ATMOSPHERES OF EARTH AND THE PLANETS

SPACE SCIENCE LIBRARY

A SERIES OF BOOKS ON THE RECENT DEVELOPMENTS

OF SPACE SCIENCE AND OF GENERAL GEOPHYSICS AND ASTROPHYSICS

PUBLISHED IN CONNECTION WITH THE JOURNAL

SPACE SCIENCE REVIEWS

VOLUME 51

PROCEEDINGS

ATMOSPHERES OF EARTH AND THE PLANETS

PROCEEDINGS OF THE SUMMER ADVANCED STUDY INSTITUTE,
HELD AT THE UNIVERSITY OF LIÈGE, BELGIUM,
JULY 29–AUGUST 9, 1974

Edited by

B. M. McCORMAC

Lockheed Palo Alto Research Laboratory,
Palo Alto, Calif., U.S.A.

D. REIDEL PUBLISHING COMPANY

DORDRECHT-HOLLAND / BOSTON-U.S.A.

Library of Congress Cataloging in Publication Data

Atmospheres of Earth and the planets.

(Astrophysics and space science library ; v. 51)
Includes bibliographies and index.
1. Atmosphere—Congresses. 2. Atmosphere, Upper—Congresses.
3. Planets—Atmospheres—Congresses.
I. McCormac, Billy Murray. II. Liège. Université.
III. Series.
QC851.A83   551.5   75–4954
ISBN-13: 978-94-010-1801-2   e-ISBN-13: 978-94-010-1799-2
DOI: 10.1007/978-94-010-1799-2

Published by D. Reidel Publishing Company,
P.O. Box 17, Dordrecht, Holland

Sold and distributed in the U.S.A., Canada, and Mexico
by D. Reidel Publishing Company, Inc.
306 Dartmouth Street, Boston,
Mass. 02116, U.S.A.

TABLE OF CONTENTS

PREFACE

This book contains the lectures presented at the Summer Advanced Study Institute, 'Physics and Chemistry of Atmospheres' which was held at the University of Liège, Belgium, during the period July 29–August 9, 1974. One-hundred nineteen persons from eleven different countries attended the Institute.

The authors and publisher have made a special effort for rapid publication of an up-to-date status of the physics and chemistry of the atmospheres of Earth and the planets, which is an ever-changing area. Special thanks are due to the lecturers for their diligent preparation and excellent presentations. The individual lectures and the published papers were deliberately limited; the authors' cooperation in conforming to these specifications is greatly appreciated. The contents of the book are organized by subject area rather than in the order in which papers were presented during the Institute. Many thanks are due to Drs Alv Egeland, Donald M. Hunten, Günther Lange-Hesse, Marcel Nicolet, Harold I. Schiff, Lance Thomas, Alister Vallance Jones, Richard Wayne, and Gilbert Weill who served as session chairmen during the Institute and contributed greatly to its success by skillfully directing the discussion period in a stimulating manner after each lecture.

Many persons contributed to the success of the Institute. Drs Alv Egeland, Donald M. Hunten, Günther Lange-Hesse, Marcel Nicolet, Harold I. Schiff, Erwin R. Schmerling, Lance Thomas, Alister Vallance Jones, Richard Wayne, and Gilbert Weill were especially helpful in preparing the technical program. Dr Jean-Claude Gerard, University of Liège, played a most important role in helping to arrange the facilities and support at the University of Liège. The assistant editor, Mrs Diana R. McCormac, checked the manuscripts and proofs, and worked hard to achieve a uniform style in this book.

Direct financial support was provided to the Institute by the Advanced Research Project Agency, U.S. Army Research Office, Defense Nuclear Agency, Lockheed Palo Alto Research Laboratory, and the Office of Naval Research.

Palo Alto   BILLY M. MCCORMAC
November 1974

PART I

INSTITUTE SUMMARY

SESSION CHAIRMEN'S SUMMARIES

H. I. SCHIFF

Faculty of Sciences, York University, Downsview, Ont., Canada

1. Optical Observations and Their Interpretation *

Optical emissions from the atmosphere may be considered to have three aspects. First, that of exploration and identification; secondly, that of understanding the processes of excitation and emission, and thirdly, that of their exploitation in the study of broader aeronomic phenomena. These phases provide a convenient framework in which to summarize the material discussed.

1.1. EXPLORATION AND IDENTIFICATION

The main emphasis is, at present, on auroral emissions. Gérard discussed the observations and measurements in the UV region where knowledge of the spectrum is still quite incomplete. Auroral EUV emission ($\sim$900 Å?) may be relatively intense and a significant source of ionization. The 2150 Å peak may be a NOγ band but the identification is still uncertain and no further spectra have been published recently. Clearly better quantitative spectra for detailed comparison with synthetic spectra are required. Progress on this task in ground-based spectroscopic measurements has been made recently by Gattinger and Vallance Jones (1974) and has resulted in the discovery of unrecognized features in the spectrum including the $v' = 2$ to 5 progressions of the atmospheric O_2 system. Measurements of system intensities have also been much improved. In the IR region some interesting radiance vs height profiles for presumed NO emission at 5.3 μm and 2.7 μm, for CO_2 at 4.3 and 15 μm and for C_3 at 9.6 μm were communicated by Stair. Some of these show striking enhancements in auroral conditions. Interesting new height profiles of the [O II] 7320 to 7330 Å emission and the [N I] 5200 Å emission were reported by Hanson from Atmospheric Explorer C.

1.2. EXCITATION PROCESSES

The AE-C profiles of [N I] 5200 Å, of [O II] 7320 to 7330 Å and of N_2 2P bands as well as rocket flights are leading to significant progress in the interpretation of dayglow excitation mechanisms.

In the aurora, the excitation model of Rees provides the most complete description to date of the maze of interacting phenomena which are involved in electron excited emissions. This type of model is of great value in assessing the validity of reaction schemes and as a guide to interpreting spectral intensity ratios (such as $I(6300)/I(4278)$) in terms of the energy parameters of the incident electrons.

---

* Presented by A. Vallance Jones.

B. M. McCormac (ed.), Atmospheres of Earth and the Planets, 3–19. All Rights Reserved.
Copyright © 1975 by D. Reidel Publishing Company, Dordrecht-Holland.

The excitation mechanism of the [O I] 5577 Å line is still uncertain although the time variation techniques discussed by Harang should provide a means of detecting indirect excitation modes. It may be necessary to combine high time resolution measurements of characteristic energy, of intensity with a realistic model to provide a definitive answer to this problem.

Gérard presented some interesting rocket observations of N_2 VK and N_2^+ 1 N bands which led to a new determination of the quenching rate constant $K_Q(O)$ for the $N_2(A\ ^3\Sigma)$ state, but since the value derived exceeds the gas kinetic collision rate some question arises as to the validity of this method. Gattinger and Vallance Jones presented evidence tending to support Cartwright et al.'s (1971) scheme for $N_2(B\ ^3\Pi)$ state excitation. Some additional excitation process now seems necessary to excite the higher v' progressions of the O_2 atmospheric bands and energy transfer from $N(^2D)$ (or energetic $O(^1D)$) was suggested. Auroral $O_2(^1\Delta)$ is still of considerable interest. The time variation measurements of Gattinger and Vallance Jones (1973) show that, in events observed from Churchill, the excitation rate is proportional to that of the N_2^+ 1 N system. The ratio, $I(O_2\ ^1\Delta)/I(4278)$, is about 85 which is comparable to the ratio of total ion production to $I(4278)$ which has a value of about 70. Thus about one $O_2(^1\Delta)$ excitation requiring about 1 eV is needed per ion pair, that is, per secondary electron. Since the average secondary electron energy is about 20 eV, it is not apparent that there is any energy problem in the observed excitation rate. Neither direct electron impact excitation nor some secondary process such as energy transfer from $N(^2D)$ or the O_2^+–NO reaction considered by Swider can be ruled out a priori. For aurora with $I(4278) = 30$ kR the steady state value of $I(O_2\ ^1\Delta)$ would be over 2000 kR and, consequently, it does not appear impossible that the values of $I(1.27\ \mu m)$ over 1000 kR reported by Noxon (private communication) are due to especially persistent IBC3 (~ 30 min) or shorter lived IBC4 substorms.

The modeling of auroral CO_2 emissions described by Kumer provides an elegant explanation of observed height profiles and a valuable indication of the concentration of vibrationally excited N_2 in aurora.

1.3. EXPLOITATION

The $\lambda 5577$ nightglow provides a powerful tool in the study of atomic O in the E region and of electron density in the F region. As shown by Donahue, height profiles obtained synoptically by satellite can yield a rich harvest of results. The ability to obtain height profiles for $n(O)$ on a synoptic basis could provide monitoring of O production and of vertical transport processes.

The measurements of $\lambda 6300$, $\lambda 5577$ and $\lambda 3914$ from Isis 2 provide a wide variety of synoptic measurements. The possibility of obtaining information over a large portion of the Earth's surface on each pass is unique to optical methods. At the same time modeling of electron and proton impact generated emissions provides an increasingly good guide to the energy spectrum and flux of the incoming particles over the wide area observed. G. Shepherd described a series of different kinds of phenomena which can be studied ranging through auroral precipitation patterns, dayside cleft emission, SAR and intertropical arcs and low latitude conjugate point photoelectron excited emission. Some previously, only partially recognized, phenomena such as the diffuse auroral zone (approximately circular about 65° N lat) and the cross polar cap Sun-aligned arcs.

Ground-based observations of intensity fluctuations in λ5577 nightglow emission were reported by Petitdidier. Under favorable conditions, observations with a set of suitably spaced fields give plausible values for wind velocities at 95 km. This kind of technique will probably be applied to other airglow emissions such as OH and Na. Noxon was reported to have detected vertically propagating temperature waves in the nightglow from rotational temperature studies of the O_2 atmospheric bands ($\sim$95 km) and the OH bands ($\sim$85 km). Another promising remote sensing technique is the inference of precipitating electron energy parameters for observations of λ4278 N_2^+ rotational temperature as reported by M. Shepherd.

1.4. Conclusions

This session covered a wide variety of studies of and conclusions from optical emission phenomena in the atmosphere. Of course, many other types of experimental observations are vital to the testing of theoretical ideas. Such techniques have been freely described in other sessions. Clearly the most effective experimental studies involve simultaneous use of all the relevant techniques.

References

Cartwright, D. C., Trajmar, S., and Williams, W.: 1971, *J. Geophys. Res.* **76**, 8368.
Gattinger, R. L. and Vallance Jones, A.: 1973, *J. Geophys. Res.* **78**, 8309.
Gattinger, R. L. and Vallance Jones, A.: 1974, *Can. J. Phys.* **52**, Nov. 15.
Noxon, J. F.: 1970, *J. Geophys. Res.* **75**, 1875.

2. Atmospheric and Ionospheric Models[*]

The study of the thermosphere and the F region is now reaching the point where the uncertainties in the chemistry are less than those in the dynamics. The introductory papers on neutral atmospheric and ionized atmospheric modeling turned away from models intended to explain effects at any one location to those associated with explaining the global behavior from first principles in a quantitative manner.

Kockarts discussed the problems associated with the development of predictive models of the thermosphere based on the simultaneous solutions of the continuity equations of number density, momentum and energy. Such models must include the effect of seasonal changes in the flows of such major constituents as atomic and molecular oxygen and helium. Major problems arise at the lower boundary associated with the interaction of absorption in the Schumann-Runge Continuum, 63 μm radiation and the effect of the changeover from eddy to molecular diffusion on the energy budget of the mesosphere. While the EUV input is uncertain to a factor of two,

[*] Presented by John Nisbet.

less is known about other sources and Joule heating, not just in the polar region, the effect of electric fields on the wind system, tides and gravity waves must all be considered. Strauss presented some preliminary results from first generation three dimensional thermospheric models which solves the three continuity equations simultaneously.

Nisbet discussed the problems associated with using such three dimensional thermospheric models in an interactive mode to develop predictive model ionospheres. This interaction is necessary because the ionosphere acts both as a damping force on the neutral winds due to ion drag and a driving force where electric fields are large. The major problems appear to be associated with the adequate representation of the neutral composition and wind and of the electric field system. The electric field system has been shown to be quite different from the E region dynamo field even at mid-latitudes at night.

Stubbe presented a detailed review of F region processes and identified the additional problems associated with reconciling the EUV fluxes with ion densities and temperatures. Detailed studies using the Atmospheric Explorer series of satellites should be useful in solving this problem. The temperature variation of the rates of the main F region loss reactions $O^+ + O_2$ and $O^+ + N_2$ has long been a problem which recent measurements by the groups of Biondi and Ferguson go far to solve. Stubbe pointed out that different behavior might well be obtained when the kinetic temperature is varied rather than the relative velocities of the particles at 300 K. Ferguson agreed but recommended that in the absence of other data they be used for the present. There remains a major discrepancy with calculated N_2^+ densities. While N_2^+ is a minor ion in the ionosphere it is one of the major ions produced and an error in its density may indicate a more fundamental problem. In the midlatitude F region the maintenance of the nighttime ionosphere and a major discrepancy in the ion balance on summer mornings remain to be quantatively explained.

Swider reviewed the current state of knowledge of the D and E regions. Basic processes in the normal midlatitude E region are understood both by day and by night. Some success has been obtained in modeling equatorial and mid-latitude sporadic E but much more work is required. There are major uncertainties in the processes resulting in the distribution of the metallic ions and about the wind fields. The minimum thickness of intense sporadic E layers may be controlled by plasma instabilities. Calculation of E region ion densities in steady aurora appears to be straightforward but there are still major problems associated with arcs and aurora varying strongly in time and space. There is no explanation for the Ziff-Donahue NO measurements in aurora.

The quiet D region is not well understood. Basic knowledge is required of the processes that convert NO^+ ions to water clusters. The NO densities are still somewhat uncertain below (80 to 85 km) and measurements are required of their geographical and temporal variability. The seasonal anomaly associated with particle events appears to be understood qualitatively but there seem to be large changes in electron densities associated with meteorological phenomena in winter such as stratospheric warmings that are not understood. The disturbed D region appears to be understood much better.

References

Detwiler, C. R., Garrett, D. L., Purcell, J. D., and Tousey, R.: 1961, *Ann. Geophys.* **17**, 9.
Hinteregger, H. E.: 1970, *Ann. Geophys.* **26**, 547.
Parkinson, W. H. and Reeves, E. M.: 1969, *Solar Phys.* **10**, 342.

3. Structure and Composition of the Neutral and Ionized Atmosphere*

Considerable discussion was given to the question of the absolute magnitudes of the fluxes in the EUV wavelength range. In the reviews by Thomas and Petit attention was drawn to studies which had indicated a need to increase the values of these fluxes by a factor of 2 from those reported by Hinteregger (1970) but it was emphasized that these studies themselves suffered from uncertainties in atmospheric composition and other input parameters. In the description of measurements carried out on board Atmospheric Explorer C satellite by Hinteregger and Heath, it was reported that Hinteregger's results seem to be consistent with his previous values whereas those of Heath supported the proposed increase. It was pointed out that preliminary studies of other measurements on board the spacecraft, such as the 7319 Å emission from $O^+(^2P)$ ions and the photoelectron energy spectra, could be reconciled with the Hinteregger measurements. No values of fluxes in the wavelength range 1300 to 1850 Å from Atmospheric Explorer C measurements were available. Thomas drew attention to the discrepancy by a factor of 2 to 3 between the (1961) and Parkinson and Reeves (1969) measurements in this range, and the corresponding discrepancy in O_2 photodissociation rates. In this connection, Strobel reported that recent work by Widing at the U.S. Naval Research Laboratory had tended to favor the lower values of Parkinson and Reeves. The need to invoke a mechanism for conducting heat downwards from 95 to 115 km, where the major part of the energy at 1300 to 1850 Å is absorbed, was discussed at length; there is some controversy as to whether turbulence would represent such a mechanism or an additional heat source. In relation to this, Petit and Walker drew attention to the information on the variation of temperature in the height range of interest provided by incoherent scatter measurements; gradients of about $12°$ km^{-1} were indicated between 108 and 123 km. Donahue mentioned that gradients larger than those shown in atmospheric models near 110 km seem to be required for the interpretation of OGO-6 satellite measurements of 5577 Å airglow emissions.

Thomas mentioned that recent measurements of positive ion composition in the D region by Kranowsky and co-workers had shown that $H^+ \cdot (H_2O)_4$ represented the major water cluster ion below about 82 km, and that ions such as $NO^+ \cdot CO_2$ were observed with substantial concentrations at greater heights; the formation of water cluster ions, $H^+ \cdot (H_2O)_n$ from NO^+ still represented one of the major problems of

* Presented by Lance Thomas.

this part of the ionosphere. He noted also that recent measurements by Ferguson's group at Boulder had indicated that the reactions of $NO^+ \cdot H_2O$ with OH and H were too slow to be effective in producing H_3O^+. In addition, they had identified two forms of NO_3^+ ion having different reactivities, and had shown that hydration seems to have little effect on some of the reactions considered in theoretical models.

Concerning neutral composition measurements, Ackerman presented a review of stratospheric measurements with particular attention being paid to odd-nitrogen constituents. The use by his group of the IR absorption technique during May 1974 had provided results on the height distributions of NO and NO_2, the former showing increases in concentrations by a factor of about 2 over their May 1973 findings. In relation to time variations, Schiff pointed out that observations by his group using the chemiluminescence method had revealed changes in NO concentration by factors of about 2 in a period of about 4 h. Ackerman also drew attention to the major discrepancies between measurements of HNO_3 concentrations obtained by the IR absorption technique and measurements of samples collected on filter paper. For greater heights Thomas reported preliminary results from the measurement of the height distributions of O concentration between 80 and 140 km at nighttime carried out by Dickinson and Bolden using the absorption and resonance scattering of $OI(^3P-^3S)$ radiation emitted from a lamp developed by Young of York University, Toronto. These results showed a rapid increase in concentration from about 10^9 cm^{-1} at 80 km to a peak value of 5×10^{11} cm^{-3} near 95 km and a decrease to 2×10^{10} cm^{-3} at 140 km. In his report of measurements at greater heights with the gas analyzer experiment on ESRO 4 satellite, von Zahn showed winter to summer concentration ratios of 20 for He and 10 for Ar at 270 km, which he attributed to dynamical and mainly temperature effects, respectively. Both the ESRO-4 and incoherent scatter results described by Petit indicated a major increase in the ratio of O/N_2 and/or O concentrations at F region heights between summer and winter. The ESRO-4 data, based on a preliminary spherical harmonic analysis, showed an increase in the O/N_2 ratio a factor of 6 at 250 km near the poles and of 3.6 near 45° latitude; the incoherent scatter results for Saint-Santin had shown an increase in the ratio of O to the weighted sum of N_2 and O by a factor of 5, this being attributed largely to a reduction in the O_2 concentration during winter.

The comparison made by von Zahn of their neutral composition results, the O^+ concentrations measurements also carried out on ESRO 4 by Raitt, and ground-based ionosonde data showed that a reduction in the ratio of O/N_2 concentration makes a major contribution to the decreases in positive ion and electron concentrations at midlatitudes during the 'negative type' storms in the F region associated with magnetic disturbances. The analysis of data for the storm which occurred during late February 1973 showed the interesting result that the reductions in the O/N_2 and in the electron and O^+ concentrations, were more pronounced at Brisbane than at the higher latitude station Hobart. This seems to be inconsistent with the model of an atmospheric disturbance propagating directly from the auroral zone to lower latitudes.

An example of a plasma drift effect at midlatitudes associated with an enhanced electric field at the time of a polar magnetic substorm was described by Blanc using data from Saint-Santin, and several examples of fields ranging in magnitude up to about 50 mV m^{-1}, as deduced from Chatanika observations, were presented by Wickwar. Comparisons of neutral winds and plasma motions observed at this high latitude site provided evidence for ion drag at both E and F region heights, and it has also been possible to examine the relative importance of Joule heating estimated from the incoherent scatter results and that due to particle precipitation during the same event. These two sources of energy input were invoked by Wickwar to explain some of the neutral wind motions observed at Chatanika, and by von Zahn to account for the heating in the polar cap and in the nightside and dayside high latitude regions as revealed by the gas analyzer results from ESRO-4 satellite.

Besides the results on 7319 Å emission from $O^+(^2P)$ ions, Hanson described measurements of photoelectron spectra and of positive ion composition carried out during the first seven months life of Atmospheric Explorer C satellite. Particularly interesting estimates of plasma drift velocities had been derived from the drift meter sensor and the retarding potential analyzer. With these two instruments, velocities of motions transverse and parallel to the satellite track had been obtained, the values ranging up to 2 km s^{-1}. It had been found that these large velocities were associated with increased ion temperatures.

4. Laboratory Measurement of Relevant Rate Coefficients*

The papers in Session 4 were clearly directed towards the chemistry of the Earth's atmosphere, yet even so the chemistry of nitrogen-containing species was only considered indirectly. Schiff's "orientation" lecture covered much of the groundwork and presented many of the important rate coefficients. In this context the following remarks are to confined the laboratory measurements in the O—H—N system; matters such as HCl, CO_2 and hydrocarbon chemistry are considered elsewhere.

Both Kaufman, and Wayne, in his introductory review, took it upon themselves to suggest possible traps lying in wait for unwary atmospheric modellers. The major point was to emphasize the danger of using simplified or incorrect kinetic models to extrapolate, to atmospheric conditions, laboratory data obtained over limited pressure or temperature ranges, even though such models adequately interpreted the *laboratory* data. It was then suggested that the transition state theory could indicate the *largest* probable pre-exponential factor for a reaction, and that experimental data giving an appreciably greater rate should be treated with suspicion. Smaller pre-exponential factors could sometimes be explained in terms of small transmission coefficients, as, for example, in the case of non-adiabatic reactions. A plea was made to use correlation rules sensibly: the basis for the arguments is to be found in Wayne's paper.

---

* Presented by Richard Wayne.

Ferguson described how the flowing afterglow technique could now be used over the temperature range 80 to 900 K, and also how the addition of a drift-tube enabled of an even wider range of kinetic energies. The curious energy dependence of reactions of the type

$$N_2^+ + O_2 \rightarrow O_2^+ + N_2$$

(Figure 4 of Ferguson's paper) gave rise to some discussion. Ferguson suggested that an energy rich complex $[N_2^+O_2]^\#$ was first formed, and that the lifetime of this complex was increased at lower temperatures. Whatever the explanation, it is clear that the form of the temperature dependence makes extrapolation to temperatures outside the laboratory range very difficult. The rate of the reaction

$$O^+ + NO \rightarrow NO^+ + O$$

increases with kinetic energy: that is, the reaction behaves as though it were endothermic even thought it is exothermic. At ionospheric temperatures the reaction is slow, and does not provide the loss process for O^+ in aurorae that might have been expected. The rate constant for

$$NO^+ + O_3 \rightarrow NO_2^+ + O_2(^1\Delta, \, ^1\Sigma, \, ^3\Sigma)$$

is small, and the reaction does not, therefore, represent an important source of $O_2(^1\Delta_g)$. On the other hand, a reaction which was expected to be slow

$$S^+ + O_2 \rightarrow SO^+ + O$$

was found to have the relatively large rate constant of 1.6×10^{-11} cm^3 s^{-1}.

With regard to metal ion chemistry, Ferguson pointed out that although the reaction

$$Mg^+ + O_3 \rightarrow MgO^+ + O_2$$

was fast $(k = 2.3 + 10^{-10}$ cm^3 s$^{-1})$, the reaction

$$Mg\,O^+ + O \rightarrow Mg^+ + O_2$$

$(k = 1 \times 10^{-10}$ cm^3 s$^{-1})$ regenerated Mg^+. A similar situation probably holds for Fe^+ and Ca^+. In contrast, a rate constant $< 10^{-11}$ cm^3 s^{-1} was found for the reaction

$$Na^+ + O_3 \rightarrow Na\,O^+ + O_2$$

suggesting that $D(Na\,O^+) < 1$ eV. The reaction

$$Si\,O^+ + O \rightarrow Si^+ + O_2$$

is fast $(k = 2 \times 10^{-10}$ cm^3 s$^{-1})$, which makes it hard to understand why $Si\,O^+$ is observed in the ionosphere. The rate constant also shows that the quoted value of $D(Si\,O^+) = 5.77$ eV must be wrong.

In the D region, the major problem is the rate of conversion of NO^+ to $H_7O_3^+$, since the straightforward route via hydration to $NO^+(H_2O)$ is too slow to answer for the observed hydrated proton concentrations. The route via CO_2 (i.e., $NO^+(CO_2)$: see Figure 6 of Ferguson's paper) is better since $[CO_2]$ is greater than $[H_2O]$ in the atmosphere. A similar route involving $NO^+(N_2)$ is speculative, since the reaction

$$NO^+ + N_2 \rightarrow NO^+(N_2)$$

is very slow at 300 K, although it might be larger at 200 K. In the troposphere, NH_3 forms water cluster ions from ammonia cluster ions.

In his discussion of negative ion chemistry, Ferguson showed that there were two kinds of NO_3^- : that formed, for example, from

$$NO_2^- + NO_2 \rightarrow NO_3^- + NO$$

and

$$NO_2^- + O_3 \rightarrow NO_3^- + O_2$$

of the structure

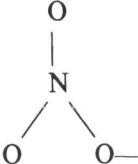

and that formed by

$$O_2^- CO_2 + NO \rightarrow NO_3^- * + CO_2$$
$$O_2^- O_2 + NO \rightarrow NO^{3*} * + O_2$$

of the structure

$$O\text{—}O\text{—}N\text{—}O.$$

A similar situation exists for NO_2^- species.

Ian Jones' paper was concerned mainly with the $O\text{—}O_2\text{—}O_3$ system and with its direct interactions with the oxides of nitrogen. The question of the quantum yields for $O(^1D)$ production from O_3 photolysis at around 3100 Å aroused some discussion, the problem being whether the published efficiency curves owed their shapes to real or to instrumental factors. The possible temperature dependence of quantum yield in the critical region was also briefly discussed.

Some of the kinetic data presented by Jones are given in Table I. Jones explained the importance of a knowledge of the rates of the reactions

$$O + O_2 + M \rightarrow O_3 + M$$

and

$$O + O_3 \rightarrow 2\,O_2.$$

The activation energy for the reaction

$$O + NO_2 \rightarrow NO + O_2$$

is now thought to be near zero, thus making the process twice as rapid in the

TABLE I

Some rate constants presented in papers on O and H atom chemistry

Reaction	Rate constant (concentration units of molecules cm^{-3}; activation energies in kcal mol^{-1})
$O + O_2 + M \rightarrow O_3 + M$	$6.6 \times 10^{-35} \exp (1.0/RT)$ for $M = Ar$ efficiency of $Ar:He:N_2 = 1.0:0.9:1.6$
$O + O_3 \rightarrow 2 O_2$	$1.1 \times 10^{-11} \exp (-4.3/RT)$
$O + NO_2 \rightarrow NO + O_2$	9.3×10^{-12} at 300 K
$O(^1D) + O_2 \rightarrow O(^3P) + O_2$	7×10^{-11} at 300 K for all pathways
$O(^1D) + O_3 \rightarrow O_2 + O_2 (\text{or } 2\,O)$	2.6×10^{-10} at 300 K
$NO_2 + O_3 \rightarrow NO_3 + O_2$	$1.2 \times 10^{-13} \exp (-4.9/RT)$
$HO + O_3 \rightarrow HO_2 + O_2$	$1.3 \times 10^{-12} \exp (-1.9/RT)$
$HO + CH_4 \rightarrow HO_2 + CH_3$	$3.8 \times 10^{-12} \exp (-3.7/RT)$
$HO + HCl \rightarrow H_2O + Cl$	$2.0 \times 10^{-12} \exp (-0.62/RT)$
$HO + HNO_3 \rightarrow H_2O + NO_3$	$0.80 - 0.95 \times 10^{-13}$; $Ea \sim 0$
$HO + HO_2 \rightarrow H_2O + O_2$	$2 \pm 1 \times 10^{-11}$ over range 200–400 K
$HO_2 + O_3 \rightarrow OH + 2 O_2$	$2.2 \times 10^{-13} \exp (-3.1 \pm 0.5/RT)$

stratosphere as formerly believed. Recent work gives an activation energy to the reaction

$$NO_2 + O_3 \rightarrow NO_3 + O_2$$

of 4.9 kcal mol^{-1}, which is also lower than earlier determinations.

Kaufman presented a number of his recent rate constant determinations involving HO and HO$_2$, which are also listed in Table I. The history of the rate constant for

$$HO + HO_2 \rightarrow H_2O + O_2$$

was described, and Kaufman offered an explanation for why Hochanadel's 1972 results may be an order of magnitude too large. Similar consideration suggests also that Hochanadel's value for the HO$_2$ + HO$_2$ reaction is too great. Kaufman also noted that the branching ratios for the reactions

$$H + HO_2 \rightarrow H_2 + O_2$$
$$\rightarrow 2\,HO$$
$$\rightarrow H_2O + O$$

were in doubt.

This section concludes with a brief personal view of what is needed next from laboratory kineticists in terms of atmospheric O$_3$ chemistry. Over the last 2 yr many of the reactions of potential stratospheric importance have been studied and re-assessed. It is probably true to say that the atmosphere can be modelled on the basis of available laboratory data and a picture consistent with the various atmospheric measurements can be obtained. Nevertheless, there seem to be several hazy areas. These include the photochemistry of NO$_3$ and its reactions with O, H and possibly HO, and, similarly, the equivalent processes involving N$_2$O$_5$. Many of the reactions involving HO$_2$ are still poorly studied. Again, while HNO$_3$ has been extensively investigated recently, and some information obtained, the same is not true of HNO_2: more knowledge is needed about the formation, photochemistry and reactions of this molecule.

Reactions involving excited species are much better understood than previously, but there are still gaps. Mention has been made of the production of $O(^1D)$ in O_3 photolysis. There are also the questions of whether it produces vibrationally excited N_2 and/or CO_2 when quenched and also what are the products of its reaction with O_3. Indeed, it is of general interest to discover to what extent vibrationally excited species play any part in atmospheric chemistry.

In every case, an attempt *should* be made to find the temperature dependence of rate, as emphasized by Kaufman. In addition, the products of a reaction should be determined wherever possible, as otherwise the right hand side of a chemical equation is largely speculative.

5. Physical Processes in the Upper Atmosphere *

The effects of the dominant physical processes acting in the upper atmosphere in connection with regular, global-scale, long-term variations were reviewed. It was demonstrated that several of the global-scale variations can be explained and predicted fairly accurately. In this brief summary only a few of the main topics discussed will be mentioned (for details refer to Kohl, Leovy, and Egeland in this volume). Some outstanding problems concerning physical processes are listed at the end of this section.

5.1. ENERGY SOURCES AND SINKS

The input of solar radiation (the dominant source), the role of hydrodynamic waves, as well as energetic particle precipitation, and electric fields were discussed in relation to the energy balance. The input of solar electromagnetic radiation is characterized by specific processes occurring within limited height regions of the Earth's atmosphere due to solar UV absorption, and it depends on solar conditions. The heat balance below, say 80 km, is now fairly well understood. Above 90 km, where the heat balance is mainly controlled by deviation from local thermodynamic equilibrium (LTE) of CO_2, there are still uncertainties.

Gravity- and Rossby-waves together with tides as possible additional energy sources in the E region and above were discussed. Their contributions to the energy budget in this region, which should be quite variable in both space and time, are far from solved. The temperature above 100 km is too large to be generated entirely by EUV radiation, tides, and atmospheric waves. Joule heating and auroral particles may be important heat sources even on a global scale. Semidiurnal tides and hydrodynamic waves, propagating laterally from the auroral zone, may consistently transfer energy into the thermosphere. The main heat source above 400 km is

* Presented by A. Egeland.

probably electron cooling. The upper atmosphere continuously loses radiation to space.

5.2. WINDS

The neutral air motions (i.e. winds) in the mesosphere, on a global scale, are persistent and fairly well understood. In the summer hemisphere the wind blows from east towards west, while the opposite is true for the winter. The average horizontal velocity is the order of 10 m s^{-1}. This wind pattern is closely related to the temperature variations in this height region. Mainly solar energy, but also atmospheric waves contribute to this wind circulation.

The tidal motion between 90 to 150 km, mainly due to solar heat input and linear gravitation, is far less understood and predictable. Even though the semidiurnal configuration of the tides is fairly regular, marked latitudinal variations are expected. Measurements have shown a prevailing horizontal wind with the order of 10 m s^{-1} as a typical speed. However, the tidal velocity will likely increase with altitudes.

Above 150 km horizontal, north-south, wind velocities of the order of 100 m s^{-1} are not unusual. The pressure variations, which cause these motions of the neutral air, are to a large extent due to absorption of solar UV radiation. This high altitude, global wind circulation can be fairly well estimated, but existing models of the F region winds cannot reproduce the low latitude observations. In addition to these periodic air movements, there is a super co-rotation of the air towards the east with angular velocity significantly larger than the Earth's velocity. This is still not explained. The connection between variations in wind (both direction and magnitude), densities, and temperature is not well understood.

5.3. ELECTRIC FIELDS

Although the plasma remains a 'minor constituent'' compared to the neutrals up to more than 1000 km, it has a large impact on the neutral gas. Due to winds, energetic particle precipitations, and magnetospheric convection fields, electric fields are important and necessary for understanding the physics of the ionosphere.

The charged particles in the E region react to the impact of the tidal wind, and an electric current flows. The magnetic effect of this is known as the Sq-variation on the ground magnetic recordings, while the lunar gravitational tides cause the magnetic L variation. The existing models of the dynamo electric fields are unsuccessful in reproducing the field observed.

In addition, marked larger, but very irregular E fields exist. Below 150 km the height-integrated Pedersen and Hall conductivities (Σ_P and Σ_H) contribute significantly to the total conductivity, while above 150 km the height-integrated parallel conductivity ($\Sigma_{"}$) completely dominates. The charged particle in the F region and above is therefore strongly coupled to the geomagnetic field lines, which can be considered as equipotentials. The consequence of this is that no significant height variation in the E field occurs. Thus, the electric field in the E region (caused by auroral particles) will extend upwards, while the electric convection field (of magnetospheric origin) will map down to the dynamo region and drive currents there. However, the electric coupling between the E and F region is not fully understood.

Even though the electrons and ions move independently of the neutrals above 150 km, the winds cause redistributions of the ionization as clearly demonstrated by the h_mF2 measurements. E fields in the F region also produce large plasma motions. In the E region, ions move with the wind, but not the electrons. The wind will still only have a minor influence on the ionization distribution here. However, in connection with E fields above say 20 mV m^{-1}, large plasma instabilities occur, which may explain much of the structures observed.

5.4. OUTSTANDING PROBLEMS

Some important, unsolved problems related to physical processes are listed below:

5.4.1. *Magnitude and Variability of Solar Flux* (< 3000 Å)

A disagreement of a factor of 2 in terms of ionizing radiation intensity seems to exist. A direct relation of solar flux parameters to thermospheric parameters, rather than ground observation (e.g. 10.7 cm solar flux) would be useful. Even a small variation (say 10%) between 1900–2100 Å will likely be important for O_3 distribution.

5.4.2. *Infrared Cooling at or above Mesopause*

More careful calculations, taking the CO_2 upper vibrational levels into account, are desirable to solve the problem of IR cooling at or above the mesopause.

5.4.3. *Flux of Gravity and Rossby Wave Energy to the Upper Atmosphere*

More theoretical work is needed on large amplitude waves to determine how they actually break and convert their energy to turbulence and heat above, say, 80 km. We need more information about the coupling (if any) between the lower atmosphere and the D region.

5.4.4. *Tidal Waves and Heating of the Thermosphere*

More experimental and theoretical work is needed to solve the heat balance and to predict the tides as a function of time and latitude.

5.4.5. *Wind Circulation in the Thermosphere*

A three-dimensional model (including time) is needed together with more data on heat input at high latitude to explain the circulation and super co-rotation. Calculation of the neutral F region wind velocity requires knowledge of the horizontal pressure gradient and this parameter is poorly known.

5.4.6. *Coupling between the Upper and Lower Ionosphere*

An interesting problem is whether this coupling (if any) would influence the weather region (i.e. below 20 km) to any large extent. The electric coupling between the E and F regions is not solved.

5.4.7. *E Fields (both DC and AC Fields) at High Latitudes*

There is a large lack of coordinated E field, energetic particle, density, and temperature measurements at high latitudes with good temporal and spatial resolutions.
The existence for parallel E fields (where and under which conditions) is far from
established. There is also a marked need for more theoretical work.

Finally, it should be stressed that only processes related to long term, global scale
variations were discussed. As the upper atmosphere undergoes more or less continuous, unpredictable changes of almost all scales of size and time, any realistic
description of the upper atmosphere must also explain the formation and the movements of such structures. These problems (i.e., physical processes acting over short
time intervals within limited areas) have barely been touched upon. As we learn more
about the general circulation, the fields, and the 'weather' of the upper atmosphere,
this region will receive an increasing interest. It is likely that this information will be
very useful for the atmospheres on other planets.

6. Planetary Atmospheres *

At this meeting we have heard papers on three planets – Venus, Mars, and Jupiter – and
two satellites – Io and Titan. I cannot summarize all this work, and will therefore
merely touch on some of the issues that are currently of concern.

Planetary research these days tends to go in a rhythm dictated by the corresponding
space missions. The most spectacular results of the Mars orbiter Mariner 9 in 1971-72
have already been published, and we didn't hear much about them. The imaging
results related primarily to geology, but there was no dearth of meteorological work,
especially on dust storms and clouds. The recent Soviet missions to Mars were
partially successful, but the results are not available yet. Late in 1973 we had the
Mariner-10 flyby of Venus and the Pioneer-10 flyby of Jupiter and Io. Those data are
still being assimilated, but we saw some interesting previews. An event that ranks
right along with these marvelous and expensive pieces of hardware was the discovery by Robert Brown of the intense Na emission from Io. The rapid rise to
prominence of Titan's atmosphere, 20 yr after its discovery, is based partly on new
data but also on a sudden awareness that here was another atmosphere that we
knew enough about for fruitful study. On the other hand, while Ganymede probably
has an atmosphere too, we know nothing quantitative about it, and it has not yet
reached a scientific take-off point. Something similar can be said about Jupiter's Red
Spot: we know a lot of facts, but they don't fit into an intellectual framework. I can't
think of it as anything more than a curiosity to be observed in the hope that some
day we may reach enlightenment.

After many years of frustration, we now have a rather firm identification of concentrated H_2SO_4 as the constituent of Venus' main cloud deck. This work, primarily

* Presented by D. M. Hunten.

that of Young, has been supplemented by the photochemical mechanism of Prinn, in which the downward flux of H_2SO_4 droplets is balanced by an upward flux of COS. As Wofsy discussed, there are some difficulties of detail, but I find the general mechanism attractive. He made the interesting point that these droplets can't be expected to survive much lower than 50 km, so that the main layer is probably no thicker than about 15 km. Above this layer there is almost certainly a thin haze, deduced for example from the cutoff of light at the horizon during a transit of Venus across the Sun. There are many suggestions of a different condensate, perhaps ice or HCl solution, at the cold temperatures of the upper stratosphere. But I see nothing unlikely about very small droplets of H_2SO_4 being lofted by eddy diffusion (or equivalent transport mechanisms) to heights far above the source. Indeed, as I discussed in my paper on eddy diffusion, Prinn has estimated cloud scale heights and attempted to set upper limits on the eddy coefficient, beyond which there would be too many particles at great heights.

Below 50 km, one or more layers of other composition may perhaps exist. For example, Lewis has suggested several Hg compounds and Hg itself. Such layers are neither required nor refuted by the photometric results from Venera 8, which measured only the downward light flux (Hansen and Lacis, 1974).

For Venus' stratosphere we are just at the point of having a satisfactory first-order description of all the known major processes. Oxidation of CO seems to be mediated by a combination of odd hydrogen and odd chlorine. One possible reaction, that of HO_2 with Cl, still needs to be pinned down. For the first time we are getting a feeling for the O_2 problem. The observational upper limit, around 1 ppm, can easily be violated simply by O_2 produced in and near the thermosphere. It must be transported down at a certain minimum rate and consumed near the cloud tops. It must supply the oxidation not only of CO, but also that of COS into SO_3, if this is indeed the source of the cloud particles. As Wofsy discussed, there is a potentially serious problem in finding an adequate supply of oxygen for the clouds.

The Venus ionosphere is a real puzzle; as McElroy commented, we simply do not have any model that fits both Mariner-5 and Mariner-10 data. McElroy has obtained reasonable fits to Mariner 5 for a pure CO_2 atmosphere, and Kumar and I recently published a model with a bit of O and a lower temperature. But the topside, which is somewhat analogous to our $F2$ region, is missing in the Mariner-10 data. Bauer and Hartle quickly suggested that it is being eliminated by solar-wind pressure. While this seems likely, their model is only semiquantitative, and by no means fully worked out. It may be that the pressure balance between the ionosphere and the solar wind is delicate, so that the height of the boundary changes drastically with the parameters of the solar wind. It's obvious that ionospheric aeronomy on Venus cannot be done without a lot of knowledge about the interplanetary medium.

Atomic hydrogen has turned out to be a surprisingly tight constraint ever since Mariner 5. The $L\alpha$ intensity is remarkably low, and was even lower at the encounter of Mariner 10. Recent work suggests that either the escape rate is orders of magnitude less than from Earth, or the atoms must be energized by some nonthermal process.

Kumar and I suggest that H_2 flows upwards and H atoms downwards, the escape flux being a small fraction of either. A different picture is suggested by Sze and McElroy, and a third version has been produced by Liu and Donahue. But we all agree that very rapid vertical transport is probably required, represented by an eddy coefficient of 10^8 cm^2 s^{-1} or more. This in turn suggests that O atoms are fairly scarce in the thermosphere, as on Mars but in great contrast to Earth. The UV emissions (1304 Å resonance triplet, 1356 Å forbidden line) seem to suggest much greater abundance, but the ratio of the two emissions is anomalous anyway. It seems that there must be additional excitation processes, or blended emissions, that are not important on Mars.

The next great event in Venus exploration by the United States will be the Pioneer-Venus orbiter and probe missions, to be launched in 1978. The orbiter will have a low and variable periapse for direct sampling of the upper atmosphere, and several solar-wind instruments too. The probes will concentrate on the clouds and measurements related to the general circulation.

Since we did not have a major review of Mars, I shall confine myself to a remark about a problem common to Mars and Venus. Calculated electron densities are too small, and calculated exospheric temperatures are too large. A change in the solar UV flux to help with one problem makes the other worse. Eddy heat conduction has been suggested by several people. One possible difficulty is that this process implies another heat source (Hunten, 1974); but perhaps it is the answer all the same.

At Jupiter we are faced with a major issue for which no solution is apparent. We have been sure for years that the visible clouds are frozen NH_3 at about 130 K and a pressure level of 1 to 2 bar. There are many lines of evidence, involving high-resolution spectroscopy in the near IR and medium-resolution radiometry in the far IR and radio region. Theoretical models of radiative equilibrium confirm the picture. The temperature profiles from the Pioneer 10 occultation experiment are wildly different: they show 550 K at 1 bar and 130 K only at 2 mb! This experiment has previously given excellent and reasonable data for Mars and Venus. Despite a great deal of thought in the last 6 mo nobody has been able to find a reason, even a wild one, why it should go wrong at Jupiter.

Then what about a cloud at the 2 mb level, as suggested by the Pioneer group? It seems unattractive, because the particles would have to be large for a high IR opacity, and with a 10 μm radius they would fall at 40 cm s^{-1}. Nevertheless, one can play the game of ignoring how implausible the particles may be, and see whether they can be made to reconcile the data. Unfortunately, we have made only a start at this process. The spectroscopic abundance of H_2 suggests that we see down to the 1-bar level. This path would have to be simulated at a much lower pressure by multiple scattering in the cloud; it seems that hundreds of scatterings would be needed. Line broadening in CH_4 would be very hard to simulate at a low pressure. Microwave emission from NH_3 would not be affected by the cloud; this may become one of the most difficult things to explain. I do not think that a workable cloud can be found, but it is worth a try.

Pioneer 10 also found a lot of fascinating things about Jupiter's magnetosphere; perhaps they will be discussed at the 1975 Institute.

The discovery of Na on and near Io appears to be linked to the remarkably dense ionosphere observed by Pioneer 10. But as Yung and McElroy emphasize, another difficulty then crops up: the ions recombine so slowly that the diurnal variation is much too small. They propose that the ions are carried down to the surface at night by a rapid current of condensing NH_3. Hopefully more observations of the Na will help disentangle the important processes. On the other hand, we may end up with a large catalog and little enlightenment – I am reminded of how I felt about Na twilight in 1955. For Io we still have quite a few interesting ideas to work out.

My paper on Titan is being published elsewhere (Hunten, 1975). There is also a NASA special publication (SP-340) with a complete summary up to late 1973. A major issue is whether or not the atmosphere contains a large amount of H_2, as strongly suggested by Trafton's spectra. The corresponding loss rate is nearly intolerable; if it were to be supplied by photolysis, nearly all the solar photons up to 2500 Å would be needed. The only molecule that seems at all possible for this photolysis is H_2S, and it is not expected to be free in the atmosphere.

A second question is whether the surface pressure is tens or hundreds of millibars. There is a lot of sentiment for the deep atmosphere, but no real proof. Unfortunately, direct probing by an entry mission would be risky and unprofitable unless the atmosphere is deep. Such probing is becoming more and more popular, for good reasons. Pioneer-Venus will be doing it, and Viking for Mars. Some very simple measurements would quickly resolve our dilemma about Jupiter, but unfortunately this planet is a forbidden target because of the huge entry velocity. Serious studies are under way for other bodies in the outer solar system, and there seems to be real hope for a mission to Uranus in the next few years. We already have a flood of new data to work with, and even better prospects for the next decade.

References

Hansen, J. E. and Lacis, A. A.: 1974, *Science* **184**, 979.
Hunten, D. M.: 1974, *J. Geophys. Res.* **79**, 2533.
Hunten, D. M.: 1975, in J. A. Burns (ed.), 'Planetary Satellites', *IAU. Colloq.*, No. 28, Cornell Univ. Press.

NEUTRAL CHEMISTRY IN THE EARTH'S ATMOSPHERE

HAROLD I. SCHIFF

York University, Downsview, Ont., Canada

1. Introduction

Our knowledge of the chemistry of planetary atmospheres has increased, almost explosively, since the last volume in this series appeared 2 yr ago. A great deal of new data has been garnered at high altitudes in the Earth's atmosphere from satellites such as OGO 6, Explorer, and ESRO 4. Some of this information is discussed in subsequent articles in this volume and will, therefore, not be treated in detail in this review. There has been a flurry of activity in stratospheric chemistry, stimulated largely by the Climatic Impact Assessment Program in the U.S.A. This program is largely directed towards the understanding of the O_3 budget in the natural stratosphere and to possible effects of human intervention by, for example, a fleet of supersonic aircraft. The stratosphere has largely been ignored by aeronomers in the past and many of the minor constituents are only now being measured for the first time.

These prodigious observational efforts have been matched by a remarkable output of good laboratory data and by increasingly sophisticated atmospheric models. Since laboratory rate data have been subjected to excellent critical evaluations (Baulch *et al.*, 1973; Hampson, 1973) no attempt will be made here to report on the large amount of new information. Instead, attention will be focussed on new measurements of atmospheric composition and its variability, and on defining some of the new problems which these measurements have raised. An attempt will also be made to discuss the salient features of the chemistry used in atmospheric models and to point out some of the inadequacies in these models and in the laboratory data which they employ.

2. The Chemical System

It will be useful to classify the components of the atmosphere according to their chemical lifetimes relative to the lifetimes for dynamical, transport, processes. Thus, Class I components can be defined as those which are so chemically inert that their distributions in the atmosphere are governed almost exclusively by physical forces such as gravitation, diffusion and winds; Class II components as those of intermediate chemical reactivity for which both chemistry and transport into and out of each region of the atmosphere must be considered; and Class III components as those which react so rapidly that photochemical steady state (PCSS) may be assumed and vertical fluxes may be ignored at any altitude regime under consideration. It will be realized, of course, that both reaction rates and transport rates will be functions of altitude so that a component which, for example, can be classified as Class I in one region of the atmosphere must be considered as Class II in a different altitude regime.

B. M. McCormac (ed.), Atmospheres of Earth and the Planets, 21–43. All Rights Reserved.
Copyright © 1975 by D. Reidel Publishing Company, Dordrecht-Holland.

2.1. CLASS I COMPONENTS

This class of components include N_2, O_2, the noble gases and CO_2. Although He is lost slowly by escape through the exosphere it has an atmospheric residence time of approximately 10^7 yr. N_2 and the other noble gases have essentially infinite lifetimes in the atmosphere. O_2 is included even though the surface of the Earth is not fully oxidized. Its concentration is controlled mainly by the familiar biological cycle and no changes have been observed over the past 100 yr. Its residence time in the atmosphere has been estimated at 10^4 yr. CO_2 can also be placed in this category although its concentration has been increasing steadily as a result of the combustion of fossil fuels. Nevertheless it is relatively unreactive in the atmosphere and its concentration is mainly controlled by ocean-air interactions.

This class of components will be uniformly mixed at any given height and the concentration of each of them will decrease with altitude, up to the turbopause, at approximately 110 km, according to the familiar barometric formula with a scale height characterized by the mean molecular weight. Above the turbopause the mean free path becomes so large that eddy diffusion is no longer effective and fractionation occurs by molecular diffusion, each constituent having its own scale height.

Above 80 km photodissociation of O_2 in the Schumann-Runge region becomes pronounced and O_2 and its dissociation product O become Class II components. The same happens with N_2 but at a much higher altitude. CO_2 has been measured mass spectrometrically (Philbrick et al., 1973) and was found to be uniformly mixed to the turbopause with a mixing ratio of about 3.2×10^{-4} and to be diffusively separated above this altitude.

Above 80 km variations in the number densities of these components with time have been observed. Seasonal and diurnal variations of densities and relative concentrations of these constituents have been studied and are discussed in detail in this volume (von Zahn, 1975; Donahue, 1975).

2.2. CLASS II COMPONENTS

H_2O, the most familiar member of this family, has a highly variable mixing ratio in the troposphere, and really behaves as a Class I constituent in this region. Its residence time in the stratosphere is, however, approximately 1.3 yr and decreases rapidly in the mesosphere. The stratosphere is very dry, having a relative humidity of about 2% or an almost constant mixing ratio of 2 to 3×10^{-6}. Its presence in the stratosphere is believed to be the result of upwelling of tropical storms and its concentration to be controlled by the tropical tropopause temperature of 190 K since the mid-latitude tropopause temperature is too high to produce so low a humidity. There is some indication of a small increase in stratospheric H_2O content during the 1960's but this trend seems to be reversing during the 1970's.

There are no direct measurments of H_2O vapor above 50 km but its concentration can be estimated from the observation of water cluster ions in the D region. Swider and Narcisi (1975) find these to be the most abundant ions below 78 km but show a sudden drop-off above this height. From the known chemistry of these cluster ions they calculate a mixing ratio of 2×10^{-6} in the 70 to 78 km region, dropping to 0.5×10^{-6} at 80 km.

2.3. O_3

This compound has long interested meteorologists and aeronomers for several reasons. Because of its strong absorption over a wide wavelength region it plays an important role in the radiation balance of the atmosphere. Because it has a relatively long lifetime in the lower stratosphere it has been widely used as a tracer to follow atmospheric reactions. Its function as a filter against lethal UV radiation is well known but has gained renewed interest in connection with the effect of supersonic transports and the increased use of halocarbons.

Ozone is relatively easy to measure and both the total column densities and its height profile are routinely measured on a global basis from the ground by balloon sondes and from satellites. At midlatitudes the concentration peaks near 25 km. In the tropics the layer is sharper and the peak occurs at a higher altitude, typically at 27 km. Conversely, at the poles a broader maximum occurs at about 18 km. There is an annual periodicity with the maximum occurring during late winter, the amplitude of the variation increasing with latitude. The maximum densities appear during polar winters with some asymmetry between the hemispheres. Daily fluctuations of as much as a factor of 2 or 3 may be observed at any station while a single ozonesonde will show considerable structure with height. Excellent reviews of all these variations are available (Dütsch, 1974).

2.4. $CH_4/H_2/CO/N_2O$

These compounds originate mainly at the Earth's surface and have sufficiently long chemical lifetimes to be transported into the stratosphere. There have been a number of recent measurements of their concentrations in the stratosphere, largely sponsored by the CIAP program and these are summarized in Figure 1 and 2. Most of the data were obtained by cryogenic sampling from balloons and rockets, although other techniques including solar IR absorption have also been used. Since considerable time is required for collecting samples, some uncertainty exists in assigning appropriate altitudes to the measurements made by rockets at the higher altitudes. Because of this and the small number of measurements made to date the trends in the height profiles must be considered to be tentative. This warning should be heeded by modellers who find the height profiles of these compounds to be attractive for deriving values for eddy diffusion coefficients in the stratosphere.

These limited results suggest that CH_4 and H_2 maintain a nearly constant mixing ratio up to 27 km and then decrease; N_2O mixing ratios decrease above 20 km while CO appears to increase in the 20 to 30 km region and to decrease at higher altitudes.

2.5. NO_x

This is the term used to describe 'odd' nitrogen compounds and include NO, NO_2,

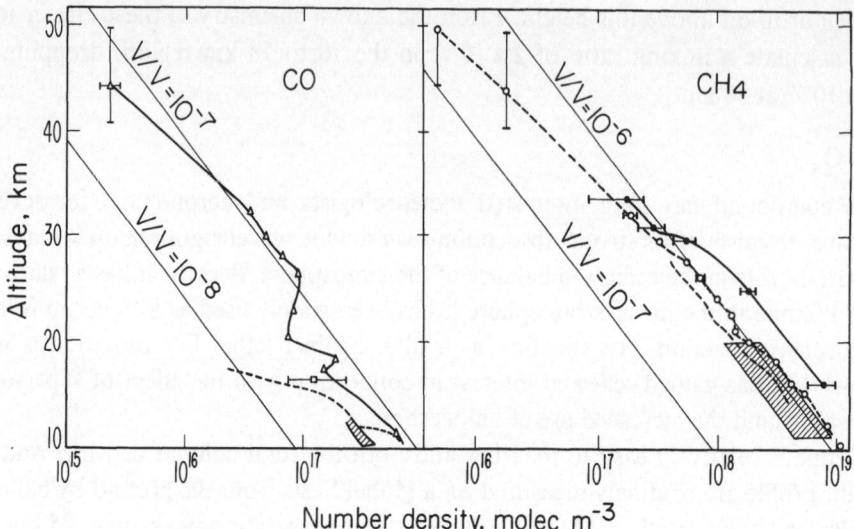

Fig. 1.   Stratospheric Measurements of CO and CH₄. CO △ – Ehhalt *et al.* (1974), cryocapture, □ – Farmer *et al.* (1974), 4.7 μm solar absorption, - - - - - Goldman *et al.* (1973), 4.7 μm solar absorption, ▩ – Seiler and Warneck (1972), hot mercuric oxide Hg 254 nm absorption. CH₄ ● – Ackerman and Muler (1973), 3.4 μm solar absorption, ○ – Ehhalt *et al.* (1974), cryocapture, - - - - - Cumming and Lowe (1973), 3.3 μm solar absorption, ▩ – Farmer *et al.* (1974), 3.3 μm solar absorption.

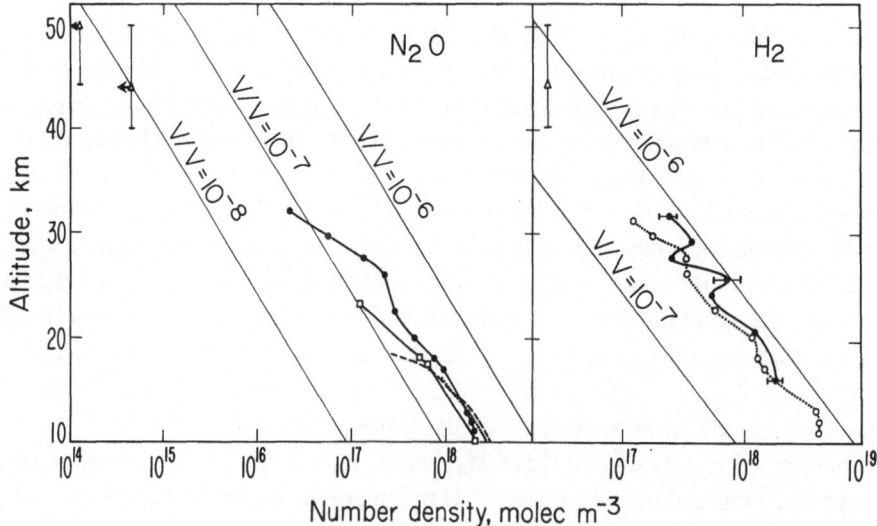

Fig. 2.   Stratospheric Measurements of N₂O and H₂. N₂O △ – Ehhalt *et al.* (1974), cryocapture, ● – Farmer *et al.* (1974), solar absorption, 3.9 and 5.3 μm, - - - - - Murcray *et al.* (1973), 4.5 μm solar absorption, ... – Harries *et al.* (1974), submillimetre thermal emission, □ – Schütz *et al.* (1970), whole air sample, gas chromatography.

NO_3, N_2O_5,* HNO_2 and HNO_3, the most abundant of which are HNO_3, NO_2 and NO. The lifetime for the conversion of one of these NO_x's to another may be con- siderably shorter than dynamic lifetimes, and may therefore be considered to be Class III components. They are included here because the chemical lifetime of *total* NO_x is longer than transport lifetimes.

As a result of the interest in the effect of NO_x on the O_3 balance a great deal of effort has been expended during the past 3 yr in measuring the concentrations of these compounds.

NO was first measured in the mesosphere by Barth (1964) and his collaborators using solar resonance fluorescence. Other measurements using this technique (Meira, 1971) have shown values which ranged from 5×10^6 to 10^8 cm^{-3} in the 65 to 110 km region. The values appear to increase with latitude and to become highly variable above 60° N. Mention should be made of the extremely high values of NO (greater than that of O or O_2) observed mass-spectrometrically by Zipf *et al.* (1970) in a number of aurora.

Ridley *et al.* (1973, 1974) were the first to measure NO in the stratosphere using an *in situ* chemiluminescence technique. These results shown in Figure 3 indicate an

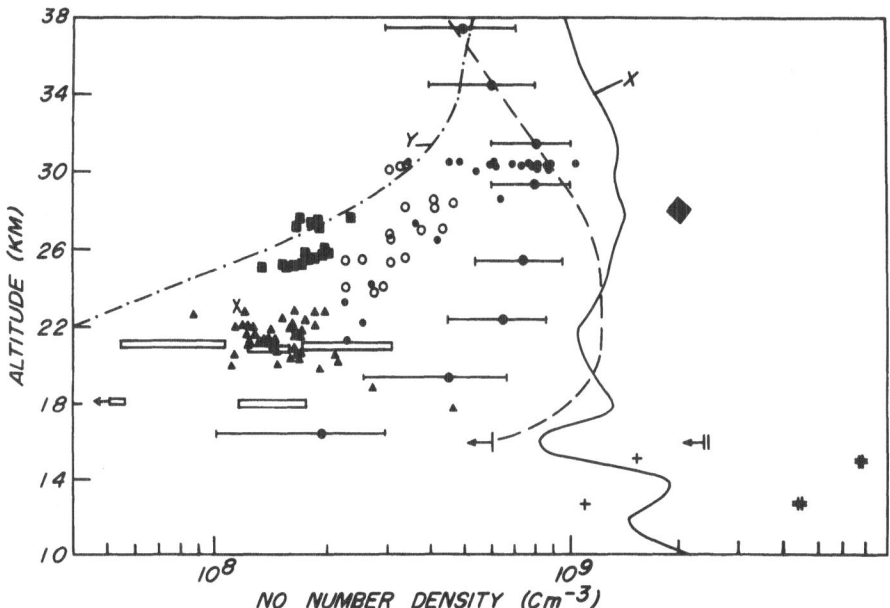

Fig. 3. Stratospheric Measurements of NO. Ridley *et al.* (1975), chemiluminescence; $\times$ – 12 Dec. 72; $\triangle$ – 16 March 73; $\blacksquare$ – 21 June 73; $\bigcirc$, $\bullet$ – 13 Nov. 73; $\vdash\!\!\bullet\!\!\dashv$ – Ackerman *et al.* (1974), 5.3 μm solar absorption, + – Farmer (1974), 5.3 μm solar absorption, $\dashv$ – Fontanella *et al.* (1974), 5.3 μm solar absorption, upper limit only, $\boxdot$ – Lowenstein *et al.* (1974), chemiluminescence, airplane, $\blacklozenge$ – Patel *et al.* (1974), spin flip laser absorption, - - - -, #, $\dashv\!\!\vert$ – data of Ackerman *et al.* (1974), Farmer (1974) and Fontanella *et al.*, respectively, converted to noon time equivalent to NO densities X, and Y are 'limiting' model calculations described in the text.

* Although N_2O_5 has an even number of nitrogen atoms, it is included because of its chemical reactivity and can be considered to be a complex of NO_2 and NO_3.

increase in number density from 18 to 30.5 km at 33° N. There is some indication of a
seasonal variation with larger values obtained in November. Short term variability
of a factor of 3 was observed at 30 km during the 4 h float period of the balloons.
Results obtained by the same technique, at lower altitudes, from aircraft platforms
were consistent with the balloon measurement but showed a much greater variability
over the length of the flight path.

Also shown in Figure 3 are data derived from remote sensing, IR solar absorption
measurements in the 5.3-μm region taken from balloons and aircraft. These measurements were made under twilight conditions to obtain long path lengths and to view as
nearly a horizontal layer as possible. The method provides an integrated column
density between the instrument and the sun. Number density profiles are obtained
from the change in column densities as a function of solar zenith angle greater than
90°. Although this technique is quite sensitive, difficulties are encountered in the
deconvolution of the data to obtain height profiles and in obtaining spectral purity
since this spectral region includes telluric CO_2 and solar CO bands. This method also
showed variability with time. Measurements made by Ackerman et al. (1974a, b)
with the same grille spectrometer at the same location exactly 1 yr apart gave values
which differ by as much as a factor of 3, and which showed quite different height
profiles.

The in situ chemiluminescence results appear to agree with the remote sensing IR
measurements at the highest and lowest altitudes but are consistently lower in the
25 km region. This difference is actually greater than would appear since the IR
measurements were taken at twilight while the chemiluminescence measurements
were made near noon where conversion from NO_2 to NO is greatest. The single
measurement reported by Patel et al. (1974) using an in situ laser absorption technique
was also taken at noon but gives a value considerably higher than those obtained by
either of the other techniques. These discrepancies may be due either to differences
in the methods or to real variability in the NO abundancies.

The NO_2 height profiles shown in Figure 4 were all obtained using remote sensing
solar absorption. The NO_2 densities above 21 km given by Ackerman and Müller
(1972, 1973) and by Murcray et al. (1974) were deduced from the same measurements
of 6.2-μm solar absorption. Differences are due to the respective methods of interpretation. Nevertheless, the values reported by these authors and those of Fontenella
et al. (1974), using a grille spectrometer and the 6.2-μm absorption band appear to
agree within the combined uncertainties.

The densities reported by Brewer et al. (1973) using visible absorption, and by
Farmer et al. (1974) using an interferometer and the 3.5 μm band appear to be
considerably higher, although the measurements of Farmer et al. (1974) were taken
under similar conditions to those of Fontenella et al. (1974).

HNO_3 has been observed with a variety of techniques and the results are summarized in Figure 5. An extensive set of measurements has been made (Murcray
et al. 1973, 1974) from balloons, using both IR solar absorption, in the 7.5-μm region
and thermal emission with a filter radiometer at 11.3 μm. Fontenella et al. (1974) have

also made IR absorption measurements and Harries *et al.* (1974) have made emission measurements from Concorde. Lazrus and Gandrud (1974) have made a series of *in situ* measurements using a filter paper sampling technique. The results all agree within a factor of 2, although those of Lazrus and Gandrud (1974) tend to be lower,

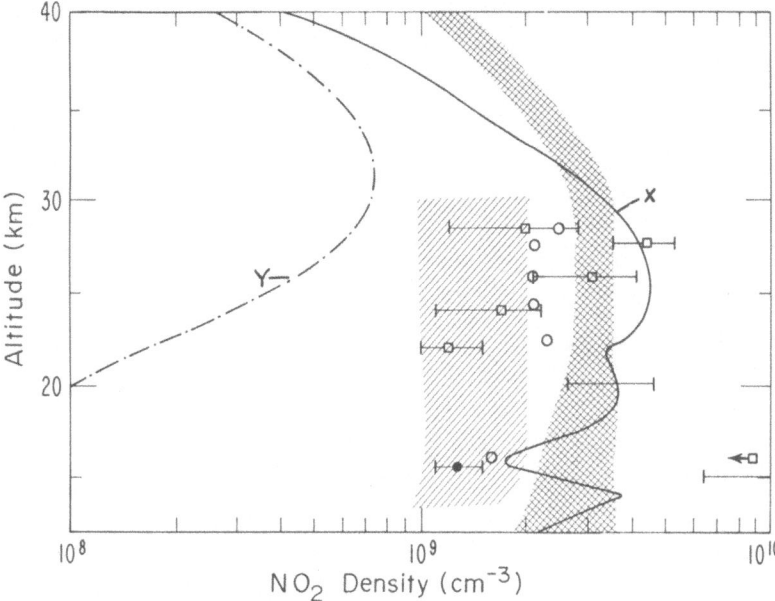

Fig. 4. Stratospheric Measurements of NO_2. $\bigcirc$ – Murcray *et al.* (1974), 6.2 μm solar absorption, $\vdash\square\dashv$ – Ackerman and Müller (1972), 3.5 μm and 6.2 μm solar absorption, ▨ $\vdash\bullet\dashv$ – Fontanella *et al.* (1974), 6.2 μm solar absorption, ▤ – Brewer *et al.* (1973), visible absorption 430–450 nm, $\vdash\dashv$ – Farmer (1974), 3.5 μm solar absorption X and Y are 'limiting' model calculations described in the text.

particularly at the lower altitudes. It has not yet been unequivocally established that the filter paper collection efficiency is 100%. The results exhibit a time variability and structure in the height profiles considerably greater than experimental error. There also appears to be a seasonal effect with the highest density occurring during the winter, in the same sense as the seasonal variability in NO indicated by the chemiluminescence measurements. Maximum densities appear to occur at lower altitudes at higher latitudes and have higher values.

2.6. CLASS III COMPONENTS

Because they are chemically very reactive the concentrations of Class III components are very low and, consequently very difficult to measure. Atomic oxygen in the ground $^3P_{1/2, 3/2}$ state has been measured in the thermosphere by mass spectrometry and the results are discussed in subsequent papers. As mentioned earlier, O becomes the major form of oxygen above 100 km and the most abundant atmospheric constituent above

160 km. Although it reaches its maximum density near 100 km the height profile in this region has continued to be remarkably resistant to accurate measurement but seems on the verge of submission. Several methods have recently been employed. It is theoretically possible to calculate $O(^3P)$ densities from the readily measured emission rates of the Oxygen Green Line on the assumption that it is formed exclusively from the Chapman reaction

$$O(^3P) + O(^3P) + O(^3P) \rightarrow O_2(^3\Sigma_g^-) + O(^1S).$$

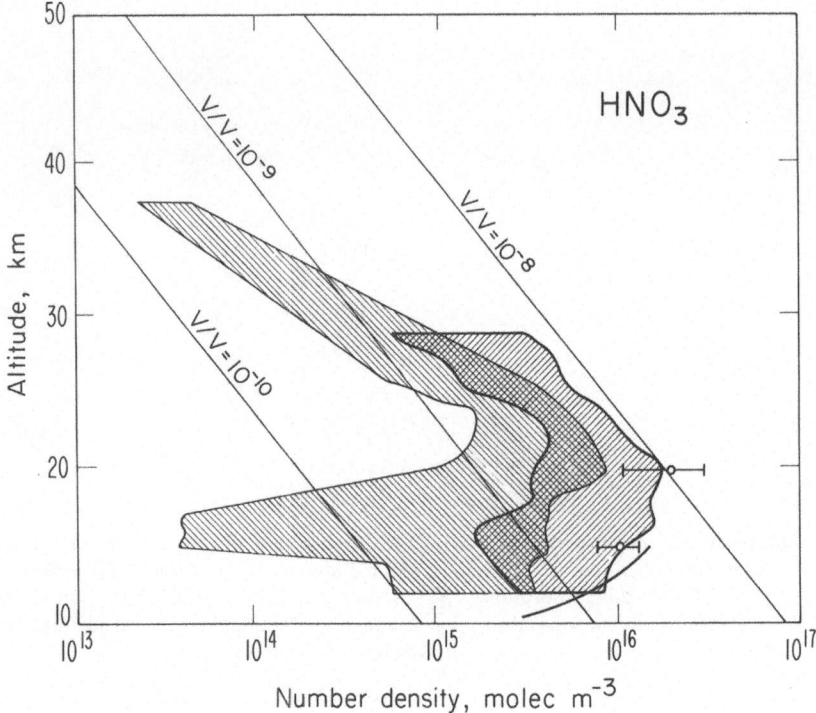

Fig. 5. Stratospheric Measurements of HNO₃. ■ – Murcray *et al.* (1973, 1974), 7.5 μm solar absorption, 11.3 μm thermal emission, ▨ – Lazrus and Gandrud (1974), filter paper capture, ⊢O⊣ – Fontanella *et al.* (1974), 7.5 μm solar absorption, —— – Harries *et al.* (1974), submillimetre thermal emission.

The difficulties inherent in this method are discussed in this volume by Donahue (1975). Henderson derived O atom densities between 89 and 94 km from the change in electrical resistance of a thin silver film resulting from the flux of O atoms to its surface (Henderson and Schiff, 1970). The most promising technique appears to be resonance fluorescence. No measurements are available below 90 km.

Hydroxyl radicals, HO, have been measured by Anderson (1971) between 45 and 70 km using solar resonance fluorescence. Densities of 4.4, 5.5 and 3.5×10^6 cm^{-3} were reported at 50, 60 and 70 km, respectively, with an estimated uncertainty of

120%. No measurements have not been reported below 50 km where HO partakes in many important chemical reactions.

3. Major Chemical Processes

3.1. O_2

Oxygen is the only Class I component which undergoes appreciable chemical reaction below 100 km. In fact, it is the photodissociation of O_2 in the UV that provides most of the chemical energy in the atmosphere. Above 70 km the strong absorption in the Schumann-Runge region

$$O_2 + h\nu(\lambda < 176\ nm) \rightarrow O(^3P) + O(^1D) \tag{1}$$

leads to one ground state atom and one excited atom, which is however, rapidly quenched to the ground state by O_2 and N_2. At these altitudes recombination occurs mainly by

$$O + O + M \rightarrow O_2 + M. \tag{2}$$

Oxygen is also dissociated by the relatively weak absorption in the Herzberg region

$$O_2 + h\nu(\lambda < 242\ nm) \rightarrow O(^3P) + O(^3P) \tag{3}$$

which occurs down to altitudes as low as 30 km. In this altitude regime recombination proceeds by the 2 step mechanism

$$O + O_2 + M \rightarrow O_3 + M \tag{4}$$

$$O + O_3 \rightarrow 2O_2 \tag{5}$$

which can repeat many times since O_3 dissociates over a wide spectral region

$$O_3 + h\nu(\lambda < 1100\ nm) \rightarrow O_2 + O. \tag{6}$$

These reactions provide qualitative explanations for the temperature profiles below 100 km and for the existence of the O_3 layer. They also predict that in order of abundance, $O > O_2 > O_3$ in the thermosphere, $O_2 > O > O_3$ in the mesosphere and $O_2 > O_3 > O$ in the stratosphere. The time required for these reactions to reach photochemical steady state is shorter than the transport lifetimes, τ_T between about 30 and 80 km but greater than τ_T at altitudes above and below this region (Nicolet, 1974). Although more than 50% of the O_3 production occurs above 30 km, more than 50% is located below 30 km. Thus O_3 behaves as a Class II or Class III component depending on altitude.

It has been recognized for some time now that other chemical reactions must affect the O_3 budget since the Chapman mechanism given above produces, on a global scale, very much more O_3 than it destroys. Since the production rate is so large the balance can only be redressed by catalytic chain processes and a number of these have been proposed involving 'odd' hydrogen and 'odd' nitrogen compounds. The most effective chains will depend on the altitude. Thus, between 45 to 100 km, where H and

O densities are relatively high the dominant chain is

A. $H + O_3 \rightarrow HO + O_2$ (7)

       $O + HO \rightarrow H + O_2$ (8)

       _____

net $O + O_3 \rightarrow 2O_2$.

Between 30 and 45 km, HO and HO_2 are more abundant than H and chain B becomes more important than A.

B. $HO + O_3 \rightarrow HO_2 + O_2$ (9)

       $HO_2 + O \rightarrow HO + O_2$

       _____   (10)

net $O + O_3 \rightarrow 2O_2$

Below 30 km $[O_3] \gg [O]$ and C become the most important chain involving 'odd' hydrogen, HO_x

C. $HO + O_3 \rightarrow HO_2 + O_2$ (9)

       $HO_2 + O_3 \rightarrow HO + 2O_2$

       _____   (11)

net $2O_3 \rightarrow 3O_2$.

Catalytic chains involving NO_x have been the subject of concern about additional amounts injected by supersonic aircraft.

The dominant chain is

D. $NO + O_3 \rightarrow NO_2 + O_2$ (12)

       $NO_2 + O \rightarrow NO + O_2$

       _____   (13)

net $O + O_3 \rightarrow 2O_2$.

The mechanism

E. $NO_2 + O_3 \rightarrow NO_3 + O_2$ (14)

       $NO_3 + h\nu \rightarrow NO + O_2$ (15)

       $NO + O_3 \rightarrow NO_2 + O_2$ (12)

       _____

net $2O_3 \rightarrow 3O_2$

may be effective at lower altitudes, including the troposphere since reaction (15) can also occur with visible light. However recent laboratory studies (Johnston, 1974) have indicated that the quantum yield for this reaction is low so that this mechanism may not be important and NO_3 will be lost mainly by

$$NO_3 + NO \rightarrow 2NO_2.$$ (16)

Current, one dimensional models (McElroy *et al.*, 1974) show that mechanisms A and B control the O_3 destruction rate above 54 km and D below 45 km. The direct Chapman reaction (5) becomes dominant only in the short altitude range 45 to 54 km.

Other chains may exist, the importance of which cannot be assessed due to the lack of reaction rate data. One example considered by McConnell *et al.* (1975)

$$HO_2 + NO_2 \rightarrow HNO_2 + O_2 \qquad (17)$$

$$HNO_2 + O_3 \rightarrow HNO_3 + O_2 \qquad (18)$$

$$HNO_3 + h\nu \rightarrow HO + NO_2 \qquad (19)$$

$$HO + O_3 \rightarrow HO_2 + O_2 \qquad (9)$$

$$\text{net} \quad 2O_3 \qquad \rightarrow 3O_2$$

may be important at low altitudes.

Another possible catalyst for O_3 destruction is Cl atoms. This possibility has been discussed recently in connection with the space shuttle (Wofsy and McElroy, 1974) which employs a fuel containing a large percentage of Cl. More recently attention has been drawn (Molina and Rowland, 1974; Cicerone *et al.*, 1974; Crutzen, 1974) to the possible effects of Freons which are finding increasing industrial use as refrigerants and in aerosol spray cans. At present the annual global production of Freons is about 1 Mt and is increasing at a rate of about 10% per annum. These Freons, mainly CF_2Cl_2 and $CFCl_3$ are all insoluble and chemically inert and, therefore, have sufficiently long lifetimes in the troposphere to be transported into the stratosphere where they will be photolyzed by UV solar radiation, presumably releasing Cl atoms, which can then react, catalytically, with O_3:

$$Cl + O_3 \rightarrow ClO + O_2 \qquad (20)$$

$$ClO + O \rightarrow Cl + O_2 \qquad (21)$$

$$\text{net} \quad O + O_3 \rightarrow O_3$$

which is the Cl counterpart of chain D for NO_x catalysis.

It should be pointed out that some atmospheric reactions can also produce O_3, and this is known to occur in polluted urban atmospheres which contain hydrocarbons and nitric oxides. The mechanisms involve oxidation of the hydrocarbon, initiated, for example by HO radicals, to peroxy compounds which, in turn, oxidize NO to NO_2. Since NO_2 is photodissociated by visible light, O_3 can be formed at low altitudes. These processes can also occur in the stratosphere starting with CH_4 and the details of the mechanisms have been discussed previously (Crutzen, 1973a).

The evaluation of the relative importance of various mechanisms requires a knowledge of the concentrations of the various NO_x, HO_x and Cl compounds involved, which are determined by the many interactions between them. Typical 1-D models invoke at least 50 such reactions, the rate constants of which are known with varying degrees of certainty. Comparison of the concentrations of these species with the measured values, when available, provides the most sensitive tests for these models. The most important reactions which are believed to determine the concentrations of Class II and III constituents in the different altitude regimes will now be summarized.

3.2. CLASS II COMPONENTS – $H_2O/H_2/CH_4$

The major sources of these hydrogeneous components are at the Earth's surface and, since they belong to Class II, flux terms must be included in the continuity equations at all altitudes. A comprehensive treatment of this system has recently been published by Liu and Donahue (1974) and by Hunten and Strobel (1974), and only the principal features will be presented here.

3.2.1. H_2O

The major loss process in the stratosphere is

$$O(^1D) + H_2O \rightarrow 2HO \tag{22}$$

where the $O(^1D)$ derives from O_3 photolysis. Above 70 km H_2O is photolysed by $L\alpha$ radiation and in the Schumann-Runge region.

$$H_2O + h\nu \rightarrow HO + H \tag{23a}$$
$$\rightarrow H_2 + O. \tag{23b}$$

There is experimental evidence (McNesby et al., 1962; Welge and Stuhl, 1967) that 25% of the photodissociation at 123.6 nm goes via channel (23b) but the branching ratio between these two channels is not known as a function of wavelength.

H_2O is also produced chemically below 50 km, mainly by the reactions

$$HO + CH_4 \rightarrow CH_3 + H_2O \tag{24}$$
$$HO + H_2 \rightarrow H + H_2O \tag{25}$$

and above 50 km mainly by the reaction

$$HO + HO_2 \rightarrow H_2O + O_2. \tag{26}$$

Reaction (26) has an uncertainty in its rate constant of about a factor of 5 and is discussed in this volume by Kaufman (1975).

3.2.2. H_2

Below 60 km the major loss processes for H_2 are reaction (25) and

$$O(^1D) + H_2 \rightarrow H + HO \tag{27}$$

which become dominant up to 110 km. Above 110 km H_2 is converted to H by the reaction sequence

$$H_2 + O \rightarrow HO + H \tag{28}$$
$$HO + O \rightarrow O_2 + H. \tag{8}$$

Although reaction (28) has an activation energy of 10 kcal mole^{-1}, it has an appreciable rate at the temperatures prevailing in the thermosphere.

Above 90 km H_2 is produced mainly by reaction (23b) and below this height by the reaction

$$H + HO_2 \rightarrow H_2 + O_2. \tag{29a}$$

There is experimental evidence that this reaction has, in addition to (29a), additional channels:

$$H + HO_2 \rightarrow HO + HO \tag{29b}$$
$$\rightarrow H_2O + O. \tag{29c}$$

The branching ratios between these channels are not well established and are also discussed by Kaufman (1975). Since this reaction assumes considerable importance, not only for the Earth's atmosphere, but for the atmospheres of Mars and Venus, further laboratory work on the rates of the three channels is clearly indicated. It may be noted that Equation (29a) results in the loss of two 'odd' hydrogen radicals, while Equation (29b) does not and Equation (29c) not only destroys the two radicals but produces an odd oxygen.

3.2.3. CH_4

There are no production processes of this compound in the atmosphere. Destruction occurs via reaction (24) and

$$O(^1D) + CH_4 \rightarrow CH_3 + HO. \tag{30}$$

The concentration is therefore expected to decrease steadily with height in keeping with observations.

3.2.4. H

Atomic hydrogen may be included in this category since flux terms must be included in modeling its behavior above 85 km. In addition to the production steps from H_2O and H_2, listed above, there is the additional stratospheric source

$$CO + HO \rightarrow CO_2 + H. \tag{31}$$
$$H + O_3 \rightarrow HO + O_2 \tag{7}$$

and

$$H + O_2 + M \rightarrow HO_2 + M. \tag{32}$$

The model of Liu and Donahue (1974) gives H as the most abundant of these compounds above 114 km, H_2 between 72 and 114 km and H_2O below 72 km. These authors have also shown that this chemistry leads to complex vertical flux patterns. At about 58 km reactions (25) and (27) provide very strong sinks for H_2 which then flows into this location both from above and from below. But the same reactions simultaneously provide a strong source for H_2O, directly by reaction (25) and indirectly by reaction (27) followed by (26). Consequently the H_2O flux is in the opposite direction to the H_2 flux, namely upwards above 58 km and downwards below this altitude.

3.2.5. $HO/HO_2/H_2O_2$

These qualify as Class III components since their chemical lifetimes are sufficiently

short that flux terms need not be considered. Their concentrations should decrease
rapidly above 80 km although the rate of decrease in the mesosphere is strongly
dependent on the branching ratios of the equivocal reaction (29).

In the mesosphere the most important production and loss terms have been discussed above in connection with the longer-lived Class II hydrogen compounds. The
major sinks are reactions (26) and (29a). Reactions (9) to (11) interconvert HO and
HO_2. In the stratosphere HO is produced by the reaction of $O(^1D)$ with H_2O, CH_4
and H_2. But in this region the chemistry becomes much more complex because of the
coupling of the HO_x and NO_x reactions. Some of these reactions (45, 45, 46) consume
HO_x, some convert one radical to the other, e.g.

$$HO_2 + NO \rightarrow NO_2 + HO \tag{33}$$

while still others (19) and (44) regenerate HO.

Conversion of HO to HO_2 will also take place through the sequence (31) and (32)
involving CO. Radical-radical recombination will, of course, provide one-way loss
processes. In addition to Equation (26) recombination of HO_2

$$HO_2 + HO_2 \rightarrow H_2O_2 + O_2 \tag{34}$$

becomes important at these altitudes and H_2O_2 can attain an appreciable concentration in the stratosphere. In fact, 1-D models indicate the order of abundance,
$H_2O_2 > HO_2 > HO$. Although H_2O_2 is a stable chemical it has a relatively short lifetime in the stratosphere because of its loss by

$$HO + H_2O_2 \rightarrow H_2O + HO_2. \tag{35}$$

The net effect of reactions (34) and (35) is the same as reaction (26). Schiff and McConnell (1973) have pointed out that the occurrence of this pair of reactions makes the
HO and HO_2 densities less dependent on the uncertainty in the rate constant of reaction (26). Nevertheless there is a wide spread in the densities of these radicals derived
by different modellers (Crutzen, 1973b; Brasseur and Nicolet, 1973; Hesstvedt, 1972;
Hudson, 1973).

3.2.6. NO_x *Chemistry*

Although the conversion rate of one NO_x constituent to another is fast (of the order of
100 s for $NO_2 \underset{O_3}{\overset{h\nu}{\rightleftharpoons}} NO_3$), production and loss of total NO_x in the stratosphere are slow.
Thus, while each individual NO_x constituent may be considered to be in photo-
chemical stationary state and thus belong to Class III, total NO_x must be treated as
belonging to Class II.

In the thermosphere and upper mesosphere NO is produced through ionization
processes. Because NO has a low ionization potential, NO^+ becomes the dominant
ion in this region as a result of such ion-molecule reactions as

$$N_2^+ + O \rightarrow NO^+ + N. \tag{36}$$

Dissociative recombination of these ions produces atomic nitrogen

$$NO^+ + e \rightarrow N + O. \tag{37}$$

The nitrogen atoms will react with O_2 to form NO

$$N + O_2 \rightarrow NO + O. \tag{38}$$

Although reaction (38) is slow for ground state $N(^4S)$ atoms it is fast for N atoms in the first excited (^2D) state, which are also formed by reaction (37). At higher altitudes N atoms are also formed by direct dissociation of N_2 by photon and electron impact. The relative importance of these processes has been discussed by Strobel (1972).

Although it was originally believed that NO produced in the mesosphere would flow downward into the stratosphere, it was later recognized that predissociation in the $\delta(0, 0)$ band:

$$NO + h\nu \rightarrow N + O \tag{39}$$

followed by the rapid reaction

$$N + NO \rightarrow N_2 + O \tag{40}$$

also provides an NO sink in the mesosphere. In fact, this sink is so strong that the NO flux is mainly in the reverse direction, namely upward from the stratosphere to the mesosphere. However, NO may be transported horizontally, in the mesosphere, to high latitudes, where it can move downwards into the stratosphere, particularly during the long polar nights.

The main production process in the stratosphere is the reaction of $O(^1D)$, from O_3 UV photolysis, with N_2O which has a biological origin at the Earth's surface and is transported upwards into the stratosphere:

$$N_2O + O(^1D) \rightarrow 2NO. \tag{41}$$

Dissociation of N_2 by cosmic rays or polar cap events followed by reaction (38) is a secondary source of NO in the stratosphere and mesosphere.

NO and NO_2 have also been observed in the troposphere and results from anthropogenic and natural sources at the surface and from electrical storms. Although they will undoubtedly be carried into the stratosphere by rising air in tropical regions there will also be transport in the reverse direction by descending air masses. Calculations indicate that the troposphere actually acts as a sink, rather than a source for stratospheric NO_x.

The other NO_x compounds all derive from NO. The reactions believed to be mainly responsible for interconversion of NO and NO_2 are

$$NO + O_3 \rightarrow NO_2 + O_2 \tag{12}$$
$$NO_2 + O \rightarrow NO + O_2 \tag{13}$$
$$NO_2 + h\nu \rightarrow NO + O. \tag{42}$$

The principal reactions used in existing models which govern the concentrations of HNO_2 and HNO_3 are

$$NO + HO \overset{M}{\rightarrow} HNO_2 \tag{43}$$
$$HNO_2 + hv \rightarrow HO + NO \tag{44}$$
$$NO_2 + HO \overset{M}{\rightarrow} HNO_3 \tag{45}$$
$$HNO_3 + hv \rightarrow HO + NO_2 \tag{19}$$
$$HNO_3 + HO \rightarrow H_2O + NO_3. \tag{46}$$

The chemistry of NO_3 and N_2O_5 is still not firmly established but it appears unlikely that they will attain appreciable concentrations in the atmosphere or play significant roles in the chemistry.

The main uncertainties in the NO_x chemistry are the quantum yields of HNO_3 photolysis, reaction (19), and the concentration of HO radicals.

3.2.7. *Chlorine Chemistry*

There are no natural sources of Cl compounds in the upper atmosphere. HCl may be injected into the stratosphere by volcanoes and, possibly, by reaction of NaCl, originating from sea spray, with aerosol H_2SO_4, but the concentrations are likely to be small. By far the greatest source of atmospheric Cl compounds is anthropogenic. Freons, CCl_2F_2 and $CFCl_3$ are being introduced into the troposphere in increasing quantities through their use as refrigerants and in aerosol spray cans, and the industrial production of other chlorinated hydrocarbons has also been increasing rapidly. These halocarbons are relatively insoluble in H_2O and chemically inert, and therefore, will be transported into the stratosphere where they will be photodissociated at wavelengths shorter than 230 nm:

$$CF_2Cl_2 + hv \rightarrow CF_2Cl + Cl. \tag{47}$$

The chlorine atoms will be converted to HCl by the reactions

$$Cl + H_2 \rightarrow HCl + H \tag{48}$$
$$Cl + CH_4 \rightarrow HCl + CH_3 \tag{49}$$
$$Cl + HO_2 \rightarrow HCl + O_2 \tag{50}$$

and reconverted by the reactions

$$HCl + HO \rightarrow H_2O + Cl \tag{51}$$
$$HCl + O \rightarrow HO + Cl \tag{52}$$
$$HCl + hv \rightarrow H + Cl. \tag{53}$$

An appreciable concentration of ClO may be present in the stratosphere. It is formed by the reaction

$$Cl + O_3 \rightarrow ClO + O_2 \tag{20}$$

and reconverted to Cl atoms by

$$ClO + O \rightarrow Cl + O_2 \tag{21}$$
$$ClO + NO \rightarrow Cl + NO_2. \tag{54}$$

It is interesting to note that reaction (54) converts NO to NO_2 and that this reaction, along with reaction (22) is equivalent to reaction (12).

Chlorine chemistry is similar to NO_x chemistry in the absence of any stratospheric sinks. The main removal process is through HCl, either by aerosols, or by transport to the troposphere and subsequent rain-out. Nothing is known about removal of other chlorinated compounds by aerosols.

4. Status of Laboratory Measurements and Models

A great deal of high quality laboratory data has become available during the past few years on rate constants and mechanisms of reactions involving ground state atoms and molecules. Some of these are reviewed in this volume and tabulations of evaluated rate data are available *(vide infra)*. The kinetic data for most of the reactions which have been used by atmospheric modellers are sufficiently well established for their purposes, with some notable exceptions. These include the rate constant for the $HO_2 + HO_2$ reaction (34); the branching ratio for $H + HO_2$ reaction (29); and for the UV photolysis of H_2O, reaction (23); the quantum yield for the photolysis of HNO_3 and NO_3. Virtually nothing is known about the photochemistry of NH_3, so that it is not possible to assess its importance or even whether it acts as a sink or a source for NO_x. Some of the reactions required for a full understanding of the photochemistry of Cl compounds are still uncertain.

Considerable progress has also been made on the chemistry of electronically excited states. Quenching rate constants for $O(^1D)$, $O(^1S)$, $N(^2D)$, $O_2(^1\Sigma)$ and $O_2(^1\Delta)$ have been measured with reasonable accuracy although the temperature dependencies have not. In many cases the products of the quenching process have not been established. One example which was discussed at length at the last Institute in connection with energy exchange in the mesosphere and thermosphere

$$O(^1D) + N_2 \rightarrow N_2^\ddagger(v > 0) + O(^3P) \tag{55}$$

now seems to be settled by the work of Slanger and Black (1974) which shows that vibrational excitation occurs for 35% of the collisions.

Although the atmospheric processes which produce most of the electronically excited species have been identified and their rates measured, two important examples may be cited where uncertainties still exist. One is the quantum yield for $O(^1D)$ production from O_3 photolysis near the long wavelength limit of the Hartley band at 310 nm. The importance of this quantity will be appreciated from the fact that $O(^1D)$ reactions (22) and (41) initiate most of the HO_x and NO_x chemistry up to about 70 km, coupled with the fact that the solar flux intensity is a rapidly increasing function of wavelength in this spectral region. The other example is the temperature coefficient for the Chapman reaction

$$O+O+O \rightarrow O(^1S)+O_2 \tag{56}$$

which has been studied only at 298 K (Felder and Young, 1972). The intensity of the green line emission ($O^1D \leftarrow {}^1S$) has been used to estimate the concentration of O atoms near 100 km but the calculations are of doubtful validity in the absence of the appropriate rate constant for the reaction and for the quenching rates at the temperature prevailing at these altitudes.

The chemical kinetics of vibrationally and rotationally excited molecules are still in a relatively primitive state. Little is known about the distribution of rotational, vibrational and kinetic energy from exoergic processes such as photolysis, e.g.

$$NO_2 + h\nu(\lambda < 398 \text{ nm}) \rightarrow NO^\dagger + O + \text{K.E.};$$

termolecular recombination, e.g.

$$H + O_2 + M \rightarrow HO_2^\dagger + M + \text{K.E.}$$

or bimolecular reactions, e.g.

$$H + O_3 \rightarrow HO^\dagger + O_2 + \text{K.E.}$$

In the last example, a large fraction of the total energy of reaction initially appears as vibrational excitation of HO. This vibrational energy can not only affect the rate of a subsequent reaction

$$HO^\dagger + O_3 \rightarrow HO_2 + O_2$$

but might even alter the nature of the products, since the production of $H + 2O_2$ become energetically possible in this example when the vibrational level of HO is greater than 1.

Although the degradation of kinetic and rotational and even vibrational energy will be rapid at low levels in the atmosphere it will not at higher altitudes where such 'hot' molecules may play significant roles in the chemistry.

Let us now consider some of the uncertainties in the models used to describe the ambient atmosphere and to predict the perturbations caused by human activity. All such models are approximations. A complete model would have to incorporate at least 50 chemical reactions, the rate constants of which are known with varying degrees of uncertainty. Furthermore, the possibility remains that one or more reactions which might prove to be very important have been overlooked. Such a model would also have to include a full description of atmospheric motion. Even in the absence of chemical reactions, if such a model existed, it would be capable of exact weather forecasting, an achievement not yet realized! Compromises must be made between atmospheric chemistry and atmospheric dynamics. The most comprehensive chemistry is usually treated in one-dimensional models in which semi-empirical, eddy diffusion coefficients are used to characterize, essentially, only vertical transport.

Attempts have been made at two and three dimensional models, but generally at the expense of the chemistry.

The major complaint which can be levelled at current models is the lack of explicit estimates of uncertainties. Frequently a modeler will select particular values for a set of rate constants, solar fluxes and eddy diffusion coefficients from which he will generate curves for the height profiles of atmospheric constituents with no indication of the error bars with which the experimentalist is familiar. In more recent models a series of profiles is given for a range of values of one of the parameters. But the uncertainties in the parameters are not additive in any linear way and one is still left with no real confidence measure in any given profile. To illustrate these problems a typical one-dimensional model for NO_x will now be explored for its sensitivity to uncertainties in some of the rate constants, to temporal variations in O_3 densities and insolation and to the inclusion of some reactions not previously considered (McConnell *et al.*, 1975).

We will take as our reference model one described by McConnell and McElroy (1973) in which the total amount of NO_x is assumed to be constant in the stratosphere. The rate constants have been updated to comply with the recommended values of the N.B.S. compilation.

Let us first consider the effect of introducing two new reactions not considered in the original model. Simonaitis and Heicklen (1974) have found evidence that the reaction

$$HO_2 + NO_2 \rightarrow HNO_2 + O_2 \tag{17}$$

occurs with a rate constant $k_{17} \geqslant 3 \times 10^{-13}$ cm^3 s^{-1}. Hampson (1974) has suggested that HNO_2 may be oxidized by O_3

$$HNO_2 + O_3 \rightarrow HNO_3 + O_2. \tag{18}$$

If these two reactions are added to the reactions previously considered in the NO_x chemistry to govern the partitioning of NO_x among its main constituents then the following ratios can be obtained from the PCSS assumption

$$\frac{[NO_2]}{[NO]} = \frac{k_{12}O_3}{J_{42} + k_{13}[O]} \equiv a$$

$$\frac{[HNO_2]}{[NO]} = \frac{a k_{17}[HO_2] + k_{43}[HO]}{J_{44} + k_{18}[O_3]} \equiv b$$

and

$$\frac{[HNO_3]}{[NO_2]} = \frac{k_{45}[HO][M] + R(k_{17}[HO_2] + k_{43}/a[HO][M]}{J_{19} + k_{46}[HO]} \equiv c$$

where R is the fraction of HNO_2 converted to HNO_3 by reaction (18) and is given by

$$R = \frac{k_{18}[O_3]}{J_{44} + k_{18}[O_3]}.$$

Figure 6 illustrates the sensitivity of the NO height profiles to the two new reactions considered. A is the reference profile which omits these reactions. If either of these two reactions are introduced separately the effect is not large. Thus, for curve B, a value of $k_{17} = 2 \times 10^{-12}$ cm^3 s^{-1} is adopted, about an order of magnitude larger than the upper limit of Simonaitis and Heicklen (1974) while reaction (18) is

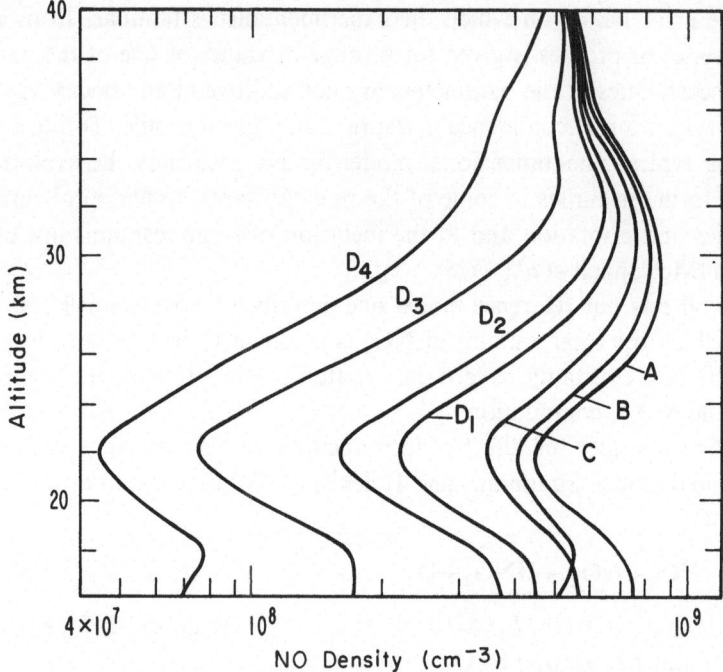

Fig. 6.   Model Calculations to Show Effect of Rate Constants.

    A – McConnell and McElroy (1973) model C, with updated rate constants, $k_{17} = k_{18} = 0$;
    B – $k_{17} = 2 \times 10^{-12}$, $k_{18} = 0$;
    C – $k_{17} = 0$; $k_{18} = 10^{-13}$;
    D_1 – $k_{17} = 2 \times 10^{-12}$; $k_{18} = 10^{-17}$
    D_2 – $k_{17} = 2 \times 10^{-12}$; $k_{18} = 2 \times 10^{17}$
    D_3 – $k_{17} = 2 \times 10^{-12}$; $k_{18} = 10^{-16}$
    D_4 – $k_{17} = 2 \times 10^{-12}$; $k_{18} = 10^{-14}$

assumed not to occur ($k_{18} = 0$). Under these conditions $b \ll 1$ and only a small fraction of NO$_x$ will be present as HNO$_2$. Similarly if only reaction (18) is introduced the effect is also small. Curve C is obtained with $k_{17} = 0$ and $k_{18} = 1 \times 10^{-13}$ cm^3 s^{-1}. But if both reactions are included a large synergistic effect occurs even if k_{18} is as small as 10^{-17} cm^3 s^{-1}. Curves D_1, D_2, D_3 and D_4 are obtained with $k_{17} = 2 \times 10^{-12}$ and $k_{18} = 10^{-17}$, 2×10^{-17}, 10^{-16} and 10^{-14} cm^3 s^{-1}, respectively. Under these conditions NO$_2$ is converted to HNO$_2$ by reaction (17) which, in turn, is converted to the relatively long-lived HNO$_3$.

Next, let us consider the effect of variations in O$_3$ densities on the partitioning of NO$_x$. It will be seen from expressions a and b that the ratios depend explicitly on O_3 which is known to have large daily and seasonal variations. There is, in addition, an implicit dependence through J_{42} and $[HO]$. Figure 7 shows height profiles calculated for model A, using average O_3 profiles given by Dütsch (1974) for Arosa, Switzerland. The differences in curves 1 to 4 are due to seasonal differences in insolation and in O_3 densities. Also shown are the differences in calculated NO_x den-

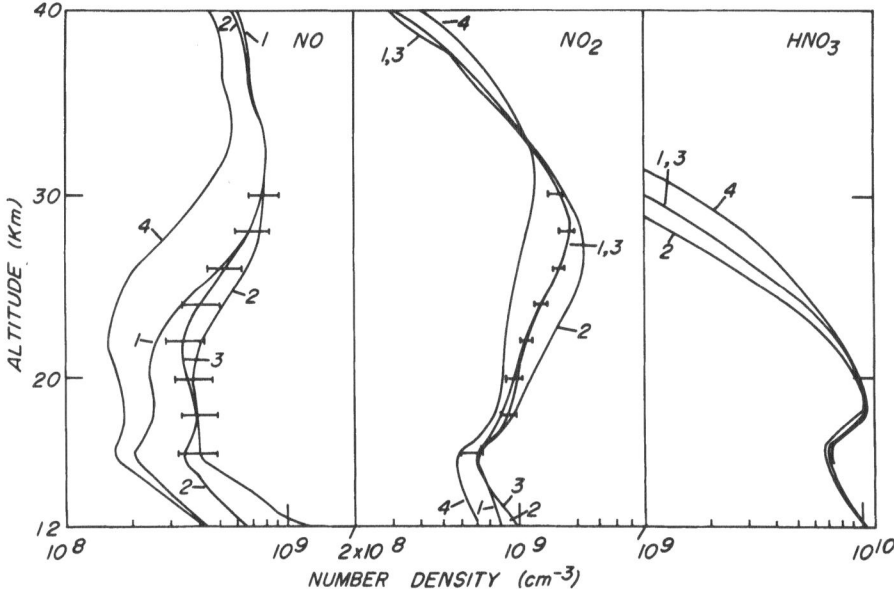

Fig. 7.   NO, NO₂ and HNO₃ densities calculated using the total NOₓ profile C from McConnell and McElroy (1973) and the O₃ densities given by Dütsch (1974) for Arosa, Switzerland. Curve 1 refers to March averages, 2 June, 3 September, 4 December. The bars which appear on curves − 3 indicate the effect on NO and NO₂ of a ±25% change in O₃, below 30 km, in a time scale such that HNO₃ does not change.

sities resulting from short term fluctuations of $\pm 25\%$ in O_3 densities for September, which are typical of daily variations. The largest effect is seen to occur with NO, which shows a change of a factor of three with season and an approximately linear relationship with daily fluctuations at 24 km. Little seasonal variation is seen at high altitudes for NO_2 or at low altitudes for HNO_3, since these are the major NO_x constituents at these respective altitudes and the model has assumed a fixed total NO_x.

Figures 3 and 4 compare existing NO measurements with some limiting model calculations. Curve X has been calculated with several parameters chosen at the limits of their uncertainties which would maximize the NO densities, viz June average O_3 profiles; quantum yield for HNO_3, $\Phi(HNO_3)=1$; $k_{17}=k_{18}=0$; $k_{34}=2\times 10^{-10}$ cm^3 s^{-1}; and NH_3 included as an additional NO source as in model A of McConnell and McElroy. Conversely, curve Y was calculated with the limits of uncertainty in these parameters chosen to minimize NO, viz: March average O_3 profiles $\Phi(HNO_3)$ as in McConnell and McElroy; $k_{17}=3\times 10^{-13}$ cm^3 s^{-1}; $k_{18}=1\times 10^{-13}$ cm^3 s^{-1};

$k_{34} = 5 \times 10^{-11}$ cm^3 s^{-1}. These 'limiting models' are capable of accommodating almost all the NO measurements as well as most of the measurements of NO$_2$ and HNO$_3$. Moreover, there is increasing evidence that each of the NO$_x$'s shows considerable long and short term variability which the current 1-D models are incapable of handling. Therefore, although the models can provide average, order-of-magnitude, NO$_x$ height profiles, they can not be used as criteria for judging the validity of any given measurement. It is therefore not possible, at present, to determine whether differences between measurements taken at different places and times by different techniques are due to real variability in the constituent or to errors in the measurements. Much more data must be obtained as a function of time and location to delineate the natural variability, ideally with simultaneous measurements of all the minor constituents, each determined by several techniques.

In summary then, although considerable progress has been made during the past 2 yr there appears to be enough work still needed to keep laboratory and atmospheric experimentalists and atmospheric modelers gainfully employed for some time in the future.

Acknowledgments

The author wishes to express his gratitude to Dr T. M. Hard, Transport Systems Centre, D.O.T. for providing the data for Figures 1 and 2 and to Dr J. C. McConnell for the model calculations.

References

Ackerman, M. and Muller, C.: 1972, *Nature* **240**, 300.
Ackerman, M. and Muller, C.: 1973, *Pure Appl. Geophys.* **106–108**, 1325.
Ackerman, M., Fontanella, J. C., Frimont, D., Girard, A., Louisnard, N., and Muller, C.: 1974a, *Aeronomic Acta A.* 133.
Ackerman, M., Frimont, D., Muller, C., Nevejas, D., Fontanella, J. C., Girard, A., Gramont, L., and Louisnard, N.: 1974b, *Can. J. Chem.* **52**, 1532.
Anderson, J.: 1971, *J. Geophys. Res.* **76**, 4634; **76**, 7820.
Barth, C. A.: 1964, *J. Geophys. Res.* **69**, 3301.
Baulch, D. L., Drysdale, D. D., Horne, D. G., and Lloyd, A. C.: 1973, Evaluated Kinetic Data for High Temperature Reactions, Butterworth and Co., London, Chemical Rubber Co., Cleveland.
Brasseur, G. and Nicolet, M.: 1973, *Planetary Space Sci.* **21**, 939.
Brewer, A. W., McElroy, C. T., and Kerr, J. B.: 1973, *Nature* **246**, 129.
Cicerone, R. J., Stolarski, R. S., and Walters, S.: 1974, *Science* **185**, 1165.
Crutzen, P. J.: 1973a, in B. M. McCormac (ed.), *Physics and Chemistry of Upper Atmospheres*, D. Reidel Publ. Co., Dordrecht-Holland, pp. 110–125.
Crutzen, P. J.: 1973b, *Pure Appl. Geophys.* **106–108**, 1385.
Crutzen, P. J.: 1974, *Geophys. Res. Letters* **1**, 205.
Donahue, T. M.: 1975, this volume, pp. 289.
Dütsch, H. U.: 1974, *Can. J. Chem.* **52**, 1491.
Farmer, C. B., Raper, O. F., Toth, R. A., and Schindler, R. A.: 1974, *Proc. Third Conf. on CIAP.*
Felder, W. and Young, R. A.: 1972, *J. Chem. Phys.* **56**, 6028.
Fontanella, J. C., Girard, A., Gramont, L. and Louisnard, N.: 1974, *Proc. Third Conf. on CIAP.*
Hampson, J.: 1974, private communication.
Hampson, R. F. (ed.): 1973, *Chemical Kinetics Data Survey*, NBSIR 73-207.
Harries, J. E., Birch, J. R., Fleming, J. W., Stone, N. W. B., Moss, D. G., Swann, N. R. W., and Neill, G. F.: 1974, *Proc. Third Conf. on CIAP.*

Henderson, W. R.: 1971, *J. Geophys. Res.* **76**, 3166.

Henderson, W. R. and Schiff, H. I.: 1970, *Planetary Space Sci.* **18**, 1527.

Hesstvedt, E.: 1972, Report DOT-TST-90-2, U.S. Dept. of Transportation, Washington, D.C.

Hudson, F. P.: 1973, *Proc. Second Conf. on CIAP*, pp. 115–129.

Hunten, D. M. and Strobel, D. F.: 1974, *J. Atmospheric Sci.* **31**, 305.

Johnston, H. S.: 1974, private communication.

Kaufman, F.: 1975, this volume, pp. 219.

Lazrus, A. L. and Gandrud, B. W.: 1974, *Proc. Third Conf. on CIAP*.

Lowenstein, M., Paddock, J. P., Poppoff, I. G., and Savage, H. F.: 1974, *Nature* **249**, 817.

Liu, S. C. and Donahue, T. M.: 1974, *J. Atmospheric Sci.* **31**, 1118.

McConnell, J. C. and McElroy, M. B.: 1973, *J. Atmospheric Sci.* **8**, 1465.

McConnell, J. C., Ridley, B. A., and Schiff, H. I.: 1975, *J. Geophys. Res.* (to be published).

McElroy, M. B., Wofsy, S. C., Penner, J. E., and McConnell, J. C.: 1974, *J. Atmospheric Sci.* **31**, 287.

McNesby, J. R., Tanaka, I., and Okabe, H.: 1962, *J. Chem. Phys.* **36**, 605.

Meira, L. G.: 1971, *J. Geophys. Res.* **69**, 3301.

Molina, M. J. and Rowland, F. S.: 1974, *Nature* **249**, 810.

Murcray, D. G., Goldman, A., Csoeke-Poeckh, Murcray, F. H., Williams, W. J., and Stocker, R. N.: 1973, *J. Geophys. Res.* **78**, 7033.

Murcray, D. G., Goldman, A., Williams, W. J., Murcray, F. H., Van Allen, J., and Schmidt, S. C.: 1974, *Proc. Third Conf. on CIAP*.

Nicolet, M.: 1974, *Can. J. Chem.* **52**, 1381.

Patel, C. K. N., Burkhardt, E. G., and Lambert, C. A.: 1974, *Science* **184**, 1173.

Philbrick, C. R., Narcisi, R. S., Good, R. E., Hoffman, H. S., Keneshea, T. J., McCleod, M. A., Zimmerman, S. P., and Reinisch, B. W.: 1973, *Space Res.* **13**, 441.

Ridley, B. A., Schiff, H. I., Shaw, A. W., Bates, L., Howlett, L. C., LeVaux, H., Megill, L. R., and Ashenfelter, T. E.: 1973, *Nature*, **245**, 310.

Ridley, B. A., Schiff, H. I., Shaw, A. W., Bates, L., Howlett, L. C., LeVaux, H., Megill, L. R., and Ashenfelter, T. E.: 1974, *Planetary Space Sci.* **22**, 19.

Schiff, H. I. and McConnell, J. C.: 1973, *Rev. Geophys. Space Phys.* **11**, 925.

Simonaitis, R. and Heicklen, J.: 1974, *J. Phys. Chem.* **78**, 653.

Slanger, T. G. and Black, G.: 1974, *J. Chem. Phys.* **60**, 468.

Strobel, D. F.: 1972, *Radio Sci.* **7**, 1.

Swider, W. and Narcisi, R.: 1975, *J. Geophys. Res.* **80**, 655.

Welge, K. H. and Stuhl, F.: 1967, *J. Chem. Phys.* **46**, 2440.

Wofsy, S. C. and McElroy, M. B.: 1974, *Can. J. Chem.* **52**, 1582.

Zipf, E. C., Borst, W. I., and Donahue, T. M.: 1970, *J. Geophys. Res.* **75**, 6371.

PART II

PHYSICAL PROCESSES

ATMOSPHERIC PHYSICS

JAMES C. G. WALKER

Arecibo Observatory, National Astronomy and Ionosphere Center, P.O. Box 995,
Arecibo, Puerto Rico 00612

1. Introduction

The atmosphere is a compressible fluid bound to the planet by the force of gravity. Pressure decreases with altitude because the weight of the overlying gas decreases. Because air is compressible the density decreases with altitude as the pressure decreases. The variation of pressure and density with altitude is described by the barometric law.

2. Barometric Law

In a static atmosphere the pressure at any level, $p(z)$, is equal to the weight of the gas in a vertical column of unit cross section above that level

$$p(z) = \int_{z}^{\infty} \varrho(h)\, g(h)\, \mathrm{d}h, \tag{1}$$

where $\varrho(h)$ is the density at height h, and $g(h)$ is the acceleration due to gravity. Differentiation of (1) yields an expression for the vertical pressure gradient,

$$\frac{\mathrm{d}p}{\mathrm{d}z} = -\varrho g. \tag{2}$$

According to the ideal gas law,

$$p = nkT = \varrho kT/m, \tag{3}$$

where n is the number density, k is Boltzmann's constant, T is the absolute temperature, and m is the mean molecular mass of the gas.

Substituting for ϱ in Equation (2) and rearranging, we obtain

$$\frac{\mathrm{d}p}{p} = -\frac{mg}{kT}\, \mathrm{d}z. \tag{4}$$

Integration of Equation (4) yields

$$p(z) = p(z_0) \exp\left[-\int_{z_0}^{z} \frac{\mathrm{d}h}{H(h)}\right], \tag{5}$$

where $p(z_0)$ is the pressure at reference level z_0, and $H = kT/mg$ is called the scale height. Given the profile of H as a function of altitude, the pressure at any level can be evaluated in terms of the pressure at the reference level by means of Equation (5).

Atmospheric scale heights are of the order of 10 km. This means that pressure (and therefore density, because pressure and density are proportional according to the ideal gas law) varies significantly over height increments of a few tens of kilometers. The variation of the gravitational acceleration (inversely proportional to the square of the distance from the center of the Earth) is very much slower; it can be neglected in many atmospheric applications.

If g is assumed to be constant (as well as m) Equation (1) can be written

$$n(z)\, H(z) = \int_z^\infty n(h)\, dh. \tag{6}$$

In words, the height integral of the number density (called the column density) above a given level is equal to the product of the number density and the scale height at that level. This result is independent of the temperature profile. The scale height is sometimes called the height of the homogeneous atmosphere because it is the thickness that the atmosphere would have if the density were constant.

2.1. INTEGRATION OF THE BAROMETRIC EQUATION

2.1.1. *Constant Scale Height*

The variation of pressure with height is particularly simple in an atmosphere with constant scale height (usually an isothermal atmosphere with constant mean molecular weight and negligible variation of gravitational acceleration). Equation (5) becomes

$$p(z) = p(z_0) \exp\left[-\left(\frac{z - z_0}{H} \right) \right]. \tag{7}$$

Pressure decreases exponentially with height, and a plot of the logarithm of pressure against height is a straight line. For this reason it is customary and convenient to use semilog plots for the profiles of pressure, density, and related quantities.

In the real atmosphere, of course, the scale height varies with altitude, and the barometric equation must usually be integrated numerically. Some simplifications are possible, however.

2.1.2. *Geopotential Altitude*

For precise work, a simple transformation of the altitude scale removes the variation of gravitational acceleration. Let

$$d\eta = \frac{g(z)}{g(z_0)}\, dz. \tag{8}$$

Then

$$\eta = \int_{R_e}^{R_e+z} (R_e/r)^2 \, dr = \frac{R_e z}{R_e+z}, \tag{9}$$

where R_e is the radius of the Earth and $r = R_e + z$. For $z \ll R_e$, there is little difference between geopotential height, η, and geometric height, z. At greater heights, the geopotential height of a given level is less than the geometric height.

Changing the variable of integration in Equation (5) yields

$$p(z) = p(0) \exp\left[-\int_0^{\eta(z)} \frac{mg(0)}{kT(\eta')} \, d\eta' \right], \tag{10}$$

and the variation of g has been removed from the integrand.

2.1.3. Linear Scale Height Variation

The barometric equation can be integrated analytically for simple variations of scale height with altitude. As an example consider a linear variation,

$$H(z) = H(z_0) + \beta(z - z_0), \tag{11}$$

where β is a constant. Then $dH = \beta \, dz$, and a change of variable in Equation (5) yields

$$p(z) = p(z_0) \exp\left[-\int_{H(z_0)}^{H(z)} \frac{dH}{\beta H} \right] = p(z_0) \left[\frac{H(z_0)}{H(z)} \right]^{1/\beta}. \tag{12}$$

This expression is convenient to use in situations where the scale height profile can be approximated by linear segments. The expression looks very different from the exponential behavior derived for the case of constant scale height. The binominal expansion can, however, be used to show that the two expressions are very nearly the same for $z - z_0 \ll H$.

3. Temperature Profile

We see that the temperature controls the rate of decrease of pressure and density with altitude. Integration of the barometric equation requires knowledge of the scale height profile and therefore of the temperature profile. A representative temperature profile shows minima of about 200 K at heights of about 15 km and 85 km. At the ground the temperature is about 300 K, and a similar value is reached at a local maximum near 50 km. Above 85 km the temperature rises rapidly at first and then more gradually until the atmosphere becomes isothermal at a temperature of about 1000 K at heights above 250 km. There are significant variations of the profile with position and time, particularly at the highest altitudes, but the pattern of temperature maxima and minima does not change.

This pattern reflects directly the sources of heat in the atmosphere provided by the absorption of solar radiation of different wavelengths. Extreme UV radiation ($\lambda < 1000$ Å) is strongly absorbed by all atmospheric constituents. The absorption rate is a maximum at levels of about 180 km. The high temperatures at the top of the atmosphere reflect this heat source. Because the solar spectrum at these wavelengths varies markedly with time and solar activity, the temperature at the top of the atmosphere varies also.

Near UV solar radiation (2000 Å $< \lambda <$ 3000 Å) is absorbed by O_3 at heights of around 50 km, leading to the temperature maximum at intermediate heights. Visible and IR solar radiation are largely absorbed by the ground, which in turn heats the bottom of the atmosphere.

The temperature minima between the levels of energy input result from the loss of heat to space by emission of IR radiation.

4. Atmospheric Nomenclature

The temperature profile with its maxima and minima provides the basis for the most common system of naming the different levels of the atmosphere. In this system, the troposphere is a region of steadily decreasing temperature extending upwards from the ground to the tropopause, at a height of about 15 km in the tropics and about 10 km at high latitudes.

At the tropopause there is a marked discontinuity in the temperature gradient, the atmosphere becoming nearly isothermal in the overlying region, the lower stratosphere. In the upper stratosphere, temperature increases to a maximum at a height of about 50 km. This maximum is identified as the stratopause. Above the stratopause is a region of decreasing temperature called the mesosphere that extends to the temperature minimum, the mesopause, at a height of about 85 km.

Above the mesopause is the thermosphere, where the temperature increases at first before leveling off at higher altitudes. Thermopause temperature is a term used for the temperature of the isothermal region at the top of the atmosphere.

As altitude increases and density decreases, the mean free path between collisions of atmospheric molecules increases steadily. At about 450 km the critical level is reached, at which the mean free path is equal to the scale height. Above the critical level is the exosphere. The gas in the exosphere does not behave strictly as a continuum fluid. Instead, individual neutral atoms and molecules follow ballistic trajectories in the Earth's gravitational field, colliding only infrequently with one another. Exospheric temperature has the same meaning as thermopause temperature.

Other names are based on the properties of the charged constituents of the atmosphere. The ionosphere is the region of the atmosphere, above about 50 km, where free electrons are sufficiently abundant to affect the propagation of radio waves. The protonosphere is the region of the ionosphere above 1000 or 2000 km in which H^+ is the predominant positively-charged species. The density of the thermal plasma (H^+ and electrons) in the protonosphere exhibits a marked decrease at heights of about 4 R_E over the equator. This decrease is called the plasmapause. Inside the plasmapause is the plasmasphere.

Energetic charged particles are trapped in the geomagnetic field at heights between about 1000 km and 10 R_E on the sunlit side of the Earth. This region is called the magnetosphere. It is bounded above by the magnetopause, where the geomagnetic field gives way to the interplanetary magnetic field. The magnetopause is the top of the atmosphere. Material inside the magnetopause travels with the Earth on its journey through space. Material outside the magnetopause does not.

Two other names are based on the altitude variation of the mean molecular mass of the atmospheric gases. In the homosphere, below about 100 km, the relative concentrations of the major atmospheric gases do not vary with height, and mean molecular mass is constant. In the overlying heterosphere light gases become progressively more abundant relative to heavy gases, and the mean molecular mass decreases with increasing height.

5. Gravitational Separation

In our discussion of the barometric law so far we have treated the atmosphere as composed of a single gas of molecular mass equal to the mean molecular mass. Then Equation (2),

$$\frac{dp}{dz} = -\varrho g = -nmg, \tag{13}$$

represents a balance between a pressure gradient force and the force of gravity. In fact, the atmosphere is composed of a mixture of gases of different masses, m_i, and gravity pulls more strongly on the heavy ones than on the light ones.

If forces are to be balanced for each constituent individually, Equation (13) becomes

$$\frac{dp_i}{dz} = -n_i m_i g, \tag{14}$$

where p_i is the partial pressure of constituent i, and n_i is the number density. Note that if we sum Equation (14) over all constituents we recover Equation (13), since $p = \sum_i p_i$, $n = \sum_i n_i$, and $m = \sum_i n_i m_i / n$. So the barometric law is satisfied in an atmosphere in which the forces on the individual constituents are balanced.

In such an atmosphere, however, we have

$$p_i(z) = p_i(z_0) \exp\left[-\int_{z_0}^{z} \frac{dh}{H_i(h)}\right], \tag{15}$$

where

$$H_i = \frac{kT}{m_i g}. \tag{16}$$

Light gases like H and He have large scale heights, while heavy gases like Ar and CO_2

have small scale heights. Therefore the densities of the heavy constituents decrease more rapidly with altitude than the densities of the light constituents, and the mean molecular mass decreases with increasing height.

This phenomenon of gravitational separation is observed in the heterosphere above about 100 km, but not in the underlying homosphere. Since gravitational forces on different constituents are as different below 100 km as above, the absence of gravitational separation in the homosphere requires some explanation.

6. Vertical Mixing

Vertical motions carry air from one level of the atmosphere to another. As it moves, the air takes with it a composition that is characteristic of the level from which it came. If enough of these vertical motions occur, any initial inhomogeneity in composition must be removed. Thus, a well-mixed layer of the atmosphere is one in which composition does not vary with altitude. Mixing is more rapid than gravitational separation in the homosphere. The reverse is the case in the heterosphere.

The ease with which vertical motions and therefore mixing can occur depends on the variation of temperature with altitude. Imagine that a parcel of air is displaced to a higher altitude by some chance perturbation. If the air in the parcel is more dense than the surrounding atmosphere at the new level, the parcel will tend to sink back down to the level from which it originated. If, on the other hand, the air in the displaced parcel is less dense than the surrounding atmosphere, buoyant forces will tend to drive the parcel still further upwards. In the first case the atmosphere is said to be stable, vertical motion is inhibited, and conservative gas properties such as composition are carried relatively slowly from one altitude to another. In the second case the atmosphere is unstable, vertical motions occur spontaneously, a phenomenon known as free convection, and mixing from one altitude to another is rapid.

As our imaginary parcel of air rises through the atmosphere it takes on, at each instant, the pressure of the surrounding gas. The temperature of the parcel may differ from that of the surroundings, however, because heat is transferred relatively slowly across the boundaries of the parcel. According to the ideal gas law, density is inversely proportional to temperature for gases at the same pressure, so whether the air in the parcel is more or less dense than the surrounding air depends on whether the temperature of the parcel is less than or greater than the temperature of the surrounding atmosphere. It is for this reason that the stability of the atmosphere depends on the rate of change of temperature with altitude.

As the parcel rises to levels of lower pressure it must expand, and common experience tells us that temperature falls as a gas expands. As it expands, the gas parcel performs mechanical work on the surrounding gas, and this work draws on the internal, thermal energy of the parcel, causing the temperature to fall. Therefore, unless the temperature of the surrounding atmosphere also decreases with increasing altitude, the parcel of displaced air will be cooler and more dense than the surrounding atmosphere; it will therefore tend to fall back to its original level.

We see that regions of the atmosphere in which temperature is either constant or increasing with altitude are stable. One such region is the thermosphere. The increase of temperature with height in the thermosphere inhibits mixing, which is part of the reason why gravitational separation occurs there.

7. Diffusion

The other part of the explanation of the difference between homosphere and hetero-sphere is that gravitational separation is rapid at high altitudes and slow at low altitudes.

Imagine an initially homogeneous atmosphere, in which all mixing ratios are independent of altitude (mixing ratio $= n_i/n$). Now, allow gravity to act on the dif-ferent constituents to move the atmosphere towards a condition of gravitational separation. The heavy constituents must move downwards while the light constituents move upwards. Thus, the approach to gravitational separation requires that the different constituents diffuse through one another. This relative motion ceases only when the pressure gradient and gravitational forces on each constituent are balanced as in Equation (14). When such a balance exists, the atmosphere is said to be in diffusive equilibrium.

The relative motion of the different constituents is inhibited by collisions between them. Collisions are frequent at low altitudes where densities are high and are in-frequent at high altitudes. Diffusion is therefore fast at high altitudes and slow at low altitudes.

Let us examine this phenomenon more quantitatively in terms of diffusion time, which is loosely defined as the time for gravitational separation to cause a significant change in the density of a given constituent at a given level of the atmosphere.

We may estimate the order of magnitude of the diffusion time for constituent i in the following way: Consider an isothermal atmosphere and suppose that the number density of constituent i, n_i, is initially independent of altitude. Constituent i must diffuse through a stationary background gas of density N much greater than n_i. Each molecule of i collides with a molecule of the background gas on average every τ seconds, where

$$\tau = \frac{1}{N\sigma v}. \tag{17}$$

In this expression, σ is the collision cross section, and v is the mean thermal speed of the i molecules. We assume that the velocities with which i molecules emerge from collisions are completely random, so the average vertical velocity of i molecules after collisions is zero.

During the period between collisions, these molecules are accelerated downwards by the force of gravity; on the average they are displaced downwards a distance of $g\tau^2/2$ between each collision. Since the average time between collisions is τ, the average vertical drift velocity of the i molecules (defined to be positive upwards) is

$$U_i = -g\tau/2, \tag{18}$$

and the flux is

$$\phi_i = n_i U_i = -n_i g\tau/2. \tag{19}$$

This downward flux of i molecules causes the density to change at a rate given by minus the divergence of the flux,

$$\frac{\partial n_i}{\partial t} = -\frac{\partial \phi_i}{\partial z} = \frac{n_i g}{z} \frac{\partial \tau}{\partial z} = \frac{n_i g}{2\sigma v} \frac{\partial}{\partial z}\left(\frac{1}{N}\right). \tag{20}$$

Since the background gas is stationary its density varies with altitude according to the barometric law, $N \propto e^{-z/H}$. Therefore

$$\frac{\partial}{\partial z}\left(\frac{1}{N}\right) = \frac{1}{NH}, \tag{21}$$

and

$$\frac{\partial n_i}{\partial t} = n_i \frac{g}{2\sigma v HN}. \tag{22}$$

As a measure of the diffusion time we may take the reciprocal of $(1/n_i)\,\partial n_i/\partial t$, or

$$\tau_D = 2\sigma v HN/g. \tag{23}$$

Collision cross sections for atmospheric gases are about 10^{-15} cm^2, and the mean thermal speed $v = \sqrt{3KT/m}$ is about 5×10^4 cm s^{-1}. Taking $H \sim 10$ km and $g = 980$ cm s^{-2} we find

$$\tau_D \sim 10^{-7} N \text{ s}. \tag{24}$$

The background gas density, N, is 2.7×10^{19} cm^{-3} at the ground, so the diffusion time is 2.7×10^{12} s or about 10^5 yr at the bottom of the atmosphere. This is very much longer than the times associated with motions in the troposphere, so we may conclude that mixing is much more important than gravitational separation in determining the variation of composition with altitude in the troposphere.

Although the diffusion time is proportional to ambient atmospheric density, N, and therefore decreases exponentially with altitude, diffusion continues to be negligible compared with mixing throughout the stratosphere and mesosphere. In the thermosphere, however, mixing is inhibited because temperature increases with altitude, and the diffusion time is short because of the low ambient density.

This is why gravitational separation becomes more important than mixing only at altitudes above about 100 km. The transition level, where mixing gives way to diffusion is frequently called the turbopause.

8. Diffusive Equilibrium

Above the turbopause we expect the atmosphere to be more or less in diffusive equilibrium. Under conditions of diffusive equilibrium, the variation with height of the partial pressure of constituent i is given by Equation (15).

It is convenient to define a dimensionless height by

$$x = \int_{z_0}^{z} \frac{dh}{H_o(h)}, \tag{25}$$

where H_O is the scale height of O. Then Equation (15) becomes

$$p_i(z) = p_i(z_0) \exp\left[-\frac{M_i}{16} x\right], \tag{26}$$

where M_i is the molecular mass of constituent i in atomic mass units. We see that the partial pressures of the different constituents in a diffusive equilibrium atmosphere are all exponential functions of x – they appear as straight lines on plots of log partial pressure against x. The slopes of these straight lines depend on the molecular weight of the constituent, but not on the temperature. Temperature enters only in the relationship between x and altitude.

Remembering that the temperature of the thermopause is variable, let us ask what will happen if the thermopause temperature changes with time, while the partial pressures in the vicinity of the turbopause, $p_i(z_0)$, do not change. The relationship between x and altitude is changed by the change in temperature, but the relationship between partial pressure and x is not affected. This important result can be expressed as follows: Changes in the temperature profile do not affect the partial pressures of the individual constituents at a given total pressure level, provided the atmosphere is in diffusive equilibrium and provided the boundary conditions, $p_i(z_0)$ do not change.

Since the partial pressure of each constituent at a given level is proportional to the column density of that constituent above the level in question, this result indicates that there are always the same number of molecules of each constituent above the altitude at which total pressure achieves a given value. This in turn indicates that changes in the temperature profile, no matter how rapid, do not cause any relative motion of the different constituents. In the absence of relative motion there can not be any departures from diffusive equilibrium. Such departures, if they occur, must be caused by processes other than changes in the temperature profile of the thermosphere.

There are simple expressions that can be used to check for departures from diffusive equilibrium in a body of data on atmospheric composition. Suppose we know the partial pressures of constituents i and j as functions of altitude. The quotient $[p_i(z)]^{M_j}/[p_j(z)]^{M_i}$ is independent of altitude if these constituents are in diffusive equilibrium. Alternatively, if we assume that the atmosphere is in diffusive equilibrium, and if we know the partial pressures at a reference altitude, this quotient can be used to derive the partial pressures of all constituents as functions of altitude from knowledge of the pressure profile of a single constituent. This procedure can provide an atmospheric model that is consistent with a rocket measurement, say, of a single constituent.

It is most usual to have data on the number densities, but not on the partial pressures of the different constituents of the thermosphere. The unknown temperature can be eliminated from the constant quotient by introducing a third constituent, k. After some rearrangement we find that the product

$$n_k^{M_j - M_i} n_j^{M_i - M_k} n_i^{M_k - M_j}$$

should be independent of altitude in an atmosphere in diffusive equilibrium. As an example, suppose that the densities of He, O, and N_2 have been measured by a satellite-borne mass spectrometer. Any variation in the value of $n(N_2) n(He)/n^2(O)$ reflects either departures from diffusive equilibrium or else horizontal variations in the partial pressures at the base of the region of diffusive equilibrium. (For purposes of illustration we are neglecting the effect of thermal diffusion on the altitude profile of He.)

9. Analytic Representation of Thermospheric Density Profiles

Our knowledge of the temperature profile in the thermosphere is frequently not good enough to justify numerical integration of the equation of diffusive equilibrium. A convenient analytic representation was introduced by Bates (1959) and developed by Walker (1965). The temperature profile is represented by a function of geopotential height which has a suitable shape,

$$T(z) = T_\infty - (T_\infty - T_0) \exp[-s\eta], \tag{27}$$

where T_∞ is the thermopause temperature, T_0 is the temperature at the reference level, s is a free parameter that controls the rate at which $T(z)$ approaches T_∞, and η is geopotential height above the reference level,

$$\eta = (z - z_0)(R_e + z_0)/(R_e + z). \tag{28}$$

With this expression for the temperature, Equation (15) may be integrated without approximation (most easily by changing the variable of integration to H_i) to give

$$p_i(z) = p_i(z_0) \left\{ \frac{T_0}{T(z)} \right\}^\gamma \exp[-s\gamma\eta], \tag{29}$$

where

$$\gamma = \frac{m_i g(z_0)}{s k T_\infty}. \tag{30}$$

This result allows for the convenient construction of models of thermospheric density profiles. A trivial extension may be made to allow for the effect of thermal diffusion (Walker, 1965).

10. Photoionization

As an application of the results derived above, let us consider the calculation of the rate of production of ionization in the thermosphere by the absorption of solar extreme UV radiation. We assume that the partial pressures at the reference altitude, $p_i(z_0)$, are constant, but that the temperature profile in the thermosphere is variable. We further assume that the sun is not close to the horizon, so the curvature of the atmosphere can be neglected (Smith and Smith, 1972).

The attenuation of the incoming beam of solar radiation at wavelength λ is described by Beer's Law,

$$dF(\lambda, z) = \left\{ \sum_i \sigma_i(\lambda)\, n_i(z) \right\} F(\lambda, z)\, \frac{dz}{\cos\theta}, \tag{31}$$

where $F(\lambda, z)$ is the flux of solar photons at wavelength λ and height z, $\sigma_i(\lambda)$ is the cross section for the absorption of these photons by constituent i, and θ is the solar zenith angle.

Integration of Equation (31) yields

$$F(\lambda, z) = F(\lambda, \infty) \exp\left[-\frac{1}{\cos\theta} \left\{ \sum_i \sigma_i(\lambda) \int_z^\infty n_i(h)\, dh \right\} \right], \tag{32}$$

where $F(\lambda, \infty)$ is the flux incident on the top of the atmosphere. If we neglect the altitude variation of gravitational acceleration, Equation (6) may be substituted for the column density in the exponent. Then

$$F(\lambda, z) = F(\lambda, \infty) \exp\left[-\frac{1}{\cos\theta} \left\{ \sum_i \sigma_i(\lambda)\, p_i(z)/m_i g \right\} \right]. \tag{33}$$

In a diffusive equilibrium atmosphere with constant reference pressures we have shown that the different partial pressures are functions only of the total pressure, p, and not of the temperature profile. So the solar flux at a given level may be expressed as a function simply of the total pressure at that level and it is convenient to use pressure rather than height as the vertical coordinate.

$$F(\lambda, p) = F(\lambda, \infty) \exp\left[-\frac{f(\lambda, p)}{\cos\theta} \right], \tag{34}$$

where f is an appropriate function.

The rate of production of ionization by photoionization at a given wavelength is equal to the rate of absorption of solar photons,

$$q(\lambda, p, \theta) = F(\lambda, p) \sum_i \sigma_i(\lambda)\, n_i(p). \tag{35}$$

The number densities, n_i, are proportional to the partial pressures, p_i, divided by temperature, and the partial pressures are functions only of the total pressure, p. So

$$q(\lambda, p, \theta) = F(\lambda, p)\, h(\lambda, p)/T(p) =$$

$$= F(\lambda, \infty) \frac{h(\lambda, p)}{T(p)} \exp\left[-\frac{f(\lambda, p)}{\cos\theta} \right], \tag{36}$$

where h is an appropriate function.

The total photoionization rate at level p is obtained by integrating this expression over all ionizing wavelengths. Our results show that this tedious integration need be performed only once at each pressure level and solar zenith angle for an atmosphere with constant reference pressures, $p_i(z_0)$. Changes in the temperature profile of the atmosphere do not significantly affect the calculation. Temperature changes cause the constant pressure levels to move up or down, but at each pressure level the photoionization rate is strictly proportional to the reciprocal of the temperature. It is therefore easy to examine the effect on photoionization rates of different thermospheric temperature profiles.

11. Conclusion

In concentrating on the effect of gravity on atmospheric composition and pressure this review has had to exclude many other important aspects of atmospheric physics. A bibliography is therefore presented below which lists a number of books on the subject.

Acknowledgment

The Arecibo Observatory of the National Astronomy and Ionosphere Center is operated by Cornell University under contract with the National Science Foundation.

References

Bates, D. R.: 1959, *Proc. Roy. Soc. London* **A253**, 451.
Smith, F. L. and Smith, C.: 1972, *J. Geophys. Res.* **77**, 3592.
Walker, J. C. G.: 1965, *J. Atmospheric Sci.* **22**, 462.

Bibliography

Banks, P. M. and Kockarts, G.: 1973, *Aeronomy*, Academic Press, New York.
Chamberlain, J. W.: 1961, *Physics of the Aurora and Airglow*, Academic Press, New York.
Craig, R. A.: 1965, *The Upper Atmosphere: Meteorology and Physics*, Academic Press, New York.
Fleagle, R. G. and Businger, J. A.: 1963, *Introduction to Atmospheric Physics*, Academic Press, New York.
Goody, R. M. and Walker, J. C. G.: 1972, *Atmospheres*, Prentice-Hall, Englewood Cliffs, New Jersey.
Green, A. E. S. and Wyatt, P. J.: 1965, *Atomic and Space Physics*, Addison-Wesley, Reading, Massachusetts.
Hines, C. O., Paghis, I., Hartz, T. R., and Fejer, J. A.: 1966, *Physics of the Earth's Upper Atmosphere*, Prentice-Hall, Englewood Cliffs, New Jersey.
McCormac, B. M.: 1967, *Aurora and Airglow*, Reinhold, New York.
McCormac, B. M.: 1971, *The Radiating Atmosphere*, Reidel, Holland.
McCormac, B. M.: 1973, *Physics and Chemistry of Upper Atmospheres*, Reidel, Holland.
Omholt, A.: 1971, *The Optical Aurora*, Springer-Verlag, New York.
Ratcliffe, J. A.: 1960, *Physics of the Upper Atmosphere*, Academic Press, New York.
Rishbeth, H. and Garriott, O. K.: 1969, *Introduction to Ionospheric Physics*, Academic Press, New York.
Whitten, R. C. and Poppoff, I. G.: 1971, *Fundamentals of Aeronomy*, Wiley, New York.

VERTICAL TRANSPORT IN ATMOSPHERES

DONALD M. HUNTEN

Kitt Peak National Observatory,
P.O. Box 26732, Tucson, Ariz. 85726, U.S.A.

1. Introduction

Two decades ago, a revolutionary paper appeared, explaining the presence of O_2 at heights above 100 km through the rapidity of molecular diffusion (Nicolet and Mange, 1954). Subsequent papers and reviews discussed this idea further (Mange, 1955, 1957; Nicolet, 1960). A second revolution took place some ten years later, with the formulation of a steady-state model incorporating eddy diffusion (Colegrove *et al.*, 1965, 1966). These papers gave the first quantitative description of the upward transport of O, between 90 and 150 km. The aeronomy to today is permeated with eddy diffusion, which for many problems is a first-order effect.

Meteorologists, who must deal every day with the details of motion fields in the atmosphere, tend to sneer at the practice of reducing them to vertical eddy transport. This viewpoint is well expounded by Lindzen (1971), who nevertheless goes on to offer some very helpful recipes. The sad fact is that fluid dynamics has nothing better to offer at present, particularly for heights above 30 km and for other planets. It is the study of the latter, and the questions raised about the exhaust products of supersonic aircraft, that have caused me to take a particular interest in eddy transport, what it means, and how to use it. The best-studied atmosphere, namely our own, naturally has the most to tell us; but the broad viewpoint is valuable too.

It is easy to lose sight of the fact that the oversimplified model that is eddy diffusion must be used with care and with a great deal of auxiliary physical insight. Well-educated handwavers often produce more useful results than the people who concentrate on the mathematical and computational aspects of the problem. The ideal is a suitable combination of the two. One way to become well-educated is to study various limiting cases that can be solved analytically. Indeed, these solutions can take us surprisingly far into many problems, sometimes rivalling what can be done with a computer. The substance of this article is a summary of the available solutions, and a critical compilation of eddy coefficients for the atmospheres of the Earth and a few other objects.

2. The Transport Equation

We shall discuss the vertical flux of an inert minor constituent. 'Inert' does not refer to the chemical behavior of the tracer, but rather to the fact that it must not affect the motion of the background atmosphere. Thus, although equations like (1) can be written for heat (usually potential temperature) and momentum, the eddy conduc-

B. M. McCormac (ed.), Atmospheres of Earth and the Planets, 59–72. All Rights Reserved.

tivity and viscosity may be, and often are, different from the eddy diffusion coefficient. I have recently discussed this question for heat (Hunten, 1974a).

The flux is assumed to be given by (cf. Strobel, 1972)

$$\phi = nw = nw_p - Kn_a \frac{df}{dz} - Dn_e \frac{dr}{dz}. \tag{1}$$

The concentrations of the tracer and the background atmosphere are n and n_a, and the mixing ratio is $f = n/n_a$. The vertical velocities are w and w_p; the latter could also include a settling velocity for aerosol particles. K and D are the eddy and molecular diffusion coefficients; n_e and r are discussed with Equation (5) below.

From the definition $f = n/n_a$, the second term in Equation (1) can be written

$$-Kn_a \frac{df}{dz} = -K\left(\frac{dn}{dz} - \frac{n}{n_a}\frac{dn_a}{dz}\right) = -K\left(\frac{dn}{dz} + \frac{n}{H_a}\right), \tag{2}$$

where H_a is the density scale height of the atmosphere. Written out still further, Equation (2) becomes

$$-Kn_a \frac{df}{dz} = -Kn\left(\frac{1}{n}\frac{dn}{dz} + \frac{m_a g}{kT} + \frac{1}{T}\frac{dT}{dz}\right), \tag{3}$$

in which m_a is the mean mass of atmospheric molecules, g is gravity, k is Boltzmann's constant, and T is temperature. One other useful form uses the density scale height H of the actual profile of the tracer:

$$-Kn_a \frac{df}{dz} = Kn\left(\frac{1}{H_a} - \frac{1}{H}\right). \tag{4}$$

It is easily seen that the eddy flux is upwards or downwards as the tracer distribution falls off faster or slower than the atmosphere with increasing height. Some of the literature relating to the troposphere uses an 'austausch' (exchange) coefficient instead of K. It is the product of K with the density (or sometimes concentration) of the tracer, and is used with the gradient of the mass mixing ratio.

Similar manipulations can be done with the third term of Equation (1), which is written out as

$$-Dn_e \frac{dr}{dz} = -Dn\left[\frac{1}{n}\frac{dn}{dz} + \frac{mg}{kT} + \frac{(1+\alpha)}{T}\frac{dT}{dz}\right], \tag{5}$$

where m is the mass of the tracer molecules, and α is the thermal diffusion factor. The diffusive-equilibrium solution n_e is obtained by setting the square bracket equal to zero, and $r = n/n_e$ plays a role analogous to f. Other transformations are discussed in some detail by Hunten (1973a, b); the first of these contains a minor inconsistency which is corrected in the second. Throughout most of the rest of this paper it will be assumed that only the second term in Equation (1) is significant: in other words, that w_p and D are negligibly small. D becomes important at about 90 km and dominant above 100 km; for light molecules, especially H_2 and H_2O, it should not be neglected at any height (Hunten, 1973b; Hunten and Strobel, 1974).

Solution of an aeronomical problem requires, in addition to Equation (1), the continuity equation

$$\frac{\partial n}{\partial t} + \frac{\partial \phi}{\partial z} = p - l, \tag{6}$$

where p and l are the production and loss rates per unit volume; sometimes l is written $l_1 n$ or n/τ, where l_1 is a loss coefficient and τ is a lifetime. If diurnal and other time variations are averaged out, rather than treated explicitly, the system may be represented by a steady state,

$$\frac{d\phi}{dz} = p - l. \tag{7}$$

Some thorough treatments require a diffusion and a continuity equation for each of 10 to 20, or even more, constituents. In most cases, it is acceptable to use grouped constituents, such as 'odd nitrogen', as discussed below. The problem is then much less formidable.

3. Remarks on Eddy Diffusion

Part of the poor reputation of eddy diffusion among fluid dynamicists is due to their use of a related concept for an entirely different purpose. The equations of motion of the atmosphere describe sound waves, and many other small-scale phenomena, as well as weather. To damp out the short-wavelength modes from a numerical calculation, an eddy viscosity is inserted. The major horizontal transports are still described dynamically, and most of the vertical transport is produced as a byproduct of horizontal motions that are not exactly horizontal. An aeronomer wishes to assimilate all the vertical transport into a single K. He includes motions on a large scale, as well as the small scale that is usually thought of as 'turbulence'. For this reason, the vertical drift velocity w_0 in Equation (1) is seldom included; if there is an upward motion somewhere it must be accompanied by a downward motion somewhere else. From far enough away the result looks like just another big eddy.

Vertical eddy diffusion is useful primarily because the atmosphere tends to be horizontally stratified. This relative uniformity of horizontal shells is due to the strong tendency of air motions to be horizontal. If a tracer has widespread sources and sinks, stratification is even more likely. The tropopause and the lower stratosphere, however, are stratified in a direction that slopes downward from equator to poles (Reed and German, 1965; Davidson *et al.*, 1966; Gudiksen *et al.*, 1968). The height of the tropopause goes from about 17 to about 10 km. The average O_3 distribution, for example, should be obtained in this sloping coordinate system if it is to be compared with an eddy treatment. The papers just cited actually use two-dimensional eddy motion, assuming uniformity around latitude circles. The tilting of the tropopause is described either literally or by use of a 2×2 eddy-coefficient tensor.

A helpful tool in visualizing eddy mixing, and in making semiquantitative estimates, is the Prandtl mixing-length hypothesis. A brief discussion is given by Lindzen (1971), but much more is to be found (for heat transport) in the literature of stellar interiors (e.g., Schwarzschild, 1958). One imagines that the mixing is due to the interchange of two large bubbles of air over a distance L, the mixing length, after which they merge into the background. While in motion, their velocity is v. We then write

$$K \sim vL. \tag{8}$$

Similarly, elementary treatments of molecular diffusion assume motion at the thermal speed $\bar{c}$ over a distance of one mean free path P, whence

$$D \sim \bar{c}P. \tag{9}$$

The coefficient in Equation (9) is actually about 0.42, but since Equation (8) is approximate anyway, and the analogy is rather loose, this factor is omitted there.

It is often stated that the idea of a K coefficient is *derived* from the mixing-length hypothesis. I see no basis for this notion. To me, K is a proportionality factor, defined by Equation (1), relating a flux to a gradient. The molecular coefficient D is defined the same way. It, too, can be derived approximately by mixing-length arguments, but they are a convenience, not a necessity.

Near the ground, and perhaps in the lower stratosphere, L should be of the order of the height above the boundary (Priestley, 1959). Elsewhere, the natural length is the scale height. This is perhaps most easily visualized in terms of downward motion: after a parcel has moved a scale height, it loses its identity by compression, if nothing else. Unfortunately, it is difficult to specify a corresponding velocity, but a couple of examples can be given. A vertical speed of 1 cm s^{-1} might be regarded as not unusual, and 100 m s^{-1}, a large fraction of the speed of sound, should be an extreme upper limit. The corresponding K's, with $L \sim 10^6$ cm, are 10^6 cm^2 s^{-1} and 10^{10} cm^2 s^{-1}. The former value is observed near the mesopause; the latter is more typical of observed *horizontal* transport, where L may be several thousand km.

Lindzen (1971) has discussed a recipe for estimation of eddy coefficients, based on the idea of interrupted wave motion. A purely coherent wave does not cause any net transport, but merely carries the tracer back and forth. If the wave motion is interrupted after a few (say N) cycles, transport will result unless N is an integer. To allow for randomness, N may be taken as the nearest integer, and a weighting factor $0.5/N$ introduced. This factor multiplies the wave velocity v, and the vertical scale is taken as $\lambda/2\pi$, where λ is the vertical wavelength. The result is

$$K \sim v\lambda/\pi N. \tag{10}$$

Lindzen gives estimates based on the diurnal tide, in which v varies inversely as the square-root of the density. K will vary the same way, as long as N does not change, and thus will tend to have a scale height

$$H_K \sim -2H_a. \tag{11}$$

Values in the range 10^5–10^7 cm^2 s^{-1} are estimated for K at various latitudes and for mesospheric heights. These ideas can be used for regions that lack tracer data.

At tropospheric and stratospheric heights, tidal motions are small, and vertical transport occurs as a byproduct of the horizontal winds, which are not strictly horizontal. This effect has been studied by Mahlman (1973) with the aid of a global circulation model in a large computer. The spread of a tracer introduced into the stratosphere of the model bears an encouraging resemblance to observation. Unfortunately, the available results do not shed much light on the difficult region just above the tropopause, because the grid is too large. This limitation should be cured in the future, with the use of still bigger computers.

4. Solutions

We shall be concerned with the solutions of Equation (1) when w_0 and D can be neglected:

$$\phi = -Kn_a \frac{df}{dz}. \tag{12}$$

The key to analytic solutions is to represent the height variation of each quantity, ϕ, K, and n_a, by exponentials. If the first two are constant, the remaining variation can be written

$$n_a = n_{a0} e^{-h} = n_{a0} \exp\left(-\frac{z-z_0}{H_a}\right). \tag{13}$$

The nondimensional height h is defined by Equation (13); subscripts 0 refer to some reference level. Taking H_a constant is an approximation since few atmospheric regions are strictly isothermal; but the approximation is acceptable for many purposes. It is common to take K independent of height due to lack of better information; we can regard any errors in Equation (13) as absorbed in our ignorance of K. With these assumptions, the solution of Equation (12) is simply

$$f = A - \int \frac{\phi}{Kn_a} dz$$
$$= A - \frac{\phi H_a}{Kn_{a0}} e^h, \tag{14}$$

where A is the constant of integration. Multiplication by n_a gives

$$n = An_{a0} e^{-h} - \frac{\phi H_a}{K}. \tag{15}$$

The solution thus has two characteristic components: one with a constant mixing ratio and an arbitrary multiplier that must be found from boundary conditions; and one determined by the flux. For downward (negative) flux, this term is positive and often is the major one, giving a density independent of height. For upward flux, the first term must be the larger, to avoid negative densities. Thus, the basic character of upward and downward flows is often very different.

If K is not independent of height, its profile may be fitted, at least over limited ranges, by an exponential

$$K = K_0 \exp\left(-\frac{z - z_0}{H_K}\right). \tag{16}$$

If K increases with height, H_K is negative. The product $K n_a$ then has the scale height

$$H_p = (H_a^{-1} + H_K^{-1})^{-1}, \tag{17}$$

and Equation (15) becomes

$$n = A n_{a0} \, e^{-h} - \frac{\phi H_p}{K}. \tag{18}$$

The second term now reflects (inversely) the height dependence of K. The case $H_K = -H_a$ requires a separate solution, in which the variation of f is linear, not exponential.

I shall not discuss the case of variable flux, except for a couple of remarks. If ϕ varies exponentially with height, its scale height can be included as another term in Equation (17). It must also be remembered that finite-difference solutions of Equation (1) and (7) are easy, reflecting the simplicity of the analytic solutions. A good scheme is given by Liu and Donahue (1974a). The case of constant flux is commoner than might be supposed, because it is often possible to use grouped constituents. An example of great current interest is odd nitrogen, often called NO_x or NO_y: the sum of NO, NO_2, and HNO_3. These molecules are mainly converted into each other, but very slowly produced and destroyed, especially in the stratosphere. If their continuity Equations (7) are added, the p and l terms almost cancel. The result for total odd nitrogen is $d\phi/dz \simeq 0$, or a nearly constant flux. Hunten and Strobel (1974) and Thomas (1974) discuss the case of total-H, which is accurately conserved above the tropopause. This happy situation is somewhat spoiled by the fact that the molecular-diffusion term in Equation (1) cannot really be neglected at any height for hydrogenous molecules, and the diffusion coefficients are not all the same. Thus, even though ϕ is accurately constant, and equal to the escape flux, Equation (18) is not the solution of the right equation and remains approximate.

Application of a boundary condition is necessary to determine the constant A in Equations (14), (15), or (18) (the value of ϕ may perhaps be regarded as another boundary condition). Examples of density distributions are shown in Figures 1 to 3 (downward flux) and Figure 4 (upward or downward flux). Figure 3 is particularly interesting, because it has a considerable resemblance to the situation around the tropopause, where $K_1 \sim 10^5$ cm^2 s^{-1} and $K_2 \sim 3 \times 10^3$ cm^2 s^{-1}. The exponential relaxation in the stratosphere towards the value $n = -\phi H_a/K_2$ takes place with the

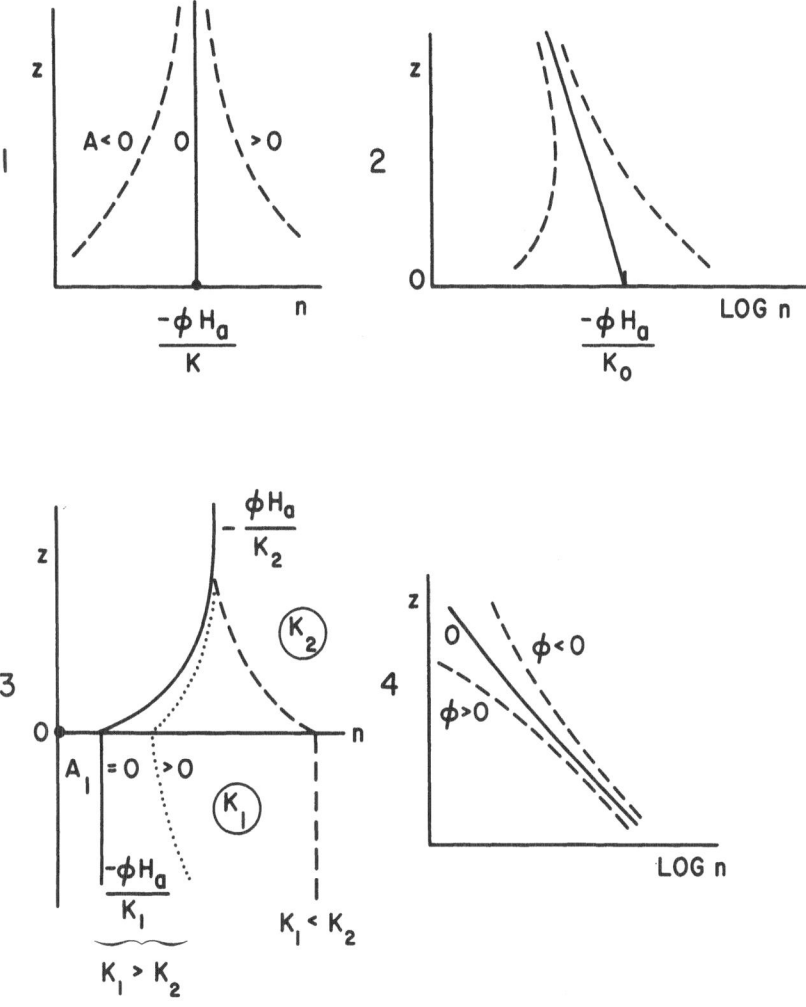

Fig. 1–4.   Vertical distributions obtained under various circumstances. (1) Constant downward flux, K constant. (2) Constant downward flux, K increasing exponentially with height. (3) Same as (1), but with two layers of differing K. (4) Constant flux, constant K.

scale height H_a, or H_p if K is variable. A more extended discussion of this case is given by Hunten (1974c), along with expressions for A in several models.

A case susceptible of a complete solution is that of a loss coefficient $1/\tau$ independent of height, and no production (Hanson and Donaldson, 1967; Hunten and Strobel, 1974). The eddy equation and the continuity Equation (7) can be combined and solved, when f is assumed to vary exponentially with scale height $H_f = -f/(df/dz)$. It is found that

$$HH_f = K\tau$$
$$H = RH_a[-1 \pm (1 + 2/R)^{1/2}], \tag{19}$$

where $R = K\tau/(2H_a^2)$. The lower sign corresponds to a downward flux. The 'diffusion time' is usually taken as $\tau_D = H_a^2/K$; thus, $R = \tau/2\tau_D$.

One case with a downward velocity w_p in Equation (1) is of considerable interest – that of particles falling under gravity. An excellent discussion by Junge *et al.* (1961) has just come to my attention; I derived some of the results a few years ago, and anoth-

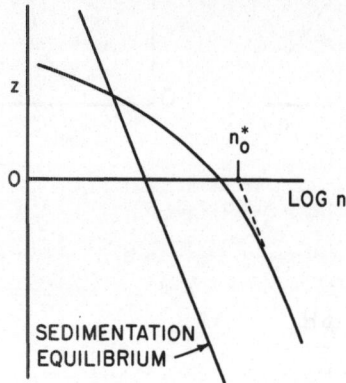

Fig. 5.   The two components of an aerosol-particle distribution, from (22). Sedimentation equilibrium corresponds to a downward flux. The other component, a 'dust bank', is supported by eddy diffusion.

er independent discussion has been given by Prinn (1974). Particles of radius a smaller than the mean free path (7 μm at 30 km, 7000 at 80) fall at a speed

$$w_p = 7.8 \times 10^{16} \, a \, (\mu\text{m})/n_a \text{ cm s}^{-1} \tag{20}$$

for a density of 2.0 g cm^{-3}. Larger particles obey a different law, but fall so fast that they are of little interest. It is convenient to choose the origin such that the value of w_p there is

$$w_0 = K/H_a. \tag{21}$$

At other heights, $w_p = w_0 e^h$. The solution of Equation (1) becomes

$$n = n_0^* \exp -(h + e^h) + \frac{\phi}{w_p}. \tag{22}$$

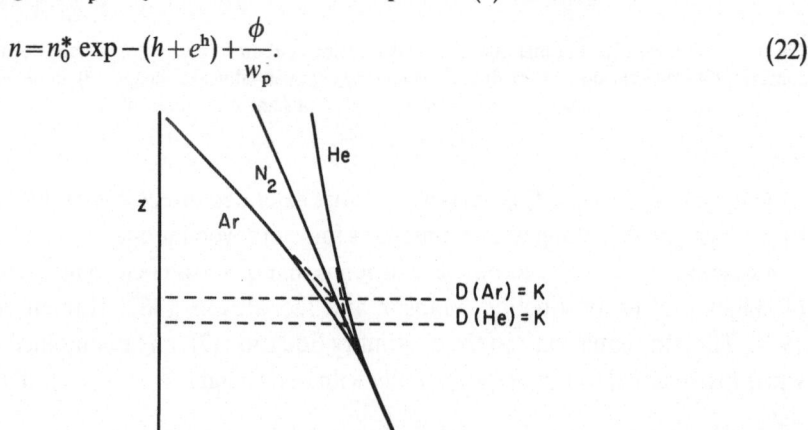

Fig. 6.   Distributions of various gases through the homopause according to (23), with K constant.

The second term depends on some imposed particle flux ϕ, and has a concentration profile parallel to that of the atmosphere; Junge *et al.* (1961) give the name 'sedimentation equilibrium' to this distribution, which does not depend on K. The first component is a 'dust bank' whose magnitude depends on a boundary condition and on the height of the origin, defined by Equation (21). Several scale heights below the origin, the asymptotic form is $n_0^* e^{-h}$; at the origin, the value is $n_0^* e^{-1}$; and at greater heights there is a rapid cutoff. The two components are illustrated in Figure 5.

The last topic of this section is the transition from mixing to diffusive equilibrium, for which we need the last two terms of Equation (1) but can set ϕ equal to zero. The procedure is outlined by Wallace and Hunten (1973), who give

$$n = A e^{-h}(1 + e^h)^{1-s}, \tag{23}$$

where $s = m/m_a$ and $D = K e^h$. The level where $D = K$ is the origin, or homopause. (A more frequent usage is *turbopause*, but it implies a mixing mechanism that may not be the whole story). Figure 6 illustrates the distributions obtained for He and Ar, with N_2 as the background gas; thus, s is 0.143 and 1.43, respectively. A is the fictitious density at the junction of the two asymptotes, which occurs at the homopause. Since D tends to be larger for lighter gases, the corresponding homopause level is higher.

Wallace and Hunten (1973) also give the distributions corresponding to constant upward and downward flux. The latter is useful to describe the motion out of the thermosphere of atoms such as O on the Earth and H on Jupiter.

5. *K* Coefficients

The definition of K is contained in Equation (12), which may be written

$$K = -\frac{\phi}{n_a(df/dz)}. \tag{24}$$

Determination of K therefore requires estimates of the flux ϕ and the mixing-ratio gradient df/dz of some tracer. Alternatively, a model of a tracer distribution may be constructed and varied until a suitable fit to observations is obtained. The two procedures are discussed and applied to the same tracer, CH_4, by Hunten (1974c) and Wofsy and McElroy (1973). The flux must usually be calculated by the continuity Equation (6) or (7) from a photochemical model, because it is difficult or impossible to observe.

Information available a decade ago, mostly from radioactive tracers, was reviewed by Bolin (1964). He concluded that K is about 10^5 cm^2 s^{-1} in the troposphere and 2×10^3 cm^2 s^{-1} in most of the lower stratosphere, rising to 10^4 in the region of the winter pole. The two-dimensional model of Davidson *et al.* (1966) worked best with vertical K's of 4×10^4 and 1×10^3 cm^2 s^{-1} for troposphere and stratosphere, and similar values were used by Gudiksen *et al.* (1968). I thus adopt a value for the troposphere of 1×10^5 cm^2 s^{-1} in Figure 7.

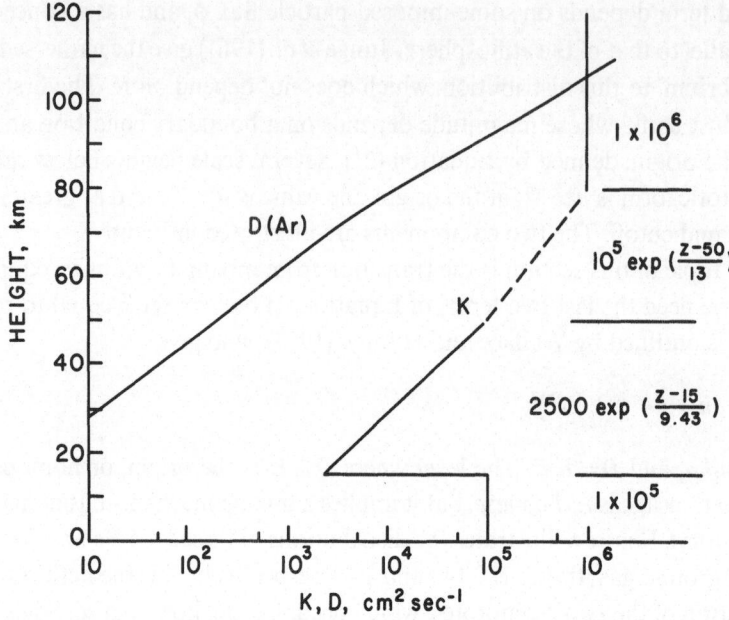

Fig. 7.   Recommended profile of eddy coefficient. The argon diffusion coefficient is shown for comparison.

Specification of K in the bottom few kilometers of the stratosphere is greatly complicated by the slope of the tropopause and by the existence of the mid-latitude 'tropopause gaps'. The work quoted above, and many other papers (e.g., Machta *et al.*, 1970) make it clear that the horizontal motion is generally parallel to the tropopause, and therefore any horizontal averaging should be done in a corresponding manner. But this is not a simple matter; the most obvious question, perhaps, is how to make a transition from these sloping co-ordinates to horizontal ones at higher altitudes. Strictly horizontal averaging, as adopted by Chang *et al.* (1973), will naturally give rather large K's at 10 km, decreasing to a truly stratospheric value at about 17 km. This procedure has the virtue of being well-defined and consistent, but may also smear out some essential physics.

Methane is an important tracer, which has recently become available because of the careful work of a group at the National Center for Atmospheric Research (Ehhalt *et al.*, 1972; Ehhalt and Heidt, 1973). The mixing-ratio gradient in the lowest strato-sphere is small, but is well resolved by the technique of sampling followed by laborato-ry analysis. Methane originates in swamps and rice paddies, and is very well mixed throughout the troposphere. As it migrates slowly up through the stratosphere, it should continue to be well mixed along the quasi-horizontal surfaces discussed above. It therefore satisfies almost ideally the condition of global or hemispheric averaging required for strict applicability of a one-dimensional eddy treatment. Methane is slowly destroyed by reaction with metastable oxygen, $O(^1D)$, and the OH radical (Wofsy *et al.*, 1972). If these processes are averaged over diurnal and sea-sonal cycles, they give a height-dependent lifetime τ that drops from 1000 yr at 15 km to a few months at 50 km. Independent analyses by Wofsy and McElroy (1973) and Hunten (1974c), as discussed above, give similar results. Figure 7 shows my version, along with a formula for the curve fit; K is 2500 cm^2 s^{-1} at 15 km and 10^5 cm^2 s^{-1} at 50 km. Data are available only below 30 and at 50 km; the K profile is thus distinctly uncertain between 30 and 50 km. The Wofsy-McElroy values are somewhat larger in this region, but do not fit the data quite as well. Moreover, the exponential segment shown here, with $H_K = -9.43$ km, fits in perfectly with Equation (18).

Despite the virtues of CH$_4$ listed above, its use as a tracer still depends on calculated densities of O(^{1}D) and OH. A systematic error would probably affect the absolute lifetimes much more than the height variation and similar remarks should carry over to the K profile. N$_2$O has similar lifetimes, controlled mainly by photolysis, and promises to be an important tracer when enough data are available.

Between 50 and 80 km, there are no usable tracer data at present; the curve shown is an interpolation going to 1×10^6 cm^2 s^{-1} at 80 km, with $H_K = -13$ km. As it happens, this shape is almost exactly in agreement with Lindzen's recipe (11). It may be possible to obtain information for the 80 km region from data on total H and especially H$_2$O. The study of Hunten and Strobel (1974) was partly directed to this end. Total H reductions of about a factor 2 were found for the smallest K's assumed, 10^5 cm^2 s^{-1} at 85 km and 3×10^5 cm^2 s^{-1} at 100 km. This reduction is probably inconsistent with the latest ideas on the hydrogen escape flux (Liu and Donahue, 1974b). The mixing ratio of H$_2$O shows a strong depletion above 75 km, below which it is essentially constant. The lifetime τ is about 3×10^5 s, due to photolysis, and if the slope of the distribution could be obtained, a K would immediately follow from Equation (19). The 'reference' model of Hunten and Strobel (1974), for example, gives an H$_2$O mixing ratio of about 0.16 ppm at 90 km. If this is not enough to support phenomena such as noctilucent clouds or hydrated ions, the K's of that model (3×10^5 cm^2 s^{-1} above 85 km) would have to be increased.

The region 90 to 110 km is represented by several tracers, and it is necessary to make a choice. Those with the longest history (heat, O, artificial trails) are full of complications and should not be given high weight. Hunten and Strobel (1974) and Hunten (1974a) discuss the details, which will not be repeated here. Diffusive separation of Ar, represented here by Equation (23), is by far the cleanest, and has been reviewed by von Zahn (1970). The homopause height is remarkably consistent at 101 km, within 1 or 2 km, and leads immediately to a K of 4×10^5 cm^2 s^{-1}. Helium is less simple because of large horizontal transports in the upper thermosphere, but they are included in the model of Reber and Hays (1973) which gives $K = 1.8 \times 10^6$ cm^2 s^{-1}. A compromise of 1×10^6 cm^2 s^{-1} is suggested, although it may not give enough weight to the Ar data.

It is difficult to wring an absolute value of K from O, but there is plenty of evidence for variability, especially from airglow observations (Donahue et al., 1973; Noxon and Johanson, 1972). It is not obvious how such results are to be reconciled with the remarkable constancy of the Ar homopause. Clearly, we have a lot to learn at all heights, even in our own atmosphere.

The next best-known atmosphere is that of Mars, essentially pure CO_2 with a surface pressure of some 6 mb. Most of the solar radiation is absorbed by the surface, and the heat transferred to the atmosphere. If the resulting vertical transport is represented by local convection (Gierasch and Goody, 1967), a K around 10^8 cm^2 s^{-1} is found for the troposphere, 0 to 15 km. Stone (1972), however, finds large-scale baroclinic waves to be more likely than local convection. Though this flow is very different, it must still carry the same amount of heat to heights where it can be radiated, and essentially the same effective value of K would be expected.

The top of the Mars atmosphere must also be strongly stirred; this is the only acceptable explanation that has been found of the remarkable rarity of O and CO in the thermosphere. These fragments are necessarily produced in CO_2 photolysis; McElroy and McConnell (1971) find that a K of 5×10^8 cm^2 s^{-1} is needed to sweep them out at the required rate.

For the stratosphere and mesosphere there are no data at all. This region has no analog of the terrestrial O_3 layer to give a high temperature and a strongly stable temperature inversion. Nor, in the absence of such a layer, is there nearly such a strong reflection of wave energy from the troposphere. The remarkably stagnant layer we find on Earth may therefore be absent. A model of the O_2 and CO_2 balance by McElroy and Donahue (1972) requires K to be 1.5×10^8 cm^2 s^{-1}, but another model is compatible with smaller values, at the expense of an assumed odd-Hydrogen content that may be unacceptably large (Parkinson and Hunten, 1972) (see also reviews by McConnell, 1973 and Hunten, 1974b).

The thermosphere of Venus seems to be in somewhat the same state as that of Mars, though O atoms may not be quite so rare (McElroy, 1969; Strickland, 1973; Strickland et al., 1973). A large K is also helpful in keeping down the abundance of H atoms. A model by Kumar and Hunten (1974) finds reasonable consistency with $K = 10^8$ cm^2 s^{-1}, but this estimate is less firm than the corresponding one for Mars. Recently Prinn (1974) has pointed out the potential of cloud and haze particles as tracers in the stratosphere and upper troposphere. He sets upper limits, $K < 7 \times 10^4$ cm^2 s^{-1} (5 to 50 mb) and $K < 2 \times 10^5$ cm^2 s^{-1} (50 to 155 mb), from an inference that the particle scale height is less than that of the gas (cf., Figure 5).

Another recent paper (Sze and McElroy, 1974) deduces a thermospheric K of 5×10^7 cm s^{-1}, slightly less than that of Kumar and Hunten. Their tracer for the stratosphere is O_2, from which a *lower* limit of 4×10^5 cm^2 s^{-1} is obtained.

Information on Jupiter's upper atmosphere comes from the downward transport of H atoms produced in the ionosphere. Eddy diffusion was introduced to this problem by Hunten (1969), and a thorough study made by Wallace and Hunten (1973). The rocket data on Ly-α discussed in the latter paper suggested a low value of K, less than 10^6 cm^2 s^{-1}. But Pioneer 10 has changed the position completely, by showing that part of the hydrogen observed from Earth was in Io's orbit rather than Jupiter's thermosphere (Carlson and Judge, 1974). The preferred value of K is now 3×10^8 cm^2 s^{-1}, with error bars of a factor 10 each way.

Another tracer, just above the cloud tops at ~ 1 bar, is NH_3. The small amount seen in rocket spectra had been a puzzle for several years, until Strobel (1973) pointed out that it is rapidly photolyzed and therefore may have a small 'photomechanical' scale height by Equation (19). He finds that $K = 2 \times 10^4 \text{ cm}^2 \text{ s}^{-1}$ satisfies the observations, giving $H = 3$ km for NH_3 while $H_a = 17$ km. Thus, we seem to have another example of a stagnant lower stratosphere. Though Jupiter has no O_3 layer, it does have a stratospheric temperature inversion (Wallace *et al.*, 1974).

6. Concluding Remarks

Some of the conceptual difficulties with the idea of one-dimensional eddy diffusion are eased if the attitude of Section 5 is adopted: K is a parameter, defined by Equation (24), to be determined from observation of tracers. Excellent tracers are known, especially CH_4 and N_2O, for which many more measurements, and better height coverage, are needed. Such measurements are not easy, but will be immensely valuable. Another interesting possibility is beginning to come to light: the following of a tracer in a dynamical model of the atmosphere, and computation of K's from the definition (24) (Mahlman, 1973). Present models lack the vertical resolution to be satisfactory near the tropopause, and also do not extend as high as the aeronomer would wish; but these deficiencies will disappear as bigger computers become available. The same technique should apply to the estimation of K's in two-dimensional models, nearly impossible from measurements in the actual atmosphere.

The solutions developed in Section 4 and illustrated in Figures 1 to 6 do not begin to exhaust the possibilities nor the published papers. They represent a technique that should be considered whenever eddy transport problems are encountered.

Acknowledgements

I am grateful to the Climatic Impact Assessment Program for the opportunity to sharpen my understanding of the stratosphere. Kitt Peak National Observatory is operated by the Association of Universities for Research in Astronomy, Inc., under contract with the National Science Foundation.

References

Bolin, B.: 1964, in H. Odishaw (ed.), *Research in Geophysics*, vol. 2, M.I.T. Press, Cambridge, Mass., p. 479.

Carlson, R. W. and Judge, D. L.: 1974, *J. Geophys. Res.* **79**, 3623.

Chang, J. S., Hindmarsh, A. C., and Madsen, N. K.: 1973, UCRL-74823, Lawrence Livermore Lab., Livermore, Calif.

Colegrove, F. D., Hanson, W. B., and Johnson, F. S.: 1965, *J. Geophys. Res.* **70**, 4931.

Colegrove, F. D., Johnson, F. S., and Hanson, W. B.: 1966, *J. Geophys. Res.* **71**, 2227.

Davidson, B., Friend, J. P., and Seitz, H.: 1966, *Tellus* **18**, 301.

Donahue, T. M., Guenther, B., and Thomas, R. J.: 1973, *J. Geophys. Res.* **78**, 6662.

Ehhalt, D. H., Heidt, L. E., and Martell, E. A.: 1972, *J. Geophys. Res.* **77**, 2193.

Ehhalt, D. H. and Heidt, L. E.: 1973, *J. Geophys. Res.* **78**, 5265.

Gierasch, P. and Goody, R.: 1967, *Planetary Space Sci.* **15**, 1465.

Gudiksen, P. H., Fairhall, A. W., and Reed, R. J.: 1968, *J. Geophys. Res.* **73**, 4461.

Hanson, W. B. and Donaldson, J. S.: 1967, *J. Geophys. Res.* **72**, 5513.
Hunten, D. M.: 1969, *J. Atmospheric Sci.* **26**, 826.
Hunten, D. M.: 1973a, *J. Atmospheric Sci.* **30**, 726.
Hunten, D. M.: 1973b, *J. Atmospheric Sci.* **30**, 1481.
Hunten, D. M.: 1974a, *J. Geophys. Res.* **79**, 2533.
Hunten, D. M.: 1974b, *Rev. Geophys.* **12**, 529.
Hunten, D. M.: 1974c, *Science*, submitted.
Hunten, D. M. and Strobel, D. F.: 1974, *J. Atmospheric Sci.* **31**, 305.
Junge, C. E., Chagnon, C. W., and Manson, J. E.: 1961, *J. Meteorol.* **18**, 81.
Kumar, S. and Hunten, D. M.: 1974, *J. Geophys. Res.* **79**, 2529.
Lindzen, R. S.: 1971, in G. Fiocco (ed.), *Mesospheric Models and Related Experiments*, D. Reidel Publ. Co., Dordrecht-Holland, p. 122.
Liu, S. C. and Donahue, T. M.: 1974a, *J. Atmospheric Sci.* **31**, 1118.
Liu, S. C. and Donahue, T. M.: 1974b, *J. Atmospheric Sci.* **31**, 1466.
Machta, L., Telegadas, K., and List, R. J.: 1970, *J. Geophys. Res.* **75**, 2279.
Mahlman, J. D.: 1973, A.I.A.A. Paper 73-528.
Mange, P.: 1955, *Ann. Geophys.* **11**, 153.
Mange, P.: 1957, *J. Geophys. Res.* **62**, 279.
McConnell, J. C.: 1973, in B. M. McCormac (ed.), *Physics and Chemistry of Upper Atmospheres*, D. Reidel Publ. Co. Dordrecht, Holland, p. 309.
McElroy, M. B.: 1969, *J. Geophys. Res.* **74**, 29.
McElroy, M. B. and Donahue, T. M.: 1972, *Science* **177**, 986.
McElroy, M. B. and McConnell, J. C.: 1971, *J. Atmospheric Sci.* **28**, 879.
Nicolet, M.: 1960, in J. A. Ratcliffe (ed.), *Physics of the Upper Atmosphere*, Academic Press, New York, p. 17.
Nicolet, M. and Mange, P.: 1954, *J. Geophys. Res.* **59**, 16.
Noxon, J. F. and Johanson, A. E.: 1972, *Planetary Space Sci.* **20**, 2125.
Parkinson, T. D. and Hunten, D. M.: 1972, *J. Atmospheric Sci.* **29**, 1380.
Priestley, C. H. B.: 1959, *Turbulent Transfer in the Lower Atmosphere*, University of Chicago Press, Chicago.
Prinn, R. G.: 1974, *J. Atmospheric Sci.* **31**, 1691.
Reber, C. A. and Hays, P. B.: 1973, *J. Geophys. Res.* **78**, 2977.
Reed, R. J. and German, K. E.: 1965, *Mon. Weather Rev.* **93**, 313.
Schwarzschild, M.: 1958, *Structure and Evolution of the Stars*, Dover, New York.
Stone, P. H.: 1972, *J. Atmospheric Sci.* **29**, 405.
Strickland, D. J.: 1973, *J. Geophys. Res.* **78**, 2827.
Strickland, D. J., Stewart, A. I., Barth, C. A., Hord, C. W., and Lane, A. L.: 1973, *J. Geophys. Res.* **78**, 4547.
Strobel, D. F.: 1972, *Radio Sci.* **7**, 1.
Strobel, D. F.: 1973, *J. Atmospheric Sci.* **30**, 1205.
Sze, N. D. and McElroy, M. B.: 1974, *Planetary Space Sci.*, submitted.
Thomas, R. J.: 1974, *Planetary Space Sci.* **22**, 175.
Wallace, L. and Hunten, D. M.: 1973, *Astrophys. J.* **182**, 1013.
Wallace, L., Prather, M., and Belton, M. J. S.: 1974, *Astrophys. J.* **193**, 481.
Wofsy, S. C., McConnell, J. C., and McElroy, M. B.: 1972, *J. Geophys. Res.* **77**, 4477.
Wofsy, S. C. and McElroy, M. B.: 1973, *J. Geophys. Res.* **78**, 2619.
von Zahn, U.: 1970, *J. Geophys. Res.* **75**, 5517.

SOME ENERGY SOURCES AND SINKS IN
THE UPPER ATMOSPHERE

CONWAY B. LEOVY

Dept. of Atmospheric Sciences, University of Washington*
Seattle, Wash. 98195 U.S.A.

1. Introduction

The overwhelmingly dominant energy source for the atmosphere is electromagnetic solar radiation, the gross disposition of which is given in Table I. The average temperature distribution in the troposphere is maintained to a large extent by the energy absorbed at the ground and transferred to the atmosphere as sensible or latent heat, together with the very effective emission of IR radiation from the upper troposphere. Less than 2% of the incident solar radiation is absorbed above 15 km, and less than $10^{-3}\%$ is absorbed above 80 km. Yet this part of the solar flux is responsible for the heating which determines the temperature structure of the stratosphere, mesosphere, and thermosphere, and for practically all of the photodissociation and ionization processes in the atmosphere. In the next two sections of this paper we shall deal with this flux and its disposition in the atmosphere. In the subsequent section we consider the role of emission of IR radiation in determining the thermal structure of the mesosphere and the mesopause region.

Since the solar energy available to heat the upper atmosphere is a very small fraction of the available incident radiation, processes other than absorption of solar electromagnetic radiation play a significant energetic role at high levels. These processes include energy deposition by particle fluxes, joule heating due to ionospheric currents, heating by dissipation of hydromagnetic waves generated in the magnetosphere, and heating by dissipation of hydrodynamic disturbances generated in the thermosphere or below. The role of hydrodynamic waves as energy sources for the thermosphere is briefly discussed in the last section.

2. Coupling between the Solar and Terrestrial Atmospheres

The monochromatic intensity of radiation, $I_\nu(\mathbf{s}, \Omega)$, is a function of position, $\mathbf{s}$, frequency, ν, and direction of travel of the photon stream confined to a narrow solid angle cone, Ω. Its spatial variations are described by the equation of transfer,

$$\mathrm{d}I_\nu = -(I_\nu - S_\nu)\,\sigma_\nu n\,\mathrm{d}s \tag{1}$$

where $\sigma_\nu(\mathbf{s})$ is the mean atomic or molecular extinction cross-section, $n(\mathbf{s})$ is the number density of atoms or molecules, $\mathrm{d}s$ is the distance increment in the direction of travel, and $S_\nu(\mathbf{s}, \Omega)$ is the source function. For a medium in which scattering is

---

* Contribution number 328, Department of Atmospheric Sciences, University of Washington.

B. M. McCormac (ed.), Atmospheres of Earth and the Planets, 73–86. All Rights Reserved.

TABLE I

Average disposition of incident solar flux, w m^{-2}

Incident solar flux	Absorbed at the ground	Transferred from ground to atmosphere	Absorbed in the atmosphere below 15 km	Absorbed in the atmosphere above 15 km	Absorbed in the atmosphere above 80 km	Reflected to space
340	163	95	68	7	<0.003	112

absent and in which local thermodynamic equilibrium (LTE) prevails, S_v is just the Planck Function,

$$S_v = \frac{2h v^3}{c^2} \left[\exp(h v / kT) - 1 \right]^{-1}. \tag{2}$$

When the atomic or molecular internal energy levels are not populated in equilibrium with the kinetic energy of the random motion of the gas particles, emission of radiation is not determined solely by the local kinetic temperature, Equation (2) is no longer valid, and local thermodynamic equilibrium breaks down (Kondratyev, 1965).

For solar radiation incident on the Earth's atmosphere, the source function can usually be neglected, except in the troposphere where scattering is important. Then Equation (1) reduces to

$$\frac{\mathrm{d}I_{v,s}}{\mathrm{d}s} = -\sigma_v n I_{v,s} \, \mathrm{d}s =$$

$$= -\sigma_v n I_{v,s} \sec \theta \, \mathrm{d}z, \tag{3}$$

where $I_{v,s}$ refers specifically to the intensity of solar radiation, and θ is the zenith angle of the sun. For monochromatic solar radiation, σ_v is nearly independent of height, so that Equation (3) can be integrated to give,

$$I_{v,s}(z) = I_{v,s}(\infty) \exp\left[-\int_z^\infty \sigma_v n \sec \theta(z') \, \mathrm{d}z' \right] \simeq$$

$$\simeq I_{v,s}(\infty) \exp\left[-\sec \theta(z) \, \tau_v(z) \right], \tag{4}$$

where

$$\tau_v(z) = \int_z^\infty \sigma_v(z') \, n(z') \, \mathrm{d}z' \simeq \sigma_v \int_z^\infty n(z') \, \mathrm{d}z'$$

is the optical depth. Zenith angle depends on height because of the sphericity of the Earth, but this dependence is very weak if $\theta(z)$ is not too large, and it has been neglected in Equation (4). The volume rate of absorption of monochromatic solar radiation, J_v, can be obtained by differentiating Equation (4). Expressing the result in terms of the pressure scale height H, one obtains,

$$J_v(z) = \cos\theta \frac{\mathrm{d}I_{v,\,\mathrm{s}}}{\mathrm{d}z} = I_{v,\,\mathrm{s}}(\infty)\,\sigma_v n(z)\,\exp\left[-\sec\theta\,\tau_v(z)\right] =$$

$$= I_{v,\,\mathrm{s}}(\infty)\,\sigma_v n(z)\,\exp\left[-\sigma_v n(z)\,H\,\sec\theta\right]. \tag{5}$$

Equation (5) describes the well-known Chapman layer distribution of absorption. It is strictly applicable to the absorption of monochromatic radiation in an exponential atmosphere, but it is approximately valid for the actual atmosphere as well. This distribution has the following properties.

(1) Maximum absorption occurs at the altitude at which

$$\sigma_v n H \sec\theta \equiv \tau_v \sec\theta = 1.$$

This maximum results from the trade-off between the increasing absorption rate because of the density increase with depth, on the one hand, and the exponential depletion with depth of the incident radiation on the other hand.

(2) The maximum rate of absorption decreases and the altitude of the maximum increases with increasing zenith angle.

(3) Because J_v decreases exponentially at heights above its maximum, and it decreases even more rapidly below the maximum, most of the solar radiation in any narrow frequency interval is absorbed within one or two scale heights of the maximum. Thus, except for sunlight incident at large zenith angles, most of the monochromatic absorption takes place near or slightly above the height at which $\tau_v = 1$, the height of unit optical depth.

The vertical distribution of contributions to the monochromatic radiation intensity leaving an emitting atmosphere exhibits a similar behavior. Applying Equation (1) to the contribution to the flux across a unit horizontal area outside of an infinite exponential atmosphere, we have

$$\cos\theta\,I_v(\infty) = \int_0^\infty \sigma_v n(z)\,\exp\left[-n(z)\,\sigma_v H \sec\theta\right] S_v(z)\,\mathrm{d}z. \tag{6}$$

Thus, the outgoing intensity is a weighted integral of the source function throughout the atmosphere. The weighting function has the same form as the Chapman layer distribution (Equation (5)), so that, unless there are extreme variations in $S_v(z)$, the emitted radiation at any frequency also originates mainly within a few scale heights of the height of unit optical depth.

Figure 1 shows the spectral distribution of the height of unit optical depth in the Earth's atmosphere and some of the associated physical processes (Hertzberg, 1965). Between 0.2 and 0.3 μm, radiation reaches the stratosphere, with particularly deep penetration near 0.2 μm. Between 0.1 and 0.2 μm, radiation is absorbed mainly near 100 km, and below 0.1 μm, most of the energy is absorbed in the thermosphere, although X-rays can again penetrate to the mesosphere.

One can compare this spectral distribution of the height at which energy is absorbed in the terrestrial atmosphere with the spectral distribution of the height from which radiation is emitted in the solar atmosphere. The photospheric absorption cross section increases with decreasing wavelength so that emitted radiation generally comes from higher and more variable levels in the solar atmosphere as wavelength decreases (Minneart, 1953). The visible and near UV continua and the Fraunhaufer lines originate at levels below the 4500 K temperature minimum. Radiation

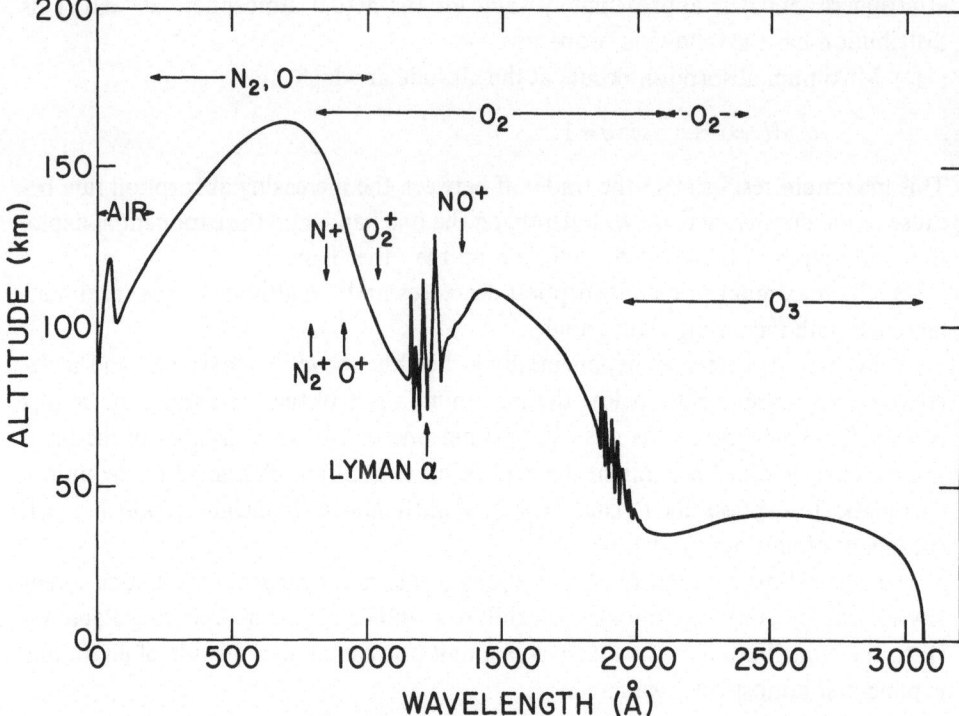

Fig. 1.   Height of unit optical depth in the Earth's atmosphere, and associated absorbing gases and ionization thresholds (from Hertzberg, 1965).

from these levels is believed to be quite constant in time, although long term absolute measurements of the solar intensity in the UV are difficult, and the actual variability is not yet known. There even remain uncertainties in the absolute magnitude of the mean solar flux below 0.21 μm wavelength amounting to a factor of 2 or 3. Below 0.15 μm the line spectrum appears in emission, indicating that these lines are formed above the temperature minimum, either in the chromosphere, the chromosphere-corona transition zone, or in the corona. In general, the lines are formed at higher levels, and the corresponding energy fluxes tend to be more variable. The coronal radiation includes a continuum portion in the X-ray region which is highly variable with the phase of the solar cycle, and with the level of solar activity (de Jager, 1964).

Thus each level and process in the Earth's upper atmosphere is tied to a specific level and a specific set of processes in the solar atmosphere. The variable part of the solar spectrum is absorbed primarily in the thermosphere above 150 km, although part of the highly variable solar X-ray flux penetrates to the D region. As a result, the composition and structure of the thermosphere and the ionization state of the D region depend on the solar cycle and state of the sun, but the neutral composition and thermal structure of the stratosphere and mesosphere are not sensitive to solar variations.

3. Disposition of the Absorbed Solar Radiation

From the stratosphere upward, most of the absorbed solar radiation initially produces dissociation, ionization, or both. Radiation at wavelengths between 0.2 and 0.3 μm dissociates O_3 in the stratosphere; radiation reaching the stratosphere in the 0.2 μm window also dissociates O_2 in the Hertzberg continuum, and, as a consequence, it is responsible for formation of the O_3 layer. Between 0.175 and 0.2 μm, the absorption spectrum shows a very complicated structure, that of the Schumann-Runge bands of O_2. The continuum peaking at 0.15 μm is the Schumann-Runge continuum; it is responsible for the rapid dissociation of O_2 which occurs in the lower thermosphere. One of the O atoms appears in the ^1D excited state. There is an absorption minimum at 0.1215 μm which allows solar Ly-α radiation to penetrate into the mesosphere. There it ionizes NO, which is present at concentrations of the order of $10^{-7}-10^{-8}$, and is responsible for formation of the upper D region under undisturbed conditions. Below about 0.1 μm, solar radiation ionizes the major atmospheric constituents, O, O_2, N_2, and is responsible for formation of the E and F regions of the ionosphere.

The extent to which the portion of the input radiation used initially for ionization or dissociation eventually appears as heat depends on subsequent processes. These include additional ionizations or dissociations by electron impact, electron impact excitation, transfer of kinetic energy between electrons or from electrons to ions and from ions to neutrals, dissociative recombination of electron-ion pairs, recombination of dissociation products and other chemical reactions involving dissociation products. The fraction of the absorbed solar energy which appears as heat in the same height range is known as the heating efficiency, and its calculation is generally very difficult because of the variety of possible paths for the input energy. If the system is in steady state, however, all of the absorbed energy eventually appears locally as heat unless there is: (1) emission of radiation from levels excited by non-thermal processes, or (2) a net flux of ionization or dissociation products out of the region in which they are produced.

Below 75 km, excited states of O and O_2 are produced, but quenching is rapid and transport is relatively slow, so that practically all of the absorbed energy appears as heat. Most of this energy is absorbed by O_3, and the height dependent absorption rate is immediately calculable upon spectral integration of Equation (5) if the O_3 concentration is known. There are still uncertainties in this concentration of a factor of 2 or so, above 40 km, and consequent uncertainties in the heating rate. The heating and cooling processes are sufficiently well known in this region, however, that the problem can be reversed; one can ask: what concentration of O_3 will lead to thermal equilibrium between 35 and 65 km at the observed temperatures? This approach can lead to useful constraints on the O_3 concentration and on the processes responsible for its destruction (Blake and Lindzen, 1973).

Two of the most intense airglow emissions originate in the region between 75 and 105 km. These are the Meinel bands of OH, and the green line of O (0.5577 μm) arising from the $^1S \rightarrow {^1D}$ transition. The average Meinel band emission amounts to some 2×10^{-3} W m^{-2}; the O emission is about 1×10^{-3}. Together they account for more than half of the energy transported downward from the thermosphere by conduction and in the form of O.

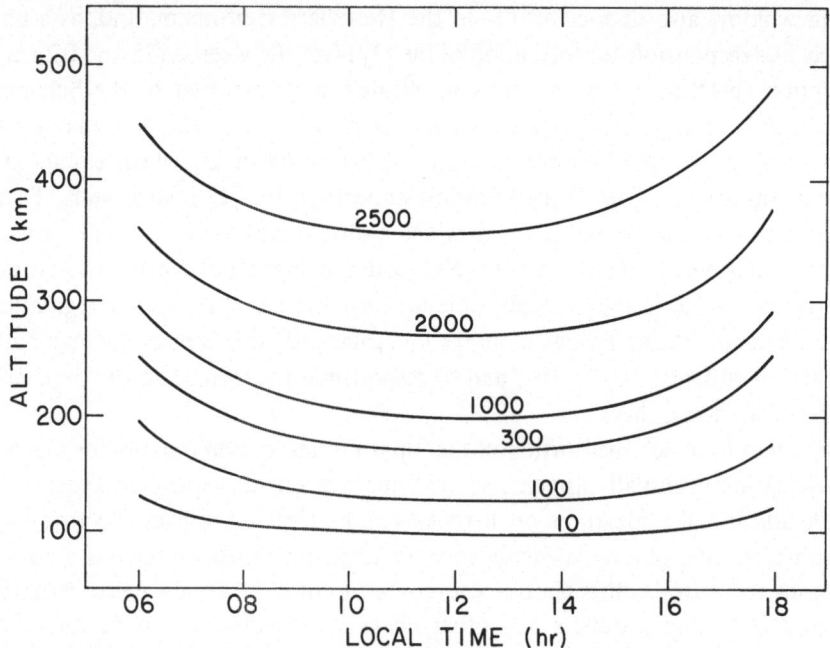

Fig. 2.   EUV heating rate (°C day^{-1}) at 30° latitude and equinox, as calculated by Lagos and Mahoney (1967) (from Dickinson and Geisler, 1968).

Above the mesopause energy is absorbed in the Schumann-Runge continuum and in the ionization continua. The heating efficiency due to the former depends mainly on the branching ratio between quenching or emission from the O^1D state, and it decreases with height (Izakov and Morozov, 1970; Izakov, 1970). Heating efficiency for the ionizing radiation is much more difficult to evaluate; it has been variously estimated at 0.1 to 0.5, but an overall average value near 0.35 gives reasonable thermal balance in the thermosphere (Lagos and Mahoney, 1967). Since the spectral distribution of input energy depends on solar conditions, the efficiency factor for the ionizing radiation must also depend on solar conditions. Figure 2 shows some estimates of the heating rate per unit mass based on the work of Lagos and Mahoney (Dickinson and Geisler, 1968).

4. Thermal Emission and Formation of the Mesopause

In order to calculate the cooling due to thermal emission of radiation, we integrate Equation (1) over frequency and solid angle making use of the plane-parallel atmosphere model (the assumption that the atmosphere is horizontally stratified). The net rate of energy gain per unit volume, $h(z)$, can then be expressed in the form

$$h(z) = \int dv \left\{ [-\pi S_v(O_-) + \pi S_v(z)] \frac{dT_v}{dz}(0, z) - \right.$$

$$\left. - \int_0^\infty dz' [\pi S_v(z') - \pi S_v(z)] \frac{d^2 T_v}{dz\, dz'}(z, z') - \pi S_v(z) \frac{dT_v}{dz}(z, \infty) \right\}. \quad (7)$$

Here $S_v(O_-)$ is the source function of the underlying surface (ground or cloud), and the frequency integration extends over the absorption bands of the radiatively active gases. The quantity $T_v(z_1, z_2)$ is the flux transmission function, defined by

$$T_v(z_1, z_2) = 2 \int_0^{\pi/2} d\theta \sin\theta \cos\theta \exp\left[-\sec\theta \left| \int_{z_1}^{z_2} \sigma_v n(z')\, dz' \right| \right] =$$

$$= 2E_3 \left[|\tau_v(z_2) - \tau_v(z_1)| \right] \quad (8)$$

so that T_v is a monotonically decreasing function of the difference in optical depth between z_1 and z_2.

The three terms on the right side of Equation (7) correspond respectively to exchange of radiation between level z and the ground, exchange of radiation between level z and all other levels, z', and exchange of radiation between level z and space. Since the thermal emission from space is negligible, the latter is a one way exchange only; the atmosphere loses radiation to space. For the stratosphere and above, exchange of radiation with the underlying surface is generally negligible because of the large optical depths and corresponding small values of dT_v/dz involved. Emission by O_3 is an exception to this; since there is very little O_3 in the troposphere, exchange with the ground remains significant at all levels. The factor $d^2 T_v/dz\, dz'$ is everywhere negative, and its magnitude decreases monotonically away from the level $z' = z$ where it is singular. Thus the exchange of radiation between levels contributes to the heating rate mainly in regions where there is strong curvature of $S_v(z)$. Under LTE conditions, curvature of S_v corresponds to curvature of the temperature profile; positive curvature contributes to heating, negative curvature to cooling.

Water vapor, CO_2, and O_3 are the main IR radiating gases. Although H_2O dominates thermal emission in the troposphere, the CO_2 concentration is at least 20 times that of O_3 and about 100 times that of H_2O in the stratosphere and mesosphere, so that CO_2 dominates in these regions. The CO_2 absorption bands are confined to a narrow spectral interval centered near 15 μm, and over this interval S_ν varies slowly with frequency, although σ_ν, which arises from hundreds of individual vibration-rotation lines, varies extremely rapidly. Under these conditions Equation (7) can be approximated by

$$h(z) \simeq \left\{ [\pi\bar{S}(z) - \pi\bar{S}(0_-)] \int \mathrm{d}\nu \frac{\mathrm{d}T_\nu}{\mathrm{d}z}(0, z) - \int \left[\int \mathrm{d}\nu \frac{\mathrm{d}^2 T_\nu}{\mathrm{d}z\,\mathrm{d}z'}(z, z') \right] \times \right.$$
$$\left. \times [\pi\bar{S}(z') - \pi\bar{S}(z)]\,\mathrm{d}z' - \pi\bar{S}(z) \int \mathrm{d}\nu \frac{\mathrm{d}T_\nu}{\mathrm{d}z}(z, \infty) \right\}, \qquad (9)$$

where $\bar{S}$ is a mean value of the source function over the narrow spectral interval. The quantity

$$\int T_\nu(z, \infty)\,\mathrm{d}\nu \bigg/ \int \mathrm{d}\nu$$

is just the probability that a photon emitted in the upward half-space from level z will escape from the atmosphere; it can be called the *escape function*. There are two main difficulties in the evaluation of the escape function, and the other spectral integrals involving T_ν in Equation (9).

(1) Even for strictly monochromatic radiation σ_ν depends on pressure and temperature because of the dependence of the line profile on these parameters. Thus, $T_\nu(z_1, z_2)$ depends on the distribution of pressure and temperature between z_1 and z_2.

(2) Lines having a range of strengths of 4 or 5 decades will contribute to the spectral integration and must be taken into account. This is because weak lines will contribute to the integration over long paths when the strong lines are saturated, but strong lines will dominate the spectral integration over short paths. Since transmission over both long and short paths must be taken into account, a very large number of lines has to be included in the integration.

Dickinson (1972, 1973) has described techniques for evaluating $h(z)$. Figure 3 shows calculated contributions to the total cooling by the fundamental vibrational transition, the overtone and combination bands (hot bands), and the fundamental transition of minor CO_2 isotopes. All of these bands make important contributions to the total cooling in some height range.

At high levels, collisions are too infrequent to maintain the populations of the vibrational levels in equilibrium, and LTE breaks down. The rates of population of the energy levels by radiation and by collision have to be taken into account. Because many levels are involved, the general problem is a very complicated one, but a qualitative picture of the major effect of the breakdown of LTE can be obtained either by ignoring the upper vibrational levels, or by assuming that they are closely coupled to the lowest excited level. In either case, only two levels, the ground state and the first excited state have to be considered explicitly. The source function is then, ap-

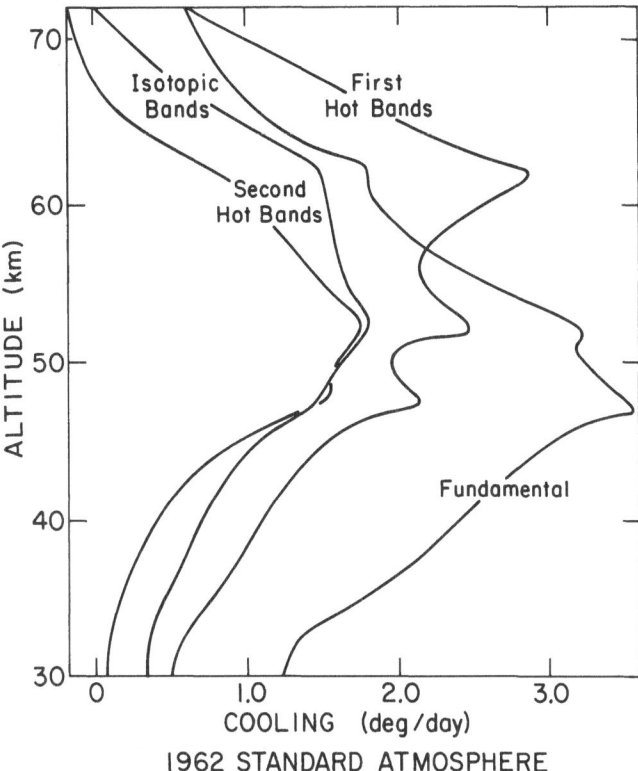

Fig. 3.   Cooling rate for the standard atmosphere due to different CO_2 bands. The small contribution by hot bands of isotopes other than the main one is not included (from Dickinson, 1973).

proximately,

$$\bar{S}(z) = \frac{n_*}{n_{*e}} \bar{B}(z),\tag{10}$$

where n_* is the number density in the lowest excited vibrational state; n_{*e} is the number density in that state under kinetic equilibrium conditions. Substitution of Equation (10) into (9) leads to an integral equation for $\bar{S}(z)$ which can be solved if the Planck function, $\bar{B}(z)$, is known. The qualitative effects of the breakdown of LTE can be seen by consideration of only the radiation to space term in Equation (9), since this term is often dominant. One obtains

$$h(z) \sim -\left\{1 + A\tau_c \int \mathrm{d}v \frac{\mathrm{d}T_v}{\mathrm{d}z}(z,\infty) \bigg/ 2n \int \sigma_v \, \mathrm{d}v\right\}^{-1} \pi \bar{B}(z) \int \mathrm{d}v \frac{\mathrm{d}T_v}{\mathrm{d}z}(z,\infty).$$

The cooling is reduced relative to that for LTE by the factor

$$\left[1 + A\tau_c \int \mathrm{d}v \frac{\mathrm{d}T_v}{\mathrm{d}z}(z,\infty) \bigg/ 2n \int \sigma_v \, \mathrm{d}v\right] \simeq \left[1 + \frac{\pi \bar{B}\tau_c}{h v n_{*e}} \int \mathrm{d}v \frac{\mathrm{d}T_v}{\mathrm{d}z}(z,\infty)\right],$$

where A is the Einstein coefficient for spontaneous emission from the lowest excited level; it is the inverse of the radiative lifetime of that level. Thus, for this simplified model, two conditions are required for the breakdown of LTE to cause large reductions in cooling rate relative to LTE conditions:

(1) $A\tau_c > 1,$

(2) $\int dv \frac{dT_v}{dz}(z,\infty) \Big/ n \int \sigma_v \, dv \sim 1.$

Radiative and collisional lifetimes are comparable and condition (1) is satisfied above about 75 km, but condition (2), requiring that the probability for escape of a photon to space be large, is not satisfied below 90 or 95 km. Above the latter height, CO_2 radiative cooling becomes very inefficient because of the breakdown of LTE, and heat absorbed at higher altitudes must be transported downward by conduction to below the 90 or 95 km level before it is lost by thermal radiation. It is this requirement which determines the altitude of the mesopause.

Detailed heat balance calculations for the mesosphere and mesopause regions have been carried out by Murgatroyd and Goody (1958) and by Kuhn and London (1969). Cooling of 12 to 15 C day^{-1} occurs in the upper winter mesosphere, but because of the very low observed temperatures in summer, some warning takes place as a result of convergence of IR radiation near the summer mesopause.

5. Dynamical Heating

Hydrodynamic waves can transport energy of organized fluid motions from one region of the atmosphere and deposit it as heat in another region. Upward propagating waves can be particularly effective; because of the decrease in density with height, their amplitudes may increase with height. The principal modes by which wave energy may be propagated upward from the lower atmosphere are listed in Table II. The rate at which energy reaches the thermosphere depends on the rate of generation and the extent to which the energy flow can be refracted away from the vertical, partially or totally reflected, or absorbed.

Gravity waves can be generated by a number of processes, including the convective motion associated with mesoscale storm systems and flow over mountains. Flow over mountains is likely to be the dominant source of gravity waves capable of propagation above the tropopause however, and an average rate of wave energy generation by this process can be crudely estimated to be in the range 10^{-2} to 10^{-1} W m^{-2}. Once these waves are produced, they face many hazards enroute to the mesopause region. These hazards can be described in terms of the group velocity, or in terms of the slope of the air parcel oscillations. The waves are essentially transverse; phase velocity and wavenumber vector are normal to the parcel oscillations, but group velocity and energy flux are parallel to the oscillations with the vertical component of energy flux opposite to that of phase propagation. When the oscillation is vertical,

TABLE II

Characteristics of hydrodynamic modes

Type of mode	Principal source	Horizontal scale (km)	Vertical scale (km)	Time scale (s)	Factors influencing reflection	Factors influencing absorption	Contribution to thermosphere heat balance
Gravity wave	Flow over mountains	50–500	5–25	10^3–10^4	Strong winds and low stability	Wind direction reversal with height; turbulence	?
Rossby wave	Global scale topography and and heating	$\sim 10^4$	10–50	$\gtrsim 10^5$	Strong winds	Easterly winds, radiative damping; turbulence	Usually small
Diurnal tide	diurnal heating	10^4	5–25	2×10^4	None	Radiative damping; turbulence	Small
Semi-diurnal tide	Semi-diurnal heating	0.5×10^4	100	10^4	None	Radiative damping	Possibly large

gravity waves degenerate to pure bouyancy oscillations which do not propagate vertically. When the oscillation is horizontal, the restoring effect of bouyancy vanishes; and again the waves cannot propagate.

These propagation properties follow from the wave equation governing the complex amplitude of the wave vertical velocity, W,

$$\frac{d^2 W}{dz^2} + m^2(z)\, W = 0. \tag{11}$$

Here,

$$m^2(z) \approx k^2 [v_B^2/(\mathbf{U} \cdot \mathbf{k})^2 - 1]; \tag{12}$$

$\mathbf{k}$ is the horizontal part of the wavenumber vector, $\mathbf{U}(z)$ is the basic state wind, and $v_B^2(z) \equiv (g/\varrho_p)(\partial s_0/\partial z)$ is the Brunt-Väisällä Frequency, which can also be expressed in terms of basic state entropy, s_0. In general, $\mathbf{U}$, v_B^2, and m^2 vary with height, but $\mathbf{k}$ does not. Since m is a local vertical wavenumber, and since the waves are transverse, the condition $m^2 \to 0$ while $\mathbf{k}$ remains fixed corresponds to nearly vertical air parcel oscillations, and the waves cannot readily propagate through a region with $m^2 \leq 0$. A deep enough layer having $m^2 \leq 0$ totally reflects the wave energy; for shallower layers partial reflections can occur (Eliassen and Palm, 1967). On the other hand, if $m^2 \to \infty$, the parcel trajectories become horizontal. In this case, both vertical and horizontal components of the group velocity vanish, and a wave group confined to a narrow range of wavenumbers and frequencies will take so long to reach such a level that its energy will be absorbed (Booker and Bretherton, 1967). Equation (12) shows that reflection is favored by small values of v_B^2 and large values of $(\mathbf{U} \cdot \mathbf{k})^2$, or by large values of the component of $\mathbf{U}$ in the direction of phase propagation, but absorption occurs if the component of $\mathbf{U}$ in the propagation direction vanishes, or, in other words, if there is a reversal of wind direction with height. When the vagaries of the background winds are considered together with the extreme time and space variability of gravity wave production, one must conclude that the flux of gravity wave energy to the lower thermosphere is at least highly variable in time and space, and fluxes large enough to make a significant contribution to the thermospheric energy balance may be quite rare.

Rossby waves, or planetary waves, are produced by flow over topography on the scale of continents, by differential heating on the same scale, or by baroclinic instability in the troposphere. The rate of energy generation, which may be of the order of $10\ \mathrm{W\,m^{-2}}$ prompted Charney and Drazin (1961) to ask why it is that vertical energy propagation and consequent thermospheric heating by these waves do not produce a geocorona at low thermospheric altitudes. The description of the dynamics reduces to that of the behavior of a single dependent variable, the stream function for the horizontal flow. The complex amplitude of the stream function for disturbances having a wave-like dependence on longitude can be denoted by $\psi(y, z)$, and it is governed by the two dimensional wave equation

$$\frac{\partial^2 \psi}{\partial y^2} + \frac{\partial^2 \psi}{\partial z^2} + M^2(y, z)\, \psi = 0 \tag{13}$$

where y is the meridional coordinate, and

$$M^2(y, z) = \frac{\beta + f(y, z)}{U} - k^2. \tag{14}$$

Here β is the meridional gradient of the coriolis parameter (positive everywhere), U is the basic state zonal wind (positive from the west), and $f(y, z)$ is a function determined primarily by the distribution of U, but it is usually smaller in magnitude than β. The quantity M^2 plays the role of a refractive index in the $(y-z)$ plane. Where $M^2 < 0$, waves will not propagate; where M^2 is small and positive, waves propagate but phase changes slowly with distance; where M^2 is large and positive, propagating waves exhibit rapid change of phase with distance. To the extent that a ray optics analogy can be applied, propagating waves will tend to be refracted away from regions with M^2 small and positive and toward regions of large positive M^2 (Dickinson, 1968). There is a partial analogy between planetary waves and gravity waves: when M^2 passes through infinity to negative values, planetary waves are also absorbed. This occurs where the basic state wind switches from westerly to easterly. Thus, such regions of easterly winds as the summer stratosphere are virtually free from planetary waves. A transition to easterly winds almost always occurs in some region of the stratosphere; for example, during winter, it occurs in the subtropics. Since planetary waves are refracted toward such a region, nearly complete absorption in the stratosphere and lower mesosphere can be expected most of the time. Under exceptional conditions, the distribution of background zonal wind may permit some energy to propagate to the mesopause. Dickinson has shown, however, that such waves will be strongly attenuated by radiative damping. Furthermore, the appearance of k^2 in Equation (14) shows that only very long waves (usually only longitudinal wavenumbers 1 and 2) can propagate even under favorable conditions. This limitation prevents baroclinically unstable waves from propagating vertically, so that this energy source is not available to the upper atmosphere, and only the upward flux due to very long planetary waves forced by differential heating and topography can be considered.

The rate of energy generation in diurnal and semi-diurnal tides exceeds 10^{-2} W m^{-2} and several of the tidal modes are not strongly subject to reflection or absorption below the mesopause (Chapman and Lindzen, 1971). The diurnal tidal modes which can propagate have short vertical wavelengths, however, and they are dissipated mainly near the turbopause where their energy makes little contribution to the local heat balance. On the other hand, the semidiurnal tide can propagate well up into the thermosphere and may make a significant contribution to heating near 150 km (Lindzen, 1970, 1971).

Although the role of upward propagating gravity waves on the heat balance of the thermosphere is unknown, there is compelling evidence from Traveling Ionospheric Disturbances (TID's) and direct satellite measurements of density fluctuations that gravity waves generated in the auroral zone thermosphere propagate laterally and provide a significant heat source for the middle and low latitude thermosphere

(Newton *et al.*, 1969). Thus, auroral heating can effectively influence the temperature of the entire thermosphere. The speed at which the energy propagates away from the auroral zone is rapid (~ 300 m s^{-1}), and as a result, the temperatures in the low latitude thermosphere can be expected to show a significant correlation with particle fluxes and geomagnetic activity indices (Klostermeyer, 1973).

Note added in proof. Recent calculations by Lindzen and Hong (1974) show that when the effects of longitudinally averaged zonal winds are taken into account, the theoretical energy flux due to the semi-diurnal is reduced by nearly an order of magnitude below the flux calculated without taking zonal winds into account. The recalculated tidal amplitudes allowing for the zonal winds are consistent with recent measurements reported by Bernard, and it now appears that the semidiurnal tide generated in the lower atmosphere does not make a major contribution to the heat balance of the thermosphere.

References

Bernard: 1974, *J. Atmospheric Terrestr. Phys.* **36**, 1105.

Blake, D. and Lindzen, R. S.: 1973, *Mon. Weather Rev.* **101**, 783.

Booker, J. and Bretherton, F.: 1967, *J. Fluid Mech.* **27**, 513.

Chapman, S. and Lindzen, R. S.: 1971, *Atmospheric Tides,*

Charney, J. G. and Drazin, P. G.: 1961, *J. Geophys. Res.* **66**, 83.

Dickinson, R. E.: 1968, *J. Atmospheric Sci.* **25**, 984.

Dickinson, R. E.: 1972, *J. Atmospheric Sci.* **29**, 1531.

Dickinson, R. E.: 1973, *J. Geophys. Res.* **18**, 4451.

Dickinson, R. E. and Geisler, J. E.: 1968, *Mon. Weather Rev.* **96**, 606.

Eliassen, A. and Palm, E.: 1961, *Geofyske* **22**, 1.

Hertzberg, L.: 1965, in C. O. Hines, I. Paghiz, C. Hart, and J. Fejer (eds.), *Physics of the Earth's Upper Atmosphere*, Prentice-Hall Inc., Englewood Cliffs, N.J.

Izakov, M. N.: 1970, *Geomagnetizm i Aeronomiya* **10**, 219.

Izakov, M. N. and Morozov, S. K.: 1970, *Geomagnetizm i Aeronomiya* **10**, 495.

de Jager, C.: 1964, in H. Odishaw (ed.), *Research in Geophysics*, vol. 1, The MIT Press, Cambridge, Mass., pp. 1–42.

Klostermeyer, J.: 1973, *J. Atmospheric Terrestr. Phys.* **35**, 2267.

Kondratyev, K. Ya.: 1965, *Radiative Heat Exchange in the Atmosphere*, Pergamon Press, New York.

Kuhn, W. and London, J.: 1969, *J. Atmospheric Sci.* **26**, 189.

Lagos, C. P. and Mahoney, J. R.: 1967, *J. Atmospheric Sci.* **24**, 88.

Lindzen, R. S.: 1970, *Geophys. Fluid Dyn.* **1**, 303.

Lindzen, R. S.: 1971, *Geophys. Fluid Dyn.* **2**, 89.

Lindzen, R. S. and Hong, S.: 1974, *J. Atmospheric Sci.* **31**, 1421.

Minnaert, M.: 1953, in G. P. Kuiper (ed.), *The Solar System*, vol. 1: *The Sun*, The University of Chicago Press, Chicago, Ill.

Murgatroyd, R. J. and Goody, R. S.: 1958, *Quart. J. Roy. Meteorol. Soc.* **84**, 224.

Newton, G. P., Pelz, D. T., and Volland, H.: 1969, *J. Geophys. Res.* **74**, 183.

WINDS AND ELECTRIC FIELDS
IN THE UPPER ATMOSPHERE

H. KOHL

Max-Planck-Institut für Aeronomie, D-3411 Lindau/Harz, Postfach 20, West Germany

1. Introduction

Winds and electric fields are in fact some of the oldest problems in the field of atmospheric physics. In 1882 Balfour Stewart suggested that an electrically conducting region may exist in the atmosphere which is moved (winds!) relative to the geomagnetic field of the Earth. In this moving conductor electromotive forces (electric fields!) will be induced which in turn will produce electric currents. Stewart introduced this hypothesis in order to explain observed variations in the geomagnetic field, although in those days nothing was known about the existence of an ionosphere nor of any motion there. Meanwhile these suggestions have been confirmed and we know that, indeed, neutral air motions, i.e., winds, and electric fields exist in the upper atmosphere. Some of their features and some theoretical aspects will be discussed now. We shall restrict ourselves to large scale phenomena, i.e., on effects that appear more or less on a global scale. Section 2 will deal with winds, where the term wind stands for movements of neutral air. First, the air circulation in the mesosphere is considered in the altitude range from about 40 to 90 km. Next, tidal oscillations of the atmosphere will be discussed. Since the atmosphere oscillates as a whole, emphasis will be placed upon the altitude range of 90 to 150 km, which is the so-called 'dynamo region'. Finally global wind systems are described that are present above 150 km up to several hundreds of kilometers height. Section 3 deals with the mechanisms that produce electric fields in the ionosphere and some of the observational evidence we have for them. In Section 4, finally, the effects of electric fields and winds on the ionospheric F layer will be briefly illustrated.

2. Winds in the Upper Atmosphere

2.1. MESOSPHERIC CIRCULATION

A large amount of data of mesospheric parameters has been gathered by various techniques, i.e., rocket-borne probes, falling spheres, chaff etc. Another very useful method is the rocket grenade experiment, where at a series of heights explosions are released from a rocket. The exact time of the explosion is determined from optical observations, while a groundbased network of microphones records the arrival times of the sound at different locations. It is then possible to derive either temperature or wind velocity from these observations, owing to the fact that the velocity of sound with respect to the moving medium is proportional to $\sqrt{T}$.

Such measurements show that the wind velocities in the mesosphere are very vari-

B. M. McCormac (ed.), Atmospheres of Earth and the Planets, 87–97. All Rights Reserved.
Copyright © 1975 by D. Reidel Publishing Company, Dordrecht-Holland.

able and have a complex morphology. Some persistent features of the observations are:

(i) the time average of the meridional velocity is small (a few meters per second);

(ii) the time average of the zonal velocity is of the order of some tens of meters per second. It shows a more regular behavior depending on season; and

(iii) The rms of the velocity data is at least of the same order as the average value.

Here only the average zonal wind will be discussed. Figure 1 shows a meridional cross section of the east-west wind for solstice conditions. It is seen that in the winter hemisphere the wind velocity is positive, i.e., from the west towards the east, while in the summer hemisphere the wind blows towards the west. Maximum wind speeds occur at about 70 km altitude (CIRA, 1965).

Wind patterns of this kind can be related to meridional temperature variations observed simultaneously. This can be seen in the following way. The equation of motion for a neutral gas is:

$$\frac{\partial \mathbf{u}}{\partial t} + (\mathbf{u}\nabla)\,\mathbf{u} + 2(\boldsymbol{\omega}\times\mathbf{u}) = -\frac{1}{\varrho}\,\nabla p + \mathbf{g} + \mathbf{F}, \tag{1}$$

where the first two terms on the left hand side represent the inertial force, and the third term is the Coriolis force ($\boldsymbol{\omega}$ is the angular velocity of the Earth). The first term on the right hand side is the force per unit mass due to a gradient in air pressure, $\mathbf{g}$ is the gravitational acceleration, and $\mathbf{F}$ means any further force, e.g., viscosity.

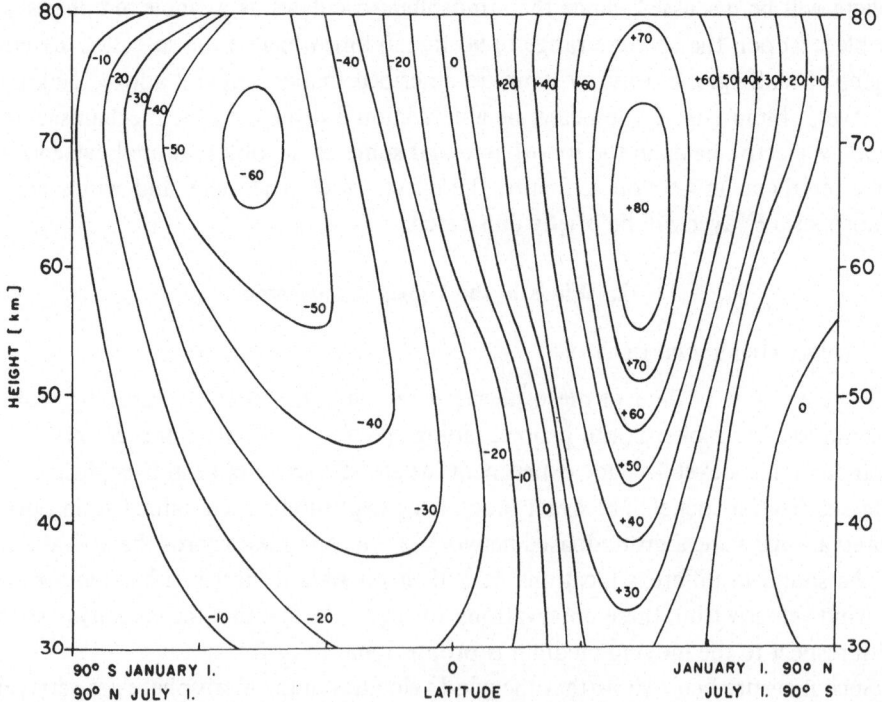

Fig. 1.   Zonal wind velocity in the mesosphere. Positive winds from W to E (after CIRA, 1965).

If we assume that the inertial force and the additional force **F** are negligible, and furthermore, if we, restrict ourselves to horizontal zonal motions, then Equation (1) reduces to

$$2\,\omega u_x \cdot \sin\varphi = -\frac{1}{\varrho}\frac{\partial p}{\partial y}, \tag{2}$$

where x and y directions are positive towards the east and north, respectively. Equation (2) means that a pressure difference in the north-south direction is balanced by a Coriolis force due to a zonal wind. It is believed that Equation (2) is valid for large scale motions in the mesosphere in most cases, since other forces, as the inertial force, are small. This kind of approximation is called 'geostrophic'.

If one differentiates Equation (2) with respect to the vertical direction z and makes use of the relation $\varrho = n\bar{m}$ ($\bar{m}$ is the mean mass of the air molecules), the ideal gas law $p = nkT$ and the barometric law $p = p_0 \exp(-\int(\bar{m}g/kT)\,\mathrm{d}z)$, then Equation (2) can be expressed as

$$\frac{\partial}{\partial z}\left(\frac{u_x}{T}\right) = \frac{q}{2\omega\sin\varphi}\frac{\partial}{\partial y}\left(\frac{1}{T}\right). \tag{3}$$

This is the so-called thermal wind equation, which relates vertical shears in the zonal wind to meridional temperature gradients.

Indeed, the observed temperature and wind variations seem to be in fairly good agreement with Equation (3) confirming the above assumptions. In particular, Figure 1 shows an increase of the magnitude of the wind velocity at medium latitudes up to about 70 km and a decrease at greater heights. From Equation (3) it can be inferred that below 70 km the temperature should decrease towards the pole in the winter hemisphere and increase towards the pole in the summer hemisphere. The opposite should be true at heights greater than 70 km. Indeed, this is just what is observed, as can be seen in Figure 2. The above discussion relates only to temperature and wind variations. It does not explain why there is such a temperature (or pressure) variation that drives the wind circulation. This is a problem of energy input into the atmosphere that cannot be considered here.

2.2. TIDAL MOTIONS

Tides are not only present in the oceans, but also in the atmosphere. Here, they are not only caused by gravitational forces of the moon and the sun, but also by the heat input of the Sun, which has primarily a diurnal variation, but, of course, includes higher harmonics. The excitation and propagation of tidal waves is a very complicated matter. We shall only mention some features of tides. The air motion due to tides is mainly horizontal. In the lower atmosphere their amplitude is small and strongly masked by meteorological effects. However, at higher altitudes the tidal velocity is expected to increase. This is a consequence of the constant energy ($\sim \varrho u^2$) in a wave, which leads to high velocities when the density becomes small.

At altitudes above 80 km tidal motions have been observed. Figure 3 shows such

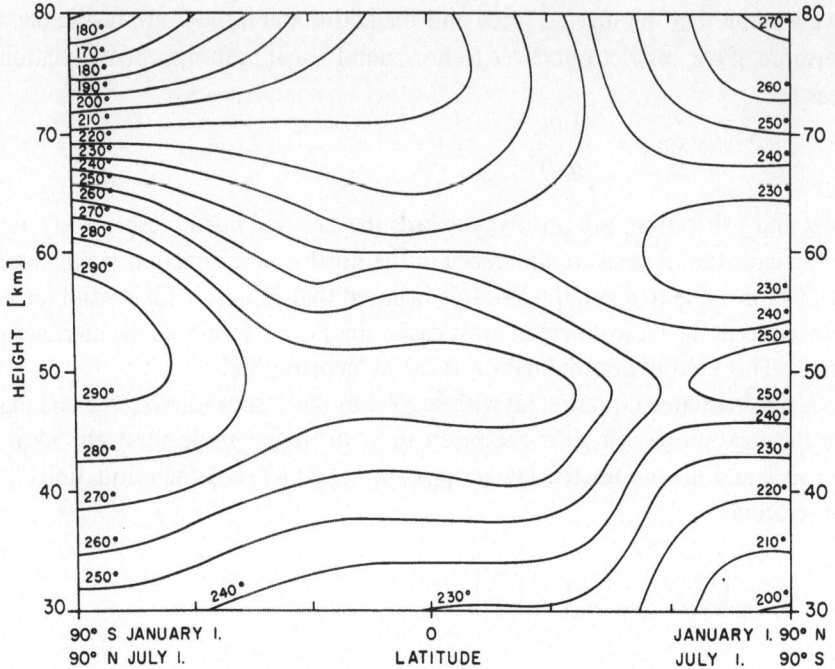

Fig. 2.   Temperature in the mesosphere as function of height and latitude
(after CIRA, 1965).

measurements made at Garchy (Spizzichino, 1968). They were obtained by the meteor
scatter technique which can be used in the height range of 80 to 110 km. When a
meteorite penetrates this range it evaporates and produces an ionized trail, which
moves with the velocity of the surrounding air. Its velocity can then be determined
by radar tracking. In the upper part of Figure 3 the observed diurnal variation of
the north-south velocity at an altitude around 100 km is shown. The lower part
presents the harmonic analysis. It is clearly seen that the semidiurnal wave is dominant, while the diurnal wave is slightly weaker. Similar observations have been made
at other locations and the results were sometimes different; this is not surprising as
tidal motions depend on the geographic position of the observing station. It should
be noted that meteor radar measurements also showed the existence of a prevailing
wind of the order of 10 m s^{-1} indicating that circulations of the kind discussed above
are also present at these heights. These prevailing winds were excluded in Figure 3.

In recent years the incoherent scatter technique has also been used for investigating
motions in the height range of 100 to 150 km. Because of the high collision frequency
at these altitude the ionization moves almost with the same velocity as the neutral
air. The velocity of the ionization, on the other hand, can be determined from the
Doppler shift of a backscattered electromagnetic wave. The results obtained are
similar to those from meteor scatter (Bernard, 1974).

2.3. GLOBAL WIND SYSTEMS IN THE THERMOSPHERE

At altitudes above 150 km the collision frequency between ions and neutrals becomes

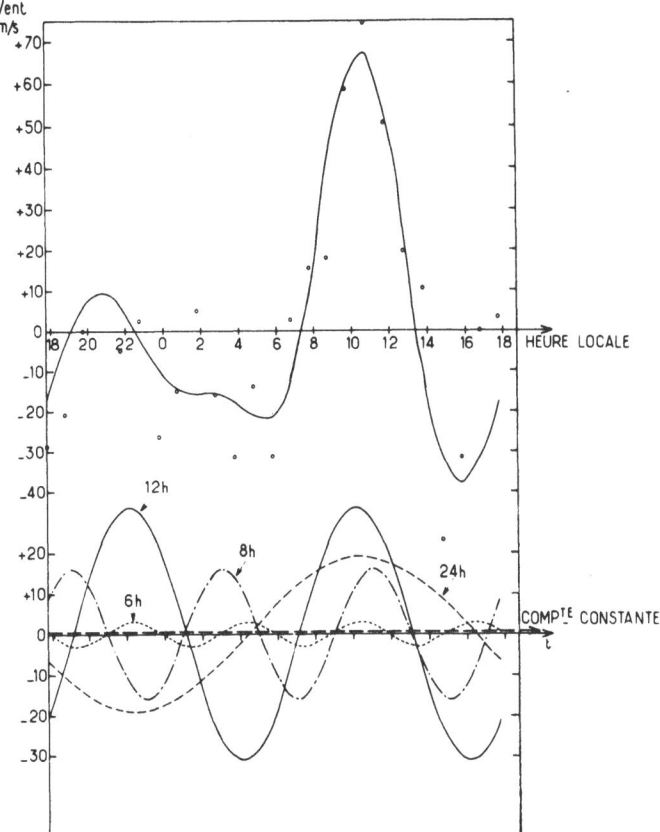

Fig. 3. Upper part: Diurnal variation of the zonal tidal velocity at 100 km altitude measured at Garchy (France) by the meteor scatter techniques. Lower part: Harmonic analysis of the velocities shown in the upper part.

so small that the ionization no longer follows the movements of the neutral air. Thus, the two methods described above and all other methods, which primarily observe ion motions, are not applicable for studying neutral air motions.

Thus vapor clouds or trails released from rockets were used for this kind of investigation. Their disadvantage is that they can only be observed during twilight conditions. Moreover, rocket flights are relatively expensive and only a limited number has been flown. The material did show that winds of the order of 100 m s^{-1} exist in the thermosphere (sometimes even several hundred meters per second in the polar ionosphere) (Stoffregen, 1972), but it was not sufficient to give a complete synopsis of thermospheric neutral air motions, in particular, because of the twilight restriction.

The observations together with theoretical considerations led to the next step. A satellite moving with a velocity of $w \approx 8$ km s^{-1} experiences a drag force from the surrounding neutral air, which is proportional to ϱw^2, where ϱ is the neutral air density. This force is small, but under thermospheric conditions it is strong enough to distort the orbit and to change the orbital period of the satellite. If the orbit is reasonably eccentric, most of the drag force is centered around the satellite's perigee where the air density is greatest. In this case the air density at the perigee can be calculated from the change of the orbital period provided that the satellite is roughly of a spherical shape, so that its effective cross section is known to be independent of any tumbling.

In this manner much air density data have been obtained at various times, geographic positions, altitudes, seasons and solar activity periods. It turned out that generally at a certain height the air density is lower on the nightside of the Earth and higher on the dayside, the density ratio being up to a factor of 2 at 300 km height. Since the temperature is also higher at the dayside, we finally have a high air pressure on the dayside and a low one on the nightside. Thus we can expect that strong winds are excited (King and Kohl, 1965).

However a more quantitative approach is possible. Using the above mentioned satellite drag data together with information on composition and temperature obtained from other measurements, and introducing physical arguments like, for instance, the barometric law, it was possible to construct models of the thermosphere which support us with data on density, composition, and temperature for any given time, height and location. This enables us to calculate wind velocities from Equation (1), in which the driving term $\varrho^{-1}\nabla p$ can be regarded as a known function.

We have, however, still to deal with the unspecified force $\mathbf{F}$ in Equation (1). For this problem an important force is the friction between ions and neutrals. It was mentioned earlier that above 150 km the ionization does not move with the neutral air. Therefore, we have to add a friction term $\mathbf{F} = v_{ni}(\mathbf{u} - \mathbf{v})$, where $\mathbf{u}$ is the wind velocity, $\mathbf{v}$ is the velocity of the ionization, and v_{ni} is an effective collision frequency describing the momentum transfer from neutrals to ions. Roughly $v_{ni} \approx 10^{-9}$ cm^3 s^{-1} N_i is the ion number density.

The ion velocity $\mathbf{v}$ can be determined from theoretical considerations. It can be shown that for low collision frequencies ($v_{in} \ll eB/m_i c$) the geomagnetic field gets control over the ion motion, in that the ion velocity perpendicular to the magnetic field $v_{\perp} \approx 0$, while for the velocity parallel to $\mathbf{B}$ it is $v_{\parallel} \approx u_{\parallel}$.

The above friction term $\mathbf{F}$ is now a function of $\mathbf{u}$ and N_i only. If the latter is given, for instance from measurements or from an F layer model, then $\mathbf{u}$ is the only unknown in Equation (1).

Figure 4 shows the solution of Equation (1) for equinox and medium sunspot activity ($F_{10.7} = 150$) at 300 km altitude using an atmospheric model worked out by Jacchia (1971). It is seen that winds are directed from the dayside equatorial region, i.e., the region of highest pressure, to the nightside equatorial region, the region of lowest pressure. The direction of the wind is roughly following grand circles. It is obvious that the geostrophic approximation is no longer valid, as the wind is directed more or less parallel to the pressure gradient rather than perpendicular. This is because the frictional force was included, which is usually stronger than the Coriolis force. Figure 4 also shows that the velocities are smaller during the day than the night. This is also caused by the friction force, which is stronger at daytime because of the higher electron densities there.

There is observational evidence that the winds in the thermosphere really behave like the calculations predict. First, releases of Ba clouds in twilight showed velocities in reasonable agreement with the above calculations. Second, the behavior of the *F*

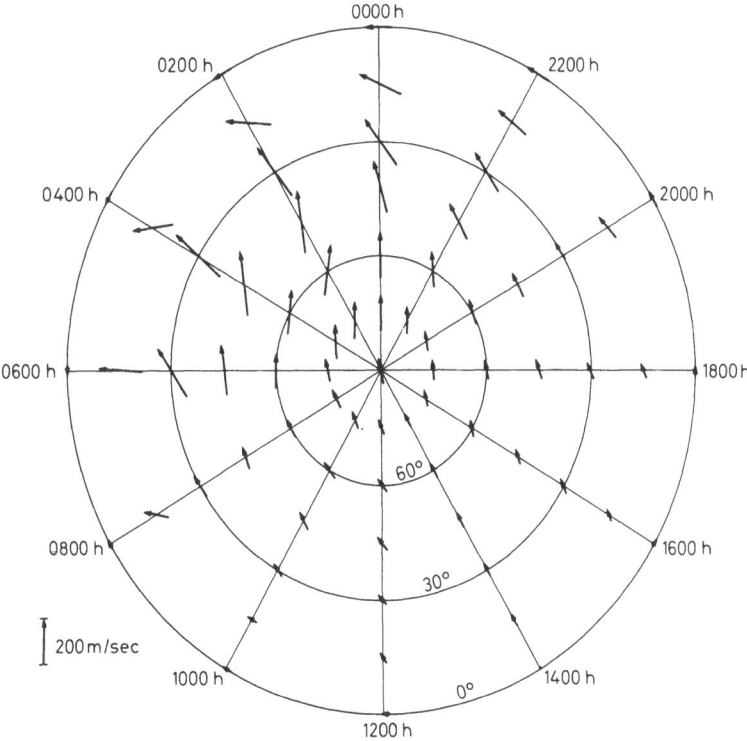

Fig. 4.   Calculated pattern of global wind system at 300 km altitude for equinox and medium sunspot activity conditions.

layer of the ionosphere is consistent with what one would expect as the effect of such winds. It was pointed out above that a wind cannot move ionization across the magnetic lines of force, but the ionization takes over the field aligned component of the wind velocity. Thus a poleward blowing wind will move ionization downwards, while an equatorward blowing wind will move ionization upwards. Observations of the height of the *F* layer show that it is about 100 km higher during nighttime than during daytime. This is in agreement with what would be expected from a wind pattern like in Figure 4.

The physical nature of the global wind system discussed here is, of course, that of a thermally excited tide, because the pressure variation exhibited by atmospheric models is produced by the diurnally varying solar energy input. A full tidal theory has to start from this excitation source, and besides Equation (1) must also take into account the conservation of mass in the tidal flow. Such complicated calculations have been undertaken (see e.g., Volland and Mayr, 1974).

It should be mentioned that besides periodic movements there seems to be also an average velocity of the upper atmosphere towards the east. It was detected by King-Hele (1964) from the changes of the inclination of satellite orbits. Satellites moving in a frame of reference which is not corotating with the Earth but is almost fixed with respect to the Sun, experience a drag force from the rotating atmosphere, which changes the inclination of the orbit. From the rate of change the angular velocity Ω of the rotating atmosphere can be determined. It turns out that Ω is about 20% higher than the angular velocity of the Earth. The reason of this surprising result is not yet clear.

3. E Fields in the Upper Atmosphere

3.1. E FIELDS IN THE DYNAMO REGION

It was pointed out earlier that above 150 km altitude charged particles cannot be moved across magnetic field lines by the action of a neutral wind, but will only take over the field aligned component of the velocity. In the lowest part of the ionosphere, say below 70 km, the collision frequency is so high that both electrons and ions move with the speed of the neutral air. The region in between is a transition region where the ions still move with the wind, while the electrons are constrained to move along magnetic field lines. Since electrons and ions respond with different velocities, the tides obviously set up an electric current. This current will usually have a divergence and, therefore, will lead to electric charge accumulation according to $\partial\sigma/\partial t =$ $= -\operatorname{div} j$ (where σ is the electric charge density). From the space charges an electric field will originate. Finally, the tides produce an electric field in the dynamo region. This field will in turn set up electric currents and modify the original current system. The modification will continue until stationary conditions are reached, where the current flow is divergence free. The so-called Sq variations of the undisturbed geomagnetic field are believed to be caused by this current system.

It is very difficult to treat the dynamo current problem quantitatively. Not only the mathematical difficulties are serious, but also the limited knowledge of the physical parameters, i.e., the height variations of electron density and tidal velocity, causes problems. Nevertheless such calculations have been performed and reasonable patterns for tidal motions and electrostatic fields have been derived that can explain the observed geomagnetic variations. The upper part of Figure 5 shows such an electrostatic field pattern calculated by Matsushita (1969). The E field is weaker than 1 mV m^{-1} at low latitudes and increases to about 4 mV m^{-1} at high latitudes. The $\mathbf{E}$ vector is rotating in a clockwise sense at latitudes above 50° and in an anticlockwise sense at latitudes below 50°. The lower part of Figure 4 shows the field $\mathbf{E} + (\mathbf{u} \times \mathbf{B})$ which is the electric field measured in a frame of reference moving with the velocity of the neutral air.

Measurements of this E field are still not sufficient. Rocket or satellite borne probes suffer from the fact that they move with high speed $\mathbf{w}$ relative to the magnetic lines of force, so that the induced field $(\mathbf{w} \times \mathbf{B})$ is usually higher than a few millivolt per

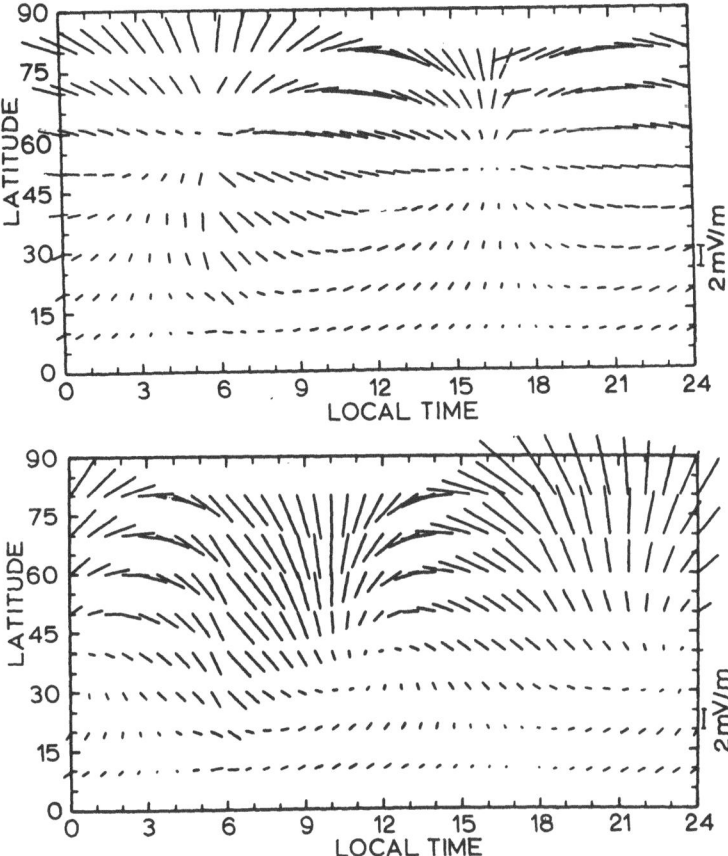

Fig. 5. Electric field pattern in the dynamo region (after Matsushita, 1969). Upper part: Electrostatic polarization field **E**. Lower part: Electric field **E**+(**u** × **B**) in a frame of reference moving with the tidal velocity **u**.

meter. Such probes are useful for measuring stronger E fields associated with magnetospheric particle fluxes. For the weaker Sq fields Ba ion clouds have been successfully used (Haerendel, 1970). The magnitude of the E field agrees reasonably well with the model prediction, however its direction is sometimes quite different.

In recent years the incoherent scatter technique was used to determine E fields in the ionosphere. At low latitudes the agreement with the model is quite good (Harper, 1971), while at medium latitudes (e.g., Taylor, 1974) the electric field shows a rather irregular behavior within time scales of a few hours. It seems that only an average over a larger amount of data can be compared with models.

3.2. ELECTRIC FIELDS FROM THE MAGNETOSPHERE

The electric fields discussed so far are produced by tides in the dynamo region, but they extend much further up. This is because the electrical conductivity of the plasma along the lines of force is very high as a result of the decreasing collision frequency. On the other hand the conductivity across the lines of force is very small, because charged particles are impeded by the magnetic field to move in that direction. Thus, the lines of force can be regarded as equipotentials, as any potential difference along these lines will be equalized quickly. As a consequence the electric field in the dynamo region will extend to the F layer and even high up into the magnetosphere.

But the reverse situation may happen as well. Electric fields may be set up in the magnetosphere by polarization of ambient plasma or of energetic particles with different sign of charge. Such fields will for the same reasons extend down to the dynamo region and set up currents there.

Measurements using probes, ion clouds and incoherent scatter have shown that in polar regions very often E fields of 30 mV m^{-1} or even more are present. So these fields are much stronger than those discussed in relation to tides. It is not clear so far, to what extent electric fields from the magnetosphere are also important at medium latitudes.

4. Effects of Winds and E Fields on the Ionospheric F Region

It was mentioned before that poleward winds will move the F layer ionization downwards, while equatorward winds will move it up. Figure 6 shows the diurnal variation of the F layer peak height h_m for two stations. The observed variations are depicted, on the right hand side, and on the left hand side the calculated variations are given including the effect of the wind (these winds are shown in Figure 4). It is clearly seen

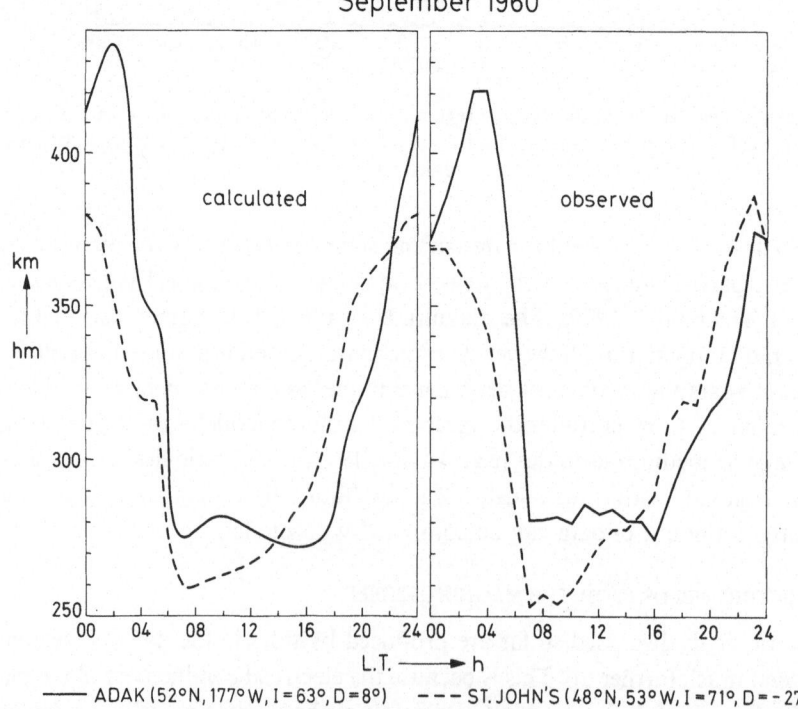

Fig. 6.　Calculated and observed diurnal variations of the height h_m of the F layer peak for two stations.

that calculations and observations are in good agreement. This graph does not demonstrate that a calculation without winds would lead to serious discrepancies. This point will be discussed in another paper on the F region. (Stubbe, this volume, p. 269.)

Electric fields in the F region also produce motions of the plasma with a velocity $\mathbf{v} = (\mathbf{E} \times \mathbf{B})/B^2$. A field of $E = 10$ mV m^{-1} will lead to a velocity $V = 250$ m s^{-1}. In the polar region such strong motions will happen. At medium latitudes E fields are weaker, as pointed out above. However, incoherent scatter measurements have shown fields up to $E = 6$ mV m^{-1} at Malvern, England.

In the meridional direction the term $\mathbf{E} \times \mathbf{B}$ can have a vertical component, so that also electric fields can move the F layer ionization up or down. From the work of Petelski (1972) it seems that irregular height variations of the F layer with time scales of a few hours should be attributed to E fields, while the more regular diurnal variations are caused by winds.

References

Bernard, R.: 1974, *Radio Sci.* **9**, 295.
CIRA: 1965, North-Holland Publ. Co., Amsterdam.
Haerendel, G.: 1970, MPI/PAE-Extraterr. 44/70.
Harper, R. M.: 1971, Thesis, Rice University, Houston, Texas.
Jacchia, L. G.: 1971, Smithsonian Astrophysical Observatory, Special Report 313.
King, J. W. and Kohl, H.: 1965, *Nature* **206**, 699.
King-Hele, D. G.: 1964, *Planetary Space Sci.* **12**, 835.
Matsushita, S.: 1969, *Radio Sci.* **4**, 771.
Petelski, E.: 1972, *J. Atmospheric Terrestr. Phys.* **34**, 1163.
Spizzichino, A.: 1968, in K. Rawer (ed.), *Winds and Turbulence in Stratosphere, Mesosphere and Ionosphere*, North-Holland Publ. Co., Amsterdam, p. 201.
Stoffregen, W.: 1972, in K. Folkestad (ed.), *Magnetosphere-Ionosphere Interactions*, Universitetsforlaget, Oslo-Bergen-Tromsö, p. 83.
Taylor, G. N.: 1974, *J. Atmospheric Terrestr. Phys.* **36**, 267.
Volland, H. and Mayr, H. G.: 1974, *Radio Sci.* **9**, 263.

AURORAL HEATING

ALV EGELAND

The Norwegian Institute of Cosmic Physics, Blindern, Oslo 3, Norway

1. Introduction

Today we can explain and predict fairly accurately several of the regular, global-scale variations in the upper atmosphere, while existing models fail to reproduce rapid structural changes which occur more or less continuously in the high latitude ionosphere. Particularly in the auroral E region, there is a wide gap between our present theoretical understanding of the physics and the experimental data.

The thermospheric temperature is too large to be generated entirely by EUV heating, winds, and hydrodynamic waves (cf. Leovy (1975) and Kohl (1975)). Auroral heating contributes a significant fraction to the global heat budget. Precipitating particles and Joule heating (cf. Section 5) are now recognized as particularly important heat sources in the auroral region (Wickwar, 1975), but they have not yet been included in any realistic ionosphere model. Two other major sources are believed to be heat conduction from the magnetosphere and acoustic-gravity waves (cf. Leovy, 1975). However, these two sources are not sufficiently known to allow any detailed discussion.

After a brief presentation of characteristic dimensions and frequencies in the ionospheric plasma (Section 2), we will discuss the basic equations needed to calculate the heat from auroral sources (Section 3). In Section 4, some electric field characteristics will be discussed and these will be related to auroral heat sources in Section 5.

2. Characteristic Frequencies and Dimensions in the Ionospheric Plasma

The Earth's magnetic field (**B**) causes the charged particles to gyrate around the force-lines at a frequency given by

$$\omega_{ck} = |q_k| \, B/m_k, \tag{1a}$$

where q_k and m_k is the charge and mass of the particle species k. The radius of the gyrating k-particle, the cyclotron radius, is

$$r_{ck} = v_\perp/\omega_{ck}, \tag{1b}$$

where $v_\perp$ is the particle velocity transverse to the geomagnetic field. Introducing the collision frequency (cf. e.g. Dalgarno, 1961) between the neutral and the charged particles, v_{kn}, the ratio ω_{ck}/v_{kn} indicates whether the motion is dominated by **B** or if collision processes play the most important role.

The magnitude of the plasma frequency (ω_{pk}) characterizes the rate at which electrostatic restoring forces eliminate deviation from neutrality. It is given by

$$\omega_{pk}^2 = N_k q_k^2 / \varepsilon_0 m_k, \tag{2}$$

where N_k is the charge particle density and ε_0 is the dielectric constant. The height variation of v_{kn}, ω_{ck}, and ω_{pk} for electrons and ions (an average mass-30-ion is assumed) corresponding to $L = 6$ is shown in Figure 1. This figure shows that the mobility of electrons is controlled by the geomagnetic field for heights above 80 km, while collisions dominate the ion motion below approximately 130 km. Thus, the plasma is strongly coupled to the neutral gas below 80 km, while above say 130 km the drift of the ionization is independent of the neutral air motion. The neutral winds will still cause redistributions of the F region ionization (cf. Kohl, 1975). In the E region, where $\omega_{ce} \gg v_{en}$ and $\omega_{ci} \approx v_{in}$, the differential streaming between ions and electrons causes plasma instabilities (cf. e.g. Egeland and Holtet, 1973).

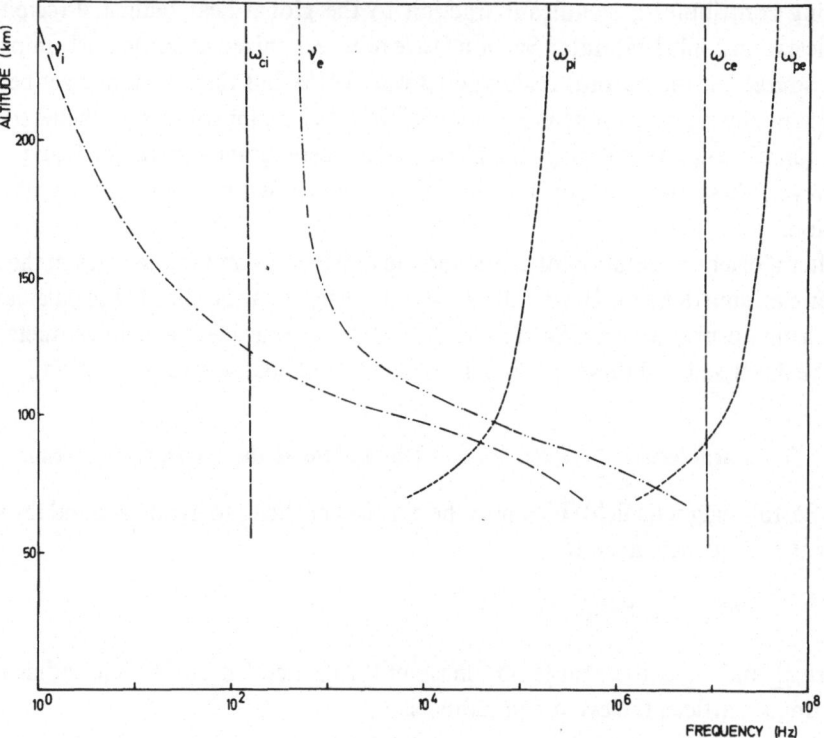

Fig. 1.   Variations with altitude of electron and ion collision frequency (v_e and v_i), electron and ion gyrofrequency (ω_{pe} and ω_{pi}), and electron and ion plasma frequency
(ω_{pe} and ω_{pi}) calculated for a mean model ionosphere.

A k particle with velocity v_k moves on the average a distance v_k/v_{kn} between each collision. This distance, called the mean free path, can be obtained from the expression.

$$\lambda_k = [\pi \kappa T_K / 2 m_k v_{kn}^2]^{1/2}, \tag{3}$$

where κ is the Boltzmann's constant and T_K is the temperature (cf. e.g., Morse, 1964).

The ratio r_{ck}/λ_k contains information about the influence of the magnetic field relative to collisions. If for instance $r_{ck}/\lambda_k \ll 1$, the trajectories of the k particles can be regarded as straight lines between consecutive collisions.

3. Particle Motions and Conductivities

An electric field $\mathbf{E}$ (cf. Section 4) will set the charged particles into motion, so the ionization will start to drift and electric currents will flow. The electric conductivity is determined by the charged particle's densities and their mobilities. Charge neutrality is assumed (i.e., $N_e = N_i$) and we neglect effects from negative ions. Mechanical forces acting in this medium are also neglected. Between collisions, the k particle is therefore only affected by $\mathbf{E}$ and $\mathbf{B}$ fields, and its equation of motion will be

$$m_k \frac{d\mathbf{v}_k}{dt} = q_k (\mathbf{E} + \mathbf{v}_k \times \mathbf{B}), \tag{4}$$

where $\mathbf{v}_k$ is the velocity relative to the neutral gas. The $\mathbf{B}$ field is directed along the z axis, while the $\mathbf{E}$ field is in the xz plane, and both are assumed homogenous. The velocity difference between the ionized and neutral gas is given by (cf. e.g. Boström, 1973)

$$\mathbf{v}_k - \mathbf{v}_n = \frac{\varepsilon_k v_{kn} \omega_{ck} (\mathbf{E}_\perp + \mathbf{v}_n \times \mathbf{B})}{(v_{kn}^2 + \omega_{ck}^2)\, B}$$
$$+ \frac{\omega_{ck}^2 (\mathbf{E}_\perp + \mathbf{v}_n \times \mathbf{B}) \times \mathbf{B}}{(v_{kn}^2 + \omega_{ck}^2)\, B^2} + \frac{q_k}{m_k v_k} \mathbf{E}_{\|}\,, \tag{5}$$

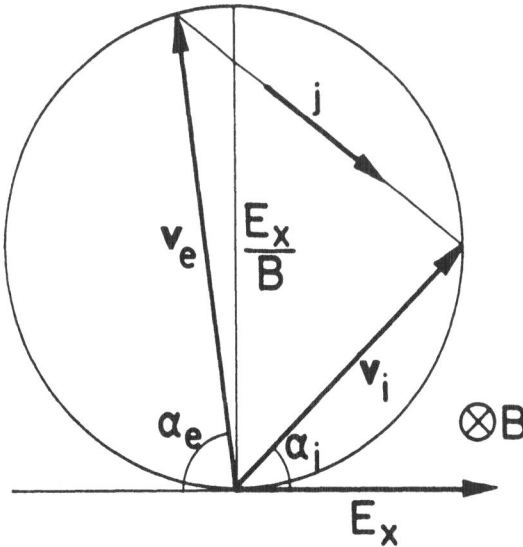

Fig. 2.   Graphical construction used to depict the directions and magnitudes of the drift velocities $\mathbf{v}_i$ and $\mathbf{v}_e$ for ions and electrons due to an electric field $\mathbf{E}_\perp$.

where $\varepsilon_k = q_k/|q_k|$, $\mathbf{v_n}$ = the velocity of the neutral gas and the indices $\parallel$ and $\perp$ refer to components parallel and perpendicular to the magnetic field.

The directions and magnitudes of the velocity component transverse to $\mathbf{B}$ are given by

$$\alpha_k = \arctan(\omega_{ck}/v_{kn}) \quad \text{and} \quad v_k = \sin \alpha_k (E_\perp/B). \tag{6}$$

Equation (6) can be represented graphically as shown in Figure 2. Here the $\mathbf{B}$ field is directed into the paper plane and the x axis is in the direction of $\mathbf{E}_\perp$. The diameter of the circle circumscribing the vectors $\mathbf{v_e}$ and $\mathbf{v_i}$ is $D = v_k/\sin \alpha_k = E_\perp/B$. Thus, by making a circle with diameter $E_\perp/B$ and drawing chords in the α_e and α_i directions, we get the magnitude and direction of the motion (cf. Figure 2). Above 150 km where $v_{kn} \ll \omega_{ck}$ we have $\alpha_e \approx \alpha_i \approx 90°$. Thus, $\mathbf{v_e} \approx \mathbf{v_i} = \mathbf{E} \times \mathbf{B}/B^2$ (cf. Equations (5) and (6)). If, on the other hand, $v_{kn} > \omega_{ck}$, as is the case below 80 km, the ions will move nearly parallel and the electrons antiparallel to the $\mathbf{E}$ field and the velocity becomes small as the collision frequency is very large (cf. Equation (5)).

The current density in this medium (i.e., $N_e = N_i$ and only singly charged ions) is given by

$$\mathbf{j} = \sigma_P(\mathbf{E}_\perp + \mathbf{v_n} \times \mathbf{B}) + \sigma_H \mathbf{B} \times (\mathbf{E}_\perp + \mathbf{v_n} \times \mathbf{B})/B + \sigma_\parallel \mathbf{E}_\parallel. \tag{8}$$

The conductivities σ_P, σ_H, and $\sigma_\parallel$ are called respectively Pedersen, Hall, and parallel conductivity. We find

$$\sigma_P = \left[\frac{v_{en}\omega_{ce}}{v_{en}^2 + \omega_{ce}^2} + \frac{v_{in}\omega_{ci}}{v_{in}^2 + \omega_{ci}^2} \right] \frac{eN_e}{B}, \tag{9a}$$

$$\sigma_H = \left[\frac{\omega_{ce}^2}{v_{en}^2 + \omega_{ce}^2} - \frac{\omega_{ci}^2}{v_{in}^2 + \omega_{ci}^2} \right] \frac{eN_e}{B}, \tag{9b}$$

$$\sigma_\parallel = \left[\frac{1}{m_e v_{en}} + \frac{1}{m_i v_{in}} \right] e^2 N_e. \tag{9c}$$

Notice that σ_P and σ_H are different from zero just because collisions occur. Furthermore, currents in the direction of the $\mathbf{E}$ field (Pedersen-current) add, whereas those perpendicular to $\mathbf{E}$ (corresponding to σ_H) subtract.

The conductivity per electron and per ion (i.e. $\sigma_P' = \sigma_P/N_k$ and $\sigma_H' = \sigma_H/N_k$ calculated from Equations (9a) and (9b)) has been computed by Sears et al. (1974) and the results are shown in Figure 3. (Concerning the model these calculations are based on, cf. Sears et al. (1974) and later in this section.) The actual Pedersen and Hall conductivity can then be estimated from Figure 3 when we knew the height profiles of electrons and ions. This figure clearly shows that the Pedersen conductivity for electrons has a negligible contribution above 100 km. Thus, σ_P in the E region is almost entirely due to the positive ions. On the other hand, the ionospheric Hall conductivity is mainly caused by the electron component of the plasma.

In the ionosphere, $\sigma_\parallel$ is much larger than the perpendicular components, and consequently $\mathbf{E}_\perp \gg \mathbf{E}_\parallel$. Often the geomagnetic field lines may be considered perfectly conductive and then $\mathbf{E}_\perp$ becomes almost height-independent. If the ionospheric

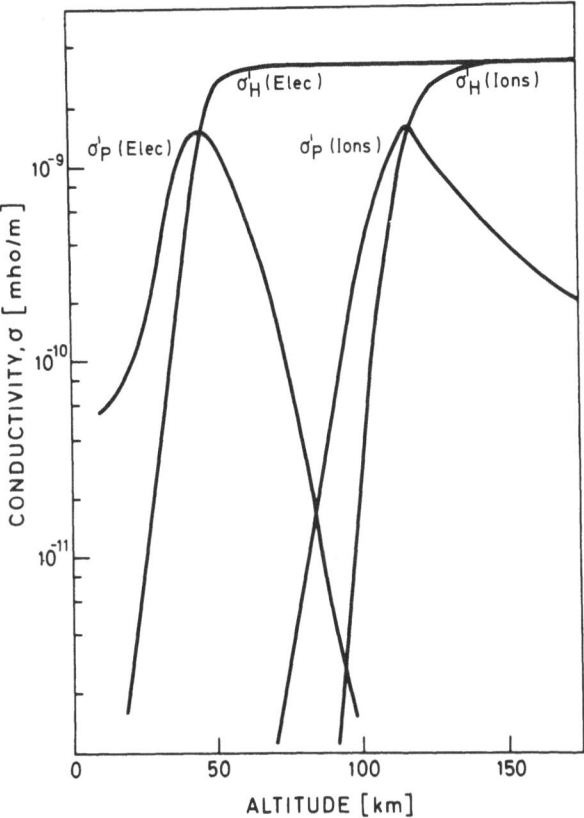

Fig. 3. Altitude profiles of Pedersen and Hall conductivity for a plasma density of 10^6 m^{-3} and an assumed average ion mass of 30 (Sears *et al.*, 1974).

layers at different heights are effectively coupled together, one may use the concept height-integrated conductivities

$$\Sigma_P = \int \sigma_P \, dh; \qquad \Sigma_H = \int \sigma_H \, dh. \tag{10}$$

The variations of σ_P and σ_H vs. altitude together with Σ_P and Σ_H, for a quiet daytime ionosphere, are shown in Figure 4. These calculations are based on the electron density profile plotted in the same figure which is measured in a rocket. Notice that the Hall-currents flow low in the ionosphere. (These currents, which are largely conducted by the ionospheric electrons, cause most of the magnetic disturbances observed on the ground.) Generally σ_P and σ_H as well as Σ_P and Σ_H are roughly of the same magnitude. In these calculations (Figure 4) it is assumed that the mean ion mass, m_i is 5.12×10^{-26} kg, and that the ion density variation with altitude is identical with the N_e profile shown in Figure 4. The ion neutral collision frequency (Banks and Kockarts, 1973) is given by $v_{in} = 7.5 \times 10^{-16} N$, where N is the neutral density m^{-3}. The atmosphere model chosen corresponds to a mesopause temperature of 190 K and a thermospause temperature of 1000 K (CIRA, 1972). By changing the mesopause temperature

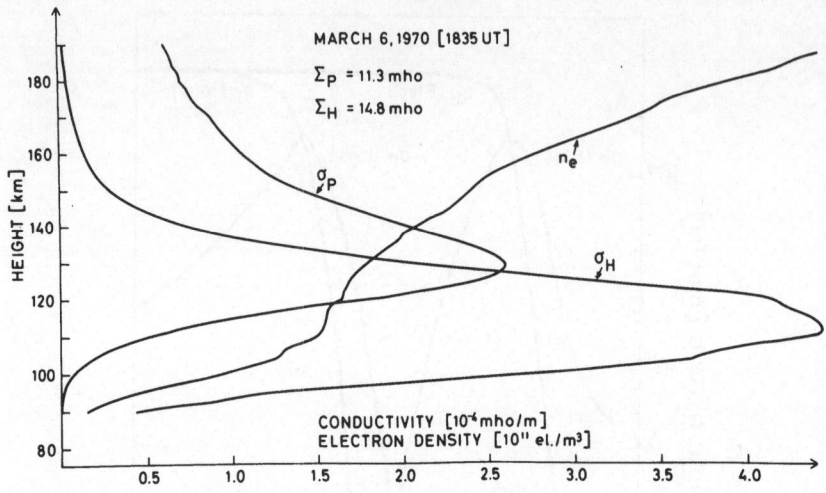

Fig. 4.   Variation of the ionosphere conductivities and electron density with
height for a quiet daytime ionosphere.

between 160 and 190 K, and the thermopause temperature between 750 and 2000 K, the Σ_H varies by 10 to 20%, while the variation in Σ_P is less than 5%. A change in ν_{in} by a factor of 2, however, will result in a 30% variation in Σ_H and less than 10% in Σ_P.

The total, height-integrated conductivity perpendicular to the **B** field in Figure 4 is close to 20 mho, which is a typical quiet daytime value. The corresponding quiet nighttime value is one order of magnitude less. During disturbed conditions, however, the day and particularly the nighttime values will often be one to two orders of magnitude larger.

4. Large-Scale Ionospheric Electric Fields

Measurements of ionospheric electric fields have only been carried out during the last decade. The main reason for this was that $\sigma_{\parallel}$ was assumed to be infinite. Thus, the E field could be obtained straightforward from ground measurements of **B**. This attitude has now changed, and it has been proved that **E** field is an important and necessary parameter to understand the physics of the ionosphere. Our knowledge about **E** fields has increased rapidly during the last years, but the resulting picture is far from unambigious.

Only a few comments concerning **E** fields will be stressed (for a review e.g. Wolf, 1973.) There is a general agreement concerning the magnitude and direction of the field. Typical values are from 20 to 50 mV m^{-1} poleward of approximately 55° Λ. Large and rapid variations often occur as shown in Figure 5, where the total **E** field reached a magnitude of almost 100 mV m^{-1}. There seems to be no systematic variation with altitude even though contradicting results have been published (cf. e.g. Maynard, 1972).

The question of what happens to **E** fields in an auroral form (which is important in relation to Joule heating), has been subject to discrepancies. The following main

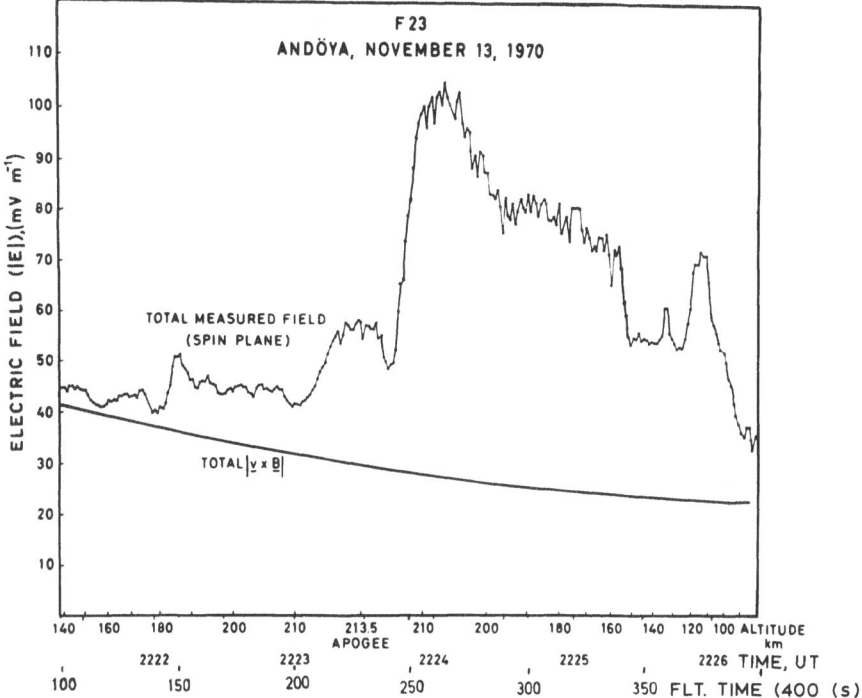

Fig. 5. Total d.c. electric field measured in the spin plane for the rocket F23. The calculated **v** × **B** contribution is also shown (cf. Måseide *et al.*, 1973).

conclusions could be drawn from some recent, *in situ* measurements (cf. Måseide *et al.*, 1973; Maynard *et al.*, 1973).

(a) The d.c. **E** field is markedly reduced within auroral arcs which are mainly produced by 10 keV electrons.

(b) In arcs where 1 keV electrons dominated, the **E** field, if anything increases.

Due to very large $\sigma_\parallel$ **E** fields will propagate effectively along the magnetic lines of force. However, the mapping (both direction and magnitude) onto the ionosphere cannot, as frequently done, be assumed to be perfect. The existence of **E** fields parallel to **B** is, however, a controversial question (e.g. Heppner, 1972).

5. Auroral Heating

Energy deposition due to Joule heating and auroral particles will be discussed. These energy sources should be compared with the thermal energy of the E region. Assuming an average temperature of 300 to 400 K or a kinetic energy corresponding to a drift velocity of 50 to 200 m s^{-1}, both correspond roughly to an energy source of approximately 10^{-2} to 10^{-3} J.

5.1. Joule heating

We will restrict ourselves to uniform **E** fields perpendicular to the geomagnetic field, and neglect gradients in density and conductivity parallel to the $\mathbf{E}$ field. Global electric fields will cause large power dissipation into the upper atmosphere (cf. Cole, 1971). Joule heating is per definition the energy deposited in the resistive ionospheric medium due to the current flow in this region. The dominating source of the ionospheric current is the convection $\mathbf{E}$ field of magnetospheric origin (cf. Section 4) propagating effectively along geomagnetic field line equipotentials (cf. Section 4). The heat input per unit volume and time can be calculated from the following equation:

$$Q' = \sigma_P(\mathbf{E}_\perp + \mathbf{v}_n \times \mathbf{B})^2 = \frac{j_\perp^2}{\sigma_C}, \tag{11}$$

where σ_C (the Cowling conductivity) is $\sigma_P + \sigma_H^2/\sigma_P$. If we integrate this equation over the actual height region (i.e., 100 to 200 km), we get

$$Q = \Sigma_P(\mathbf{E}_\perp + \mathbf{v}_n \times \mathbf{B})^2 = \Sigma_P(\mathbf{E}_\perp + \mathbf{E}_W)^2, \tag{12}$$

where $\mathbf{E}_W$ is the field due to the neutral wind. (Since $\mathbf{B}$ is assumed homogenous and $\mathbf{v}_n$ is uniform, $\mathbf{v}_n \times \mathbf{B}$ is a potential field.) Notice that Q is proportional to Σ_P while Q' is inversely proportional to σ_C. Maximum heating is therefore found in the region where the ionosphere is most conductive. By comparing Figure 3 together with Equation (12) it can be concluded that only the ions will be of significant importance for the ionospheric Joule heating. However, it should be stressed that this conclusion is based on a simple model. If polarization $\mathbf{E}$ fields and currents due to drifting irregularities are included, Σ_H will also be important for the Joule heating (see later).

Assuming $v_n = 30$ m s^{-1} (cf. Kohl, 1975) and $B = 0.6 \times 10$ T (auroral zone), we find that $E_W \approx 1.5$ mV m^{-1} when the angle between $\mathbf{v}_n$ and $\mathbf{B}$ is 70°. This is the averaged global $\mathbf{E}$ field for the S_q current system. The Joule heating due to $\mathbf{E}_W$ (Equation (12)) will not contribute with more than 10^{-5} W m^{-2}, more or less equally distributed over the globe, above say 100 km altitude. Comparing this value with the solar EUV-input (Leovy, 1975), it can be concluded that S_q currents will not have any significant influence on the heat budget.

During disturbed conditions, Σ_P-values up to 20 mho are not unusual, and $\mathbf{E}$ fields of 40 mV m^{-1} across a latitude belt of several degrees are often observed. These parameters will cause a Joule heating of 3×10^{-2} W m^{-2}, which is larger than that due to solar UV absorption during daytime and order of magnitude larger than during nighttime. Within more limited regions during large magnetic storms the heat due to Joule dissipation can be one or even two orders of magnitude larger still. In the auroral zone, Joule heating during night will often exceed that in the day by a factor of 2.

Even for fairly quiet conditions Joule heating plays an important part in the dynamic and thermal structure above 100 km altitude. This is due to the widespread (poleward of say 55° Λ) magnetospheric, convection $\mathbf{E}$ field of magnitude a few 10 mV m^{-1}. Furthermore, parts of the auroral electrojet flow at middle latitudes as return current (cf., Akasofu and Chapman, 1972). The averaged Joule heating amounts to a significant fraction (20% or more) of the global EUV heat input between 100 and 200 km during at least 50% of the time, and it should therefore be included in any realistic

atmosphere model. The significance of this heating increases as the magnetic disturbances increase, because the fields increase and the area of the globe directly heated
becomes greater at these times.

Within an auroral form, as well as in the auroral electrojet, significant polarization
electric fields occur. In addition, drifts of ionization irregularities (cf. e.g. Bostrøm,
1973) cause enhanced currents. The result of this is an effective conductivity within the
auroral form of $\Sigma_P + \Sigma_H^2/\Sigma_P = \Sigma_P'$ which is markedly larger than outside. Because
$E = J_P/\Sigma_P'$ (Cole, 1971) and Σ_P' is maximum within the arc, while J_P (the height integrated Pedersen current) stays constant, the E field inside the auroral form will be
reduced (cf. Section 4). The volume rate of Joule heating will therefore be smaller
inside auroral forms than outside. Heating due to parallel E fields may be important,
but as we have no reliable information of such fields and the area they cover, it is not
possible to give any realistic estimates.

Thus, a significant amount of energy (mainly created by the Pedersen current) is
deposited in the ionosphere by Joule heating, but the distribution mechanisms of this
energy are far from known. According to Cole (1971) the global effect of this heat input
will likely last several hours after the onset of the disturbances, but the storing and the
time constant for dissipation is not fully understood. Both thermodynamical and
kinetic energy partitions have been considered by Sears et al. (1974). They give the
following equation for the (center of mass) kinetic energy W per ion, in a weakly
ionized gas. (Only positive ions are considered as they almost entirely are responsible
for the Joule heating.)

$$W = 0.5\, m_i v_d^2 + 1.5\, \kappa T. \tag{13}$$

Here v_d is ion drift velocity, m_i is average ion mass, and $1.5\, \kappa T$ is the thermal energy
of the ion. Typical ion drift velocities for an E field of $30\,\mathrm{mV\,m^{-1}}$ are a few hundred
meter per second in the ionosphere between 100 and 200 km. However, the ion kinetic
energy gain is small in the E region, even though it exceeds the thermal energy.
Furthermore, such motion of heat waves should give rise to significant pressure
gradient which are not included in the theory. In general, the problem of energy
transport to lower latitudes is not solved satisfactorily.

5.2. HEATING DUE TO ENERGETIC PARTICLE PRECIPITATION

The average power in the solar wind is well documented to be $5 \times 10^{-4}\,\mathrm{W\,m^{-2}}$
$(0.5\ \mathrm{erg\ cm^{-2}\ s^{-1}})$, and it should therefore in general have a negligible effect in the
heat budget of the upper atmosphere. However, the solar wind impinges on a cross
section of approximately $10\,R_E$, while the Earth's atmosphere is roughly $0.5\,R_E$. Thus,
if the solar wind is focused on our atmosphere there is enough energy ($\sim 0.2\,\mathrm{W\,m^{-2}}$) to
power most of the ionospheric phenomena of interest.

The energy deposition associated with intense auroral PCA events can be very
important. During strong PCA's, the energy deposition over the polar caps can go up
to $1\ \mathrm{W\,m^{-2}}$ for more than a day (Hultqvist, 1973). However, such intense PCA events

occur fairly infrequently, and will therefore not be important in an average global heat budget.

The auroral particles precipitating into the atmosphere normally cover a broad energy spectrum (0.1 to 100 keV). However, both the particle spectrum and the fluxes vary considerably from one event to another. Some characteristic values for different zones and time sectors are given in Table I (cf. Mende, 1974) for relative quiet magnetic conditions. An estimate of the energy carried by these particles into the ionosphere is also shown in the last column of Table I. As approximately 30 to 40% of this initial

TABLE I

Auroral particle precipitation characteristics for relatively quiet magnetic conditions
($K_P \leqslant 3$) (Mende, 1974)

Region	Sector	Invariant latitude	Particle	Energy	Flux $(m^{-2} s^{-1})$	Energy $(W m^{-2})$
Auroral oval	Dusk	74°–76°	Electrons	~10 keV	~5×10^{11}	8×10^{-4}
	Midnight	67°–71°	Electrons		~2×10^{12}	3×10^{-3}
	Dawn	70°–74°	Electrons		~2×10^{11}	3×10^{-4}
	Noon	75°–77°	Electrons		~2×10^{11}	3×10^{-4}
Nightside proton aurora	Dusk	68°–71°	Protons	~10 keV	~2×10^{10}	3×10^{-5}
	Midnight	64°–69°	Protons		~4×10^{10}	6×10^{-5}
	Dawn	70°–76°	Protons		~2×10^{10}	3×10^{-5}
Nightside soft zone	Dusk	76°–82°	Electrons	~0.5 keV	~10^{12}	8×10^{-5}
	Midnight	71°–79°	Electrons	~0.5 keV	~10^{12}	8×10^{-5}
	Dawn	74°–82°	Electrons	~0.5 keV	~10^{12}	8×10^{-5}
Dayside soft zone	9–15 h	76°–83°	Electrons	~0.15 keV	~3×10^{11}	7×10^{-6}
			Protons	~0.5 keV	~3×10^{11}	3×10^{-5}

particle energy will go into heat (cf. Rees, 1975), it can be concluded that even during relatively quiet conditions the auroral particles will be of some importance for the heat budget of the ionosphere.

It is also possible to obtain some information of the particles' heat contribution from visible aurora. In order to produce aurora visible to the naked eye (IBC I equivalent to 1 kR), a particle influx of about 10^{-3} W m^{-2} is needed (cf. Deehr and Egeland, 1971). The corresponding figures for aurora class IBC II, III, and IV are respectively 10^{-2}, 10^{-1}, and 1 W m^{-2}. Again roughly 40% of this energy will go into heat. It should, however, be noticed that mainly electrons between 1 and 15 keV will contribute significantly to the optical emissions, while the total particle energy influx is important for the heat budget.

During intense auroral events the particle flux will increase by a factor of 10 to 100. However, these disturbed events are fairly limited both in areas and times.

Recent satellite measurements show that auroral particles (0.1 to 100 keV) exist near the Earth, mainly at geomagnetic latitudes above 55°, but are also observed at lower latitudes. Even though a lot of information of these energetic particles has been accumulated during the last years, we still do not have a complete picture covering the whole energy spectrum of precipitated particles (cf. Hultqvist, 1973). In a global scale averaged over the Earth, it looks like the particle precipitation will contribute less than the Joule heating to the heating budget in the ionosphere.

Acknowledgment

The author would like to thank Dr B. M. McCormac for valuable suggestions and comments.

References

Akasofu, S.-I. and Chapman, S.: 1972, *Solar-Terrestrial Physics*, Clarendon Press, Oxford.
Banks, P. M. and Kockarts, G.: 1973, *Aeronomy*, Academic Press, New York.
Bostrøm, R.: 1973, in A. Egeland, Ø. Holter, and A. Omholt (eds.), *Cosmical Geophysics*, Universitets-forlaget, Oslo, p. 151.
Cole, K. D.: 1971, *Planetary Space Sci.* **19**, 59.
Dalgarno, A.: 1961, *Ann. Geophys.* **17**, 16.
Deehr, C. S. and Egeland, A.: 1971, in B. M. McCormac (ed.), *The Radiating Atmosphere*, Reidel, Dordrecht, Holland, p. 125.
Egeland, A. and Holtet, J.: 1973, *European Research At High Latitudes*, ESRO, SP-97, Paris, p. 61.
Heppner, J. P.: 1972, *Geophys. Publ.* **29**, 105.
Hultqvist, B.: in A. Egeland, Ø. Holter, and A. Omholt (eds.), *Cosmical Geophysics*, Universitetsforlaget, Oslo, p. 161.
Kohl, H.: 1975, this volume, p. 87.
Leovy, C. B.: 1975, this volume, p. 73.
Måseide, K., Holtet, J., Egeland, A., and Maynard, N. C.: 1973, *J. Atmospheric Terrestr. Phys.* **35**, 1833.
Maynard, N. C.: 1972, in K. Folkestad (ed.), *Magnetosphere-Ionosphere Interactions*, Universitetsforlaget, Oslo, p. 155.
Maynard, N. C., Bahnsen, A., Egeland, A., and Lundin, R.: 1973, *J. Geophys. Res.* **78**, 3976.
Mende, S.: 1974, private communication.
Morse, P. M.: 1964, *Thermal Physics*, Benjamin Inc., New York.
Rees, M. H.: 1975, this volume, p. 323.
Sears, R. D., Evans, J. E., and Varney, R. E.: 1974, Report DNA 3293F, Lockheed Palo Alto Res. Lab., Palo Alto, Calif. 94304.
Wickwar, V.: 1975, this volume, p. 111.
Wolf, R. A.: 1973, in B. M. McCormac (ed.), *Magnetospheric Physics*, Reidel, Dordrecht, Holland, p. 167.

CHATANIKA RADAR MEASUREMENTS

VINCENT B. WICKWAR

Stanford Research Institute, Menlo Park, Calif. 94025, U.S.A.

1. Introduction

The auroral ionosphere often differs from the mid- and low-latitude ionosphere due to the influx of energetic particles (primarily electrons) and the mapping of convection electric fields down the magnetic field lines from the magnetosphere. The auroral ionosphere can significantly affect the neutral atmosphere above about 100 km. This paper presents Chatanika Radar measurements related to the physics and chemistry of the atmosphere.

1.1. LOCATION

The Chatanika Radar (Leadabrand *et al.*, 1972) is an incoherent-scatter radar located about 30 km north of Fairbanks, Alaska operated by Stanford Research Institute. Measurements began in July 1971. Its location is unique among incoherent-scatter radars in that measurements can be made in the plasmasphere, auroral oval, or polar cap, depending on the geomagnetic activity and local time. Figure 1 shows the position of the radar at dipole latitude 65° ($L = 5.6$) with respect to the auroral oval (Feldstein, 1963) as a function of magnetic local time for three values of Q [approximately equal to K_p (see Eather and Mende, 1971)]. During the large magnetic storm of August 1972, the oval expanded even further, to the extent that the radar was in the polar cap, as determined by the energetic proton precipitation (Watt, 1974).

At the radar site and in the vicinity, other facilities and instruments help provide additional information about the geophysical phenomena being studied. On the radar

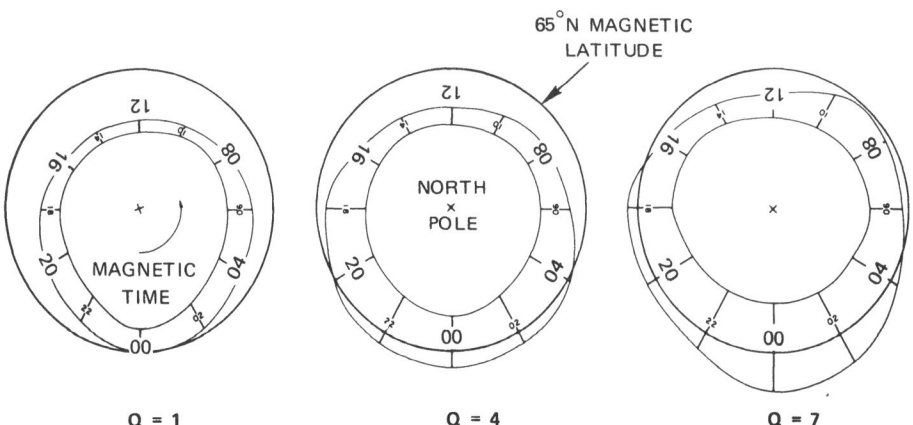

Fig. 1. Chatanika and the auroral oval. – The position of Chatanika (on the 65° N magnetic latitude circle) relative to the auroral oval is indicated as a function of local magnetic time and geomagnetic activity (Baron, 1973).

B. M. McCormac (ed.), Atmospheres of Earth and the Planets, 111–124. All Rights Reserved.
Copyright © 1975 by D. Reidel Publishing Company, Dordrecht-Holland.

dish there is a bore-sited, low-light-level TV. About 3 km away is the Poker Flat rocket range, operated by the Geophysical Institute of the University of Alaska. At a variety of nearby locations there are all-sky cameras, auroral photometers, magneto-meters, riometers, an all-sky TV, an ionosonde, and both NASA and ESRO satellite tracking and receiving stations.

1.2. Measured and Derived Parameters

The principles of incoherent scatter and the determination of profiles of electron density, electron and ion temperature, ion velocity, and ion composition are discussed by Petit (1975) and Evans (1969). Specific reference to the Chatanika radar is given by Baron et al. (1970), Rino (1972), Doupnik et al. (1972). Rino et al. (1974), and Wickwar (1974a). Additionally, the radar observations (sometimes combined with other obser-vations) have been used to determine other important geophysical parameters: integrated Hall and Pedersen conductivities, currents, and Joule heating (Brekke et al., 1974b); effective D and E region recombination coefficients (Baron, 1973; Watt et al.. 1974); and information about the total energy, mean energy, and pitch-angle distribution of auroral primaries (Baron, 1973, 1974).

2. Particle Precipitation

The incident particle fluxes produce not only discrete and diffuse optical emissions (the visual aurora), but also D, E, and F region ionization that can be measured by the radar. In this section, electron density profiles are examined to see what can be learned about the energetic particle input, effective ionospheric recombination rates, and in-cident energy input from energetic electrons.

2.1. Survey of Particle Precipitation

The highest-energy particle precipitation was observed during the PCA event that accompanied the large magnetic storm of August 1972. Two resultant representative electron density profiles are shown in Figure 2a. The unusual features are the well defined D layer on the earlier profile and the significant ledges below the E layer peak on the later profile. To reach 85 km, electrons require 80 keV and protons 2 MeV; to reach 65 km, protons require 10 MeV.

Electron density profiles more typical of auroral conditions are shown in Figure 2b. There are no layers or ledges below the E region maximum. In this example, the maximum density shifts from 125 km when weak to 100 km when strong, but may vary between about 90 and 130 km. Electrons require 50 keV and 2 keV, respectively, to penetrate to these altitudes. The rapid time variation in these profiles is characteristic of observations during periods of high magnetic activity when discrete auroral features are present. It reflects the temporal variability of the auroral source and the passage of discrete features through the radar beam. At other times, the profiles may be stable for tens of minutes. Baron (1973, 1974) has classified the auroras into discrete and diffuse categories on the basis of differences in the E region profile, such as temporal

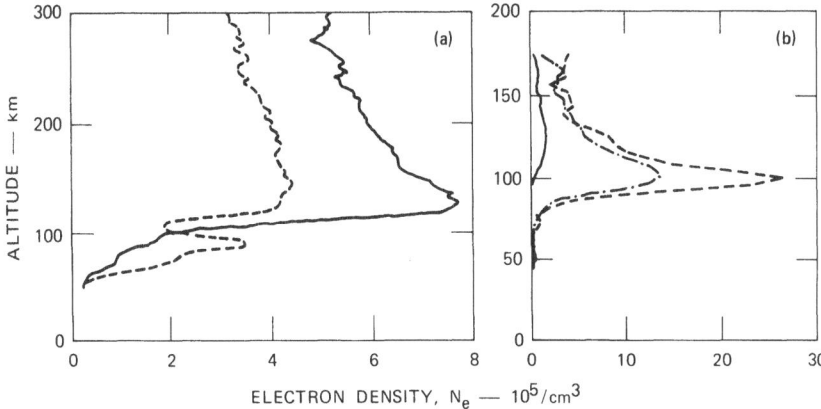

Fig. 2. Electron density profiles. – The measurements in part (a) are from Aug. 4, 1972, during a large PCA event – 2210–2220 UT for the dashed line, and 2240–2249 UT for the solid line. The steady electron and proton precipitation enabled 10 min averages to be used. The measurements in part (b) are from March 21, 1973. The solid curve, 0804:37–0807:12 UT, is during diffuse auroral conditions shortly before a large discrete electron precipitation event. The dot-dashed, 0808:51–0808:57 UT, and dashed lines, 0808:57–0809:03 UT, show the rapid time variation and large size of the disturbance. [Note that the scales in parts (a) and (b) differ. Alaska Standard Time (AST) = UT – 10 h.]

behavior, spatial distribution, peak electron concentration, height of layer maximum, and several deduced parameters about the incident flux. These categories presumably are closely related to the discrete and diffuse visual aurora.

There is evidence for even lower-energy electron precipitation. Banks *et al.* (1974a) show what may be interpreted as a wall of ionization extending from the E region to at least 400 km. If that interpretation is correct, the E region densities indicate particle precipitation and the large F region densities indicate a substantial component of the flux with energy less than 400 eV. In a combined radar and 6300 Å photometer experiment during a very quiet period, there is possible evidence for electrons of still lower-energy which give rise to 6300 Å emission but not ionization detectable with the radar (Wickwar, 1974b).

2.2. Effective Recombination Coefficient

The effective recombination coefficient can be derived from the radar data or a combination of radar data and production-rate information. When negative ions are present (below about 80 km), the continuity equation neglecting transport is

$$(1+\lambda)\frac{dN_e}{dt} = p(t) - (1+\lambda)(\alpha_d + \lambda\alpha_i) N_e^2 - N_e \frac{d\lambda}{dt} \qquad (1)$$

where N_e is the electron density concentration, λ is the ratio of negative ions to electrons, $p(t)$ is the electron production rate, α_d is the recombination coefficient for electrons with ions, and α_i is the recombination coefficient for negative ions with positive ions. The effective recombination coefficient is

$$\alpha_{eff} = (1+\lambda)(\alpha_d + \lambda\alpha_i). \qquad (2)$$

Without negative ions the continuity equation reduces to

$$\frac{dN_e}{dt} = p(t) - \alpha_d N_e^2 \tag{3}$$

and α_{eff} becomes simply α_d.

When the incident particle flux is measured by rocket or satellite the electron production rate can be calculated. Then, for steady state conditions, the effective recombination coefficient is

$$\alpha_{eff} = p(t)/N_e^2. \tag{4}$$

Figure 3 shows a number of altitude profiles of α_{eff} obtained in this fashion. Below 93 km the three curves were obtained by Watt and Reagan (1974) for the August 1972 PCA event. The electron densities were measured by the radar, and the particle fluxes by the Lockheed 1971-089A satellite. The production rate was calculated using programs developed by Walt et al. (1969) for the incident electrons and by Francis (1974) for the incident protons and alphas. The bump near 80 km is believed to be real. It is probably related to the transition form NO^+ and O_2^+ ions to water-cluster ions (Thomas, 1971, 1974). Above 95 km, there is a curve from Watt et al. (1974) derived in the same fashion, except that the data were obtained during an active discrete aurora. The dotted line above 95 km (Ulwick and Baron, 1974) was obtained by the same procedure during a discrete aurora, but $N_e(h)$ was obtained with an AFCRL rocket from Poker Flat [and agreed with $N_e(h)$ measured simultaneously by the radar], and the production rate was calculated with the Rees (1963) method.

The effective recombination rate can be calculated using only radar data by a

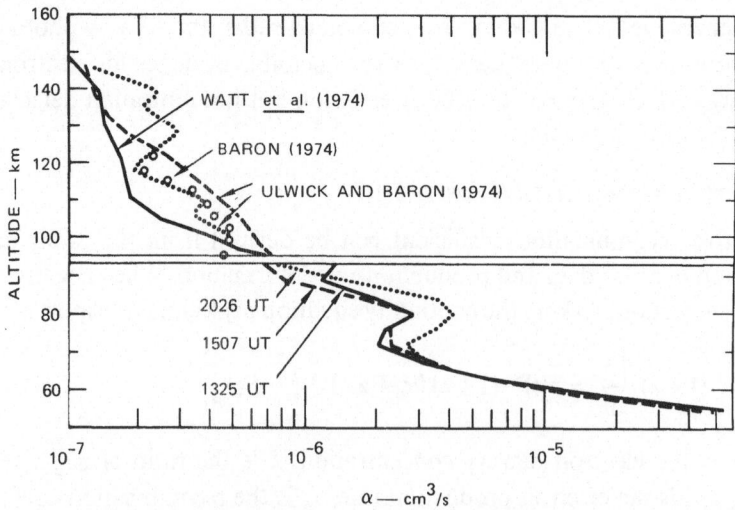

Fig. 3. Effective recombination coefficients. – The E region determinations were made during discrete auroras. The curve from Baron (1974) is for February 24, 1972, that from Watt et al. (1974) is for January 27, 1972, and the two from Ulwick and Baron (1974) are for March 16, 1972. The D region determinations were made during the PCA event of August 4, 1972 (Watt and Reagan, 1974).

method developed by Baron (1973). When production ceases – i.e., $p(t)$ is zero – the electron density decreases due to recombination. The solution for Equation (3) then becomes

$$N_e = N_0/(1 + \alpha_{eff} N_0 \Delta t), \tag{5}$$

where N_0 is the electron density at time t_0 and N_e at time $t_0 + \Delta t$. This equation may then be solved for α_{eff}. With this method, Baron (1973, 1974) obtained the curve of small open circles in Figure 3 for a discrete aurora. At high altitudes transport effects limit the applicability of this technique.

The remaining curve above 95 km is obtained using the laboratory recombination rates of Biondi (1969), the NO^+ and O_2^+ densities measured on the above-mentioned AFCRL rocket flight, and the CIRA (1965) model atmosphere (Ulwick and Baron, 1974).

It is noted that these E region profiles obtained for conditions of discrete aurora are in good agreement, and match remarkably well at 93 km with the PCA derived results for the D region.

2.3. ELECTRON ENERGY INPUT

There are two procedures for determining the incident energy flux due to auroral electrons. For both, the production rate is determined from the electron density profile by using Equation (4) and the effective recombination coefficient. Since 35 eV is required per ion pair created (see Rees, 1963) the energy flux can be determined directly from the ionization profiles. Alternatively, the maximum production rate and its altitude can be compared to calculated profiles in Rees (1963) or Banks *et al.* (1974b) to find the characteristic energy and flux of the incident electrons. A discussion of the derived energy flux is delayed until Section 4 where it is compared with the energy input due to Joule heating.

3. Convection Electric Fields

Another feature that distinguishes the region observed by the Chatanika Radar is the presence of convection electric fields mapped from the magnetosphere into the iono-sphere (Cauffman and Gurnett, 1971; Haerendel, 1972; Mozer, 1973). These fields produce large electron and ion velocities that in turn affect E and F region neutral winds, help form the F region trough, produce the auroral-zone electrojets, add energy to the auroral atmosphere through Joule heating, and modify the atmospheric chemistry.

3.1. WINDS

With a steerable monostatic radar, ion vector velocities can be determined by com-bining three line-of-sight velocities (Doupnik *et al.*, 1972) or by combining line-of-sight velocities obtained during a continuous azimuth scan (Hagfors and Behnke, 1974). Figure 4 shows ion velocities determined by the first method at three altitudes for two

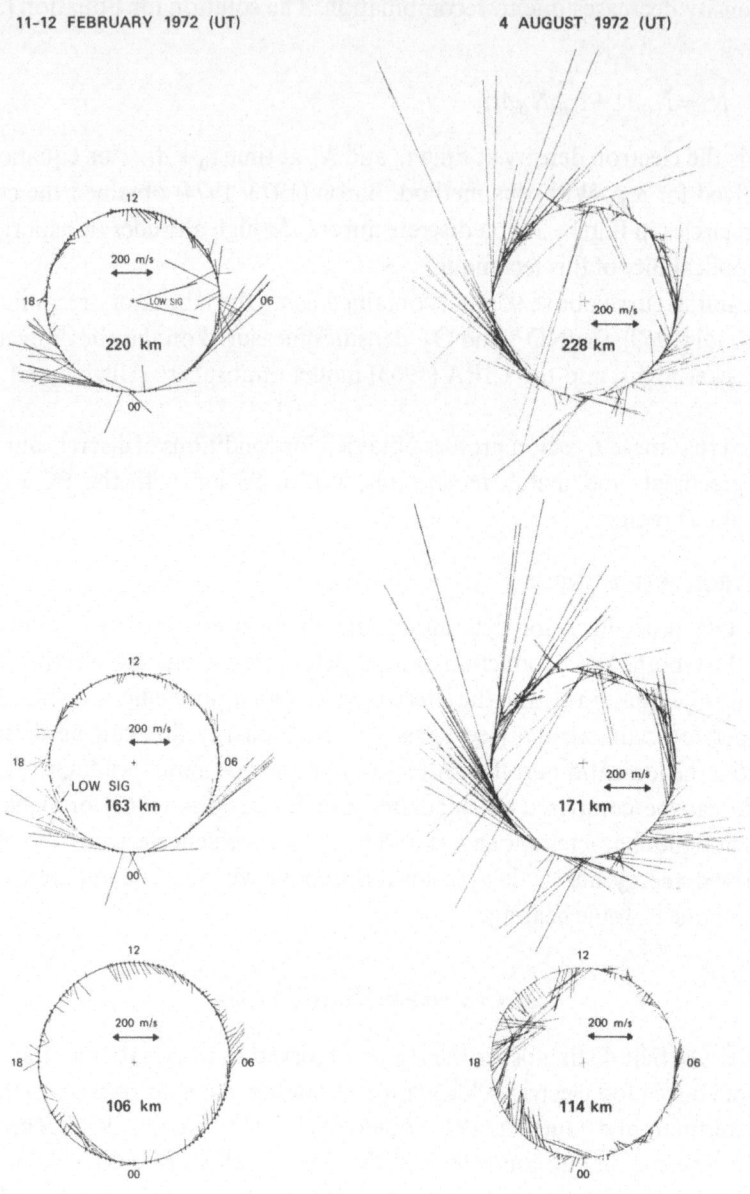

Fig. 4. Ion velocities. – The velocities are given at three altitudes as a function of AST for each of two days. The lowest altitude is affected by ion-neutral collisions and neutral winds. The upper two are free of these complications. A vector toward the center of a circle represents a velocity to the magnetic north. A vector tangent to a circle and pointing toward increasing time represents a wind to the magnetic east. (The data for February 11–12, 1972, are from Doupnik *et al.*, 1972.)

24 h periods: a quiet period ($K_p \leqslant 3$) and an active period ($4 \leqslant K_p \leqslant 9$). These velocities are related to electric fields, ion neutral collisions, and neutral winds. When pressure gradients and gravity are neglected, the equation of motion of the ions is

$$m_i \frac{\partial \mathbf{v}_\perp}{\partial t} = e(\mathbf{E} + \mathbf{v}_\perp \times \mathbf{B}) + m_i v_i (\mathbf{u} - \mathbf{v}_\perp), \tag{6}$$

where m_i is the ion mass, $\mathbf{v}_\perp$ is the ion velocity perpendicular to the magnetic field $\mathbf{B}$, $\mathbf{u}$ is the neutral-wind velocity, $\mathbf{E}$ is the perpendicular electric field in the frame of the radar (assumed constant along a magnetic field line), and v_i is the ion-neutral collision frequency. For steady-state conditions (which are established by the ion-neutral collision time – of the order of 10^{-2}s), the equation of motion becomes

$$(\mathbf{E} + \mathbf{v}_\perp \times \mathbf{B})/B + (\mathbf{u} - \mathbf{v}_\perp)\, v_i/\Omega_i = 0, \tag{7}$$

where Ω_i is the ion cyclotron frequency, eB/m_i. Above 150 km, $v_i/\Omega_i \ll 1$; hence,

$$\mathbf{E} = -\mathbf{v}_i \times \mathbf{B} \quad \text{or} \quad \mathbf{v}_i = \mathbf{E} \times \mathbf{B}/B^2. \tag{8}$$

In agreement with the above assumptions the velocity measurements centered at 165 and 225 km are similar, whereas those centered at 110 km are very different due to ion-neutral collisions and neutral wind. The very high velocities at 165 km and above found in the evening and early morning are due to the convection electric field (Doupnik *et al.*, 1972; Banks *et al.*, 1973). It is these large velocities that distinguish auroral latitudes from lower latitudes (e.g., see Evans, 1972).

These ion winds have a large effect on the neutrals because of ion drag. Although above 150 km ion-neutral collisions have only a small effect on the ion velocity compared with the electric field, the collisions have a significant effect on the neutrals. The effect has been seen by Nagy *et al.* (1974), who compare *F* region ion drifts measured with the radar and neutral drifts determined with a Fabry-Pérot interferometer measuring the Doppler shift of the 6300 Å emission line. In the evening sector there is a strong westward motion of the neutrals, which is unexpected from the wind systems predicted on the basis of solar heating alone (Kohl, 1975).

There can also be *E* region winds – e.g., they have been detected in 5577 Å Doppler observations (Meriwether, 1974). From Equation (7) it is apparent that, given an ion-neutral collision-frequency profile and the **E** field derived from measurements above 160 km, the neutral velocity can be found with the radar. Due to the finite length of the radar pulse, the quantity actually found is the neutral wind averaged over the E region (Brekke *et al.*, 1973). The behavior of the neutral wind is summarized in Brekke *et al.* (1973, 1974a). During the day, between 09 and 16 LT, the wind is directed nearly parallel to the noon-midnight line as would be expected from solar-induced pressure gradients. But during the night, between 16 and 09 h, when Chatanika is closer to the oval, the wind is directed southward in the evening and eastward in the morning sectors; between 20 and 02 h its direction is variable. The nighttime wind pattern most probably reflects the influence of ion drag at the evening and morning, and local auroral heating in between. The *E* region neutral wind for March 13 to 14, 1972, is shown later (Figure 5c). The value in m s^{-1} is 20 times the induced electric field in mV m^{-1}. Typical velocities are between 50 and 150 m s^{-1}.

During large magnetic storms, strong equatorward neutral winds have been observed in midlatitudes at F region altitudes (Hays and Roble, 1971) and in lower latitudes at E region altitudes (Smith, 1968). These winds are believed to result from large E region heating in the auroral zone due to particle bombardment and Joule heating. The F region winds would start as vertical winds, as implied by Hays et al. (1973). Confirmation of the theory is still being sought.

The large F-region ion velocities to the west in the evening sector appear to play a large and perhaps dominant role in the formation of the nighttime trough (Belon et al., 1972). The stronger the northward electric field (westward ion drift), the faster the F region decays. This mechanism and others are considered theoretically by Knudsen (1974).

3.2. CURRENTS

Since the ion velocity above 150 km is negligibly affected by ion-neutral collisions, it is essentially equal to the electron velocity and there is negligible current. But, still assuming that the magnetic field lines are equipotentials, at lower altitudes the ion velocity is progressively impeded by collisions, while the electron velocity is unaffected above about 90 km. Hence, in the E region the convection electric field drives horizontal currents.

Additionally, the E region neutral winds give rise to currents. In the rest frame of the ions the neutral wind effects are equivalent to an electric field $\mathbf{u} \times \mathbf{B}$. The current perpendicular to B is given by

$$\mathbf{j}_\perp = \sigma_P \mathbf{E}' - \sigma_H \mathbf{E}' \times \mathbf{B}/B \tag{9}$$

and the height-integrated current by

$$\mathbf{J}_\perp = \Sigma_P \mathbf{E}' - \Sigma_H \mathbf{E}' \times \mathbf{B}/B, \tag{10}$$

where

$$\mathbf{E}' = \mathbf{E} + \mathbf{u} \times \mathbf{B}. \tag{11}$$

In the above, σ_P and σ_H are the Pedersen and Hall conductivities (Egeland, 1975), and Σ_P and Σ_H are the height-integrated Pedersen and Hall conductivities. As in the case of neutral winds the finite length of the radar pulse has so far made it necessary to work with height-integrated currents (Brekke et al., 1974b).

Figure 5 (Brekke et al., 1974b) shows the quantities in Equations (10) and (11) measured on March 13 to 14, 1972. Using a model of the collision frequencies, and the electron densities measured by the radar, the integrated conductivities are shown in Figure 5a, the convection electric field is shown in 5b, and the neutral wind induced field is shown in 5c. The smooth variation of Σ_P between 1600 and 0400 UT is typical of the solar-induced daily variation. The Σ_H variation shows some auroral disturbance in addition to the solar variation. In the nighttime period, between 0400 and 1600 UT,

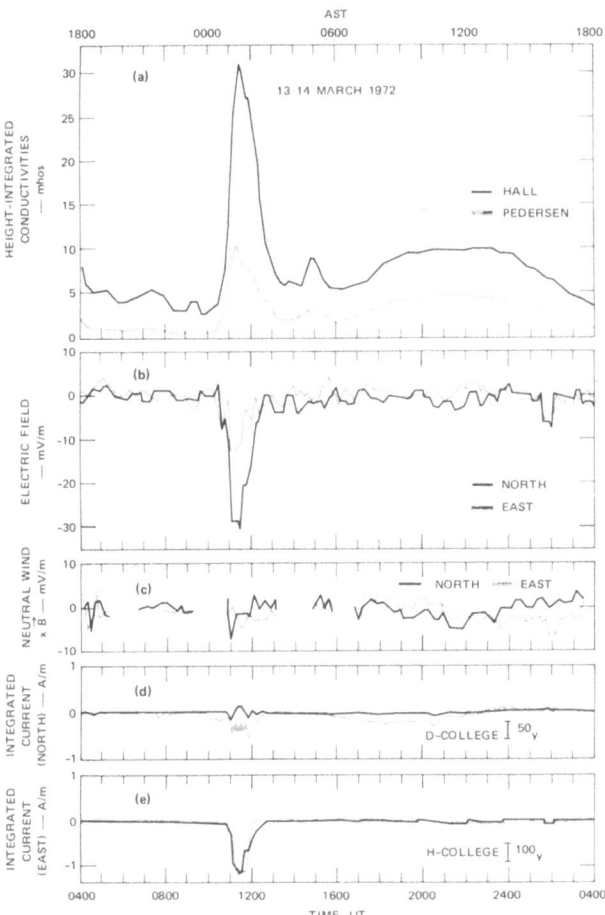

Fig. 5. Integrated conductivities, electric field, neutral wind, and magnetic-field disturbances for March 13–14, 1972 (see text). – The gray curves in parts (d) and (e) are the magnetometer traces, while the black curves are the currents. (Figure from Brekke *et al.*, 1974b.)

both Σ_P and Σ_H show considerable auroral effects, especially between 1030 and 1230 UT during a negative bay. Despite variations in the integrated conductivities, Brekke *et al.* (1974b) show that the ratio Σ_H/Σ_P varies very little, between about 1.5 and 4. The data gaps in the neutral-wind-induced electric field occur due to low signal-to-noise ratio. An important observation is that on occasion, such as between 2000 and 0500 UT in the example, the neutral wind is more effective in causing current flow than the imposed electric field.

Once the ionospheric currents are known, their effect on the magnetic field can be obtained. Variations in the D and H components are proportional to the currents, as follows:

$$\Delta D: \; -(J_N \sin I + J_B \cos I) = -(0.97\, J_N + 0.03\, J_B) \approx -J_N \tag{12}$$

$$\Delta H: J_E \tag{13}$$

where J_N and J_E are the integrated currents perpendicular to $\mathbf{B}$ in the geomagnetic north and east directions, and are defined by

$$J_N = \Sigma_P E'_N - \Sigma_H E'_E$$
$$J_E = \Sigma_P E'_E + \Sigma_H E'_N. \tag{14}$$

J_B is the field-aligned current, and I is the dip angle.

Figure 5d and 5e illustrate comparisons of the measured D and H components of the magnetic field with deflections expected from the ionospheric currents (Brekke *et al.*, 1974b). The D component compares well during portions of the day, but departs badly between 1100 and 1230 UT, and again between 1600 and 2200 UT. The first departure is consistent with a strong upward current north of Chatanika as might be expected, since all-sky photographs show discrete aurora north of the radar and the auroral electrons may carry a net current (Anderson and Vondrak, 1975). The second departure is a typical morning feature, with a less obvious interpretation. The H component shows good agreement, even during the negative bay, which appears to be typical.

Furthermore, from the measurements of currents and electric field the energy flux from Joule heating can be computed by

$$Q_j = \Sigma_P E'^2 = \mathbf{J}_\perp \cdot \mathbf{E}'. \tag{16}$$

The Joule heating is examined and compared with the energy flux from energetic electrons in Section 4.

3.3. CHEMISTRY

In addition to the above effects, theory predicts that the convection electric fields can affect the chemistry by affecting the reaction rates. The large ion velocities can become an appreciable fraction of the ion thermal velocity, thereby upsetting the average over a Maxwellian distribution that occurs in reaction-rate calculations (Schunk and Walker, 1973). The case of $O^+ - O$ charge exchange has been treated by Horwitz and Banks (1973) and applied to the determination of neutral temperatures (Watkins and Banks, 1974). Further complicating the situation, St.-Maurice and Schunk (1974) suggest that the ion velocity distribution may depart significantly from Maxwellian. Joule heating is expected to significantly increase the ion temperature (Rees and Walker, 1968; Schunk and Walker, 1971; Fedder and Banks, 1972), which in turn increases the recombination rate for $O^+ + N_2$. This increase, combined with other factors, leads to the prediction that there are conditions when NO^+ could be the dominant F region ion in the auroral zone (Banks *et al.*, 1974c).

4. Energy Input

The energy flux from energetic electrons and from Joule heating has been mentioned

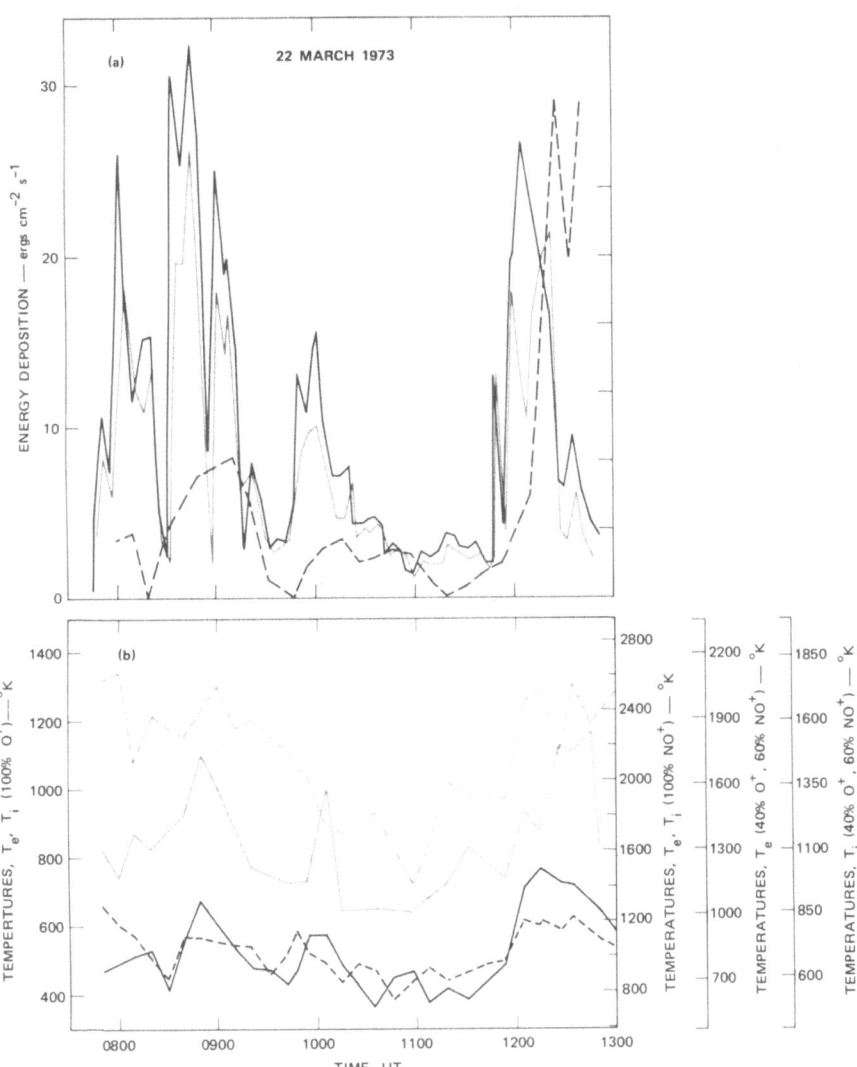

Fig. 6.   Energy inputs and temperatures. – In part (a) the energy flux of energetic electrons is obtained from the radar data in two different ways. The black line is the first and the gray line the second method described in the text. The Joule heating (dashed curve) is calculated according to Equation (16). In Part (b) the black curves are the temperatures for 177 km, and the gray curves for 233 km. The dashed curves are T_e, and the solid curves T_i. Temperature scales are given for three combinations of ion composition: 100% O⁺, 40% O⁺ and 60% NO⁺, and 100% NO⁺.

in the previous two sections. Figure 6(a) shows energy inputs during the midnight sector for March 22, 1973 – two radar curves for energetic electrons and a curve for Joule heating. There is good agreement in both time and magnitude on the energetic electron input. However, the second radar method, using Rees (1963), is smaller by increasingly large factors as the characteristic energy of the incident electrons becomes greater. This behavior can be understood qualitatively because the fluxes in Rees are monoenergetic, while real fluxes are not. The greater the characteristic electron energy, the larger the number of lower-energy electrons that have not been included.

The curves show increases in the energy input from Joule heating and auroral electrons occurring at about the same time (within the 30 min time resolution of the Joule heating determination). Given an $E_\perp$, a close time correspondence is to be expected because the auroral electons produce enhancements in the E layer ionization, and hence enhancement in the Pedersen conductivity, current, and Joule heating. The energy input from auroral electrons is seen to vary from about one-third to four times that from Joule heating. However, since at least the electron input is subject to time variations of the order of a few seconds, it should be noted that the 5 min averaging reduces the peak input, and the 30 min smearing of Joule heating undoubtedly has a similar effect. The base level for the combined auroral energy inputs is about 3 erg cm^{-2} s^{-1}, or comparable to the energy absorbed above 80 km from the overhead sun (Leovy, 1975). During the two large events the input is about an order of magnitude larger. So far, the maximum electron input seen has been about 100 erg cm^{-2} s^{-1}.

As a result of the large energy inputs, increases in electron and ion temperatures are expected (Schunk and Walker, 1973). Figure 6b shows temperature measurements by the radar indicating qualitative agreement with the theory. During each of the four events involving increases in the particle precipitation and Joule heating in Figure 6a, (around 0800, 0900, 1015, and 1215 UT) there are increases in both electron and ion temperatures at 177 km. The changes in the two temperature curves occur at the same time. During the two largest events, centered at 0900 and 1210, there are increases in the electron and ion temperatures at 233 km. However, in contrast with the lower altitude, the electron temperature is elevated for longer periods and there is less convincing evidence of heating during the two smaller events.

In discussing the temperature changes above, it is difficult to assign numerical values because the ion composition has to be known (Petit, 1968; Wand and Perkins, 1970; Wickwar, 1974a). The ion composition changes in this region under quiet conditions from mostly a mixture of NO^+ and O_2^+ (the NO^+ scale) near 120 km, to O^+ near 250 km. But, under auroral conditions it may well vary at a given altitude. There are theories for intermediate altitudes that predict either an increase (Jones and Rees, 1973) or a decrease (Banks et al., 1975) in the ratio of O^+ to heavy ions. Furthermore, the data represent a weighted average over about 60 km of this rapidly changing region. Hence, the three temperature scales are included to show the possible range. Note that if the composition were almost entirely NO^+ at 177 km, there would be occasions when T_i is greater than T_e. However, if the composition were 40% O^+, then the temperature ratio would be unity on these occasions.

Improvements in the data-acquisition system since these data were taken have improved the time resolution for Joule heating, and improvements presently being incorporated will significantly improve the altitude resolution. Hence, it is hoped that more definitive results on temperatures will be forthcoming.

5. Conclusions

The Chatanika Radar has brought the powerful technique of incoherent scatter to the auroral zone during the last 3 yr. Considerable ionization occurs between about 60 and at least 350 km from energetic particle bombardment. The resultant electron density profiles have been observed, and those from the D and E regions have been used to find the effective recombination coefficient. The energy flux of incident auroral electrons has been determined. Large ion velocities resulting from magnetospheric convection have been observed. From them the E region neutral winds have been derived. When combined with integrated conductivities based on measured electron densities, the ion and neutral winds enable the auroral electrojet currents, Joule heating, and magnetic-field changes to be found. Comparison of ion and neutral winds shows the importance of ion drag and auroral heating upon the neutral winds. The evening-sector ion velocities appear to be important in the development of the F region trough. The energy inputs from energetic electrons and Joule heating are compared. Significant energy input is found from both Joule heating and energetic electrons, compared to the daytime solar input in the same altitude region.

Acknowledgments

I would like to thank members of the incoherent-scatter groups at SRI and UCSD for helpful discussions during the preparation of this paper. In particular I would like to thank T. Watt of SRI and J. Reagan of Lockheed for allowing some of their data to be presented prior to publication. This research has been supported in part by Contract DNA001-74-C-0167 from the Defense Nuclear Agency and Grant GA-36095 from the National Science Foundation to Stanford Research Institute.

References

Anderson, H. R. and Vondrak, R. R.: 1975, *Rev. Geophys. Space Phys.* **13**, 243.
Banks, P. M., Doupnik, J. R., and Akasofu, S. I.: 1973, *J. Geophys. Res.* **78**, 6607.
Banks, P. M., Rino, C. L., and Wickwar, V. B.: 1974a, *J. Geophys. Res.* **79**, 187.
Banks, P. M., Chappell, C. R., and Nagy, A. F.: 1974b, *J. Geophys. Res.* **79**, 1459.
Banks, P. M., Schunk, R. W., and Raitt, W. J.: 1974c, *Geophys. Res. Letters* **1**, 239.
Baron, M. J.: 1973, Ph.D. thesis, Stanford University, Stanford, California.
Baron, M. J.: 1974, *Radio Sci.* **9**, 341.
Baron, M. J., de la Beaujardiere, O., and Craig, B.: 1970, 'Radar Readiness Achievement Program, Part A – Data Processing and Analysis', report, Stanford Research Institute.
Biondi, M. A.: 1969, *Can. J. Chem.* **47**, 1711.
Belon, A. E., Bates, H. F., Hunsucker, R. D., Hays, P. B., and Wickwar, V. B.: 1972, *Trans. Am. Geophys. Un.* **53**, 1070.
Brekke, A., Doupnik, J. R., and Banks, P. M.: 1973, *J. Geophys. Res.* **78**, 8235.
Brekke, A., Doupnik, J. R., and Banks, P. M.: 1974a, *J. Geophys. Res.* **79**, 2448.
Brekke, A., Doupnik, J. R., and Banks, P. M.: 1974b, *J. Geophys. Res.* **79**, 3773.
Cauffman, D. P. and Gurnett, D. A.: 1971, *J. Geophys. Res.* **76**, 6014.
CIRA: 1965, North-Holland, Amsterdam.
Doupnik, J. R., Banks, P. M., Baron, M. J., Rino, C. L., and Petriceks, J.: 1972, *J. Geophys. Res.* **77**, 4268.
Eather, R. H. and Mende, S. B.: 1971, *J. Geophys. Res.* **76**, 1746.

Egeland, A.: 1975, this volume, p. 99.

Evans, J. V.: 1969, *Proc. IEEE* **57**, 496.

Evans, J. V.: 1972, *J. Atmospheric Terrestr. Phys.* **34**, 175.

Fedder, J. A. and Banks, P. M.: 1972, *J. Geophys. Res.* **77**, 2328.

Feldstein, Y. I.: 1963, *Geomagnetizm i Aeronomiya* **3**, 183.

Francis, W.: 1974, private communication.

Haerendel, C.: 1972, in B. M. McCormac (ed.), *Earth's Magnetospheric Processes*, D. Reidel Publishing Company, Dordrecht-Holland, p. 246.

Hagfors, T. and Behnke, R. A.: 1974, *Radio Sci.* **9**, 89.

Hays, P. B. and Roble, R. G.: 1971, *J. Geophys. Res.* **76**, 5316.

Hays, P. B., Jones, R. A., and Rees, M. H.: 1973, *Planetary Space Sci.* **21**, 559.

Horwitz, J. and Banks, P. M.: 1973, *Planetary Space Sci.* **21**, 1975.

Jones, R. A. and Rees, M. H.: 1973, *Planetary Space Sci.* **21**, 537.

Knudsen, W. C.: 1974, *J. Geophys. Res.* **79**, 1046.

Kohl, H.: 1975, this volume, p. 87.

Leadabrand, R. L., Baron, M. J., Petriceks, J., and Bates, H. F.: 1972, *Radio Sci.* **7**, 747.

Leovy, C.: 1975, this volume, p. 73.

Meriwether, J.: 1974, private communication.

Mozer, F. S.: 1973, *Rev. Geophys. Space Phys.* **11**, 755.

Nagy, A. F., Cicerone, R. J., Hays, P. B., McWatters, K. D., Meriwether, J. W., Belon, A. E., and Rino, C. L.: 1974, *Radio Sci.* **9**, 315.

Petit, M.: 1968, *Ann. Geophys.* **24**, 1.

Petit, M.: 1975, this volume, p. 159.

Rees, M. H.: 1963, *Planetary Space Sci.* **11**, 1209.

Rees, M. H. and Walker, J. C. G.: 1968, *Ann. Geophys.* **24**, 193.

Rino, C. L.: 1972, *Radio Sci.*: **7**, 1049.

Rino, C. L., Baron, M. J., Burch, G. H., and de la Beaujardiere, O.: 1974, 'A Multipulse Correlator for the Chatanika Radar', report, Stanford Research Institute.

Schunk, R. W. and Walker, J. C. G.: 1971, *J. Geophys. Res.* **76**, 6159.

Schunk, R. W. and Walker, J. C. G.: 1973, in C. A. Rouse (ed.), *Progress in High Temperature Physics and Chemistry*, Pergamon Press, New York, p. 1.

Smith, L. B.: 1968, *J. Geophys. Res.* **73**, 4959.

St.-Maurice, J.-P. and Schunk, R. W.: 1974, *Planetary Space Sci.* **22**, 1.

Thomas, L.: 1971, *J. Atmospheric Terrestr. Phys.* **33**, 157.

Thomas, L.: 1974, *Radio Sci.* **9**, 121.

Ulwick, J. C. and Baron, M. J.: 1974, 'Simultaneous Rocket-Probe and Incoherent-Scatter Measurements During an Aurora', submitted for publication.

Walt, M., MacDonald, W. M., and Francis, W. E.: 1969, in R. L. Carovillano (ed.), *Physics of the Magnetosphere*, D. Reidel Publishing Company, Dordrecht-Holland, p. 534.

Wand, R. H. and Perkins, F. W.: 1970, *J. Atmospheric Terrestr. Phys.* **32**, 1921.

Watkins, B. J. and Banks, P. M.: 1974, *J. Geophys. Res.* **79**, 5307.

Watt, T. M.: 1974, private communication.

Watt, T. M. and Reagan, J.: 1974, private communication.

Watt, T. M., Newkirk, L. L., and Shelley, E. C.: 1974, *J. Geophys. Res.* **79**, 4725.

Wickwar, V. B.: 1974a, *Analysis Techniques for Incoherent-Scatter Data*, Interpretation in the 100- to 300-km Region, report, Stanford Research Institute.

Wickwar, V. B.: 1974b, *Planetary Space Sci.* **22**, 1297.

COORDINATED OPTICAL AND RADAR MEASUREMENTS OF AURORAL MOTIONS AND IONOSPHERIC PLASMA DRIFT

ROBERT D. SEARS and JOHN E. EVANS

Lockheed Palo Alto Research Laboratory, 3251 Hanover Street, Palo Alto, Calif. 94304, U.S.A.

1. Introduction

Optical measurements of the motion of auroral emission features in the O_I, 5577 Å line were made with a three beam photometer in close coordination with incoherent scatter radar measurements at Chatanika, Alaska, during several periods in 1973. The motion of auroral emission irregularities is compared to the ionospheric plasma drift motion which is derived from the radar Doppler frequency data. We find that the chaotic auroral emission irregularities present during periods of high activity have velocities which often correspond to the radar-derived plasma drift velocities. We conclude that the magnetospheric E fields, which cause the plasma drift, also cause the motion of irregularities in the precipitating particle flux which produces the observed 5577 Å variations.

2. Optical Measurements and Analyses

The three beam photometer technique for describing the horizontal motions of wavelike irregularities in upper atmospheric emissions was described by Sears and Evans (1974). This technique is the optical analog of the radio spaced receiver method for measuring ionospheric drifts. Optical measurements of auroral O_I, 5577 Å emission irregularities were made at Chatanika, Alaska during February and March 1973. Table I summarizes the characteristics of the three beam photometer as utilized for the measurements described herein.

For much of the experimental period in 1973, the photometer observed the 5577 Å line during active auroral conditions. Three sources of excitation for the O_I, 5577 Å line are considered:

airglow excitation by the Chapman mechanism,
$$O + O + O \rightarrow O_2 + O(^1S),$$
direct excitation by energetic precipitating particles
$$O + e \rightarrow O(^1S) + e,$$
and delayed excitation by the dissociative recombination of O_2^+
$$O_2^+ + e \rightarrow O + O(^1S).$$

Detailed examination of photometric intensity data and the all sky camera films obtained during the experiments clearly shows that most of the time the auroral excitation is more important than the background airglow. Likewise, the production of

B. M. McCormac (ed.), Atmospheres of Earth and the Planets, 125–130. All Rights Reserved.
Copyright © 1975 by D. Reidel Publishing Company, Dordrecht-Holland.

TABLE I

Three beam photometer description

Field of view (each beam)	1°
Beam separation	2°
Filter wavelength	5577 Å
Sensitivity	~10 R
Approximate useful intensity range	100 R to 10 kR
Approximate altitude range	100–120 km
Beam separation distance at altitude	3.9 km
Orientation toward geomagnetic zenith	Elevation angle 78
	Geographic azimuth 209°

$O(^1S)$ by dissociative recombination of O_2^+ is also shown to be unimportant from consideration of the electron density profile data, the known rate of production of $O(^1S)$ by dissociative recombination, and from other experimental considerations. Therefore, the intensity observations reported herein relate principally to direct excitation by precipitating electrons. Consequently, the motions of irregularities in emission intensity are directly related to the motions of equivalent irregularities in the precipitating particle flux.

Evans and Sears (1974) have described the analytical method for determination of the horizontal phase velocity of upper atmospheric emission irregularities. The computational procedure involves cross correlation and cross spectral analyses of the three data time series which are the photon count rates from each of the three photometric fields of view. The principal output of interest to this experiment is the plot of the horizontal phase velocity spectra, as seen in Figure 1. Two types of spectra are observed: the most predominant form during the presence of auroral activity is illustrated in Figure 1a, where an inverse dispersion type of behavior is present. This spectral form is the result of the effect of an emitting region of finite thickness. This effect is discussed in more detail by Sears (1975). The other form of velocity spectrum, illustrated in Figure 1b, shows a true dispersive character and may well indicate the presence of acoustic mode waves in the emitting region (see for example, Francis, 1973).

In order to describe the horizontal motion of the emission irregularities, hence of the precipitating particle flux irregularities, we select the minimum value of horizontal phase velocity which satisfies the statistical criteria described by Sears (1975). These criteria involve the accuracy with which the spectral variables have been computed. They include the spectral coherence between data series, and the requirement that the direction of propagation at a selected wave frequency cannot vary by more than 0.2 rad for at least two of the three redundantly computed values. In addition, the propagation velocity and direction vs frequency must present a smooth curve through the value selected. In most cases where the phase velocity curve is of the form of Figure 1a, the low frequency or long period portion of the curve contains either a true minimum velocity or an asymtotic value which satisfies the above criteria.

The result of selecting the minimum horizontal phase velocity for a consecutive

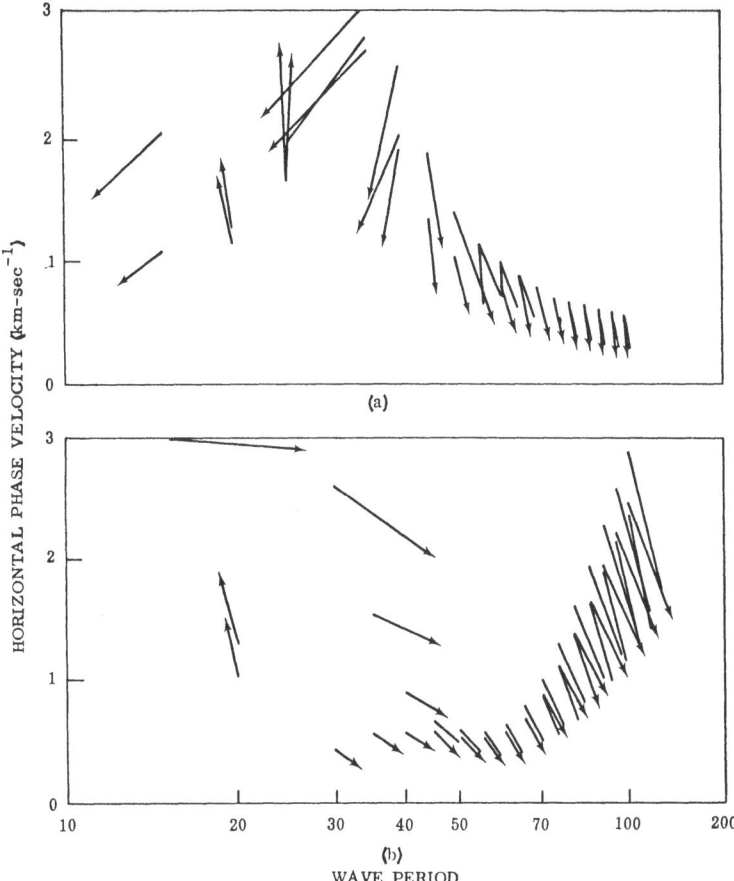

Fig. 1. Horizontal phase velocity of auroral motions vs. period; (a) The effect of finite vertical wave-number for wave propagation is shown. (b) The data were taken during increasing activity and show a quasi-dispersion effect. Times are 1112 to 1117 UT, March 21, 1973.

series of 5 min data intervals is presented in Figure 2. Here, the variability of the N-S and E-W components is illustrated for a period of very active auroras during a magnetic storm. The relationship of the structured features seen in the auroral emission as described by such a plot to all-sky photographs and other detailed information is beyond the scope of this paper. Sears (1975) discusses the significance of the auroral emission phase velocity structure and compares it with other geophysical data obtained at the same location.

3. Radar Measurements

Simultaneous measurements of electron density profile and the line of sight plasma velocity were made by the Chatanika incoherent scatter radar. Techniques for interpreting the spectral shape of radar return pulse in terms of the Doppler velocity of the ionospheric plasma are described by Wickwar (1975), and the radar itself is described by Leadabrand *et al.* (1973). *F* region plasma motion is generally controlled by the

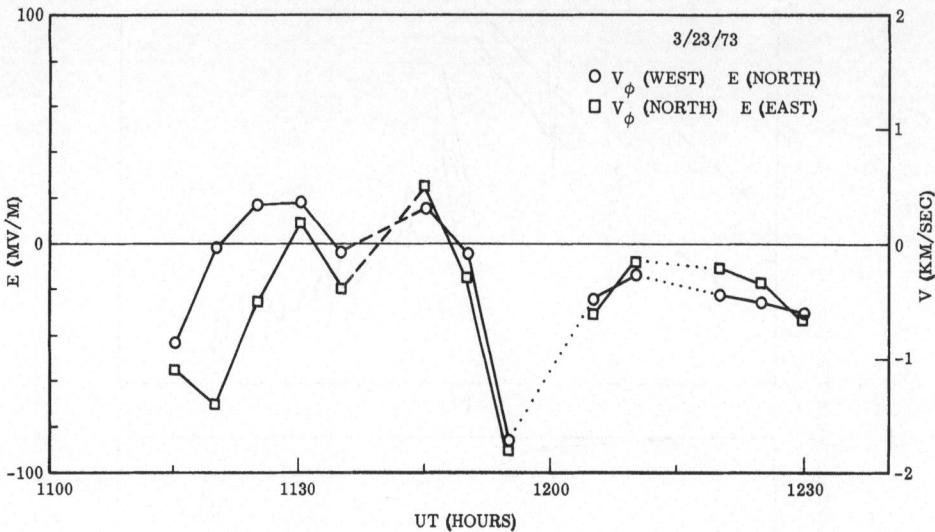

Fig. 2.   Plot of the velocity components of auroral motions for successive 5 min periods of observations
and the inferred ionospheric E fields. Geomagnetic coordinates are used.

E fields which are imposed by the magnetosphere, or generated by neutral winds in
the dynamo region. As altitude decreases in the ionosphere, the effects of neutral
winds upon the plasma motion become more important, such that the E region plasma
motion may be influenced by either E fields or by neutral winds or by both. We assume
herein that the F region plasma drifts are a direct map of the E fields at ionospheric
heights. Furthermore, we assume that the horizontal E fields are essentially constant
with height and may be mapped along geomagnetic equipotentials.

The incoherent scatter radar was operated in two different modes during the optical observations: one, a three position mode allowed determination of the line of
sight plasma drift velocity for a 10 min integration period in each position, from
which a 30 min average resolved velocity vector could be determined each 10 min.
The second mode of operation involved 10 min integrations with the radar looking
either to the north or to the east at 45° zenith angle. This mode produced a series of
plasma velocity component values every 10 min, either along the N-S or E-W geomagnetic coordinates. In the three position mode, the radar's effective spatial resolution is controlled by its 13.5° zenith angle. This corresponds to a horizontal resolution
element of about 50 km in the auroral emission region at 110 km, or to about 150 km in
the F region. In the 45° position, the radar plasma velocity measurements are spatially
far removed from the magnetic zenith where the photometric observations are made.

4. Comparisons of Plasma Drift and Auroral Motions

Auroral motions observed with the three beam optical technique and the plasma
drifts observed with the incoherent scatter radar are compared in Figures 3 and 4.
Figure 3 illustrates the N-S, and E-W components of electric field inferred from

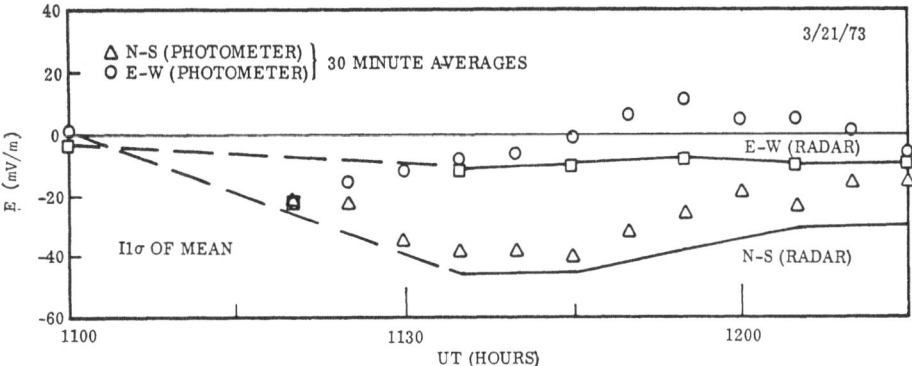

Fig. 3.   Comparison of E fields inferred from photometer measurements of the emission irregularity statistical properties, with E fields inferred from radar measurements of plasma drifts.

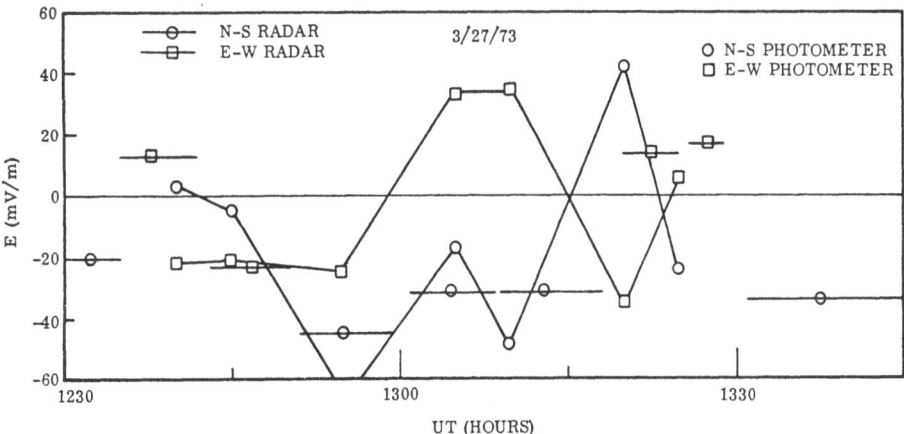

Fig. 4.   Comparison of E fields inferred from photometer measurements with E fields derived from radar measurement of plasma drifts.

auroral motions and plasma drifts. The photometer data were smoothed by a 30 min running average in order to allow comparison with the radar data on an equivalent time base. This figure shows that for at least part of the time, the N-S inferred E fields agree well in temporal behavior, but not so well in magnitude. The E-W E fields vectors agree part of the time, but show an opposite direction part of the time which is smaller in total magnitude than the orthogonal component. This behavior suggests, but does not prove, that the motion of auroral emission irregularities is controlled in part by the same E field forces that cause plasma drifts in the F region. Further comparisons of auroral motion and plasma drift velocities are presented in Figure 4, where a mixed series of northward and eastward radar observations is compared with the photometer data. Again, temporal behavior of the N-S E fields observed by the two techniques agree reasonably well quantitatively, whereas the orthogonal E-W components do not agree as well.

These data, plus other comparisons not shown here, suggest the following behavior:

both auroral motions and F region plasma drifts in the E-W directions appear to be controlled by the N-S component of magnetospheric electric field but the degree of equivalence between the two effects cannot be determined fully as yet. E fields used in these plots are related to the auroral motion and plasma drift velocities by the proportionality $20 \, (\text{m s}^{-1})$ per $1 \, (\text{mv m}^{-1})$.

Relatively poorer agreement exists for the comparison of N-S velocities and E-W E field components. This may be a result of inaccuracy in the photometric analysis technique, or it may be caused by the differences in spatial scale and locations observed by the two methods. Mozer and Manka (1971), for example, conclude that the highly variable components of ionospheric E fields may have a spatial scale of 100 km or less. Hence, separation of the optical and radar fields of view by even a few tens of kilometers may result in drastically different E fields inferred from the auroral motions and plasma drifts. The significant question appears to be the degree of control which the magnetospheric E fields may have on the visible auroral motions. Further study of coordinated measurements such as these may yield this information.

Acknowledgments

The research reported herein was supported by Defense Nuclear Agency and by the Lockheed Independent Research Program. The Chatanika incoherent scatter radar is operated by Stanford Research Institute under Defense Nuclear Agency and National Science Foundation sponsorship. The authors are indebted to Drs M. J. Baron and V. Wickwar of SRI for their continuing assistance and interest, both in the field and during the analysis, and for providing the radar data. Mr D. E. Hillendahl provided valuable assistance in preparation and fielding of the apparatus.

References

Evans, J. E. and Sears, R. D.: 1974, *EOS Trans. Amer. Geophys. Union* **55**, 364.
Francis, S. H.: 1973, *J. Geophys. Res.* **78**, 2278.
Leadabrand, R. L., Baron, M. J., Petricks, J., and Bates, H. F.: 1972, *Radio Sci.* **7**, 747.
Mozer, F. S. and Manka, R. H.: 1971, *J. Geophys. Res.* **76**, 1697.
Sears, R. D.: 1975, *J. Geophys. Res.* **80**, 215.
Sears, R. D. and Evans, J. E.: 1974, *EOS Trans. Amer. Geophys. Union* **55**, 364.
Wickwar, V.: 1975, this volume, p. 111.

PART III

STRUCTURE AND COMPOSITION OF THE
NEUTRAL AND IONIZED ATMOSPHERE

EARLY AERONOMY RESULTS FROM THE
SATELLITE ESRO 4

U. von ZAHN

Physikalisches Institut, Universität Bonn, 53 Bonn 1, West Germany

1. Introduction

After 510 days in Earth orbit the polar orbiting satellite ESRO 4 re-entered the atmosphere on April 15, 1974. Within its experimental payload it carried 3 instruments with particular aeronomic relevancy: An ion spectrometer (ESRO designation: S45) provided by the University College London (principal investigator: W. J. Raitt), a gas analyzer (S80) provided by the University of Bonn (principal investigator: U. von Zahn), and an auroral particle spectrometer (S94) provided by the Kiruna Geophysical Observatory (principal investigator: P. Christophersen). Although the wealth of data received from these instruments is large enough to keep us busy analyzing them for a couple of years to come, I feel that some of the early results obtained so far meet with sufficient interest to present them to you on this occasion.

I will divide this presentation into four parts – first a brief description of the spacecraft and its experiments, second, results with respect to the neutral atmosphere, third, interactions of the neutral atmosphere-ionosphere at midlatitudes, and last, results with respect to high latitude processes.

2. The Spacecraft and Its Experiments

2.1. THE SPACECRAFT

The general configuration of ESRO 4 is shown in Figure 1. The mass of the spacecraft was 113 kg. It was attitude stabilized by a spin of about 1 rev s^{-1}. Very important for our gas analyzer was the fact that there was an active attitude control system onboard using a magnetorquer. This device consisted of a couple of coils inside the spinning satellite which could be energized with electric current thereby creating a magnetic field which interacted with the Earth's magnetic field. If used properly a torque could be created sufficient to turn the satellite spin axis over in a controlled manner. A typical change of attitude by 90° took about three days to accomplish. It is a comparatively simple system but it has worked very nicely. Very important, furthermore, was the capability of onboard data storage by tape recording, where the tape recorder had a capacity of storing the data of 1.2 orbits.

2.2. THE ORBIT

Selected parameters of the ESRO 4 orbit are given in Table I. With an inclination of 91° ESRO 4 is an almost true polar orbiting satellite. This allows an intensive study of the upper atmosphere and ionosphere at high latitudes and across the polar cap

B. M. McCormac (ed.), Atmospheres of Earth and the Planets, 133–157. All Rights Reserved.

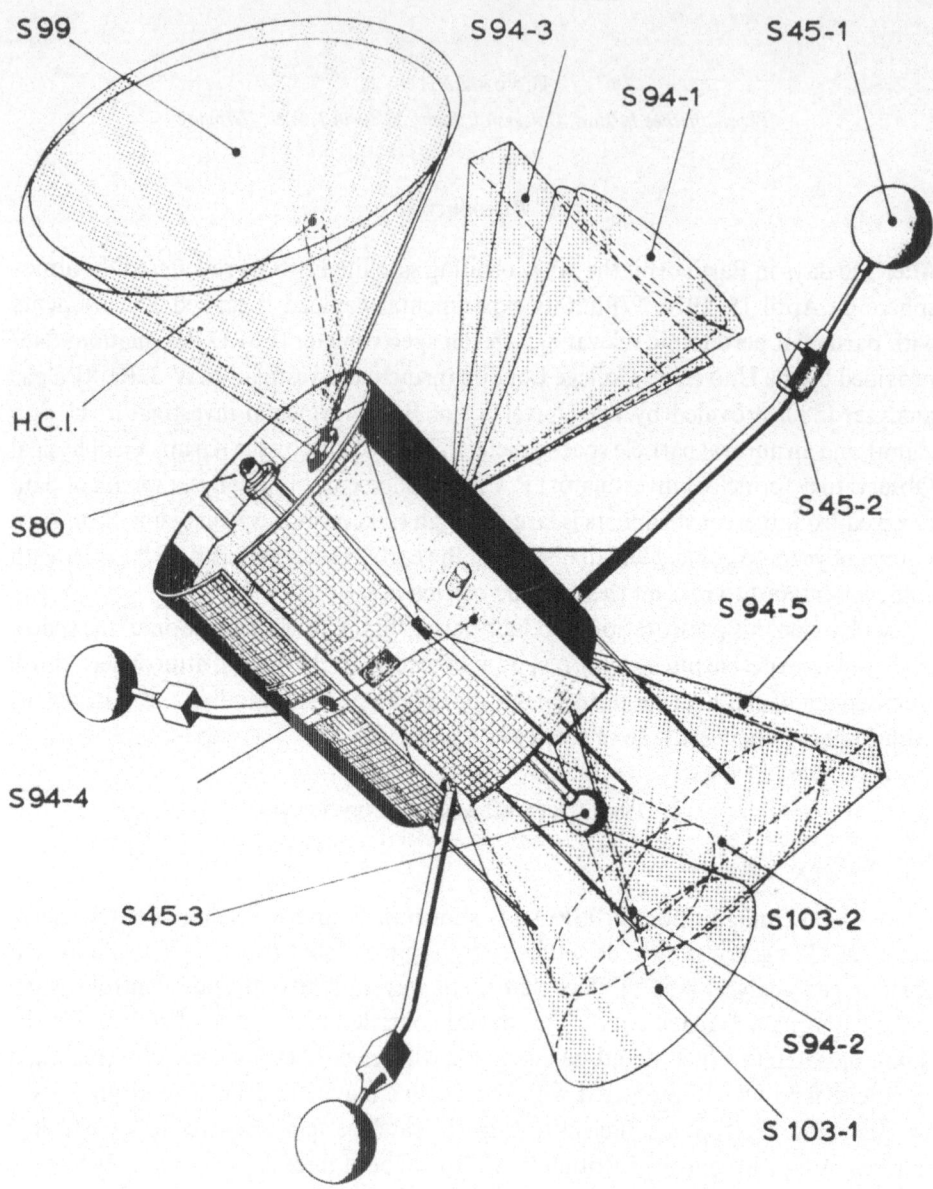

ESRO IV SATELLITE

Fig. 1.   The ESRO-4 satellite showing the viewing cones of the various experiments.

TABLE I

The orbit of ESRO 4

Initial parameters	245×1177 km
Inclination	$91.1°$
Mean motion of perigee	$-3.7°$ day^{-1}
Number of perigee passes	6 north pole passes
across polar caps	5 south pole passes
1. pass of perigee across	Dec. 16, 1972
north pole	at 0235/1435 LT
3. pass of perigee across	July 5, 1973
north pole	at 1532/0332 LT
Mean change of local time	-3.4 min day^{-1}
for ascending node	(28.6 h in lifetime)

areas which are very important to our understanding of the high atmosphere. The polar orbit is in addition particularly well suited for the investigation of the seasonal-latitudinal variations in the neutral atmosphere which among other things are indicators for important global circulation systems. The low perigee allows measurements to be taken closer than before to the source region for many dynamical processes acting in the upper atmosphere, that is the 100 to 150 km altitude region. In addition, extensive mapping of $F2$ layer parameters can be performed and direct comparison of *in-situ* measurements with ground based ionosonde data is greatly facilitated. Last but not least the comparatively high number densities at the 250 km level ease the measurement job for the gas analyzer.

Why did we not choose exactly 90° inclination? We wanted to eliminate the diurnal variation in our most basic measurements of the seasonal-latitudinal variations in composition and density of the upper atmosphere. Thus we wanted the first perigee pass across the north pole in deep winter to occur at the same local times as the later pass in the high summer season. This was accomplished by choosing an inclination of 91°. By further selecting a proper local time for the launch of the spacecraft the conditions during these two perigee pole passes were such that measurements were taken close to the local times of maximum and minimum densities in the neutral atmosphere (see also Table I).

2.3. THE MISSION

ESRO 4 was launched on November 22, 1972, and stayed in orbit for 510 days (7686 orbits). Figure 2 indicates that the perigee during the orbital lifetime moved 6 times across the north pole and 5 times across the south pole. The very last ESRO 4 pass over an active telemetry ground station occurred at 115 km altitude above Spitzbergen a little bit more than an hour before disintegration of the spacecraft. At this time ESRO 4 was still at 100% performance. The flawless performance of the spacecraft including its tape recorder throughout its lifetime and efficient operations planning by the European Space Operations Center (ESOC) formed the basis for an excellent rate of data recovery from the instruments.

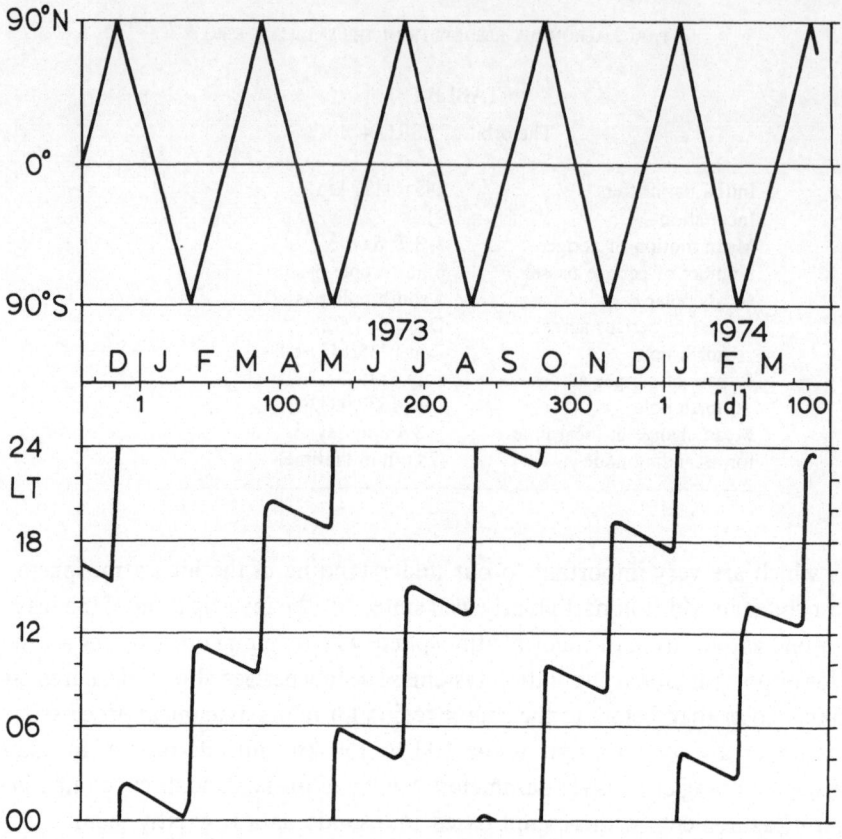

Fig. 2. The position of the ESRO-4 perigee in geographic latitude (upper panel) and local solar time (lower panel).

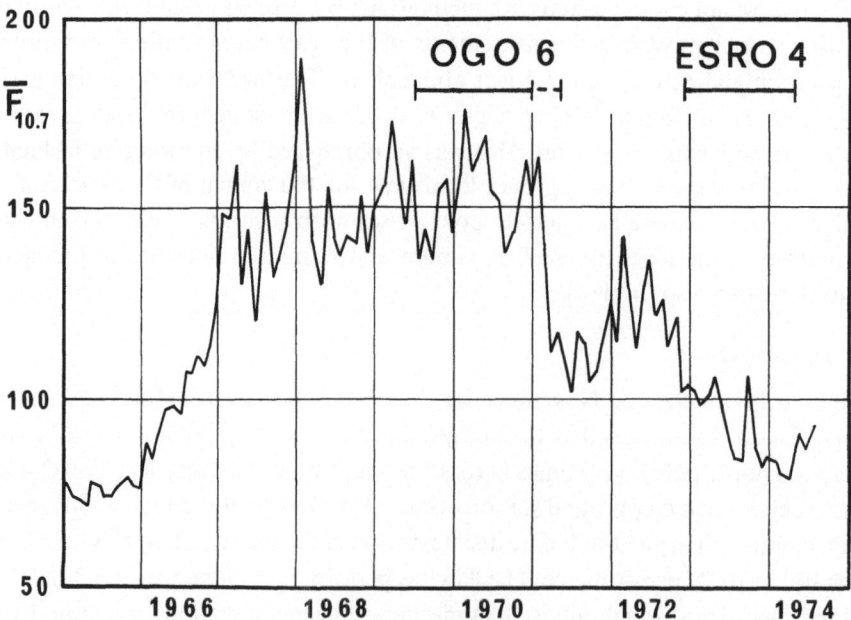

Fig. 3. Plot of the monthly mean value of the 10.7 cm solar flux with the operational periods of OGO 6 and ESRO 4 indicated.

The aeronomic importance of the ESRO 4 mission rests also in the fact that it occurred at a level of low solar activity. The mean solar flux at 10.7 cm over the orbital lifetime was 91.5×10^{-22} W m^{-2} Hz^{-1}. Figure 3 shows that the ESRO 4 measurements taken at low solar activity should nicely complement the data collected by the satellite OGO 6 at medium and moderately high solar activity.

During the orbital lifetime of ESRO 4 the maximum geomagnetic activity occurred on April 1, 1973 with $K_p = 8 +$ for 3 h.

2.4. THE INSTRUMENTS

The ion spectrometer (S45) consists of a spherical gridded Langmuir probe mounted on a boom (sensor S45-1 in Figure 1). The mode of operation of this probe is such that the velocity of the satellite through the ionospheric plasma enables mass identification of the major positive ion species. Once this has been done the number density and temperature of the ion species can also be determined. In addition there is a small probe (S45-2) for measuring electron density, electron temperature, and the spacecraft potential with respect to the ambient plasma.

The gas analyzer (S80), as shown in Figure 4, consists of a so-called antechamber, an

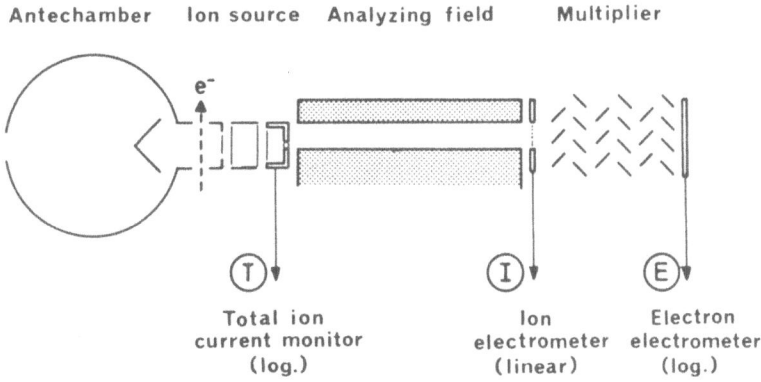

Fig. 4.   Schematic plan of the ESRO-4 gas analyzer (S80).

electron impact ion source, a monopole analyzing field and a triple current detection system (Trinks and von Zahn, 1975). The purpose of the spherical antechamber is to force full thermal accommodation of the incoming gas particles to the temperature of the ion source walls and to recombine the atmospheric O to O_2 before it either enters the ion source proper or leaves the antechamber through the front orifice. The instrument performs one gas analysis every 9 s (corresponding to 0.6° in latitude) and uses a total bit rate of 90 bps. Typical measurement ranges are below 600 km for He, below 500 km for O, below 400 km for N_2, and below 300 km for Ar. The instrument is mounted parallel to the spin axis. The attitude of the spacecraft was readjusted regularly to maintain an inflow of ambient gas into the antechamber at perigee. The instrument was operated with about 50% duty cycle each orbit, with the active period centered about perigee.

The auroral particle spectrometer (S94) consists of a complement of 14 various detectors to measure the energy spectrum and pitch angle distribution of electrons and protons in the energy range 0.2 to 150 keV. One group of sensors looks radially outward through the waistband of the satellite (Figure 1), the others look parallel to the spin axis through the bottom of the spacecraft. The full instrument needed, however, a bit rate of 10.2 kbps which substantially exceeded the storage capability of the tape recorder. This mode therefore was used only on ground command in real time over a few selected ground stations. In order to secure at least a minimum amount of information about the particle fluxes encountered throughout the orbit the outputs of 3 selected detectors were recorded with decreased bitrate. Two of them looked outward radially and monitored the differential electron flux at 0.21 keV and the integral electron flux above 44 keV. A true pitch angle distribution can not be analyzed in this low speed mode due to the reduced bit rate. Nevertheless, the following discussions will use those two channels only as their data processing is most advanced.

2.5. Data processing for the gas analyzer

The present state of processing for the data received from our gas analyzer is as follows: all 135 tapes (7 track) with raw data received from ESOC have been converted and re-formatted to 24 high-density tapes (9 track). All the raw data on ion currents representing He, O, N_2, and Ar have been plotted on microfilm in the form of 'telemetry volts'. That means the data are without corrections for temperature variations, gain changes of the multiplier, background gases, and so on. We also developed microfilm routines which put the results of the experiments (S45 and S80, or S94 and S80) (see later Figures) on one picture.

This is about the present state and all the results which I am going to present are extracted from these early microfilm plots.

The next essential steps will be first the conversion of the ion currents to absolute number densities in the ion source and second conversion to absolute ambient number densities through application of the ram effect corrections and the appropriate model for the chemistry taking place at the ion source surfaces. The chemistry corrections appear to be negligible for all data on He and Ar and for the data on O and N_2 below 300 km. At altitudes above 300 km the measurements of O and N_2 become affected by surface reactions inside the ion source. The effects which we observe are, however, notably different from those described by Hedin *et al.* (1973) for the OGO 6 mass spectrometer. Also their ion source model cannot describe the behavior of our ion source with what I consider physically reasonable surface parameters. We are therefore investigating the possibility of explaining our ESRO 4 observations in terms of either one or a combination of the following two effects: first a considerably higher surface density of O adsorbed to the ion source surfaces (possibly beyond one monolayer) than assumed in the OGO-6 model (3 to 5% of a monolayer). Second a significantly increased gasatom-adatom recombination coefficient for the incoming O as long as they possess their approximately 5 eV kinetic energy with respect to the ion source surfaces. Only future work will show the success or shortcomings of these working hypotheses.

3. Neutral Atmosphere Results

Due to its polar orbit ESRO 4 is particularly well suited for the investigation of seasonal-latitudinal variations in upper atmosphere composition and density, as well as the reaction of the atmosphere to geomagnetic activity. In this section I will deal with the first of these two aspects, the next two sections will discuss selected observations related to geomagnetic activity.

3.1. THE HELIUM BULGE

The He bulge is a phenomenon which consists of much higher He densities over the winter pole of the Earth than over the summer pole. Since its discovery (Keating and Prior, 1967) it has attracted great interest because it is considered an indicator for important transport processes taking place in the upper atmosphere. (For recent reviews the reader is refered to Johnson (1973) and Kockarts (1973)).

Before we can use He bulge observations for a quantitative evaluation of global circulation systems we need to know, however, at least its amplitude, that is the variation of He number densities at a given altitude between the winter and summer pole. We also should expect that this amplitude depends on the altitude, on the mean solar activity, and on the level of geomagnetic activity. As shown in Table II there exists

TABLE II

Variation of He number density between winter and summer poles

	$\dfrac{n(\text{winter})}{n(\text{summer})}$	Altitude
Keating and Prior (1968)	2.5 to 4.4	120 km
Jacchia and Slowey (1968)	4.5[a]	700 km
Bitterberg et al. (1970) (summary of rocket-borne mass spectrometer data)	10 (to 50)	150 km
CIRA 1972	4.5	independent
Hedin et al. (1974) (OGO-6 model)	12	450 km
von Zahn et al. (1973) (ESRO-4 data)	20	270 km
Keating et al. (1974) (drag improved OGO-6 model)	21	450 km

[a] derived from the stated temperature variation for Explorer 19, on the assumption that the effective height of the fitted densities was 700 km

presently little agreement as to the true amplitude of the He bulge and its altitude dependence. The early studies of the bulge amplitude through satellite drag data analysis possibly led to an underestimate of the amplitude due to the difficulty of applying the right corrections for the remaining O and temperature effects at exospheric heights. Bitterberg *et al.* (1970) summarized results from rocket-borne mass spectrometers obtained in the years 1965 until 1968. At 150 km altitude absolute He number densities varied by a factor of 10, whereas the ratio $n(\text{He})/n(\text{N}_2)$ changed seasonally by as much as of a factor 50. The global empirical model of thermospheric composition based on OGO 6 mass spectrometer measurements of Hedin *et al.* (1974) puts the amplitude of the He bulge to 13 for 450 km altitude and mean solar flux of $F_{10.7\ cm} = 150$. Keating *et al.* (1974) on the other hand have critically compared drag measurements of satellites over the solar cycle through 1973 with the OGO-6 model atmosphere. Their analysis indicated a factor of 21 polar variation in He concentration, again normalized for an altitude of 450 km and $F_{107\ cm} = 150$. A factor of about 20 was also found for the altitude of 270 km from a preliminary analysis of ESRO-4 gas analyzer data by von Zahn *et al.* (1973). In summary all the direct *in-situ* measurements by mass spectrometers confirm the comparatively high He bulge amplitude of about 15 to 20 for the variation between winter and summer pole conditions in the middle thermosphere. For even higher altitudes one would on the other hand expect a decrease of this factor due to the increasing and smoothing influence of lateral diffusion and exospheric outflow of He.

Another interesting question is the actual distribution of He across the geographic poles. Since a sample of OGO-6 observations was published by Reber *et al.* (1971) it has generally been assumed that the maximum He concentration does not occur over the winter pole itself but rather at about 53° magnetic dipole latitude. More recently Reber and Hedin (1974) have re-interpreted the variation in latitude of the peak in the winter He distribution as being due to a persistent mid- to high-latitude heating phenomenon which also depends on UT time.

The observations of our ESRO-4 gas analyzer appear just to add a little more complexity to this question. We find that under quiet geomagnetic conditions the peak of the He distribution can indeed be found on many occasions very close to the *geographic* pole. Figure 5 gives two examples of our raw data plots showing He distributions which peak right on top of the geographic pole even though the perigee for both orbits has a slight offset from the true pole. Example 5b was selected to prove that this can happen even during moderately disturbed conditions ($K_p = 4_o$). The slight dip in the He distribution of Figure 5b is caused by a local disturbance due to a magnetospheric cusp event.

Figures 6a and 6b on the other hand give examples of a pronounced He density minimum over the poles even at low geomagnetic activity, similar to the observations of Reber *et al.* (1971). From this we conclude that in the future it will be necessary, even for so-called 'quiet time' atmosphere models, to distinguish between various forms of high latitude He distributions if a serious attempt is to be made to give a realistic description of the behavior of the He bulge.

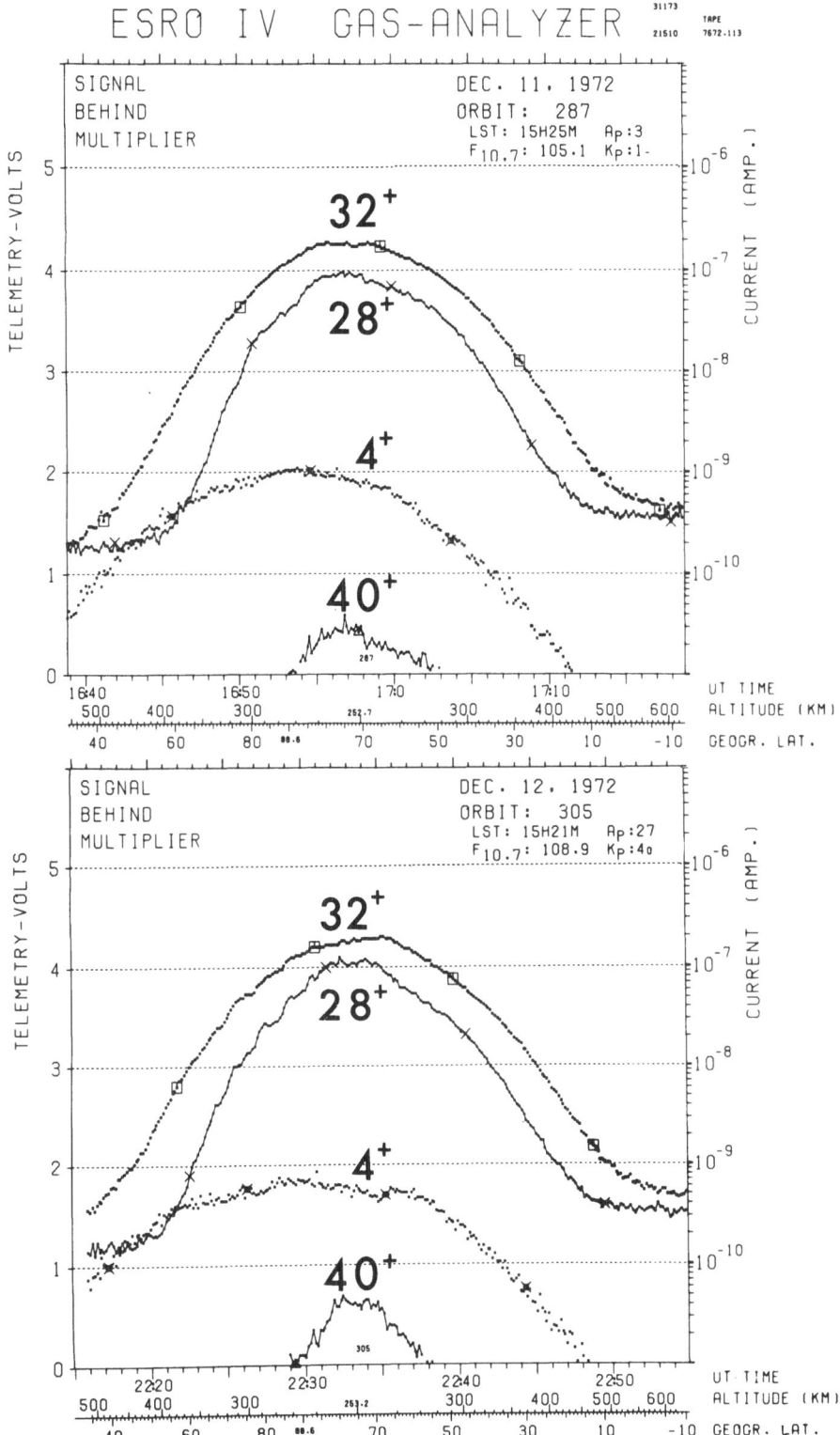

Fig. 5. Telemetry signals received during two north pole passes of ESRO 4 with an approximate scale for the measured ion currents on the right ordinate axis. Bold faced numerals indicate the measured amu, 4^+ representing He for example. From the bottom scales one can see that the peak He densities were observed over the geographic pole rather than at perigee.

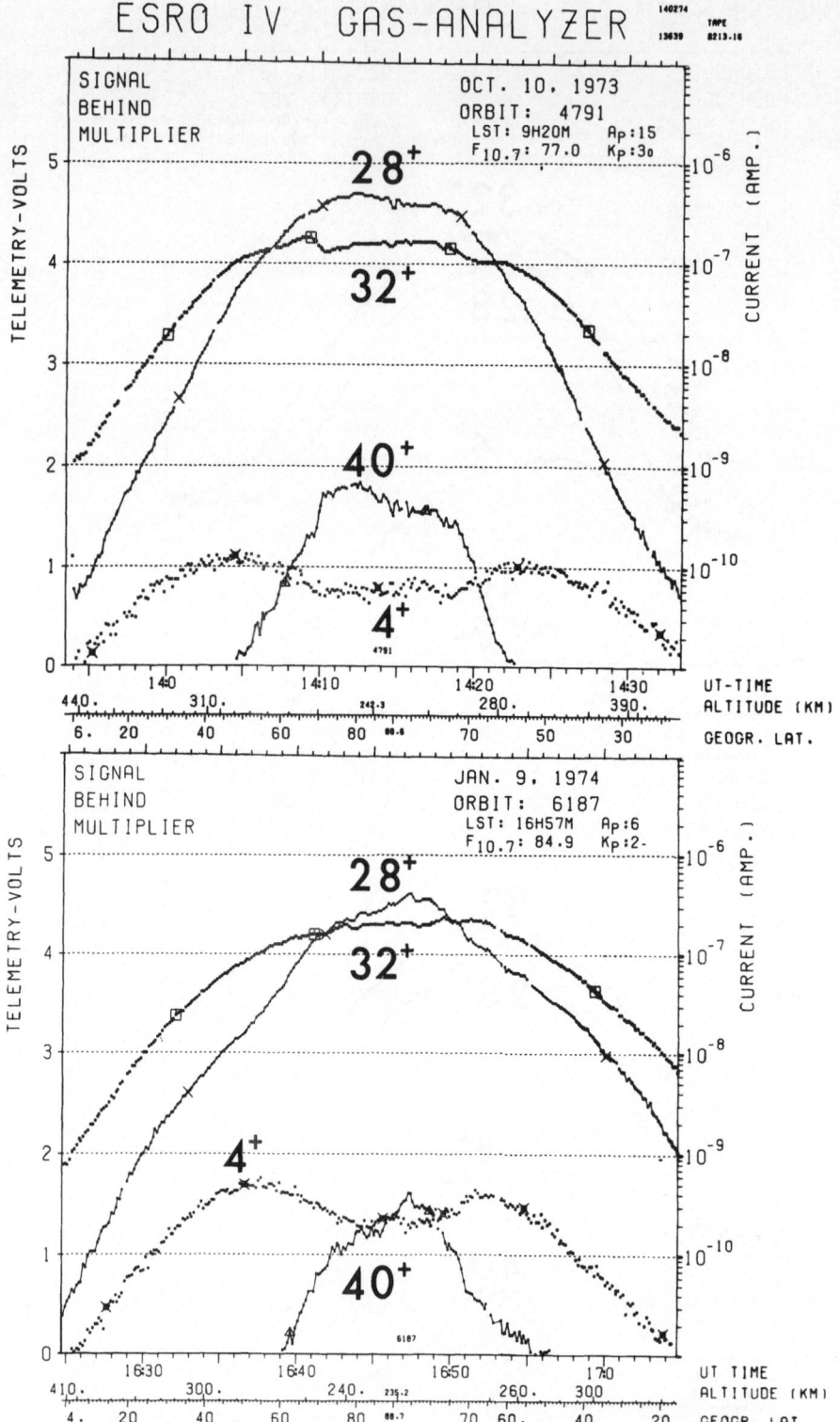

Fig. 6. Telemetry signals received during two north pole passes in moderately disturbed conditions. The He distributions are characterized by pronounced minima over the geographic poles.

Another interesting aspect of the He bulge is its apparent north-south asymmetry (Keating *et al.*, 1973). Our ESRO-4 observations will certainly contribute to this question, but only after quantitative analysis of much more of our data.

3.2. THE ARGON BULGE

Due to the opposite deviation of their atomic mass from the mean molecular mass of thermospheric air, He and Ar react in opposite ways to many dynamical processes occurring in the thermosphere like changes in turbopause altitude, vertical winds, and so on. As long as we believe that the winter He bulge is basically caused by dynamical processes one would anticipate that there should be a corresponding summer Ar bulge. Therefore when the ESRO-4 gas analyzer performed routine Ar measurements at satellite altitudes for the first time it was not too surprising that we immediately found a summer Ar bulge (von Zahn *et al.*, 1973). Comparing winter solstice conditions over the north pole with late summer conditions over the south pole, Ar densities turned out to be larger by a factor of 10 in late summer than in winter at 270 km altitude and under geomagnetically quiet conditions.

A closer look at this factor 10, however, reveals that the relationship between the He bulge and the Ar bulge may not be as close as one could assume from first sight. One of the basic differences in the behavior of these two constituents rests in the rather different reaction of their altitude profiles to changes in exospheric temperature. In the case of He the effect of thermal diffusion very nearly cancels any dependence of the He density in the 300 km level on the exospheric temperature (see for example CIRA 1972). On the other hand Ar densities above, say 150 km altitude, react quite strongly to temperature changes. Since the OGO-6 mass spectrometer measurements indicate very considerable variations in exospheric temperature in going from the winter to the summer pole (Hedin *et al.*, 1974) it becomes clear that a large part of the summer Ar bulge may be due to the temperature effect. This would then make the Ar bulge a different phenomenon than the winter He bulge which undoubtedly is entirely controlled by dynamics.

To evaluate this line of thought further, we can calculate the summer Ar bulge amplitude resulting from the temperature difference between the summer and winter pole with the help of a pure static diffusion model with fixed lower boundary conditions (like CIRA 1972) thereby neglecting any additional dynamics effects. Hedin *et al.* (1974) derive a mean temperature difference between the poles at solstice conditions of somewhat more than 400 K for $F_{10.7\,cm} = 150$. Let us therefore assume that during the measurement period of ESRO 4 with $F_{10.7\,cm} = 91.5$ the polar temperature difference was 300 K, or more specific varied from 700 K to 1000 K. Such temperature change alone would create an Ar bulge amplitude of a factor of 12, whereas our preliminary analysis of ESRO 4 observations indicates a factor of 10.

Admittedly this comparison is somewhat ambiguous and in addition the predicted Ar bulge amplitude turns out to depend strongly on the minimum temperature reached over the winter pole. The point I want to make, however, is that the Ar bulge is strongly temperature controlled and it is left to a much more quantitative analysis of the now available Ar measurements to determine what role is left for dynamical influence in the Ar bulge phenomenon.

3.3. THE RATIO $n(O)/n(N_2)$

Köhnlein *et al.* (1974) investigated the density ratio $n(O)/n(N_2)$ as obtained from the ESRO-4 gas analyzer under geomagnetically quiet conditions. Constructing a global model by use of a spherical harmonic expansion the variations of O/N_2 were analyzed for annual, semiannual (both symmetric and asymmetric), and diurnal terms at 250 km altitude and for the time period December 1972 to December 1973. However the number of coefficients solved for was much less than in the OGO-6 model of Hedin *et al.*

Figure 7 shows the latitudinal-seasonal variations of O/N_2 with the yearly mean on

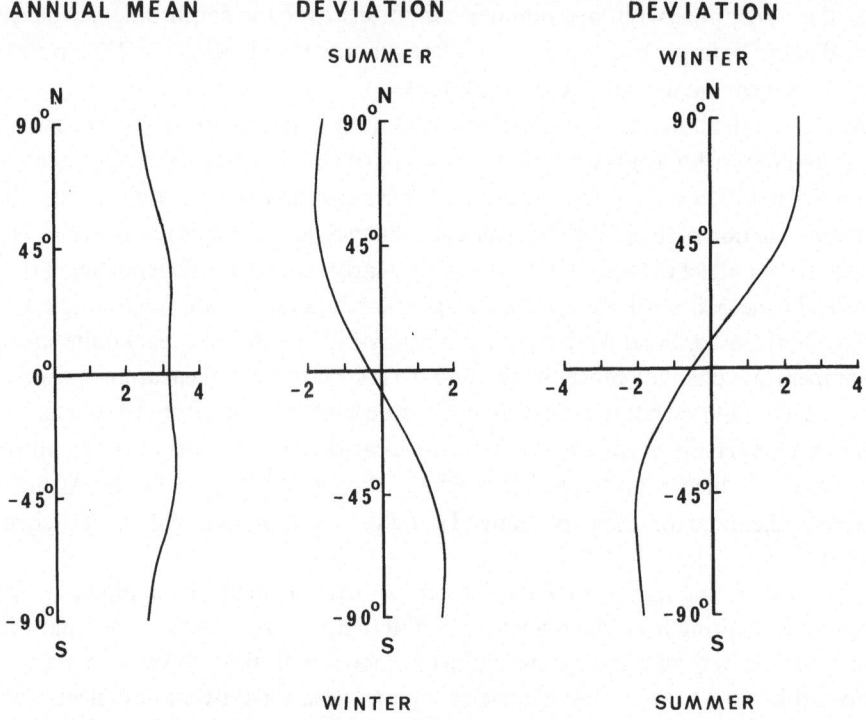

Fig. 7.   The ratio $n(O)/n(N_2)$ at 250 km altitude as derived from a preliminary model fitted to the ESRO-4 gas analyzer data covering the time period December 1972–December 1973 and quiet conditions only.

the left to which the deviations on the right side of the figure must be added in order to obtain the true O/N_2 number density ratio. The values refer to $F_{10.7 \, cm} = 100$, $a_p = 5$, and midnight local time. The increase in the ratio O/N_2 in going from summer to winter conditions amounts to a factor 3.6 at 45° latitude and about 6 right over the poles. The direction of this change in composition is correct for an explanation of the $F2$-layer seasonal anomaly. On the other hand the amplitude of the change, in particular under the comparatively quiet conditions of ESRO 4, is considerably higher than predicted from ionospheric observations.

The analysis of Köhnlein *et al.* showed another significant fact: the number density ratio O/N_2 measured by the ESRO-4 gas analyzer is in close agreement, sometimes surprisingly close, with the predictions made from the OGO-6 model for 250 km altitude (Hedin *et al.*, 1974). For example at the equator and under equinox conditions ESRO 4 measured a ratio of 3.8 at the above stated standard conditions, which compares with a value of 3.7 calculated from the OGO-6 model. An immediate consequence of this apparent close agreement between the ESRO-4 and OGO-6 derived O/N_2 ratios is that it lends support to the conclusions drawn by Taeusch and Carignan (1972) that CIRA 1972 somewhat overestimates the O/N_2 ratio in the thermosphere.

4. Neutral Atmosphere – Ionosphere Interactions at Mid-Latitudes

The complex behavior of the ionospheric F layer during geomagnetic disturbances has long been a subject of intensive studies. The data collected by ESRO 4, in connection with ground based observations of the ionosphere and the geomagnetic field, provide an excellent means to study more closely the interactions of the neutral atmosphere and ionosphere, as well as the spatial and temporal development of the observed disturbances. So far we have concentrated our efforts of analyzing storm time effects mostly on a storm period in late February 1973 at which time the perigee of ESRO 4 was located near 30° S, at 10:45 LST, and an altitude of about 253 km. Therefore many of the following observations pertain to dayside and mid-latitude conditions.

Our studies have given evidence for two distinctive features during this storm period: first, a very strong correlation between changes in the F region plasma density and changes in the neutral atmosphere composition, and second a complex spatial structure of the storm effects in both the neutral atmosphere and the ionosphere (Prölss and von Zahn, 1974; Trinks *et al.*, 1974; Prölss *et al.*, 1974). Let me discuss these observations in turn.

4.1. CORRELATION BETWEEN CHANGES IN THE O/N_2 RATIO AND THE PLASMA DENSITY

During the second half of February 1973 a series of moderate and moderately strong magnetic storms occurred as indicated by the magnetic indices K_p and D_{st} in Figure 8. The D_{st} index started to drop rapidly at 1800 UT of Feburary 21, 1973, indicating the onset of the main phase of a geomagnetic storm. Before the D_{st} index had fully recovered from this event, another storm occurred in the afternoon of February 22. After this second storm the D_{st} index did not recover, and the magnetosphere was continuously disturbed until the end of February.

In Figure 9 we have plotted the $F2$ layer critical frequency during the period February 20 to 24 as observed at eight ionosonde stations in the southern hemisphere. It is noticeable that the reaction to the disturbances is quite different for the ionosphere above the various stations as indicated by the difference between the observed f_0F2 and the monthly means of the hourly values of f_0F2 (dotted lines). The depression of

f_0F2 during the storm is most pronounced for the stations at mid-latitudes in the African (Hermanus) – Australian (Mundaring, Salisbury, Brisbane) sector. For these stations we have correlated the observed changes in upper atmosphere composition with observed changes in plasma density. For each day of the time period February 17 to 25 the measured O/N_2 ratio was taken from the ESRO-4 gas analyzer data when-

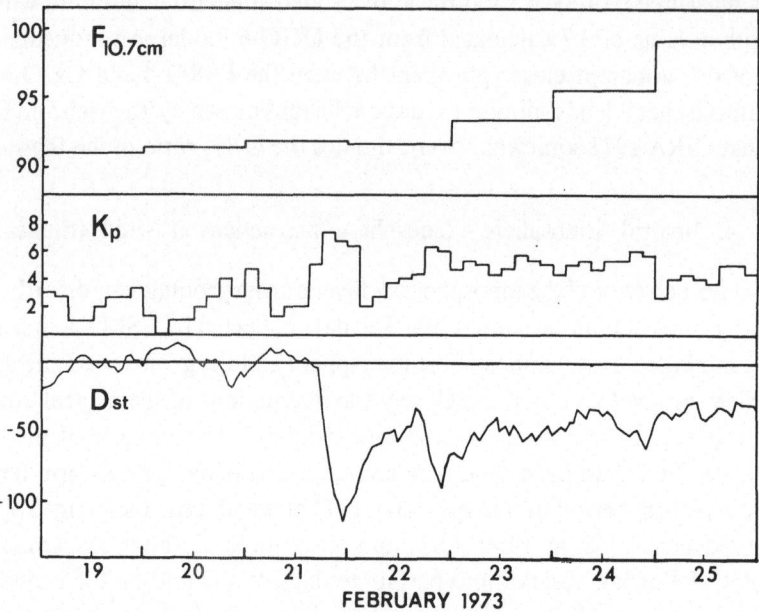

Fig. 8.   The 10.7 cm solar flux, K_p geomagnetic index, and the D_{st} ring current index for the storm period in late February 1973.

ever the satellite was closest to the particular station. In the same way the *in-situ* O^+ density was collected from the ESRO-4 ion spectrometer data. Finally the maximum electron densities of the $F2$ layer were determined from ionosonde observations once every day for the time of closest approach by the ESRO-4 satellite. The observations were normalized by calculating the ratio of parameters under disturbed over un-disturbed conditions, referring to either February 20 or 21 as the undisturbed reference day depending on when the satellite came closest to the ionosonde station. The time variation of these normalized ratios is shown in Figure 10. The correlation between the 3 parameters is indeed very strong throughout this storm period. This clearly supports previous theoretical studies that delineate neutral composition changes as the major cause for negative ionospheric storm effects.

The close correlation between changes in neutral atmosphere composition and $F2$ plasma density can be used to study the temporal and spatial development of storm effects in the neutral atmosphere through investigation of the strongly correlated observations of f_0F2 from many ionosonde stations. An example for this concept will be given in the next subsection.

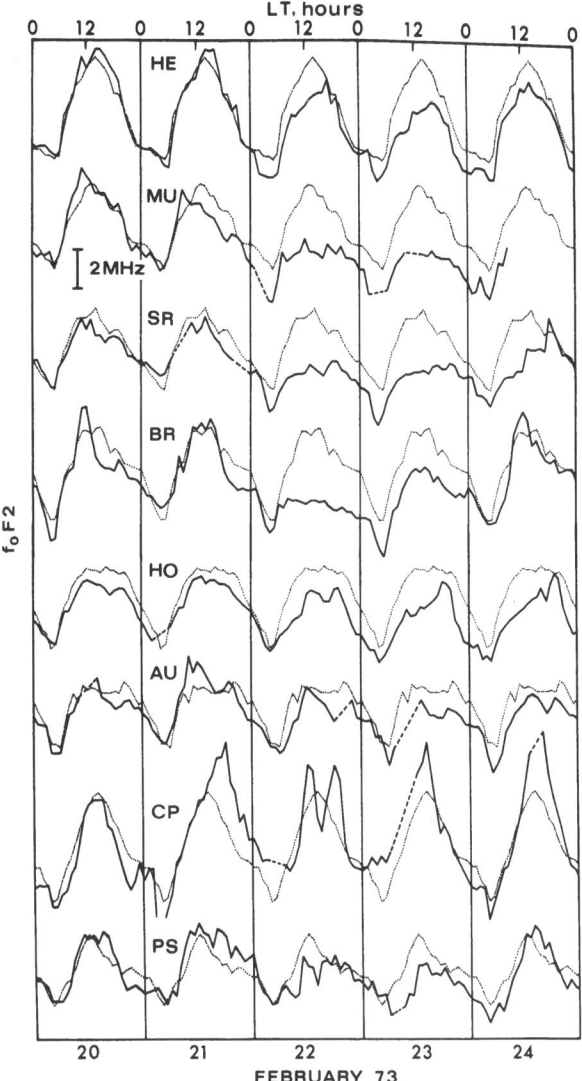

Fig. 9.   Time variation of the $F2$ layer critical frequency as observed at representative ionosonde stations in the southern hemisphere (solid lines). Also shown are the respective monthly mean values of f_0F2 for February 1973 (dotted lines). The ordinate scale is linear and the bar indicates the magnitude of a 2 MHz change in f_0F2. HE: Hermanus at 34S/19E; MU: Mundaring at 32S/116E; SR: Salisbury at 34S/138E; BR: Brisbane at 27S/152E; HO: Hobart at 42S/147E; AU: Auckland at 37S/175E; CP: Concepción at 36S/73W; PS: Port Stanley at 52S/58W geographic position.

An additional detail in Figure 10 is that the *in-situ* measurements of $n(O^+)$ fall somewhat below the other two curves under the most disturbed conditions for 3 out of the 4 stations investigated. This could be explained by the fact that the height of the $F2$ layer maximum rose sufficiently during these days to have ESRO 4 actually sample its data at the bottomside of the $F2$ layer. This rise in $F2$ layer height is most likely the result of increased loss processes for the ionization due to the changed neutral atmospheric composition, with strong equatorward winds contributing to this height increase as well.

4.2. SPATIAL STRUCTURE OF STORM EFFECTS

The magnitude of the disturbances observed in both the neutral atmosphere and ionosphere exhibits a complex spatial structure with strong longitudal control and also an unexpected latitudinal dependence. Because I consider the latter fact the more interesting one, I will discuss it first.

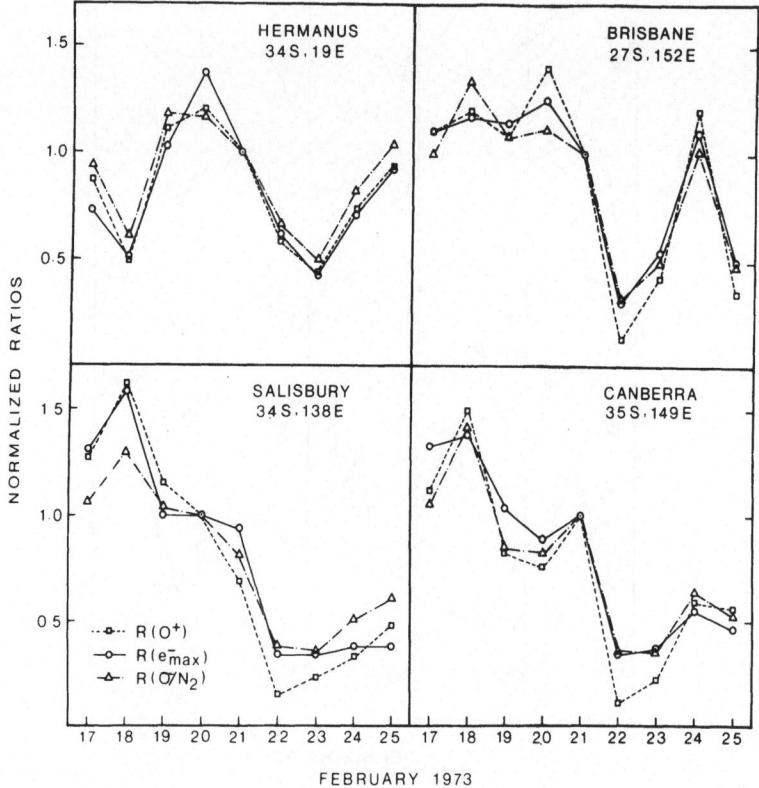

Fig. 10.   Time variations of the ratios of disturbed over undisturbed parameters for the O^+ ion density, $R(O^+)$, the O/N_2 ratio, $R(O/N_2)$, measured by ESRO 4 and the maximum electron density, $R(e_{max}^-)$, measured above four individual ionosonde stations. The ESRO 4 data are taken for the points of closest approach to the particular station.

Although substorm activity in the southern hemisphere started at about 1600 UT on February, 21, 1973, and the D_{st} index started to drop at 1800 UT it was not until about 2340 UT that the satellite ESRO 4 entered a strongly disturbed region at mid-latitudes and near 170° E longitude. This disturbance is indicated in Figure 11 at about 48° IN Lat by a significant 'hole' in the O^+ density, a regional deficiency of O and an increase in N_2 and Ar. There exists, however, a well defined second local disturbance close to 70° IN Lat which is separated from the mid-latitude disturbance zone by an almost undisturbed region centered at 64° IN Lat. This pattern was evident for three consecutive orbits with minimum disturbances at about 64° IN Lat each time. Furthermore, this pattern was again very clearly developed during the initial phase of the geomagnetic storm of February 22, for which Figure 12 gives the evidence. Here the mid-latitude disturbance is just developing, again at 48° IN Lat and the undisturbed region is centered at 65° IN Lat.

From ESRO-4 observations alone we cannot decide how persistent in time this 'minimum reaction zone' was. Figure 9 on the other hand allows a very interesting comparison of the state of the ionosphere above Salisbury (45° S geomeg. lat.) and Hobart (52° S geomag. lat). Salisbury, which is very close to the center of the mid-latitude disturbance zone identified by ESRO 4, shows a very strong depression in $f_0 F2$ for more than 2 days (note that this figure is ordered by LT!). Hobart, on the other hand, lies poleward (!) of Salisbury, but is about half way to the center of the minimum reaction zone found by ESRO 4. The considerably weaker reaction of $f_0 F2$ to the storm events at Hobart throughout February 22 and 23 is taken as a clear indication that the minimum reaction zone did in fact persist for at least these two days with only insignificant smoothing.

As will be discussed in the next section the higher latitude disturbance zone is most likely positioned at the location of the intersection of the magnetospheric cusp with the upper atmosphere. Both particle precipitation and Joule heating may contribute to the observed reactions in the neutral atmosphere and ionosphere. The sources for the mid-latitude disturbance zone on the other hand, are more difficult to identify. In particular the observations in the earliest storm period (as in Figure 12) suggest a fairly localized source even for the mid-latitude zone. The position perhaps coincides with the footprint of the plasmapause or with the inner edge of the outer radiation belt. Preliminary studies of simultaneous measurements of the ESRO-4 particle spectrometer (S94) do not encourage the assumption that hard particle precipitation is the major source for this phenomenon. Five hours later and 73° westward in longitude the satellite finds, still at the same *local* time, the mid-latitude disturbance to be a rather broad and structureless feature which extends from about 15° IN Lat to about 60° IN Lat. This much wider latitudinal range of reaction could more easily be associated with the dissipation of gravity waves that originate in high latitudes, or with storm related Joule heating. It should be clear that the cause for the mid-latitude disturbance may actually be a multitude of different processes.

As was mentioned before, this latitudinal structure develops and exists together with a longitudinal dependence of the storm effects in the atmosphere and ionosphere. For illustration of the *longitudinal* dependence, the ESRO-4 observations of $n(O)/n(N_2)$ and $n(O^+)$ along the 30° S parallel are shown in Figure 13 for the time period February 17 through 25, 1973. There is a weak 24 h periodicity noticeable in the $n(O)/n(N_2)$ ratio before February 22 which becomes very pronounced with the onset of the actual storm sequence. In order to distinguish between a possible 24 h or 360° longitudinal periodicity we turn again to the ionospheric records of Figure 9. All the stations in the African-Australian sector close to 30° S confirm a *persistent* decrease in F2 plasma

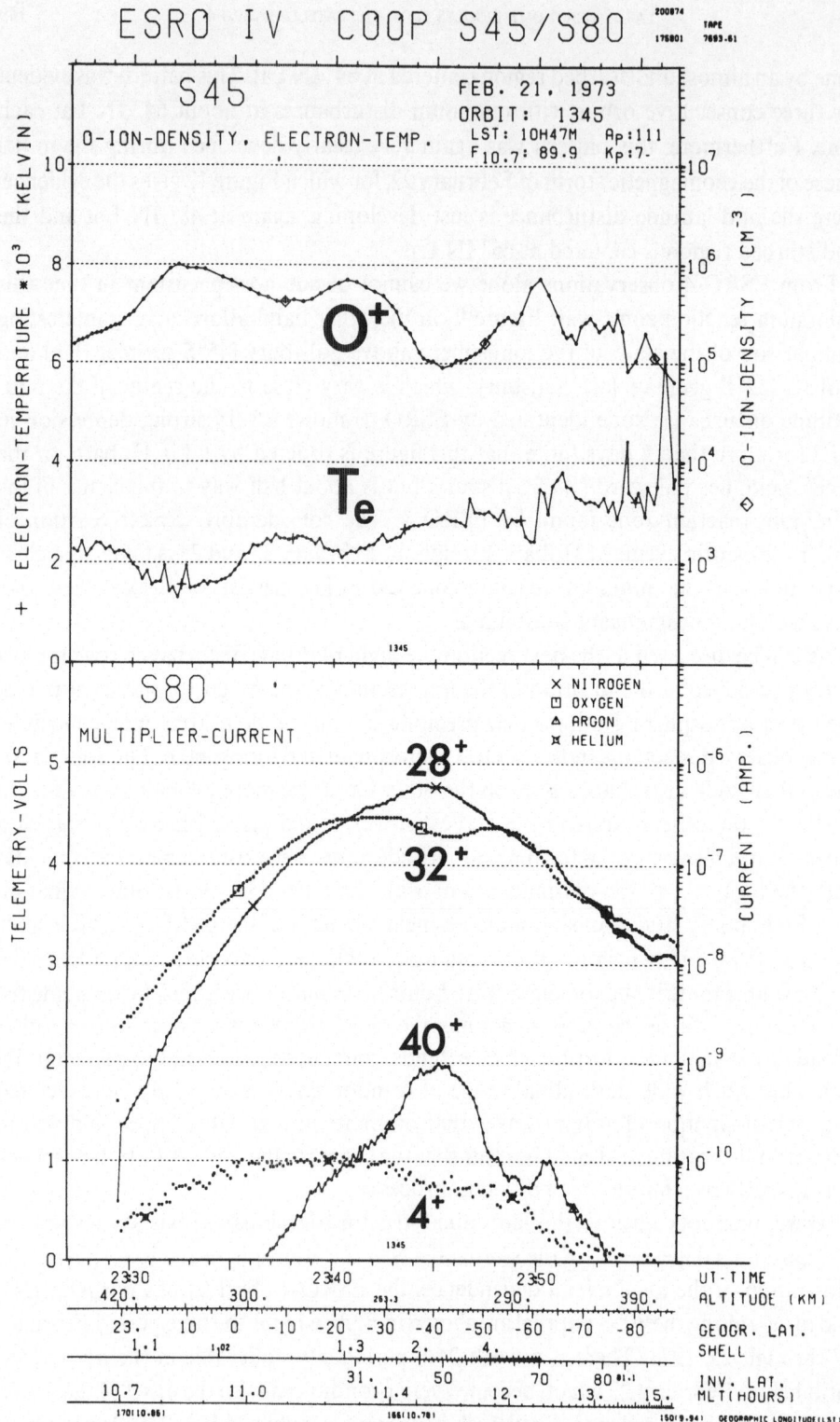

Fig. 11. O⁺ ion density (O⁺), electron temperature (T_e), from the ESRO-4 ion spectrometer and approximate currents measured by the ESRO-4 gas analyzer during the early phase of the February 21, 1973 storm. Note the two latitudinally separated zones of disturbance most clearly indicated in the Ar profile.

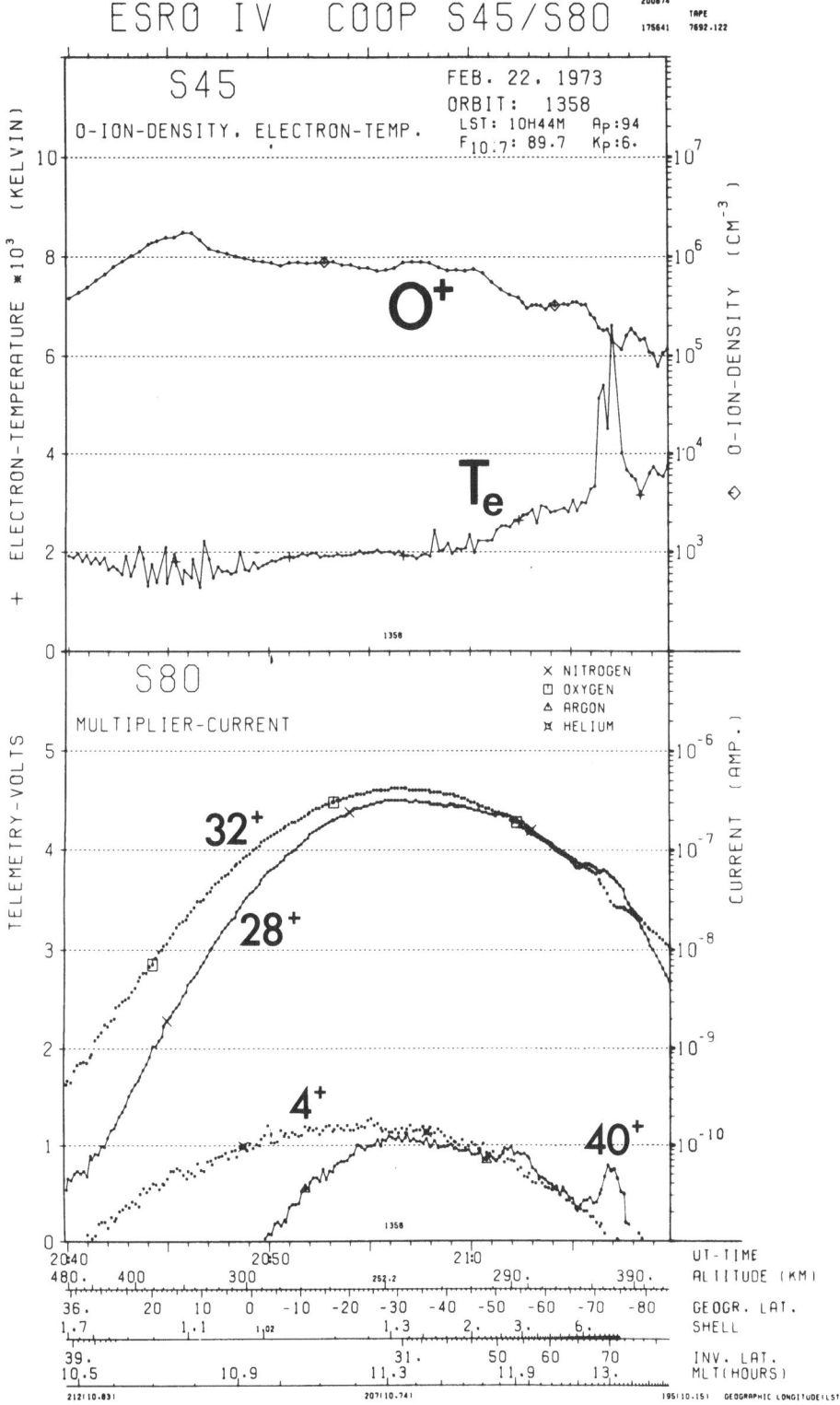

Fig. 12. Parameters as in Figure 11, only 22 h later. The two zones of disturbance can again clearly be identified. Notice that a significant increase in electron temperature has occurred in the high latitude disturbance zone.

density on February 22 which lasted even until February 24 at Salisbury. ESRO 4, on the other hand 12 h after having passed over West-Australia repeatedly probes the South American sector of the southern hemisphere and finds virtually undisturbed conditions at 30°S during the whole storm sequence. We conclude therefore that the deep minima in the $n(O)/n(N_2)$ and $n(O^+)$ curves of Figure 13 are the result of ESRO 4 moving through a spatially fixed disturbance region in the African-Australian sector of the southern hemisphere. Further, this mid-latitude disturbance zone had a very sharply defined eastward boundary, which was located close to 180° longitude and persisted for at least 3 days.

I conclude from this combination of ESRO-4 and ground based observations that even for the described rather complex storm patterns the very close relationship between neutral atmosphere changes and plasma density changes holds. The existence of such complex spatial structures during storms may in the future become a nightmare for atmospheric modelers. Last, but not least the observed complex storm pattern must lead to a similarly complex wind field at F region height because the neutral air pressure, the amount of vertical air movements and the ion drag force all become strongly spatially structured. Hence we may not expect that the developing wind system can be described by a simple circulation cell with rising air at high

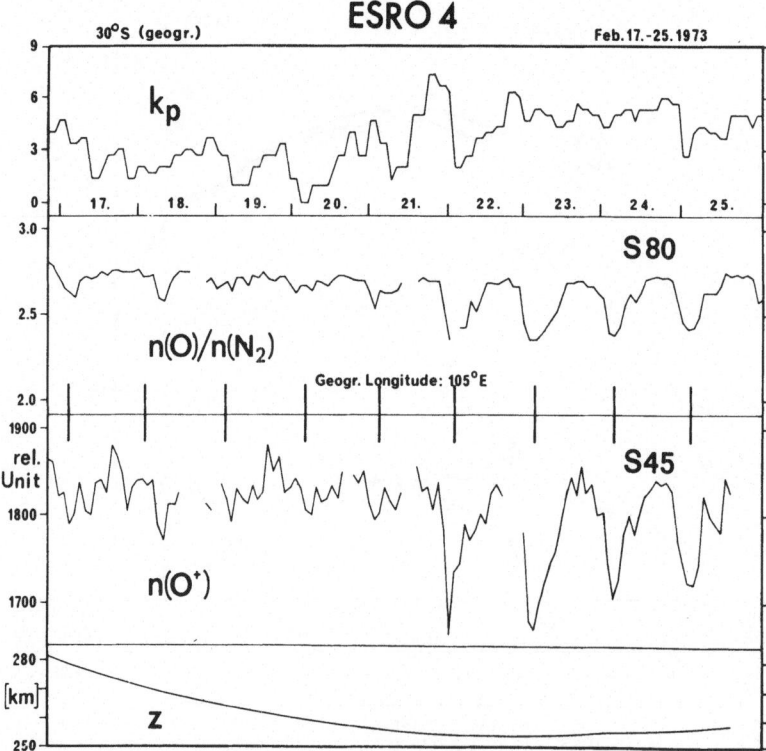

Fig. 13. The K_p geomagnetic index, the O/N_2 ratio and the O^+ ion density as measured by ESRO 4 at 30°S geographic latitude during the February 1973 storm period. Note the 24 h periodicity in the data, which is mainly due to a strong longitudinal dependence of the measured parameters O/N_2 and $n(O^+)$.

latitudes, lateral winds blowing from the poles towards the equator and subsiding air at low and equatorial latitudes.

5. Atmosphere-Ionosphere-Magnetosphere Interactions at High Latitudes

5.1. HEATING ZONES FOR THE NEUTRAL ATMOSPHERE DURING GEOMAGNETIC DISTURBANCES

A correlative study of ESRO-4 *in-situ* observations of neutral gas composition, ionospheric plasma density, electron temperature, and auroral particle fluxes at high invariant latitudes during magnetic disturbances revealed the existence of three zones of heating in the thermosphere (Raitt *et al.*, 1975): over the nightside auroral oval, over the polar cap, and in the interaction region of the dayside magnetospheric cusp with the upper atmosphere. Identification of these three zones is possible through their different characteristics.

The nightside heating zone can be identified by a fairly wide bulge in the N_2 distribution between about 50° and 80° IN Lat during geomagnetic disturbances. The bulge boundaries correlate fairly well with the boundaries of enhanced hard particle fluxes as indicated by the > 44 keV electron flux measurements of experiment S94. The equatorward edge of the soft particle fluxes as determined by the 210 eV electron channel of experiment S94 closely follows the poleward boundary of the ionospheric plasma density trough. This latter correlation also holds for the dayside as will be shown later. The relative location of the observed N_2 enhancements, the particle fluxes and the plasma trough are shown in Figure 14 for the period December 12 to 15, 1972. In general all the different boundaries move towards lower latitudes with increasing K_p.

The trans-polar heating zone for the upper atmosphere is a frequent and very noticeable feature in the ESRO-4 observations. It is most apparent in an enhancement of N_2 and Ar, and a decrease in He over a wide range of latitudes on either side of the pole. The difference between undisturbed and disturbed periods can be seen clearly in Figure 5 and Figure 6, respectively. During the moderate geomagnetic disturbances studied so far the O density in general changed only slightly. This observation is consistent with a source of heat at altitudes of around 150 km rather than the 100 km level where the energy of 10 to 50 keV particles would be dissipated (Mayr and Volland, 1972). In addition the 210 eV electron fluxes over the polar cap did not greatly differ between undisturbed and disturbed conditions during the period December 12 to 15, 1972 for which this effect was analyzed. We are thus led to the conclusion that the major heat input into the trans-polar upper atmosphere is via Joule heating caused by electric current systems flowing in the polar cap, with soft particle precipitation a contributing, but secondary source of heating. It is also to be noted that the plasma density in this region is only weakly controlled by the neutral atmospheric composition, contrary to our observations at mid-latitudes.

The dayside high latitude heating zone is characterized by a narrow band ($< 10°$) of enhanced N_2 and Ar, increased T_e, and decreased plasma density at invariant latitudes in the range 65° to 80° (Fricke *et al.*, 1974; Wulf-Mathies *et al.*, 1974). It can easily be recognized in Figure 12 at 70° IN Lat. The position coincides closely with the equatorward edge of the soft electron fluxes and poleward boundary of the hard particle fluxes as shown in Figure 15 (same orbit as Figure 12). Because the boundary between these two particle populations is considered indicative of the location of field lines

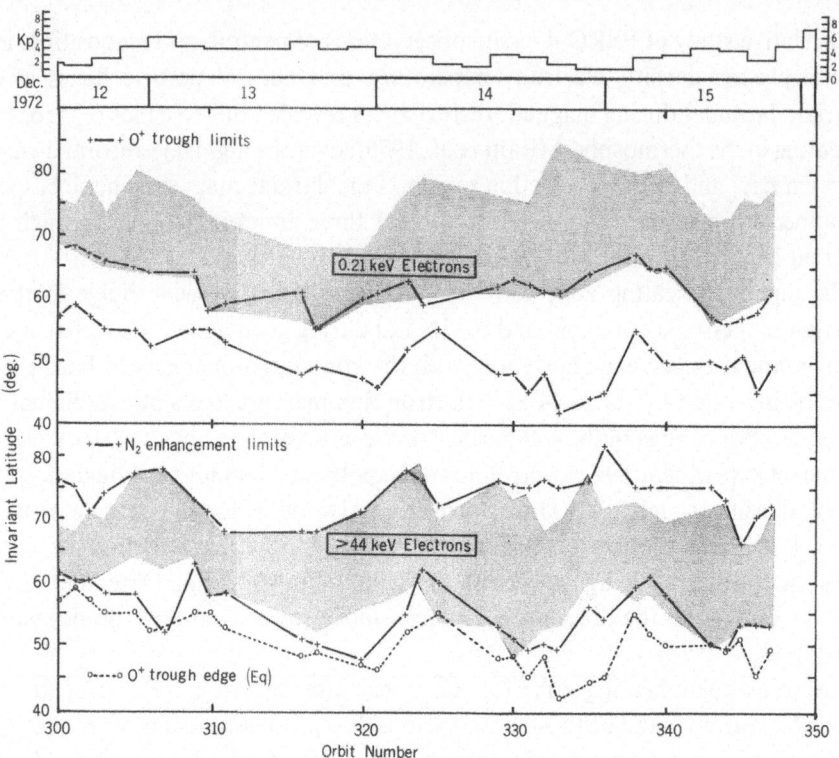

Fig. 14.   Plot showing the local correlation of regions with soft (0.2 keV) electron fluxes, hard (>44 keV) electron fluxes, nitrogen enhancements, and the boundaries of the O^+ ion trough on the nightside.

connected to the magnetospheric cusp we suggest that the disturbances observed in the neutral atmosphere are occurring at the intersection between the cusp field lines and the upper atmosphere. Both corpuscular precipitation and Joule heating could be responsible for the observed effects in the neutral atmosphere.

5.2. NEUTRAL ATMOSPHERE DAYSIDE HIGH LATITUDE DISTURBANCES

A statistical analysis was made by Fricke *et al.* (1974) of the spatial and temporal distributions of the dayside disturbances in the neutral atmosphere, which were described in the last section. This investigation concentrated on local N_2 enhancements occurring poleward of 30° IN Lat and in the time sector defined by 0600 < < MLT < 1800. In 1588 orbits leading at the appropriate MLT through the high latitudes 1881 events of local N_2 enhancements were found. Approximately 30% of

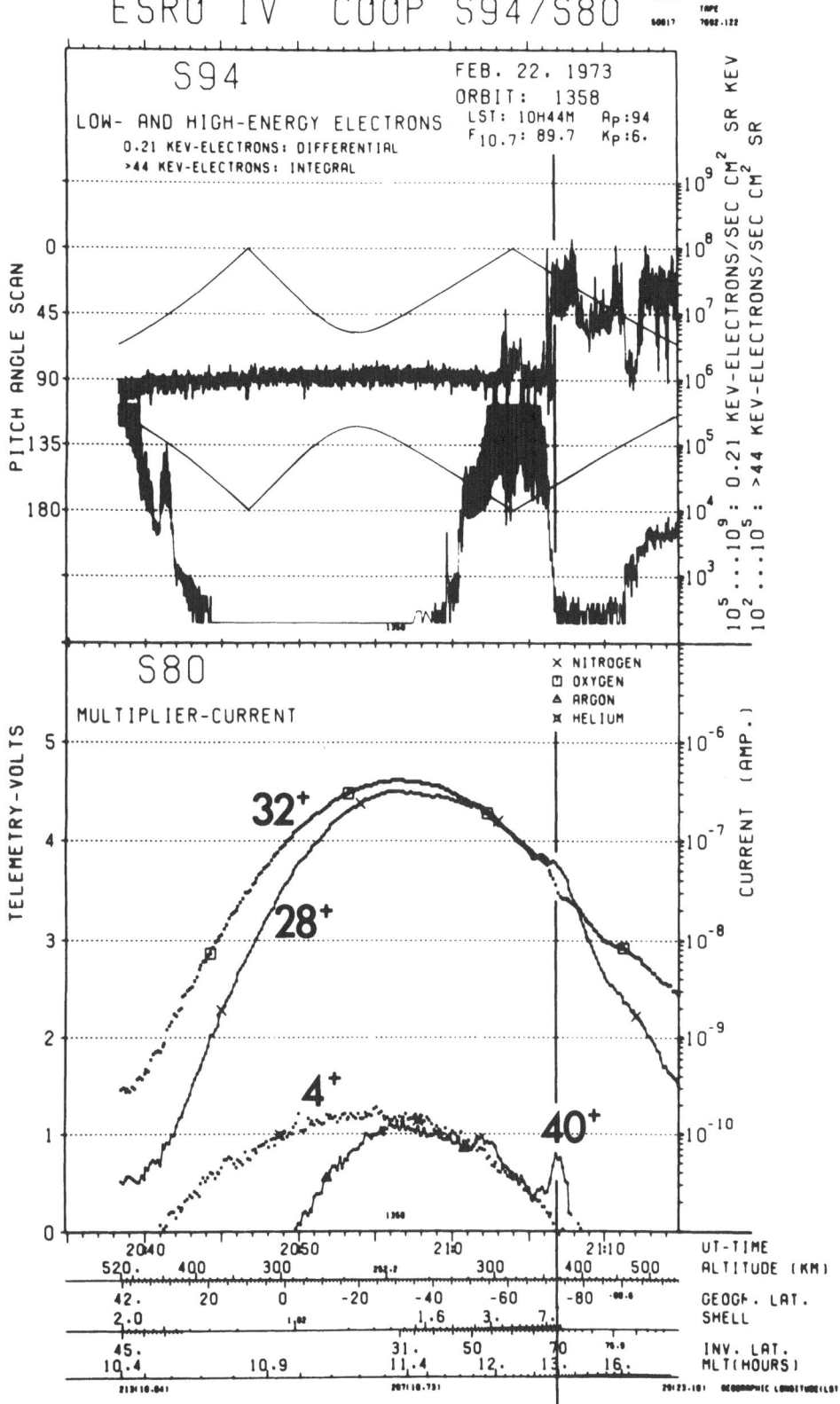

Fig. 15.   Auroral particle fluxes and approximate gas analyzer currents measured by ESRO 4 during a pass through the low altitude extension of the magnetospheric cusp. The neutral atmosphere high latitude disturbance occurs at the poleward edge of the hard particle fluxes (lower trace of upper panel) and the equatorward edge of the soft particle fluxes as indicated by the vertical bar.

those orbits showed no event, 40% showed one event per orbit, and 30% showed more than one local disturbance per orbit. We thus can conclude that these localized day-side heating phenomena are a very common, although not really continuous feature in the neutral atmosphere. The peak in the latitudinal distribution lies close to 80° IN Lat. If the latitudinal distribution is analyzed for various levels of geomagnetic activity a noticeable shift of the most probable position towards lower latitudes is observed with increased geomagnetic activity. An investigation of the relative abundance at various magnetic local times gave the result that the abundance appears to be fairly independent of MLT. The lack of a pronounced peak of events at noon time or in the early afternoon somewhat contradicts the predictions made by Olson (1972) and Olson and Moe (1974), who studied the motion of charged particles from the magnetospheric cusp region toward the dayside auroral atmosphere. This mild disagreement, which we do not want to overemphasize due to the preliminary nature of our own investigation, may on the other hand indicate that in fact Joule heating is significantly involved in causing these local disturbances which was not considered by Olson (1972).

Our statistical analysis of the above described effects is incomplete yet because the various distributions obtained so far still need to be folded with the probability distribution for the spacecraft to reach in its orbital motions a given invariant latitude and magnetic local time. We do not expect that the conclusions drawn in the previous paragraph become invalidated by this next analysis step. At the present time we wish to refrain, however, from quoting precise numbers for the various distributions.

Blum et al. (1974) suggested a first order theoretical interpretation of the observed local disturbances. Starting from the observation that some of the disturbances continued over several satellite orbits with relatively minor changes in amplitude they suggest that neither longitudinal nor latitudinal interactions have a major influence on the behavior of these events. Therefore a quasi-steady state of the thermosphere can be reached even under the disturbed conditions. These quasi-steady events lend themselves to a description by a one dimensional static diffusion model similar to CIRA 1972. Blum et al. (1974) thus attempted to fit four independent observables, these are the four ratios of number densities during undisturbed and disturbed conditions for He, O, N_2, and Ar, with a CIRA 1972 type static diffusion model allowing for only two free parameters, these are the height of the turbopause and the increase in exospheric temperature under disturbed conditions. The result of this fitting procedure is that many of the observed compositional changes can be described in terms of an additional heating of the thermosphere amounting to between 75 K and 150 K increase in exospheric temperature and a height increase of the turbopause level by 1.5 to 5 km.

Acknowledgments

A spacecraft mission becomes a success only through the contributions of a large group of able and dedicated people and their supporting institutions. This is also true for ESRO 4 and is gratefully acknowledged here.

With reference to the subject of this paper I wish to express my gratitude to Dr K. Lenhart from the European Space Operations Center (ESOC), who was responsible for the well prepared and most efficiently performed job of processing the raw data of ESRO 4 at ESOC.

References

Bitterberg, W., Bruchhausen, K., Offermann, D., and von Zahn, U.: 1970, *J. Geophys. Res.* **75**, 5528.

Blum, P., Wulf-Mathies, C., and Trinks, H.: 1974, Paper, XVII Plenary Meeting of COSPAR.

CIRA 1972, *COSPAR International Reference Atmosphere 1972*, Akademie-Verlag, Berlin.

Fricke, K. H., Trinks, H., and von Zahn, U.: 1974, *EOS* **55**, 370.

Hedin, A. E., Hinton, B. B., and Schmidt, G. A.: 1973, *J. Geophys. Res.* **78**, 4651.

Hedin, A. E., Mayr, H. G., Reber, C. A., Spencer, N. W., and Carignan, G. R.: 1974, *J. Geophys. Res.* **79**, 215.

Jacchia, L. G. and Slowey, J. W.: 1968, *Planetary Space Sci.* **16**, 509.

Johnson, F. S.: 1973, *Rev. Geophys. Space Phys.* **11**, 741.

Keating, G. M. and Prior, E. J.: 1967, Paper, 48th Annual Meeting of the AGU.

Keating, G. M. and Prior, E. J.: 1968, *Space Res.* **8**, 982.

Keating, G. M., McDougal, D. S., Prior, E. J., and Levine, J. S.: 1973, *Space Res.* **13**, 327.

Keating, G. M., Prior, E. J., and McDougal, D. S.: 1974, Paper, 17th Annual Meeting of COSPAR.

Kockarts, G.: 1973, *Space Sci. Rev.* **13**, 723.

Köhnlein, W., Trinks, H., and Volland, H.: 1974: Paper, XVII Plenary Meeting of COSPAR.

Mayr, H. G. and Volland, H.: 1972, *Planetary Space Sci.* **20**, 379.

Olson, W. P.: 1972, *Space Res.* **12**, 1007.

Olson, W. P. and Moe, K.: 1974, *J. Atmospheric Terrestr. Phys.* **36**, 1715.

Prölss, G. W. and von Zahn, U.: 1974, *J. Geophys. Res.* **79**, 2535.

Prölss, G. W., von Zahn, U., Raitt, W. J., and Christophersen, P.: 1974, *EOS* **55**, 369.

Raitt, W. J., von Zahn, U., and Christophersen, P.: 1975, *J. Geophys. Res.*, in press.

Reber, C. A., Harpold, D. N., Horowitz, R., and Hedin, A. E.: 1971, *J. Geophys. Res.* **76**, 1845.

Reber, C. A. and Hedin, A. E.: 1974, *J. Geophys. Res.* **79**, 2457.

Taeusch, D. R. and Carignan, G. R.: 1972, *J. Geophys. Res.* **77**, 4870.

Trinks, H. and von Zahn, U.: 1975, 'The ESRO 4 Gas Analyzer', *Rev. Sci. Instrum.*, in press.

Trinks, H., Fricke, K. H., von Zahn, U., and Prölss, G. W.: 1974, *EOS* **55**, 369.

Wulf-Mathies, C., Blum, P., and Trinks, H.: 1974, Paper, XVII Plenary Meeting of COSPAR.

von Zahn, U., Fricke, K. H., and Trinks, H.: 1973, *J. Geophys. Res.* **78**, 7560.

INCOHERENT SCATTER RADAR RESULTS

M. PETIT

Centre National d'Études des Télécommunications Département RSR, 3 Avenue de la République
92131 Issy-les-Moulineaux, France

1. Incoherent Scatter Sounding

1.1. PRINCIPLES OF THE METHOD

Incoherent scattering is the Thomson scattering of electromagnetic waves of very high frequency (e.g., several hundred MHz) by the free electrons in the ionosphere. In any scattering experiment, the received signal is proportional to a spatial Fourier component of the scatterers density (here the electron density); the relevant wave vector **k** is mainly determined by the frequency and, through a trigonometric factor, by the geometry of the experiment. Incoherent scattering occurs even when the medium is macroscopically homogeneous; this phenomenon is due to the randomness of the position of each electron, which results from the thermal motion. In the absence of Coulomb · interactions, between electrons, each particle would be independent from the others and the total received power would be the sum of the power scattered by each particle and the power spectrum would be Gaussian in shape, reflecting the electron velocity distribution. This truly incoherent scattering actually occurs when the wave used has a wavelength λ much smaller than the Debye length λ_D of the plasma. This condition does not apply to the present experiments and one should speak rather of quasi-incoherent scattering.

When the opposite condition applies $(\lambda \gg \lambda_D)$, the electrons cannot be displaced without carrying along the ions and the power spectrum is narrowed to the characteristics of the ion gas even though the electrons are the only really effective elements in the scattering phenomenon. In particular, the size of the central part of the spectrum roughly corresponds to a Doppler effect tied in with the thermal velocity of the ions rather than the electrons. In other words, the scattered energy is concentrated in a much narrower band of frequency and the energy density per Hertz is therefore greater, which facilitates experimental study by improving the signal-to-noise ratio. This is one of the reasons why all ionospheric studies have been carried out with $\lambda \gg \lambda_D$. In addition, the Debye length in the ionosphere is less than 1 cm at altitudes less than 1000 km, while very high power radars have been built in longer wavelengths.

The power spectrum of the scattered wave, in addition to the ionic spectrum which occupies the immediate neighborhood of the incident frequency, also includes two lines called electronic, situated on either side of the incident frequency at a separation equal to the plasma frequency. These lines are greatly enhanced by the presence of a small amount of suprathermal photoelectrons.

B. M. McCormac (ed.), Atmospheres of Earth and the Planets, 159–175. All Rights Reserved.
Copyright © 1975 by D. Reidel Publishing Company, Dordrecht-Holland.

1.2. RAW PARAMETERS DEDUCED FROM INCOHERENT SCATTERING

Although interesting studies have been performed from the 'plasma lines', the bulk of data has been gathered from the 'ionic spectrum'.

The expression for the power spectrum of the received wave is function of a certain number of parameters: plasma density, temperatures, etc. Starting from an experimentally obtained spectrum, we can find 'these parameters by adjusting their values in such a way that the theoretical spectrum is fitted, in the sense of 'least squares', to the experimental spectrum (Petit, 1968).

The errors of measurement are essentially of a statistical type, so that they would tend to zero if one could allow the time T during which the received signal is integrated before recording its power spectrum to approach infinity. The residual fluctuations are in most cases due to receiver noise, very often more powerful than the received signal, which always remains small because of the small Thomson scatter cross section. Under such conditions, the error in the measured parameters varies as the noise to signal ratio. This implies that the D region and nighttime E region are poorly explored by this technique.

1.2.1. *Electron Density*

To a first approximation, the electron density enters the expression for the spectrum only as a factor of proportionality, upon which, indeed, the total received power depends. However, certain technical characteristics of the experiment, such as the emitted power and antenna gain, also have a direct influence on the power received. Therefore, the error in measurement of the electron density is not mainly statistical, but comes from imprecision in the determination of these technical characteristics, some of which can change with time. Since these variations are especially important over long periods, it is often useful to normalize the observed vertical profile, that is to multiply the values obtained at all altitudes by the same factor, in such a way that the maximum density is equal to that which can easily be deduced from a classical ionosonde. A relative precision of the order of 1% can thus be obtained (although the error in absolute value can reach 10%), depending on the care with which the power calibration is carried out.

1.2.2. *Bulk Motion of the Ions*

If the plasma has a bulk motion, it is evident that the Doppler effect shifts the entire spectrum in frequency. It has been shown that this effect corresponds to the bulk motion of the ions rather than that of the electrons if they are different. The precision that can be obtained is of the order of several meters per second. Here, the only cause of error is normally statistical, the problem of frequency stability having been resolved in a satisfactory way by the techniques currently in use. Of course, with only one emitter and receiver only one component of the vector mean velocity of the ions (in the direction of the vector $\mathbf{k}$) can be measured at one ionospheric location, and three receiving stations must be used in order to reconstruct the vector in space. Assuming that the ion velocity is the same over large horizontal scales, this difficulty may be overcome by tilting the beam to get **k** vectors of different orientation.

1.2.3. *Electron Temperature, Ion Temperature*

When the ionic composition is known, one can easily measure the electron temperature T_e, and the ion temperature T_i. Only statistical errors need be taken into account, and under the best conditions precision of the order of 1% can be achieved.

Actually the hypothesis of known ionic composition applies well below 120 km, where there exist only molecular ions, O_2^+ of mass 32 and NO^+ of mass 30, masses close enough that they need not be distinguished from each other. Similarly, between 250 and 500 km of altitude, it is known that one species, the O^+ ion of mass 16, largely dominates. On the other hand, in the transition region between the molecular ions and the atomic O^+ ion, the ionic composition cannot be considered as known. It is the same at very high altitudes, where the proportion of H^+ becomes important.

When the ionic composition is unknown, it is harder to deduce the electron and ion temperatures from a given spectrum. In fact, it becomes very difficult to distinguish the effect of this new parameter from that of the temperatures. The errors are greatly increased, to the point where they sometimes destroy the significance of the measurement. When this happens, it is impossible to deduce N, T_e, T_i and p from the observed spectrum if an outside source of information about one of these parameters is not available. Nevertheless, we should not conclude that there are no circumstances under which we can extract all of these parameters from a single observed power spectrum. All the errors can be reduced by increasing the signal to noise ratio and the time of integration. However, experimental conditions must be particularly favorable if N, T_e, T_i and a parameter of ionic composition are to be extracted altogether from a unique incoherent scattering spectrum.

1.2.4. *Ionic Composition*

We have already presented the essentials of the problem. However, measurements of the ionic composition of the region of transition between molecular and atomic ions (120 to 220 km) have been obtained, using assumptions, for example about the ion temperature. In fact, as we shall see below, the ion temperature in this region is practically equal to that of the neutral particles. One can use a model for the neutral atmosphere in order to interpolate, without large error, between the two regions, below and above, where T_i is measured without problems, the ionic composition being known. This indirect procedure has allowed us to obtain very interesting results about the variation of composition in this transition region.

1.2.5. *Density of the Neutral Atmosphere*

When collisions between ions and neutral particles are sufficiently numerous, they influence the form of the 'ionic' part of the spectrum of the scattered waves. Thus ion-neutral collision frequency can be measured directly by incoherent scattering at an altitude of about 100 km. This permits study of the variations of the density of N_2 molecules which are the main constituents of the atmosphere at this altitude.

These raw parameters are routinely deduced from the observations in each incoherent scatter observatory. Some of these data are deposited in World Data Centers and all the data can be obtained on request from the observatories. The simultaneous availability of several accurately measured ionospheric parameters permits one to solve equations describing basic physical laws and to deduce very interesting features on structure and composition of the atmosphere. Although the ionized species are mainly involved in the incoherent scattering process, properties of the neutral atmosphere can be determined as well. We shall examine in turn a few typical studies which have been recently performed along these lines.

2. Solar EUV Fluxes and Incoherent Scatter Data

The solar EUV flux is responsible for the ionization process which creates the ionosphere. Therefore, observed electron density profiles depend directly upon this solar flux. The same holds for the electron temperature: the basic heat input to the electron gas comes from the kinetic energy of the photoelectrons. Therefore, both the observed electron densities and temperatures can be interpreted in terms of solar EUV fluxes.

2.1. ELECTRON DENSITY INTERPRETATION

In the lower F region production and recombination determine the electron density value N_e, the transport term being usually negligible (Walker and McElroy, 1966). All the calculations of production and recombination rates require a model of neutral densities. Diffusive equilibrium is generally assumed with the exponential temperature profile of Bates (1959) which depends upon three parameters. A fourth parameter permits one to determine the density profile of any neutral constituents i.e., the value of the density at some boundary level (usually 120 km). The recombination rate of electrons depends upon the chemical nature of the ions. To overcome the lack of measurements of the ion composition Swartz and Nisbet (1973) assume photochemical equilibrium and express the recombination rate in terms of the ion production ratios instead of the absolute values of the individual ion production rates. Scialom (1974) uses the ion composition results of Giraud et al. (1971) based on simultaneous use of incoherent scatter and rocket-borne mass spectrometry techniques. When comparing the production rates based on Hinteregger's (1970) solar fluxes with the recombination rates deduced from the density measurements at Arecibo, Swartz and Nisbet (1973) find that the agreement is good (Figure 1), provided the solar EUV has been increased by a factor of 2. Scialom (1974) using Saint-Santin data, reached a similar conclusion, the scaling factor being of 3.

2.2. THERMAL BALANCE OF ELECTRONS

In the daytime F region, the electron temperature T_e exceeds the ion and neutral particles temperature as a result of photoelectron heating. Thermal balance of the

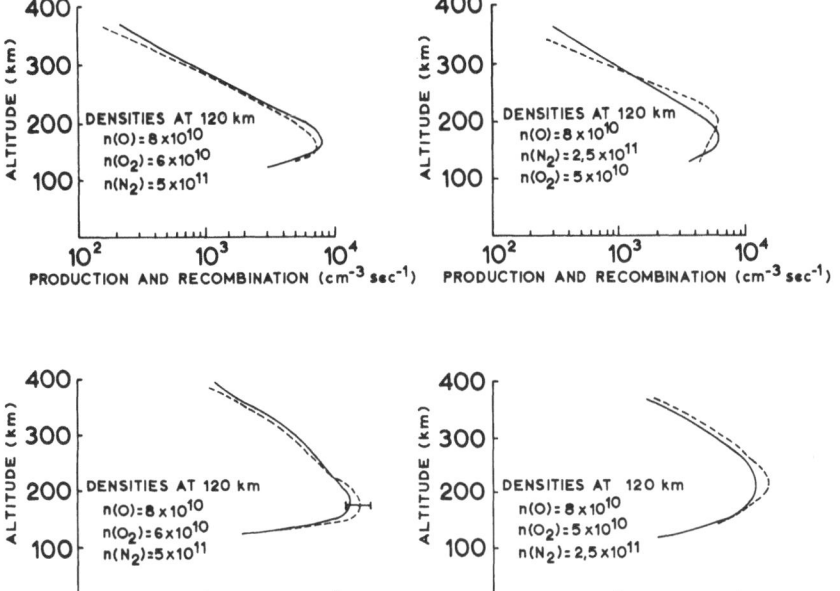

Fig. 1. Comparisons of the production and recombination rates (top figures) and of the heating and cooling rates (bottom figures). Left hand figures concern Arecibo, June 26, 1968, at 12.25 and right hand figures Arecibo, December 24, 1968, at 12.17.

electrons is achieved by transfering energy to the ions and to the neutral particles. If the cross section relative to the rate for transferring energy to the ions is well established on plasma physics theoretical grounds, the cross sections relative to transfer to the neutral particles are more questionable. Their determination has been progressively improved; a major step in this direction has been taken by Dalgarno and Degges (1968), who stressed that the most efficient loss process had been neglected, namely the fine structure excitation of the 3P levels of O. In the incoherent scatter experiment, electron density, electron temperature and ion temperature are measured. It has been shown that the ion temperature does not differ insignificantly from the neutral temperature in the region where cooling to the neutral particles is important. If a model is adopted for the neutral particles density (see Section 2.1 above), the cooling rates implied by the experimental data can be estimated. To satisfy thermal balance of the electrons, these cooling rates must be equal to the heating rates which can be estimated from the kinetic energy of photoelectrons. The main difficulties in this estimation are to properly consider photoelectron transport of energy away from the region in which the electrons originated and to account for the quantified discrete steps of energy loss via inelastic collisions with neutral particles. Various techniques have been developed for this study and lead to consistent results.

Cooling and heating rates are compared on Figure 1 taken from Swartz and Nisbet (1973). To obtain a satisfactory fit, the solar EUV flux has been again multiplied by a factor of 2. To explain both the electron density profiles and the electron temperature

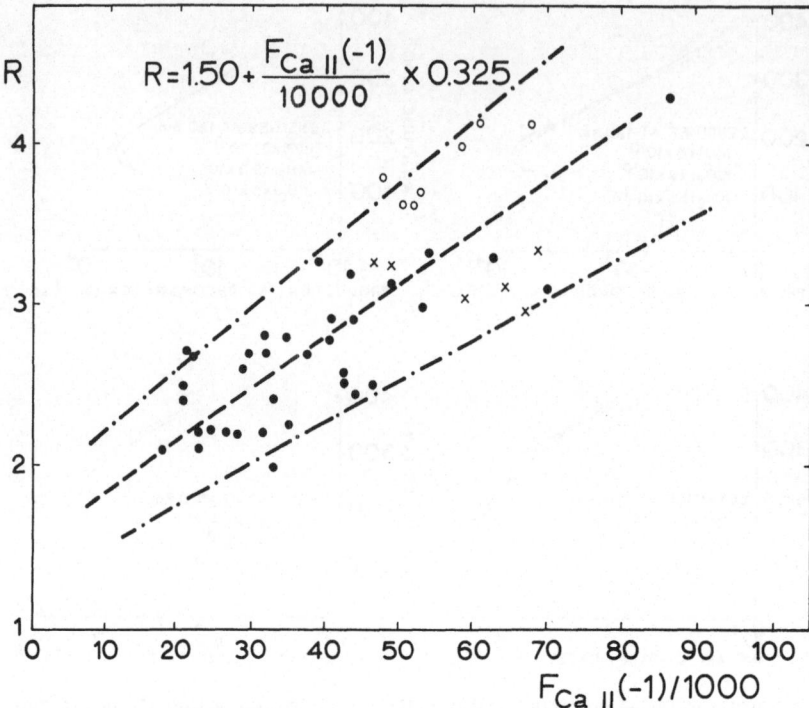

Fig. 2.   Scaling factor R as a function of the Ca index of the preceding day. A seasonally varying O has been taken into account ($\bullet$ equinox relative concentration $r_0 = 1$; $\circ$ winter $r_0 = 1.4$; $\times$ summer $r_0 = 0.7$).

profiles measured simultaneously at Arecibo, the scaling factor of 2 for the solar flux is necessary. A similar conclusion has been derived by Lejeune (1973) from the electron temperatures measured at Saint-Santin. Figure 2 shows scaling factor as a function of the Ca index of the preceding day. A quite good correlation is clearly seen with a minimum value of R of about 2.

2.3. CONCLUSION

Electron density and temperature measured by incoherent scattering suggest that the solar flux in the ionizing part of the spectrum (300 to 1000 Å) is currently under-estimated by a factor of about 2 to 3. Extrapolating this result to larger wavelengths would certainly create problems with the thermal balance of the neutral atmosphere. In this respect, the question remains open.

3. Long Term Variations of Thermospheric Temperature and Composition

3.1. ATOMIC OXYGEN CONCENTRATION AND TEMPERATURE ABOVE 200 KM

Solving the heat balance equation for the ions in the $F2$ region allows for a determination of the neutral temperature profile and the oxygen concentration at an arbitrary level, under the assumption of diffusive equilibrium (Bauer *et al.*, 1970). Energy is transferred via Coulomb collision from the electrons to the ions. This heat input to the ions is balanced by a cooling to O which is the major neutral constituent. At low altitude when the neutral density is high enough, this balance is achieved by a small undetectable increase in T_i above the neutral temperature T_n. At higher altitude $T_i - T_n$ becomes larger and measurable from 300 km up. In this range T_n is not altitude dependent and the observed variations of T_i vs height, in addition to T_e and N_e (which permits to compute the heating rate of the ions) permit to determine the O concentration, with a limited accuracy of above 15%. This method is obviously restricted to daytime, when T_e exceeds T_i. On the contrary, T_n is available in day and nighttime with an accuracy of about 2%.

3.2. Molecular concentration at 200 km

Up to 200 km the incoherent scatter experiments yield a determination of the ratio p of the molecular ion concentration. Cox and Evans (1970) developed a procedure based on photochemical equilibrium to deduce the ratio $(O)/(\gamma M_2)$ of the O concentration to the weighed sum of the N_2 and O_2 concentrations.

3.3. Results about long term variations

In addition to the quantities which can be deduced from the data through the procedure described in Section 3.1 and 3.2, neutral density around 100 km is directly deduced from the power spectrum of the scattered signal (see Section 1.2.5). All these data can be combined to give an overall picture of the thermosphere behavior.

3.3.1. Molecular Nitrogen

Salah et al. (1974) at Millstone Hill found a small annual oscillation of about 13% of the mean with a winter maximum (November) for the altitude region 110 to 120 km. At Saint-Santin Alcaydé et al. (1974) also found a weak annual effect with winter (January) N_2 number densities about 20% higher than the equinoxes and summer densities. When the annual variation of the observed temperature profile is taken into account to compute the density variation at 200 km under the assumption of diffusive equilibrium, one obtains the clear variation shown on Figure 3.

3.3.2. Atomic Oxygen

The O concentration (Alcaydé et al., 1974) exhibits a small winter to summer variation at 200 km with winter maximum. Considering again the observed temperature profile, the result leads to a much larger variation at 120 km with a winter to summer ratio of 1.5, consistent both with earlier experimental results (Waldteufel, 1970; Perret, 1971) and with theoretical models including circulation (Johnson and Gottlieb, 1973).

3.3.3. Molecular Oxygen

The method presented under Section 3.2 permitted Alcaydé et al. (1974) to obtain a very clear seasonal effect from the Saint-Santin data (Figure 4) on the ratio $(O)/(\gamma M_2)$ of O to molecular constituents. This result is in qualitative agreement with Cox and

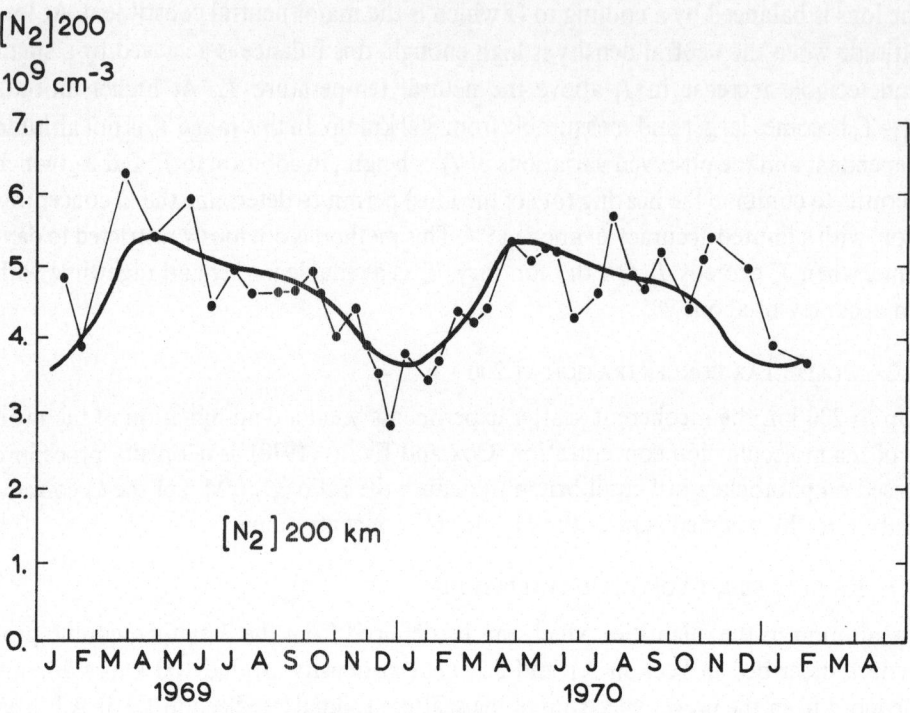

Fig. 3.   Molecular N number density at 200 km over Saint-Santin. The continuous line is a running average of the data obtained over 1969 and 1970.

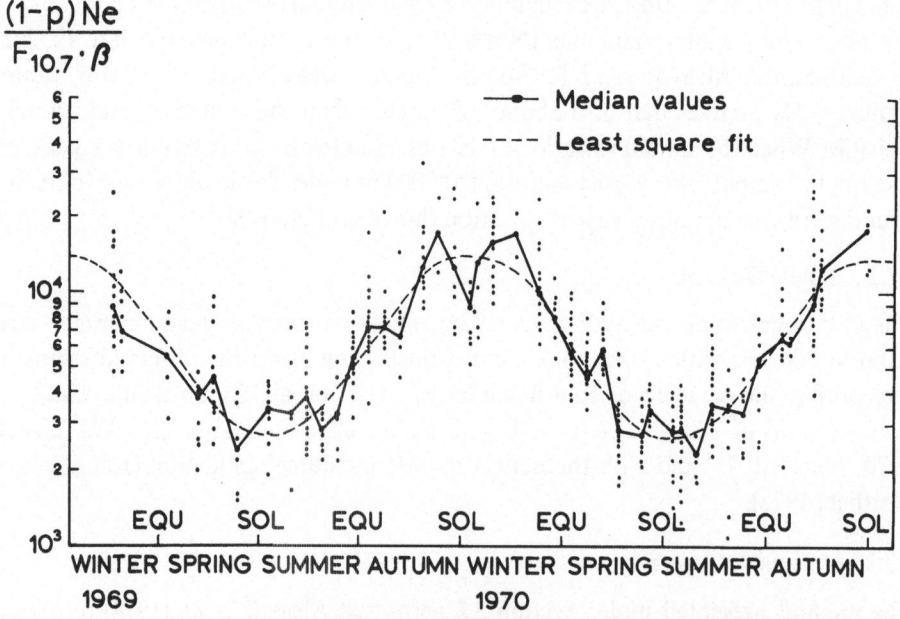

Fig. 4.   Variation of the $(O)/(\gamma M_2)$ ratio in 1969 and 1970. The dashed line corresponds to the least square fit to an annual component.

Evans (1970). The variations of (O) and (N_2) are not large enough to reproduce this $(O)/(\gamma M_2)$ seasonal trend. This suggests the existence of a large change in the O_2 concentration at 200 km, with a summer to winter ratio of about 6.

The results confirm the theoretical predictions of Rishbeth and Setty (1961) that neutral compositional changes at constant pressure level are responsible for the winter F region anomaly and suggest that the dominant role is played by (O_2).

4. Thermospheric Tides

Neutral temperature does not differ very much from the ion temperature and is in any case (see Section 3.1) easily deduced from T_i measurements. In the lower thermosphere, the measured parallel to the magnetic field B ion drift is practically equal to the projection of the meridional neutral wind along B. Above 225 km, ambipolar diffusion velocity becomes significant and the measured ion drifts must be corrected from the diffusion effect before being interpreted in terms of neutral winds. Neutral temperatures and winds measured in this way show the presence of large tidal oscillations throughout the thermosphere.

4.1. E REGION

Measurements on individual days clearly show an oscillation with a period of 12 h. When the data are averaged over many periods of observation, statistical errors and fluctuations caused by gravity waves with shorter periods are reduced and one gets a picture of the type obtained by Salah and Wand (1974) for the Millstone Hill data on Figure 5. Saint-Santin data exhibit a similar behavior. The altitude behavior of the observed semi-diurnal amplitude and phase is in good agreement for the two midlatitude stations, as shown on the harmonic dial from Salah *et al.* (1974) in Figure 6. Distance from the origin gives the percentage amplitude while the polar angle gives the time of maximum. Bernard (1974) analyzed simultaneously temperature and wind data over Saint-Santin gathered over 3 years. His conclusion, similar to Salah *et al.*'s (1974), is that the mode S_2^4 is dominant in the E region, providing a good match to the observations, not only for wavelengths, but also for the maximum amplitude of the wind and temperature variations, their phase and amplitude relationship.

4.2. F_2 REGION

The dominant tidal component in the upper thermosphere is the diurnal one, although the semi-diurnal one is still present as shown on Figures 7 and 8 (Alcaydé, 1974; Amayenc, 1974). These authors analyzed temperature and winds at Saint-Santin in terms of a mean value, a diurnal and semi-diurnal component. The phase of the diurnal component is constant throughout the year both for winds and temperature. Salah and Evans (1973) drew the same conclusion from the Millstone Hill temperature data (the hour of the maximum is 14.00 at Millstone Hill and 15.00 at Saint-Santin). The semi-diurnal component of temperature is observed to vary widely with seasons. The time variations of the semi-diurnal component are different for temperature and wind:

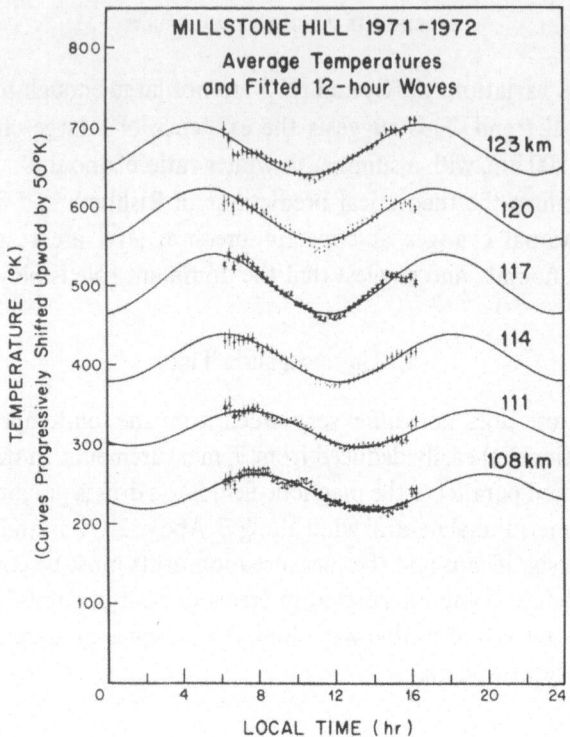

Fig. 5. Average temperatures measured at (a) Millstone Hill interpolated at 15 min intervals and fitted with 12 h sinusoids. For clarity, each curve above the lowest has been progressively shifted upward by 50 K.

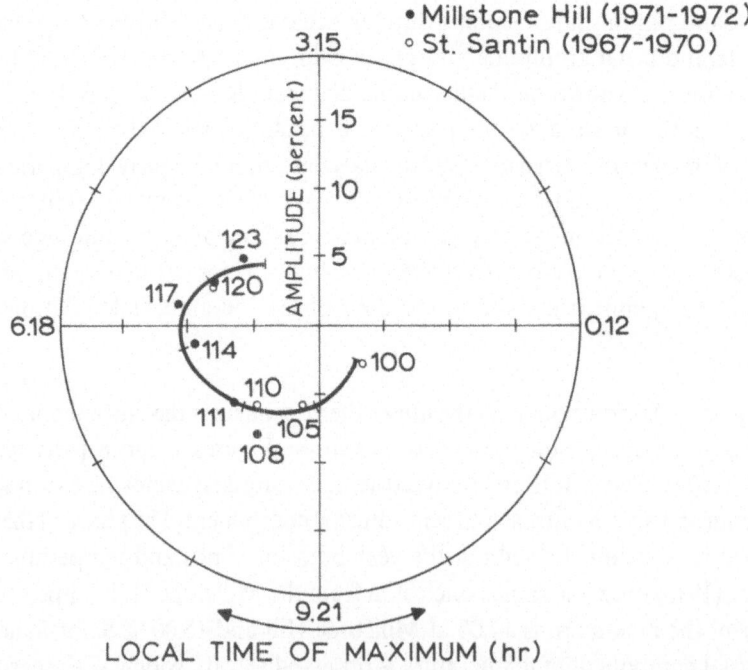

Fig. 6. Harmonic dial showing amplitudes and phases of the semi-diurnal waves fitted to the average temperatures. A smooth curve with tick marks at 5 km intervals from 100 to 125 km is also given.

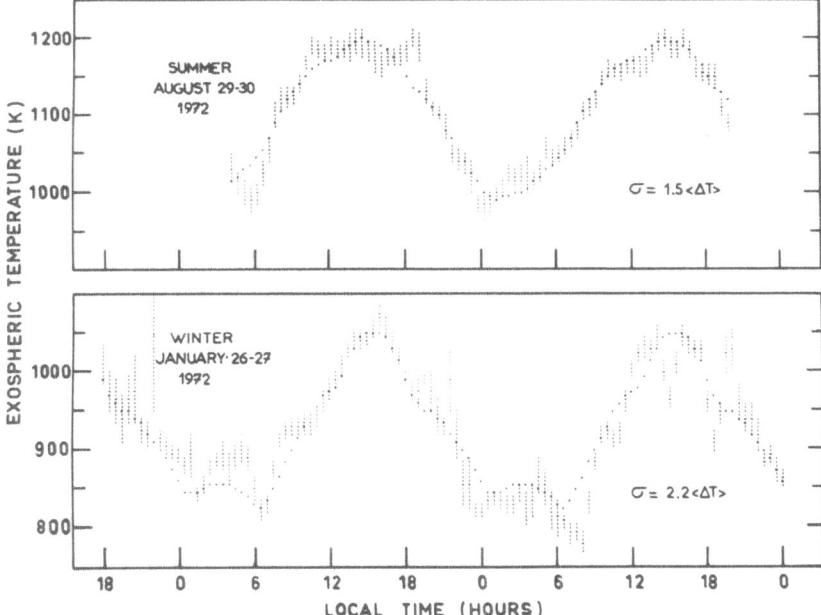

Fig. 7. Daily variation of the exospheric temperature for a typical summer case (top) and a typical winter case (bottom). σ is the standard deviation expressed as a function of the mean error bar (ΔT).

while the wind results indicate a clear dominating semi-annual variation around a mean value the temperature results seem more scattered and, in any case, suggest an annual variation rather than a semi-annual one. A terdiurnal oscillation is clearly present for the winds, while it is insignificant for the temperature.

5. Middle and Low Latitude Effects of Auroral Disturbances

5.1. DISTURBANCES IN THE NEUTRAL ATMOSPHERE

5.1.1. Atmospheric Waves

Many observations using ionosonde network, radio waves backscattering and electron content measurements, have shown that during magnetic storms large disturbances in electron density travel with sonic velocity from the pole towards the equator. Several theoreticians suspected that these phenomena are in fact the manifestation of the passage of large scale gravity waves generated in the auroral region. This view has been confirmed by incoherent scatter data on ion velocity and temperature which can be easily interpreted (see above) in terms of neutral velocity and temperature. Figure 9 shows an example of large scale gravity waves observed at Saint-Santin (Testud, 1973). Similar observations have been performed at Arecibo (Harper, 1971). Testud (1973) developed a calculation, including dissipative processes to determine first the free wave spectrum launched by the source and then the propagation from the auroral

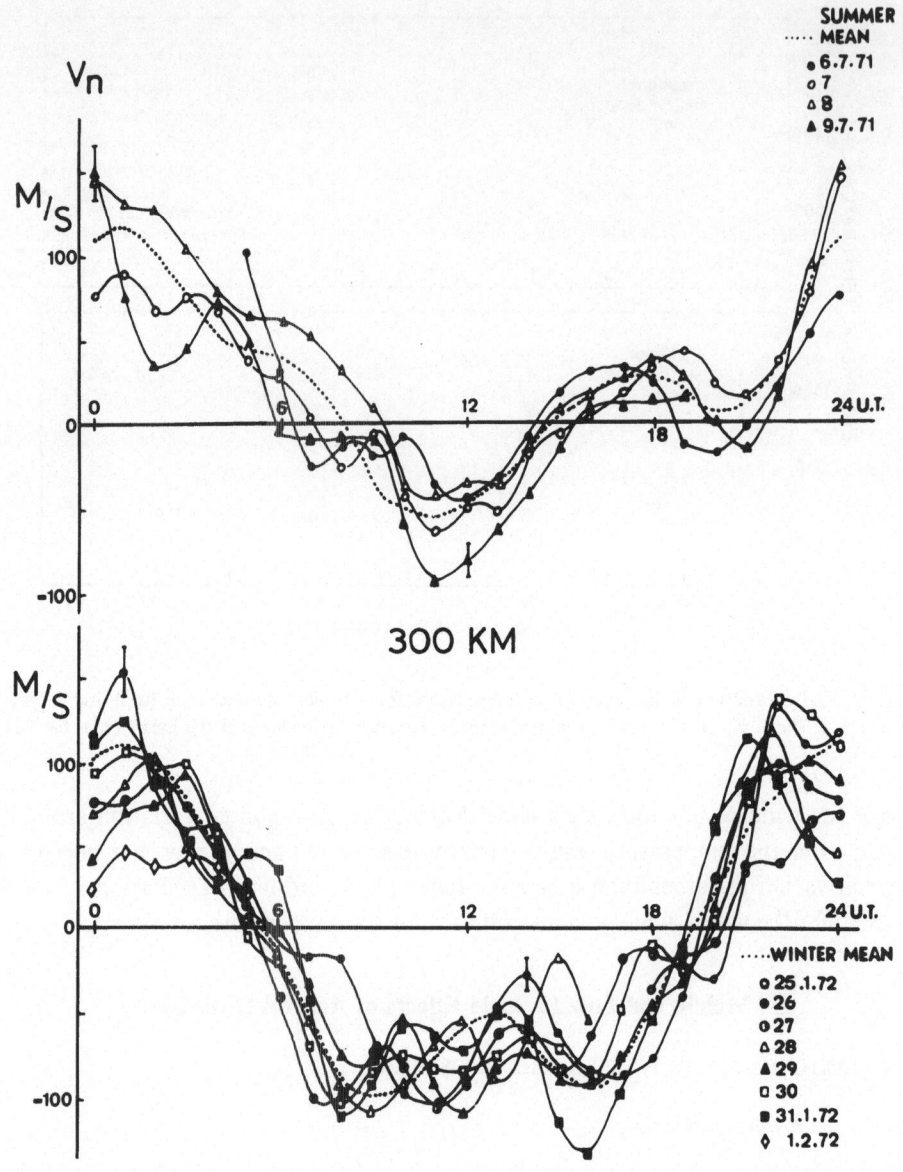

Fig. 8.   Meridional neutral wind at 300 km altitude, deduced from St-Santin-Nançay observations during two long periods of measurements: July 6 to 9, 1971, and January 25 to February 1, 1972. The seasonal averages (for the years 1971–1972) are indicated by dotted lines. Apart from some short-period oscillations, the variability from day to day is not large.

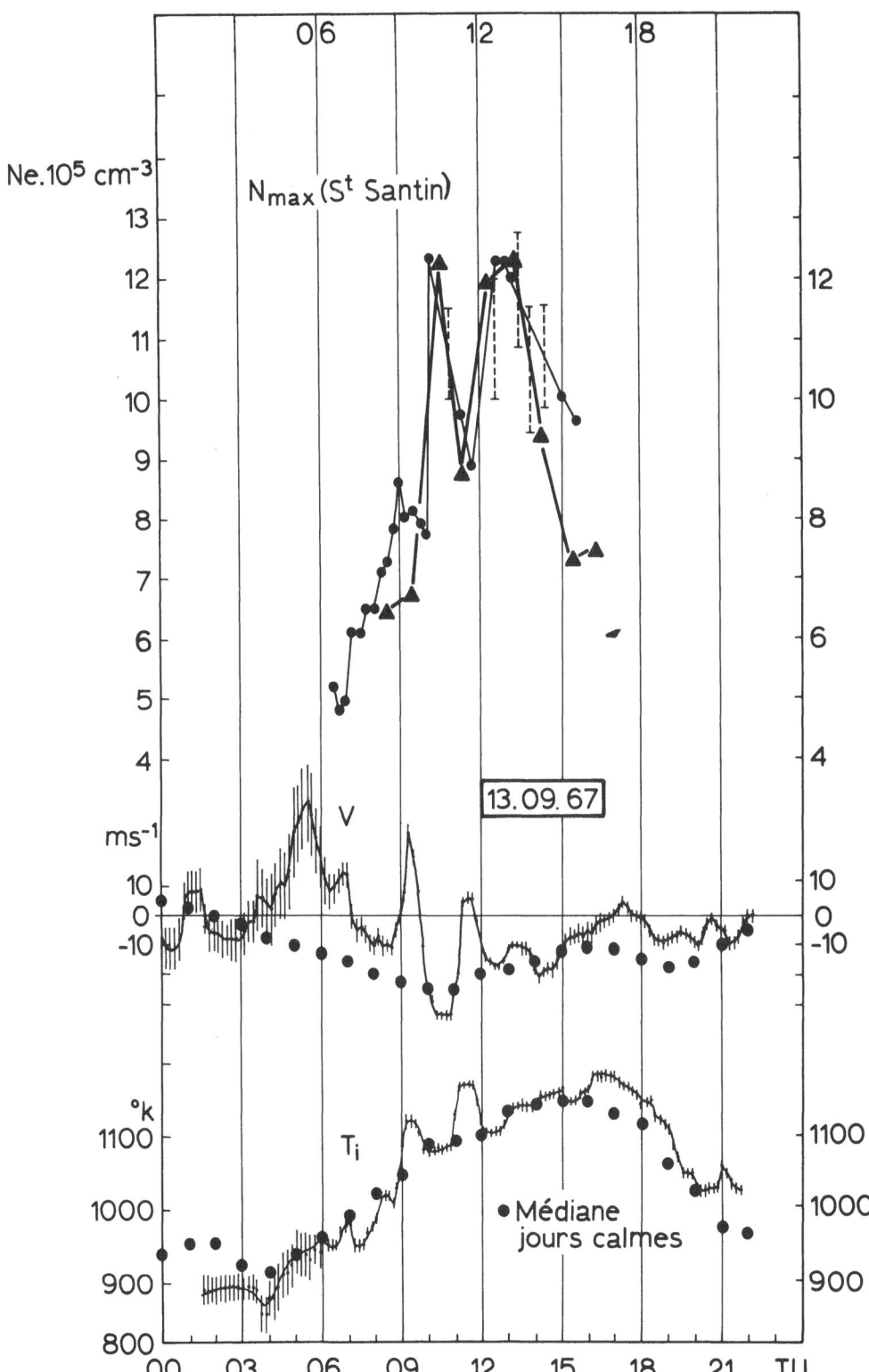

Fig. 9. Ion temperature T_i (averaged over 40 mn and between 275 and 350 km altitude), ion velocity $V = V_i \sin I$ (averaged over 40 mn and between 200 and 400 km altitude) and peak electron density (triangles correspond to incoherent scatter data, circles t_0 ionosonde data). Black circles superimposed on curves for temperature and velocity are the hourly median values of 6 quiet equinox days.

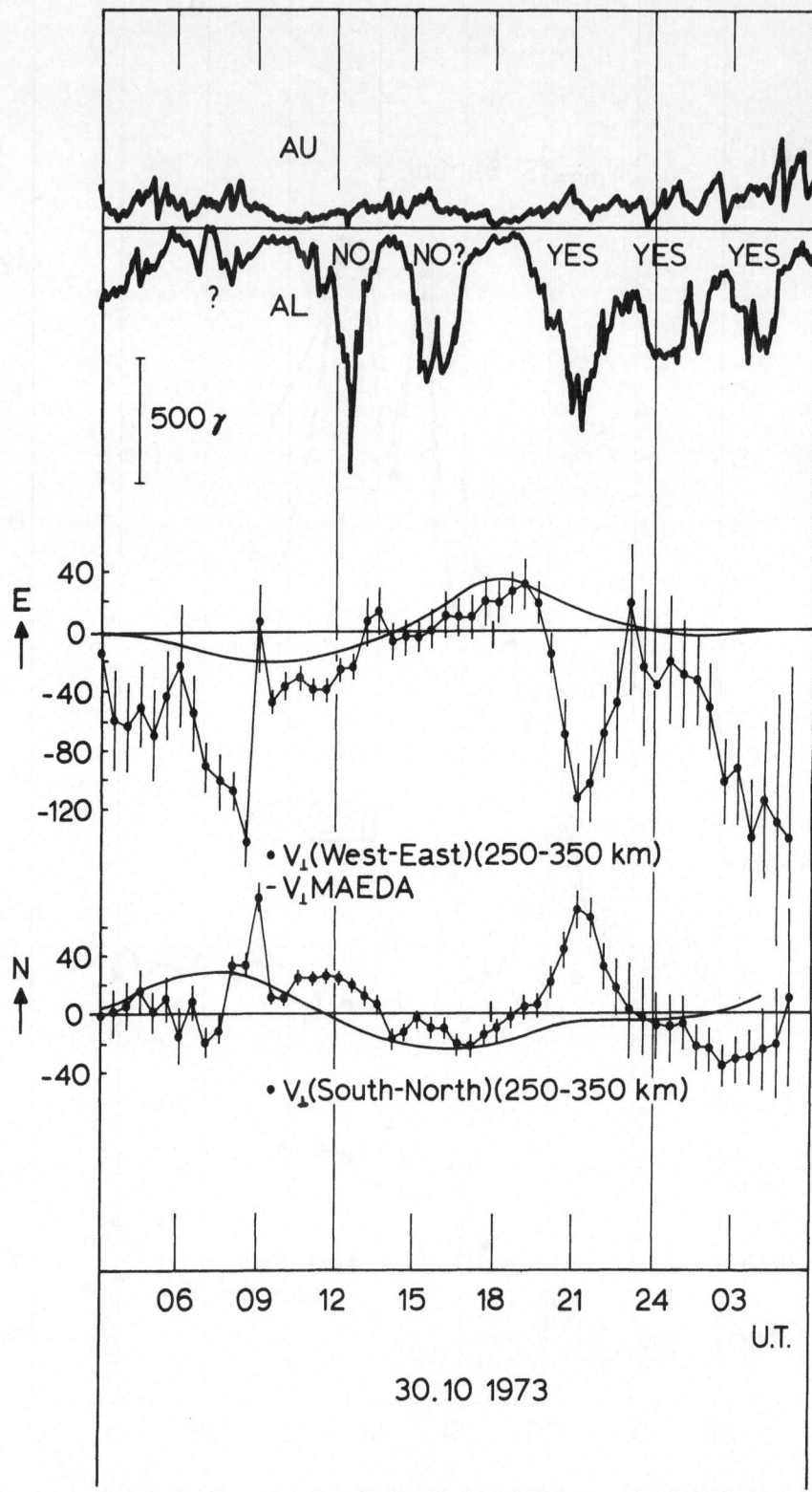

Fig. 10. Comparison between the perpendicular velocities data and auroral indices.

source region to the point of observation. According to this theory, the response of the atmosphere does not appear as a wave train but as a single pulse. This might seem to be in contradiction with the observations of Figure 9. However auroral magnetograms reveal that two events took place in the same longitudinal sector as Saint-Santin at 07.40 and 10.05 h. These events are linked with magnetic reversals of the interplanetary magnetic field observed by Explorer 33 and appear as a manifestation of a substorm growth phase. So the oscillations of Figure 9 can be interpreted as a succession of two positive impulses starting, 70 and 65 min respectively, after the two auroral events, a propagation time in good agreement with the theoretical value (75 min).

Larger period atmospheric waves have also been observed and could be the result of excitation of global modes of oscillation (Volland and Mayr, 1972).

5.1.2. *Global Modification of the Neutral Atmosphere*

A good statistical correlation between magnetic activity increase and exospheric temperature increase has been evidenced (Waldteufel, 1970; Alcaydé *et al.*, 1974; Salah and Evans, 1973; Waldteufel and Cogger, 1971).

Modifications of ionization transports in F region observed at Millstone Hill (Evans, 1973) can be partly interpreted by enhancement of equatorward neutral winds. The most significant circulation alteration of neutral gas has been pointed out by Reddy (1971) in the E and F_1 regions at Saint-Santin. The existence of a meridional circulation cell during the daytime is observed during a highly magnetically disturbed winter day. In this cell equatorward winds blow above 120 km and poleward winds below 120 km. This supplementary cell is superimposed to the quiet time circulation.

The OGO-6 mass spectrometer measurements have shown large alterations of the ratio $(O)/(N_2)$. Although incoherent scatter permits it, it does not seem that any study of the storm-time modification of this ratio has been attempted.

5.2. Disturbance in the Electric Field

5.2.1. *Penetration at Mid-Latitude of the Substorm Electric Field*

On some occasions at Millstone Hill (Evans, 1972, 1973) and at Saint-Santin (Testud *et al.*, 1974), the measured electric field components are strongly and suddenly increased up to amplitudes much greater than the usual values. The duration of this perturbation and its time of occurrence are the same as those of a particular polar magnetic substorm. This feature is illustrated on Figure 10 which shows a strong enhancement up to 120 m s^{-1} of the westward component of the perpendicular ion drift measured at Saint-Santin. A similar enhancement up to 80 m s^{-1} affects the northward component. The electron density exhibits a specific variation quite different from the one observed under the influence of a gravity wave: the F layer is lifted up by 100 km, without change in the maximum value of density.

A theoretical and quantitative interpretation of these observations, taking into account the time variation of the substorm electric field, is still needed.

5.2.2. *Motion of the Mid-Latitude Trough*

The location of the presently existing incoherent scatter facilities does not permit frequent observations of overhead passage of the mid-latitude trough. However, on November 7, 1970, Taylor (1973) observed a trough passage at Malvern ($L=2.6$). At 01.32 LT, the profile is normal with a maximum density of 2×10^5 cm^{-3} around 350 km altitude. From 02.17 to 03.08, the density decreases by a factor 4. Then from 03.20 to 04.16, a replenishment occurs, starting at high altitudes and propagating downward in such a way that the peak of the layer presents an apparent downward motion. Taylor's interpretation is as follows: the apparent inward motion of the density maxima is related to an actual horizontal motion of the oblique F region plasma tubes, along which diffusive equilibrium prevails.

More observations of this type would certainly help our understanding of the mid-latitude trough and its relationship with the plasmapause.

5.3. EQUATORIAL EFFECT OF MAGNETIC STORMS

Woodman *et al.* (1972) observed a great storm in March 1970 at the Jicamarca incoherent scatter observatory (2° dip latitude north). Severe perturbations in the vertical ionic drift appeared in the F region. They were associated with severe changes in the electron concentration and temperatures; they are highly correlated with the variation of the H component of the Earth's magnetic field. The behavior of the electron concentration and temperature is a consequence of the vertical drift perturbation, while the H variation can be interpreted as a mere effect of the electric field disturbance. An important question remains open: which mechanism creates this electric field disturbance?

References

Alcaydé, D.: 1974, *Radio Sci.* **9**, 239.
Alcaydé, D., Bauer, P., and Fontanari, J.: 1974, *J. Geophys. Res.* **79**, 629.
Amayenc, P.: 1974, *Radio Sci.* **9**, 281.
Bates, D. R.: 1959, *Proc. Roy. Soc.* **A253**, 451.
Bauer, P., Waldteufel, P., and Alcaydé, D.: 1970, *J. Geophys. Res.* **75**, 4825.
Bernard, R.: 1974, *J. Atmospheric Terrestr. Phys.* **36**, 1105.
Cox, L. P. and Evans, J. V.: 1970, *J. Geophys. Res.* **75**, 6271.
Dalgarno, A. and Degges, T. C.: 1968, *Planetary Space Sci.* **16**, 125.
Evans, J. V.: 1972, *J. Geophys. Res.* **77**, 2341.
Evans, J. V.: 1973, *J. Atmospheric Terrestr. Phys.* **35**, 593.
Giraud, A., Scialom, G., Pokhounkov, A., Poloskov, S., and Toulinov, G.: 1971, *Space Res.* **11**, 1057, Akademie-Verlag, Berlin.
Harper, R. M.: 1971, Ph.D. Thesis, Rice University, Houston.
Hinteregger, H. E.: 1970, *Ann. Geophys.* **26**, 547.
Johnson, F. S. and Gottlieb, B.: 1973, *Planetary Space Sci.* **21**, 1001.
Lejeune, G.: 1973, Ph.D. Thesis, University of Paris, Paris.
Perret, D.: 1971, M.S. Thesis, University of Paris, Paris.
Petit, M.: 1968, *Ann. Geophys.* **24**, 1.
Reddy, C. A.: 1971, private communication.
Rishbeth, H. and Setty, C. S. G. K.: 1961, *J. Atmospheric Terrestr. Phys.* **20**, 263.
Salah, J. E. and Evans, J. V.: 1973, *Space Res.* **13**, 267, Akademie-Verlag, Berlin.

Salah, J. E. and Wand, R. H.: 1974, 'Tides in the Temperature of the Lower Thermosphere at Mid-Latitudes, to be published.

Salah, J. E., Evans, J. V., and Wand, R. H.: 1974, *Radio Sci.* **9**, 231.

Scialom, G.: 1974, *Radio Sci.* **9**, 253.

Swartz, W. E. and Nisbet, J. S.: 1973, *J. Geophys. Res.* **78**, 5620.

Taylor, G. N.: 1973, *J. Atmospheric Terrestr. Phys.* **35**, 647.

Testud, J.: 1973, Ph.D. Thesis, University of Paris, Paris.

Testud, J., Amayenc, P., and Blanc, M.: 1974, *J. Atmospheric Terrestr. Phys.*, to be published.

Volland, H. and Mayr, H.: 1972, *J. Atmospheric Terrestr. Phys.* **34**, 1745.

Waldteufel, P.: 1970, Ph.D. Thesis, University of Paris, Paris.

Waldteufel, P. and Cogger, L.: 1971, *J. Geophys. Res.* **76**, 5312.

Walker, J. C. G. and McElroy, M. B.: 1966, *J. Geophys. Res.* **71**, 3779.

Woodman, R. F., Sterling, D. L., and Hanson, W. B.: 1972, *Radio Sci.* **7**, 739.

PART IV

LABORATORY MEASUREMENTS OF
RELEVANT RATE COEFFICIENTS

LABORATORY MEASUREMENT OF RATE COEFFICIENTS: INTRODUCTORY REVIEW

R. P. WAYNE

Physical Chemistry Laboratory, South Parks Road, Oxford, OX1 3QZ, England

1. Introduction

This review lays the foundations for the experiments that later papers will discuss. I shall emphasise certain features of elementary kinetics which seem open to doubt or where assumptions are frequently made without circumspect consideration of their validity. I shall then go on to discuss the laboratory methods available for the kinetic study of reactions of atmospheric significance.

2. Reaction Kinetics

2.1. DESCRIPTIVE MATTERS

A generalised reaction

$$aA + bB + \cdots \rightarrow \text{products} \tag{1}$$

is found to proceed with a rate proportional to $[A]^\alpha [B]^\beta \ldots$. The sum of the exponents, $\alpha + \beta + \cdots$, is the overall *order* of the reaction, and any individual exponent, $\alpha, \beta, \ldots$, is the order in that particular reactant to which it applies. The *molecularity* of the process is the sum $a + b \ldots$ as expressed by the chemical equation. If the reaction is an elementary process (i.e., occurs in a single chemical step) then the molecularity and order may be assumed numerically equal; if the reaction takes place *via* several steps, then the two quantities may well not be the same, and, in particular, the order may no longer be integral.

The constant of proportionality between rate of reaction and concentration terms is the *rate constant* or *rate coefficient*. In order to define the rate constant, it is necessary to know how the rate itself is defined. This is not so trivial as it seems, since if one or more of the molecularity factors $a, b, \ldots$ is not unity, the rate of loss differs for the various reactants. The mathematically rational procedure is to define a rate by the equation

$$\text{Rate} = -\frac{1}{a}\frac{d[A]}{dt} = -\frac{1}{b}\frac{d[B]}{dt} \cdots = k[A]^\alpha [B]^\beta \ldots \tag{2}$$

Unfortunately, such a definition of rate may be chemically or physically unsatisfactory. Thus in a reaction of the type

$$A + A \rightarrow A_2 \tag{3}$$

B. M. McCormac (ed.), Atmospheres of Earth and the Planets, 179–195. All Rights Reserved.

it is probably more sensible to define the rate constant by the equation

$$-\frac{d[A]}{dt}=2\frac{d[A_2]}{dt}=k[A]^2 \tag{4}$$

(see, for example, Wayne, 1969a, p. 195). This point is emphasised not so much because it has any profound theoretical value as because it can lead to obvious numerical errors if the person quoting and the person using a rate constant have opposed ideas about the definition. It pays to be very cautious with rate constants for reactions like (3).

The rate of a chemical reaction usually varies with temperature, and Arrhenius (1889) first suggested that the equation

$$k=A\exp(-E_a/RT) \tag{5}$$

would describe the temperature dependence of any rate constant k. E_a is referred to as the *activation energy* and A as the *pre-exponential factor*. In most cases (but see Section 2.3.3 below) E_a is positive, so that the rate constant (and thus rate) increase with increasing temperature. The theories of kinetics to be presented in Section 2.3 frequently predict forms for the temperature dependence of k different from the Arrhenius expression (5). The reproducibility of experimental data for the gas reactions of interest here is rarely good enough for the critical distinction between one mathematical form and another.

2.2. MULTISTEP PROCESSES

A reaction may proceed *via* two or more discrete chemical steps. Consider the formal two-step process

$$A+B\rightarrow C+X \tag{6}$$
$$X+D\rightarrow E. \tag{7}$$

The species X is a *reaction intermediate*, and the rate of reaction (7) depends, *inter alia*, on the instantaneous concentration of X. In general, therefore, it is necessary to solve the several simultaneous differential equations in order to predict the rates of either the overall process or individual reactions. Analytical solution of these equations is sometimes, but by no means always, possible. Numerical integration procedures may of course be used.

Fortunately, a great simplification may frequently be achieved by adoption of the *stationary state hypothesis*. The concentration of intermediate is assumed to build up with time to a steady-state value and then remain invariant. That is, the rates of production and loss of X become equal. During the stationary phase of the reaction we may write $d[X]/dt=0$. This assumption is always approximate (since $[X]$ only reaches its steady state value at infinite time) but the errors introduced by making it may be small. Whether or not the hypothesis may be used depends qualitatively on two factors. The time taken for $[X]$ to approach its steady-state value must first be small compared with the time to which the calculation applies, and secondly be short enough that other parameters controlling the rate (reactant concentrations, light intensities, temperatures, etc.) are sensibly constant. These conditions taken together mean (especially for the laboratory situation where reactants may be readily consumed) that, other things being equal, the rate of loss of intermediate must be high. That is, the intermediate must be highly reactive, and its concentration consequentially small, for the stationary state hypothesis to be applicable. A *quantitative* test of applicability can (and should) be made by comparing the predictions of stationary and non-stationary kinetic analyses. The information to be obtained from stationary and non-stationary systems is discussed in Section 3.1.

2.3. THEORIES OF KINETICS

2.3.1. *Bimolecular Reactions*

Two species must clearly approach each other closely if they are to react. Furthermore they must bring to their interaction sufficient energy to enable the necessary molecular re-arrangements to occur. Molecular dynamics involves solution of the equations of motion subject to the potential energy (hyper)surface representing the energy as a function of the coordinates describing the position of each of the atoms present. A study of the number of trajectories successful in bringing about reaction can then lead to a prediction of rate of reaction as a function of kinetic (translational) and internal energy of the reactants. If the dynamical calculations allow for quantum effects, they should predict accurately rate constants for any chosen reaction. Unfortunately, quite apart from any difficulty associated with the actual computations, potential surfaces are not at present calculable *ab initio* except in the very simplest of cases. It is therefore usual to employ one of two less rigorous treatments which are nevertheless of some predictive value.

The *Collision Theory* (McC-Lewis, 1918) assumes that the reaction partners are hard spheres, that for reaction to occur the reactants have to collide with an energy exceeding a critical value, E_c, and that the species possess a Maxwell distribution of velocities. The theory predicts a rate constant, k, given by

$$k = Z' \exp(-E_c/RT), \tag{8}$$

where Z' is the collision frequency factor (concentration independent), given itself by

$$Z' = \left(\frac{\sigma_A + \sigma_B}{2}\right)^2 \left(\frac{8\pi kT}{\mu}\right)^{1/2} \tag{9}$$

(σ_A, σ_B are the collision diameters of reactants A and B, and μ is their reduced mass).

Equation (8) is satisfyingly close to the experimental Arrhenius form (5). It is true that Z' contains a $T^{1/2}$ term not present explicitly in Equation (5), but since the major dependence of rate on temperature normally derives from the exponential term, this is, in itself, no particular problem. More serious are the discrepancies between experimentally determined Arrhenius pre-exponential factors, A, and Z' (for ordinary small reactant species, Z' is around 3×10^{-10} cm^3 molec^{-1} s^{-1}). It is necessary, therefore, to modify Equation (8) by inclusion of an experimental 'probability' factor, P, whose value is almost always less than (and often much less than) unity. The failures of the theory arise from the assumptions: in particular, since reactants are *not* hard spheres, some collision orientations may be more favorable to reaction than others, and energy of collision may be dissipated in internal motions rather than utilised in reaction. Further, probability of reaction is not likely to show the zero to unity step-function dependence on both impact parameter ('closeness of approach') and energy of impact assumed by simple collision theory. The theory quite expressly does not allow for a non-Maxwellian distribution of energy or for contribution to the activation energy from excitation in internal (vibrational and rotational, and possibly electronic) modes. Again, charged particles may experience long-range attractions which will cause the rate of collision to greatly *exceed* the gas kinetic value for non-interacting particles.

The somewhat damning list of problems associated with collision theory led Eyring (1935) and Evans and Polanyi (1935) to develop the *Transition State Theory* (also known as *Absolute Rate Theory* – an overconfident name, for reasons that will appear later). This theory considers that the reactants A and B are in 'equilibrium' with an activated complex, or transition state, $AB^\dagger$, lying at the position of maximum (free) energy on the pathway connecting reactants and products. The activated complex is then taken to yield products at a calculable frequency, $v^\dagger$. The equilibrium may be treated by the methods of either classical or statistical thermodynamics. After appropriate manipulation (see, for example, Wayne, 1969a, pp. 220–223) the statistical treatment yields the final expression for the rate constant

$$k = \frac{kT}{h} \frac{Q'_{AB\dagger}}{Q_A Q_B} \exp(-E_c/RT), \tag{10}$$

where the Qs represent total partition functions for A, B and $AB^\dagger$. The significance of the prime in $Q'_{AB\dagger}$ is that one degree of freedom (that leading to product formation) is omitted. Q_A and Q_B are certainly calculable, so given that the molecular parameters of $AB^\dagger$ at the transition state are known and that the correct degree of freedom is factorized out, the pre-exponential factor of the rate constant can be determined.

The need to calculate $Q'_{AB\dagger}$ means that a knowledge is required of the potential energy surface in the region of the transition state. At least a reasonable guess can often be made of $Q'_{AB\dagger}$, so that the transition state theory can predict pre-exponential factors which allow for molecular complexity and internal degrees of freedom. The temperature dependence of the pre-exponential term will now depend on the temperature terms in the various total partition functions taken together with the first power (kT/h) term.

A major cause for concern with the transition state theory centers on the 'equilibrium assumption'. A position of free energy *maximum* is certainly *not* one of thermodynamic equilibrium. However, it may be that the algebraic expressions derived on the basis of equilibrium are still acceptable. Laidler and Polanyi (1964) give a number of arguments in favor of this view for classical reactions. Even so, severe doubts must be entertained about the applicability of an equilibrium hypothesis for very fast reactions (where the more energetic reactants are consumed faster than equilibrium can be regained), or for non-equilibrated reactants. Unfortunately, one or other of these situations often obtains in systems of atmospheric importance.

Both collision theory and transition state theory make use of the 'critical energy' concept, the transition state theory equating the energy E_c with the potential energy of the activated complex. In the absence of the potential energy surface for the reaction, it is impossible to predict this critical energy. Although a number of empirical and semi-empirical rules exist for estimating activation energies, for most practical purposes, it is essential to measure this quantity. Naturally, if the room temperature rate constant for a *simple* reaction approaches the gas kinetic frequency factor, Z', then E_c is likely to be near zero. In other cases where it is impossible to measure the experimental temperature dependence of rate, it may be necessary to combine a rate constant measured at a fixed temperature with a pre-exponential factor predicted from the transition state theory to provide an estimate of the activation energy.

The collision and transition state treatments both implicitly assume that a continuous potential energy surface links reactants with products: that is, that the reaction is *adiabatic*. If no such surface exists, then there must be a crossing between surfaces, and the pre-exponential factor for the reaction is then likely to be considerably lower than otherwise predicted. The first test of whether a reaction can proceed adiabatically is to discover whether it is even possible to construct the same symmetry species from both reactants and products. This kind of argument was first used by Wigner (1927) to derive his *spin conservation rules*.

Quantum restrictions can be most important in determining the probability of a reaction, but two features deserve consideration. First, the restrictions will only be of importance to the extent that the quantum numbers employed are realistic. Thus, if there is extensive spin-orbit coupling in the isolated reactants, S has little meaning, and the collision encounter is not likely to be heavily governed by spin correlation. Secondly, although the rules may show that a reaction is *able* to proceed adiabatically, the adiabatic path may pass through high energies. In this case, reaction may actually proceed more probably *via* crossing lower energy surfaces: the decrease in activation energy may more than offset the decrease in pre-exponential factor.

2.3.2. Unimolecular Reactions

An excited species AB* can dissociate (or isomerize) to form products with kinetics that are both unimolecular and first order. If the species is polyatomic, then a number of different products may be energetically accessible. The unimolecular process is analogous to radioactive decay; AB* may be excited in a number of different ways. Absorption of light is one, and formation in an exothermic chemical reaction is another. What is not immediately obvious is how thermal activation in collisions can also lead to overall first-order kinetics, since the activation step is so clearly bimolecular. Lindemann (1922) first suggested the sequence of processes

$$A + A \overset{k_1}{\rightarrow} A^* + A \qquad \text{(activation)} \qquad (11)$$

$$A^* + A \overset{k_2}{\to} A + A \qquad \text{(deactivation)} \qquad\qquad (12)$$

$$A^* \overset{k_3}{\to} \text{products} \qquad \text{(reaction)}. \qquad\qquad (13)$$

Application of the stationary state hypothesis for [A*] leads to the result

$$-\frac{d[A]}{dt} = \frac{k_1 k_3 [A]^2}{k_2 [A] + k_3}. \qquad\qquad (14)$$

Thus, so long as $k_2[A] \gg k_3$, the system obeys a first-order kinetic law. At pressures sufficiently low that the inequality is no longer true, the process becomes second order.

While the rate law (14) gives a qualitative explanation of thermal unimolecular reactions, there are some quantitative problems. First, it would seem that, knowing the energy needed in A*, k_1 could be calculated from the collision frequency factor modified by the exponential activation energy term, as in Equation (8). In fact, measured values of k_1 for polyatomic species often greatly exceed the value calculated in this way. Hinshelwood (1927) proposed that, in a polyatomic molecule, the energy possessed in the various vibrational modes could contribute to the total energy needed for the reaction step (13).

A second problem is that k_3 has been assumed energy-independent, so long as A* possesses more than the critical excitation value. If k_3 is a function of energy, ε, then we must also consider the rate of excitation to, and deactivation from, that specific energy. Several approaches to this problem have been made, the best known being those of Rice and Ramsperger (1927), Kassel (1928) and Slater (1939). Rice, Ramsperger and Kassel (RRK) assume that there is a completely free flow of energy between the different vibrational modes of the molecule, and that the arrival of sufficient energy for dissociation of the critical bond can be calculated statistically. The more energy in the molecule, the less time taken for the critical amount to appear in the reacting bond, and the larger k_3. Slater, on the other hand, adopts a strictly harmonic model, in which there is no flow of energy: dissociation follows the build up of a critical amplitude of vibration in the dissociating bond as a result of a specific phase relationship between the various normal modes. So far as the experimental evidence is concerned, the free flow theories are favored (see, for example, Wayne, 1969a, pp. 278–287). Both RRK and Slater theories put an upper limit on the magnitude of k_3 around the value for molecular vibration frequencies (ca. 10^{13}–10^{14} s^{-1}). Some experimental data require k_3 up to 10^{18} s^{-1}. Marcus (1952) has developed a transition state modification of RRK theory (RRKM theory!) in which the interconversion of rotational and vibrational degrees of freedom in the activated complex can account for these otherwise anomalously high rate constants.

2.3.3. Termolecular Reactions

A number of processes of atmospheric interest can be kinetically third order and be represented by a termolecular chemical equation. Although some molecular reactions of NO (e.g., $2NO + O_2$) fall into this category, most of the processes are recombinations of atoms or small radicals, and we may write the formalized equation

$$A+B+M \rightarrow AB+M; \tag{15}$$

M is a 'third body' which does not become consumed. It is required to stabilize the product AB molecule. For example, in the recombination of two atoms, the newly formed molecule clearly possesses sufficient energy for redissociation even if the atoms initially had no relative translational kinetic energy. The 'hot' molecule must therefore be stabilized within about one vibrational period if it is to survive. In a process involving more complex fragments – and thus a polyatomic product – the energy of recombination can be dissipated by flow into vibrational modes other than that first excited (the converse of the flow process leading to dissociation in unimolecular reactions).

These ideas may be expressed mathematically by considering the simple model

$$A+B \overset{k_4}{\rightarrow} AB* \tag{16}$$

$$AB* +M \overset{k_5}{\rightarrow} AB+M \tag{17}$$

$$AB* \overset{k_6}{\rightarrow} A+B. \tag{18}$$

Reactions (17) and (18) are now directly equivalent to reactions (12) and (13). Solution of the steady state equations in [AB*] leads to the result

$$\frac{d[AB]}{dt} = \frac{k_4 k_5 [A][B][M]}{k_5[M]+k_6} \tag{19}$$

so that, if $k_5[M] \ll k_6$,

$$\frac{d[AB]}{dt} = \frac{k_4 k_5}{k_6}[A][B][M] \tag{20}$$

that is, the kinetics are third order, and the rate depends on the efficiency with which M stabilizes AB*. For the atom + atom situation, $k_6 \sim 10^{14}$ s^{-1}, and $k_5 \ngtr 3 \times 10^{-10}$ cm^3 molec^{-1} s^{-1} (the collision frequency factor), and the recombination remains third order at all sensible pressures. On the other hand, k_6 decreases rapidly with increasing atomicity, and, as an example, if $k_6 \sim 10^7$ s^{-1}, the reaction is likely to be second order at all pressures above a few Torr: the rate of recombination is determined now by the rate of reaction (16).

In many cases of atmospheric importance, the termolecular process can change from third to second order over the range of pressures encountered. The full form of Equation (19) must then be employed. Although, in principle, sufficient data can be extracted from high and low pressure limiting laws $(k_4, k_5/k_6)$ to use Equation (19), unfortunately the same problems are encountered here as described for unimolecular reactions. The rate of reaction (18) may be dependent on the energy of AB*, and process (17) may only serve to reduce the energy in AB* rather than completely deactivating it. Of course, fragmentation to the initial reactants $A+B$ is impossible at energies below the dissociation threshold, but there may nevertheless be alternative, lower energy, dissociations or eliminations which compete with formation of fully stabilized AB.

A final feature for discussion concerns the temperature dependence of the rate of termolecular reactions. Processes (16) and (18) are affected by temperature in so far as the kinetic energy of A and B, and hence internal energy of AB*, is affected. Process (17) is likely to become less efficient with increasing temperature. Both effects lead, therefore, to an anticipated *decrease* in overall rate with increased temperature: that is, the termolecular processes are associated with a *negative activation energy*.

3. Experimental Techniques

The kinetics of a chemical reaction are, in essence, studied by mixing the reactants in a time short compared to the time constant of the reaction, and sampling the mixture at appropriate intervals. Most of the reactions of atmospheric interest are too *fast* and the reactants (e.g., atoms, radicals, excited species) too *labile* for study by classical ('hand' mixing, direct observation as a function of time) means, and special techniques have to be adopted. Secondary reactions of the products, and competitive interfering reactions (e.g. wall loss) often pose problems for the laboratory worker. Unless the secondary reactions are deliberately permitted (as, for example, in a steady-state experiment) the conditions must be chosen to obviate their occurrence, or at least make sure that the correct allowance can be made for them.

Ideally, one would wish to study the reactivity of isolated reaction partners possessing selected velocity, internal excitation and orientation, in order to parallel the 'ideal' theoretical calculations outlined in Section 2.3.1. Experiments of this kind can, indeed, be performed using crossed molecular beam techniques, but – largely because of the lack of suitable sources, and more particularly detectors – the systems to which these techniques have so far been applied are rarely of direct atmospheric importance. The techniques with which we are concerned normally allow thermalizing collisions (homogeneous and heterogeneous) on the reaction time scale, so that the translational (and probably rotational) energies have a near-Boltzmann distribution appropriate to the temperature of the surroundings. On the other hand, it may still be possible to work with species in which vibrational and/or electronic relaxation has not occurred.

3.1. TIME RESOLUTION

The information to be obtained from stationary, non-stationary and pseudo-stationary systems is well illustrated by a consideration of the lifetime and quenching of fluorescence. The general scheme

$$A + h\nu \overset{I_{abs}}{\rightarrow} A^\dagger \qquad \text{absorption} \qquad (21)$$

$$A^\dagger + M \overset{k_q}{\rightarrow} A + M \qquad \text{quenching} \qquad (22)$$

$$A^\dagger \overset{k_f}{\rightarrow} A + h\nu \qquad \text{fluorescence} \qquad (23)$$

applies to quenched fluorescence ($A^\dagger$ is electronically excited A). In the steady state, we may show that

$$I_{\text{fluorescence}} = k_r [A^\dagger] = \frac{k_r I_{\text{abs}}}{k_r + k_q [M]}. \tag{24}$$

Thus the most we can hope to obtain from a measurement of fluorescence intensities in the steady state is the ratio k_q/k_r. Immediately after shutting off the exciting radiation, the fluorescence intensity decays according to

$$I^t_{\text{fluorescence}} = I^{t=0}_{\text{fluorescence}} \exp - (k_r + k_q [M]) t \tag{25}$$

so that measurement of $I^t_{\text{fluorescence}}$ as a function of t yields both k_r and k_q if $[M]$ is varied. A pseudostationary system may also provide information about individual rate constants. If a modulated excitation source is employed, the fluorescence intensity reaches a maximum after the excitation intensity, and similarly decays after the exciting light decays: that is, there is a phase lag. For sinusoidal modulation, it may be shown that the phase lag, ϕ, is related to the lifetime $\tau (= (k_r + k_q [M])^{-1})$ by the relation

$$\tan \phi = 2\pi v \tau, \tag{26}$$

where v is the frequency of modulation.

3.1.1. Flash Photolysis

The non-stationary method with perhaps the widest applicability is flash photolysis, developed by Norrish and Porter (1949). Figure 1 shows the basic apparatus. In essence, photochemically active reactant is placed in a vessel which is illuminated by a flash lamp. The flash is fired, and the concentrations of reactant, product, or intermediate monitored as a function of time. The apparatus shown in the figure is, in fact, set up to measure concentrations spectrographically. A second small flash, triggered after a predetermined delay by the photolytic flash, is used as a source lamp to obtain an absorption spectrum (recorded photographically). Several sep-

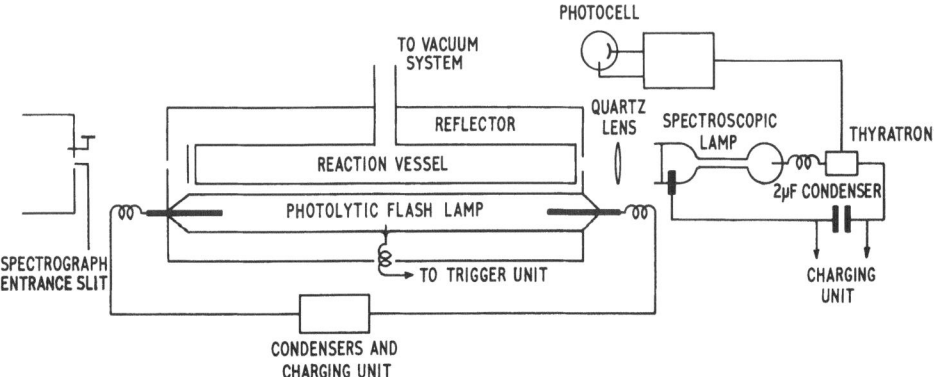

Fig. 1.   Schematic diagram of flash photolysis apparatus.

arate experiments are then conducted at different delays to obtain a time-resolved concentration profile. Thus these experiments combine the required features: rapid initiation and rapid measurements with time resolution. Norrish and Thrush (1956) give a more extensive review of the technique and some of its earlier applications.

Two variants of the initiation process have been employed. In the more usual, *isothermal* flash photolysis, a small quantity of absorber is diluted with a large amount of buffer gas to prevent any temperature rise. The initiation of reaction is then purely photochemical. The second variant, *adiabatic* flash photolysis, deliberately dispenses with the buffer, and the rapid temperature rise following absorption, by a sensitizer, of the flash can be used to initiate thermal reactions. This latter technique is clearly of value in the study of high temperature combustion processes (Erhard and Norrish, 1956).

The available time resolution is determined both by the speed of the detection method, which will be discussed in Section 3.2, and by the duration of the photolytic flash. It is clearly convenient to make the flash short compared with the reaction time to obviate difficult (or impossible) deconvolution procedures. The early flash lamps had lifetimes of the order of milliseconds, but progressive improvements in technique have yielded lifetimes of less than a microsecond. Using pulsed lasers, 'lamp' lifetimes of nanoseconds or even picoseconds are accessible. To some extent, reduced lamp lifetimes have been accompanied by reductions in the available energy input for a single flash. However, suitable application of modern signal averaging methods permits use of many repeated low intensity flashes rather than one single large one.

3.1.2. *Shock Tubes*

The shock tube technique provides another example of adiabatic heating. A driver gas at high pressure is separated from the lower pressure sample by a diaphragm (Figure 2). On rupturing the diaphragm, a shock wave travels along the tube, supersonically with respect to the cool gas. There is a very thin transition zone between high and low pressure zones, and the reaction sample is rapidly heated (up to thousands of degrees K), as the shock front passes, thus initiating thermal reactions. Observations made on the reaction mixture at different times after the shock passes then yield rate data. Kinetic spectrometry has been used to measure concentrations in the same way as for flash photolysis (Campbell and Johnson, 1957; Schott and Davidson, 1958). The temperature in the shock may be perturbed by the occurrence

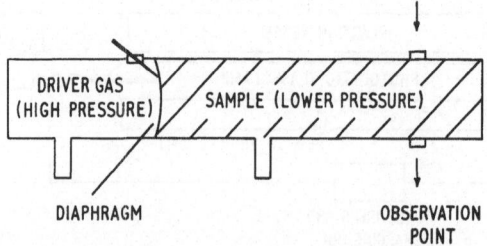

Fig. 2.   Simplified diagram of shock tube.

of exo- or endo-thermic reaction, but if a buffer gas is used, then the temperature may be estimated by techniques such as atomic line reversal (cf. Gaydon, 1957, Ch. VIII), or spectroscopic measurement of vibrational temperature in gases such as N_2. Timing also offers some difficulty, since transport of gas, from high to low pressure ends of the shock tube, provides an elapsed time contribution to be added to the time measured since the shock front passed the observation port.

A useful and more detailed discussion of the chemical applications of shock tubes is given by Bauer (1963).

3.1.3. *Flow Methods*

Although stopped and stirred flow techniques have been used with success for certain reactions, continuous flow systems are of wider application, and alone will be discussed here. Figure 3a shows diagrammatically a typical flow tube for gas phase studies. A vacuum pump establishes a flow of gas from the left to right, and it is this flow which provides the time resolution. In a typical experiment one reactant might be present in the main gas flow, and a second reactant added at a jet. Assuming good mixing at the jet, distance downstream corresponds to reaction time. Linear velocities are normally in the range 1 to 10 m s^{-1}, so that if measurements can be made with a *spatial* resolution of 1 cm, the time resolution is of the order of milliseconds. Much slower and much faster flow velocities have been used, but in the former case axial diffusion may be a problem, while in the latter turbulence must be avoided.

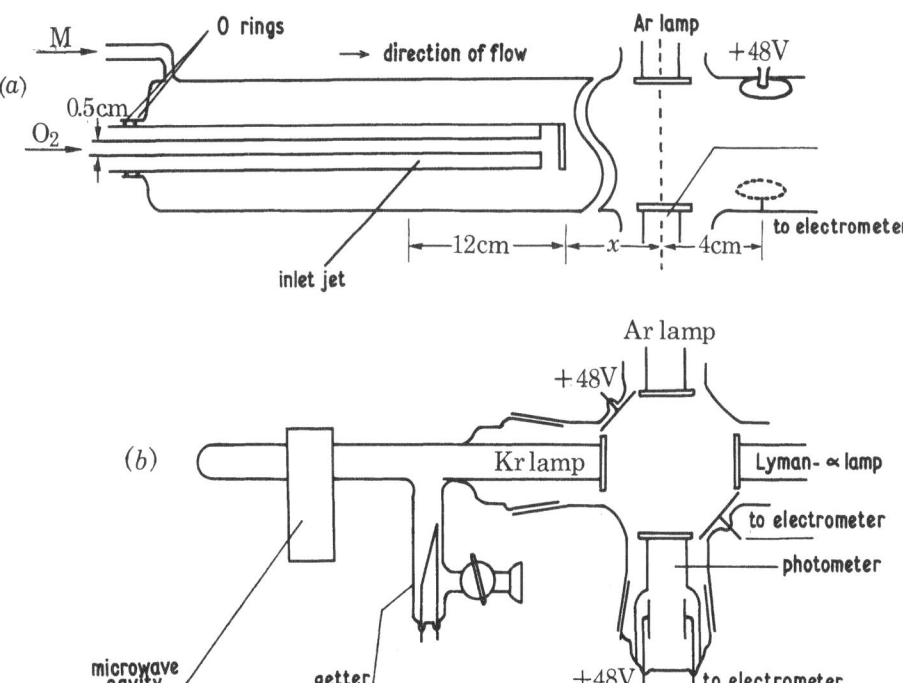

Fig. 3.   A typical flow system: (a) the flow tube showing inlet arrangements; (b) radial section showing the photoionization detector.

The detail of the execution of the experiments often depends largely on the method used for concentration measurement. If the 'detector' must be fixed in position, then differing contact times may be achieved by adding the second reactant either at a series of jets or *via* a movable inner jet. Alternatively, if the detector can be moved, then it may be positioned at various points downstream from a single second-reactant jet.

A special feature of continuous flow methods is that a steady concentration of all species is normally maintained at any point in the tube by virtue of the supply of new reactants and removal of products. Thus, although the technique is fully time resolved at the millisecond level, the method used for concentration measurement does not need to have a correspondingly rapid response. In a sense, the argument presented earlier that steady state systems only provide information about ratios of rate constants still holds, but here the competing process is the pumping operation, whose speed may be directly measured.

3.1.4. *Pseudostationary Methods*

Regular repetition of a non-stationary experiment may produce a time-independent change in some parameter from which kinetic information can be extracted. The use of intermittent radiation, produced by a rotating sector to intercept the light beam, was first applied to gas phase reactions by Briers and Chapman (1928). For a photo-chemical reaction terminated by second order loss of radical intermediates, the rate can be shown to vary with the sector speed. With a 1:1 mark to space ratio, the rate increases by a factor of $\sqrt{2}$ as the illumination period becomes comparable to and less than the radical lifetime in the system. Comparison of predicted and experimental rates over a range of sector speeds can yield a numerical value for the lifetime; the rate constants for the radical reaction then follow (cf. Wayne, 1970, pp. 207–211). As it stands, the rotating sector method requires both that the intermediates are lost in a second (or higher) order process, and that the lifetime fall within attainable sector speeds. However, these limitations are removed if concentrations of intermediates can be followed. The use of phase shift measurements for the determination of fluorescence lifetimes has been described already: here the fluorescence intensity is proportional to the excited state ('intermediate') concentration. A particularly successful recent 'chemical' application of the modulation technique is to the HO_2 radical (Johnston *et al.*, 1967; Paukert, 1969). A square-wave chopped photolytic source was used, and the HO_2 concentration followed *via* optical absorption measurements in the IR (1095, 1390 and 3410 cm^{-1}) or UV (ca. 2100 Å). The resultant modulation of the absorption signal first of all permits observations of much smaller absorptions than would be possible with a d.c. system, and, further, of small modulated absorptions in the presence of strong unmodulated ones. Secondly, the measured phase-shift provides the desired kinetic information. One further very interesting possibility arises out of this work. Paukert (1969) shows that first order decay of radicals yields an absorption modulation waveform containing only *odd* harmonics (from the original square wave photolysis beam); if, however, second order loss occurs, then *even* harmonics are introduced. It seems, therefore, as though simultaneous first and second order processes could readily be studied by use of suitable harmonic filters.

3.2. MEASUREMENT OF CONCENTRATION OF INTERMEDIATES

The labile species with which we are concerned are mainly atoms, radicals, and vibrationally and electronically excited atoms and molecules. In many respects the methods available for the investigation of ions follow closely those used for neutrals, although the presence of charge may necessitate special design.

3.2.1. *Optical Spectroscopy*

Optical spectroscopy, from the IR to the vacuum UV, is perhaps the method of choice both for the identification and estimation of intermediates. The spectrum should allow positive identification to be made, and the optical technique produces little disturbance to the system. Spectroscopy can be used in one of two modes: absorption or emission. Excited fluorescence (especially resonance fluorescence) can be regarded as falling within the category of emission spectroscopy.

Photographic recording of an absorption spectrum, using a continuous source, is probably the best way of *identifying* intermediates. This method is the one depicted in Figure 1 for the flash photolysis-kinetic spectroscopy experiment. For *concentration* measurements of a specific species a monochromator-photomultiplier combination is more satisfactory than photography and plate photometry, although both methods have provided valuable information. Higher sensitivity may be obtained by using a source whose characteristics match the absorption. Thus DelGreco and Kaufman (1962) describe the use of an OH resonance lamp for the measurement of [OH] in a flow system. Hydroxyl lamps have been used frequently since these experiments, and a typical application is that of Morley and Smith (1972) in a flash photolysis-kinetic spectroscopy experiment. The narrowness of atomic spectral lines makes the use of a matched source particularly valuable.

Absolute concentration measurements require a knowledge of the extinction coefficient or oscillator strengths for the observed transition. Direct measurement is, of course, impossible for transitions from excited states or involving chemically labile species, so that indirect methods must be used (Wayne, 1969b). Most obviously, the oscillator strength for the absorption may be calculated from a measured radiative lifetime for the reverse emission. Bennett and Dalby have measured oscillator strengths for a variety of species such as OH, NH, CN and CH; their experimental technique is described in their first paper (Bennett and Dalby, 1959). Consideration of the chemical stoichiometry may sometimes give absolute concentrations (e.g. Lipscomb *et al.*, 1956). It is worth remembering that *relative* concentrations are sufficient to obtain kinetic data in cases where the intermediate decays in a first order fashion.

Emission spectroscopy is obviously of value where excited species are to be studied. As an example, emission (although electric dipole forbidden) from the singlet states of O_2 has provided both for their identification and for concentration measurements

(Wayne, 1969c). Again, any study of fluorescence or of a chemiluminescent reaction relies implicitly on the optical emission.

Where ground state species are involved, they may be excited to give rise to emission. Two main methods are used: chemical excitation (i.e., chemiluminescence) and optical excitation (i.e., fluorescence). As an example of the first method, if a trace of NO be added to O, the processes

$$O + NO + M \rightarrow NO_2^\dagger + M \tag{27}$$

$$NO_2^\dagger \rightarrow NO_2 + h\nu \text{ ('air afterglow')} \tag{28}$$

$$O + NO_2 \rightarrow NO + O_2 \tag{29}$$

produce the green 'air afterglow' emission with no net consumption of NO, nor, at low enough [NO], of substantial amounts of O. Thus the intensity of the air afterglow may be used to monitor relative [O], and, if needed, an absolute scale can be established either against another technique (e.g. titration – see Section 3.2.4) or by measurement of the absolute intensity of emission (Fontijn et al., 1964). Excitation of *fluorescence* is particularly valuable if the resonance source is employed, since then any observed emission identifies the presence of the specific absorber-emitter. Further, the sensitivity of the fluorescence technique is largely limited by the amount of stray scattered light, so that good matching of source and absorber characteristics improves signal to 'noise' (i.e., scatter) ratios. Resonance fluorescence techniques are currently highly popular because of their sensitivity and specificity: they are applicable to both molecular (e.g. OH) and atomic (e.g. H, N, O, C) intermediates. It may be noted in concluding this section that some caution should be exercised in the interpretation of resonance absorption and fluorescence data on three counts: (i) reversal of source and mismatch of line profile with absorber, (ii) fluorescence in an absorption experiment, and (iii) radiation trapping.

3.2.2. *Electron Paramagnetic Resonance (EPR)*

Net electronic magnetization can be detected by EPR methods. Although the ultimate sensitivity is lower than for transitions in the optical region, the exceptionally high resolution to some extent offsets this factor. In the gas phase, both spin and orbital momentum can produce paramagnetism, so that species such as $O_2(^1\Delta_g)$ have an EPR spectrum in the same way as more conventional 'free radicals'. Species such as H, O, N, halogen atoms, $O_2(^1\Delta_g)$, OH, SH, SeH, SO, NF_2, ClO, BrO and $NS(^2\Pi_{3/2})$ have been detected by EPR, and in several cases kinetic studies have been made (cf. Westenberg and de Haas, 1964).

A cylindrical cavity, used in TE_{011} mode, has proved popular for gas phase studies since it can be employed for simultaneous studies of magnetic and electric dipole interactions. Spectrometer calibration may be achieved using a stable paramagnetic gas such as nitric oxide or molecular oxygen.

3.2.3. *Mass Spectrometry and Related Methods*

Mass spectrometry offers some obvious advantages of identification, but there are two major difficulties associated with its use for the estimation of labile intermediates. First, sampling offers considerable problems since there is usually a large pressure differential between the kinetic reaction system and the spectrometer. In achieving the pressure reduction, the necessary surfaces may cause recombination of radicals or deactivation of excited species, so that the sample reaching the stage of mass analysis may not be representative of the sample in the reaction system. Probably the most satisfactory sampling system is a molecular beam arrangement, which provides that the species entering the spectrometer proper shall have made no collisions.

The second problem is fragmentation of species during ionization. A molecule AB can produce the ion A^+ if there is sufficient energy to dissociate AB and ionize A. Thus the investigation of A in the presence of AB (a very common problem – CH_3/CH_4, H/H_2, etc.) can be confused by the production of fragment ions. With electron impact ionization, electron energies of about 70 eV are often used to obtain high ionization cross sections. However, 70 eV vastly exceeds the energy needed to fragment and ionize almost every molecular species, so that it is necessary, wherever A^+ could be formed from A or AB, to use much lower energy electrons and accept the reduction in ionization efficiency. Unfortunately, the thermal spread of energies of electrons from a hot filament may still make it difficult to detect small amounts of A in the presence of excess AB. Photoionization (and to some extent field ionization) overcomes this problem satisfactorily. With suitable precautions (e.g., avoidance of large electric potential gradients which could accelerate photoelectrons), photoionization can be achieved with an energy spread near that of the photon source. Atomic line sources (e.g., Ar, Kr, O) are essentially monochromatic and the wavelengths lie in a range corresponding to ionization thresholds. For example, Ly-α radiation (10.2 eV) can ionize NO (IP $\sim$ 9.5 eV), but cannot produce NO^+ from NO_2 (total energy ca. 12.6 eV), so that small amounts of NO can be measured in the presence of excess NO_2. Similarly, excited states can be investigated even when the ground state species is present. Thus $O_2(^1\Delta_g)$ (excitation energy 0.98 eV, IP $\sim$ 11.2 eV) can be ionized by Ar-resonance radiation (11.6 eV) while ground state $O_2(^3\Sigma_g^-)$ (IP $\sim$ 12.2 eV) cannot.

Various types of mass filter have been employed in kinetic experiments: conventional magnetic instruments, time of flight, Bennett-type RF devices and so on. However, for small species, the quadrupole is at present the device of choice. With large enough poles the transmission is high and the resolution adequate. The direction from which ions arrive is not particularly critical: an important feature where unstable or reactive species are to be introduced to the filter.

The energy discrimination provided by photoionization may, in fact, be used to identify and estimate suitable species even without any mass analysis. The excited O_2 species described above have been investigated by photoionization within a flow system (e.g. Clark and Wayne, 1969), the ionization being observed in terms of an overall ion current. Figure 3b shows the photoionization detector arrangement used with the flow system of Figure 3a.

3.2.4. *Chemical Methods*

A wide variety of chemical methods has been described for the identification and estimation of intermediates; a number of these are discussed by Wayne (1969b). In each case, a foreign substance is added to the reaction system which is therefore perturbed to a greater or lesser extent. The greatest value of the chemical methods is thus in *calibrating* absolutely one of the physical techniques. For the purposes of illustration here, we will consider the 'titration' techniques. The chemiluminescent emission of the air afterglow, produced by NO in the presence of O, was described in Section 3.2.1. If NO_2, rather than NO, is used as reactant, then reaction (29) converts NO_2 to NO and the air afterglow is again seen. However, O is consumed in reaction (29), so that while the intensity of chemiluminescence at first increases with increasing $[NO_2]$, it subsequently decreases, and is sharply extinguished when all the O has been removed. At the 'end-point', the $[NO_2]$ added is equal to the $[O]$ originally present. An exactly similar titration is available for H, the HNO chemiluminescence being used as 'indicator'. For N, the reactions

$$N + N + M \rightarrow N_2^\dagger + M \tag{30}$$

$$N + NO \rightarrow N_2 + O \tag{31}$$

are used. The first of these produces the yellow chemiluminescence of 'active nitrogen', but reaction (31) is very rapid, and addition of the stoichiometric quantity of NO extinguishes the chemiluminescence.

References

Arrhenius, S.: 1889, *Z. physik. Chem.* 4, 226.
Bauer, S. H.: 1963, *Science* 141, 867.
Bennett, R. G. and Dalby, F. W.: 1959, *J. Chem. Phys.* 31, 434.
Briers, F. and Chapman, D. L.: 1928, *J. Chem. Soc.* p. 1802.
Campbell, C. E. and Johnson, I.: 1957, *J. Chem. Phys.* 27, 316.
Clark, I. D. and Wayne, R. P.: 1969, *Proc. Roy. Soc.* A314, 111.
DelGreco, F. P. and Kaufman, F.: 1962, *Disc. Faraday Soc.* 33, 128.
Erhardt, K. H. L. and Norrish, R. G. W.: 1956, *Proc. Roy. Soc.* A234, 178.
Evans, M. G. and Polanyi, M.: 1935, *Trans. Faraday Soc.* 31, 875.
Eyring, H.: 1935, *J. Chem. Phys.* 3, 107.
Fontijn, A., Meyer, C. B., and Schiff, H. I.: 1964, *J. Chem. Phys.* 40, 64.
Gaydon, A. G.: 1957, *The Spectroscopy of Flames*, Chapman and Hall, London.
Hinshelwood, C. N.: 1927, *Proc. Roy. Soc.* A113, 230.
Johnston, H. S., McGraw, G. E., Paukert, T. T., Richards, L. W., and van den Bogaerde, J.: 1967, *Proc. Natl. Acad. Sci.* 57, 1146.
Kassel, L. J.: 1928, *J. Phys. Chem.* 32, 225, 1065.
Laidler, K. J. and Polanyi, J. C.: 1964, *Progr. Reaction Kinetics* 3, 3.
Lindemann, F. A.: 1922, *Trans. Faraday Soc.* 17, 598.
Lipscomb, F. J., Norrish, R. G. W., and Thrush, B. A.: 1956, *Proc. Roy. Soc.* A233, 455.
Marcus, R. A.: 1952, *J. Chem. Phys.* 20, 359.
McC-Lewis, W. C.: 1918, *J. Chem. Soc.* p. 471.
Morley, C. and Smith, I. W. M.: 1972, *Trans. Faraday Soc.* 68, 1016.
Norrish, R. G. W. and Porter, G.: 1949, *Nature* 164, 658.
Norrish, R. G. W. and Thrush, B. A.: 1956, *Quart. Revs.* 10, 149.
Paukert, T. T.: 1969, Ph.D. Thesis, Univ. California, Berkeley (UCRL-19109).

Rice, O. K. and Ramsperger, H. C.: 1927, *J. Amer. Chem. Soc.* **49**, 1617.

Schott, G. and Davidson, N.: 1958, *J. Amer. Chem. Soc.* **80**, 1841.

Slater, N. B.: 1939, *Proc. Cambridge Phil. Soc.* **35**, 56.

Wayne, R. P.: 1969a, in C. H. Bamford and C. F. H. Tipper (eds.), *Comprehensive Chemical Kinetics*, Vol. 2, Elsevier, Amsterdam.

Wayne, R. P.: 1969b, in C. H. Bamford and C. F. H. Tipper (eds.), *Comprehensive Chemical Kinetics*, Vol. 1, Elsevier, Amsterdam.

Wayne, R. P.: 1969c, *Advan. Photochem.* **7**, 311.

Wayne, R. P.: 1970, *Photochemistry*, Butterworths, London.

Westenberg, A. A. and de Haas, N.: 1964, *J. Chem. Phys.* **40**, 3087.

Wigner, E. P.: 1927, *Gött. Nachr.* **4**, 375.

ION CHEMISTRY

ELDON E. FERGUSON

Aeronomy Laboratory, NOAA Environmental Research Laboratories, Boulder, Colo. 80302, U.S.A.

1. Introduction

The development of the laboratory capability of measuring ionospheric reaction rate constants has occurred almost entirely in the past decade. The progress in this field in the past decade has been enormous and there are now a number of groups actively measuring ionospheric ion-neutral reaction rate constants by a variety of laboratory techniques and a tremendous reservoir of data now exists on many hundreds of reaction rate constants. Recently a tabulation of binary ionospheric rate constants was published (Ferguson, 1973).

There are several aspects of laboratory ion-molecule reaction studies to be considered. First is simply the reaction rate constant at thermal energy for a given ion-neutral combination. The measurement precision required by aeronomers varies considerably from reaction to reaction and of course increases with time as a consequence of the ever increasing sophistication of ionospheric measurements. Second is a knowledge of the reaction products. Often this follows simply and unambiguously from the reactants but there are many important aeronomic cases where it does not. In particular a number of important ionospheric reactions branch, i.e., they produce more than one set of products, and it is usually much more difficult to determine this branching ratio precisely than it is to measure the rate constant. Next is the matter of the energy dependence of rate constants. For the binary positive ion reactions of the E and F region one would like temperature dependences from 300 K up to around 2000 K and in some cases it would be desirable to have energy dependences for non-Boltzmann distributions of vibrational, rotational and translational energy. There is a great deal of current progress in this area. For the three-body positive and negative ion reactions of the D region, the temperature dependences of the rate constants from 300 K down to ~ 120 K are very critical and not in general well known.

An area in which very little data exist is the product state distributions, both vibrational and electronic, of the ion-neutral reactions. This is of great importance but such experiments are more difficult than usually can presently be done. Finally, reactions of electronically excited ions are of substantial importance but very little studied so far.

The principal techniques which have been used for laboratory studies of ionospheric ion reactions are: (i) the Flowing Afterglow method which has the advantage of wide chemical versatility at thermal energy. The temperature range studied has

B. M. McCormac (ed.), Atmospheres of Earth and the Planets, 197–210. All Rights Reserved.
Copyright © 1975 by D. Reidel Publishing Company, Dordrecht-Holland.

been from 80 to 900 K, although most reactions have been measured only at 300 K so far. Some excited state reactions have been measured. (ii) Beam studies have also demonstrated a wide chemical versatility. Generally they have been restricted to relative kinetic energies above ~ 1 eV. (iii) Drift tubes have recently been applied to aeronomical problems with the advantage of a continuous control of relative kinetic energy from thermal to over 1 eV. (iv) Recently a combined Drift Tube-Flowing Afterglow has combined the energy capability of the former with the chemical versatility of the latter. (v) High Pressure Mass Spectrometry has been of particular value in measuring D region clustering reactions and determining cluster bond energies and equilibrium constants. (vi) Ion Cyclotron Resonance has recently proven to be a very powerful technique for the study of ion chemistry. As yet it has been little applied to aeronomy but the potential is clearly very great. In addition, conventional low pressure mass spectrometers and other techniques such as stationary afterglows have provided useful data in spite of their limitations in versatility.

2. E and F Region (Non-Metallic) Ion Chemistry

The ion chemistry of the upper ionosphere is dominated by a relatively small number of binary atmospheric positive ion-neutral reactions so that it is feasible to present current data on these reactions and discuss them individually.

2.1. $O^+ + H \rightleftarrows H^+ + O$

This important reaction, which is the major source of H^+ in the F region, is near resonant and the rate constants are large in both directions in the ionosphere so that

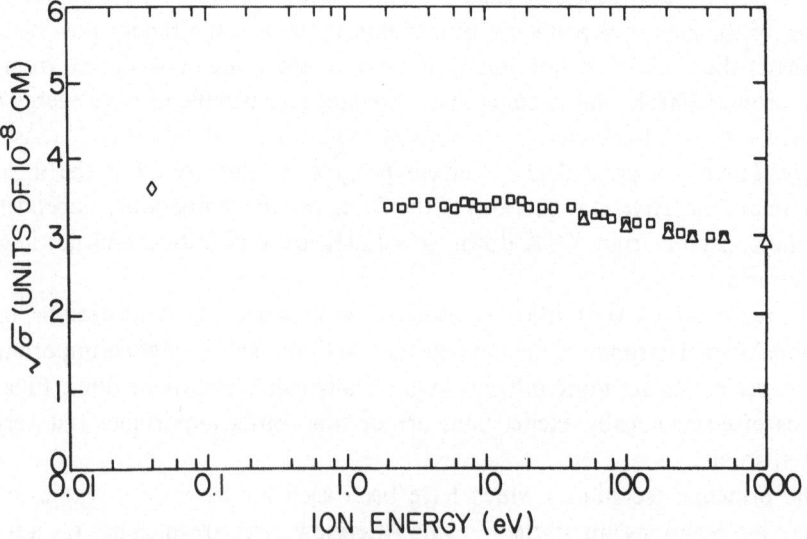

Fig. 1.   Square root of the cross section for the reaction $H^+ + O \rightarrow O^+ + H$ as a function on ion kinetic energy. $\diamondsuit$ represents the thermal 300 K measurement of Fehsenfeld and Ferguson (1972), $\square$ is the beam data of Rutherford and Vroom (1974), and $\triangle$ is the beam data of Stebbings et al., (1964).

equilibrium can sometimes be assured and used to determine the H concentration from O^+, H^+ and O measurements. Figure 1 shows the cross section as a function of energy as determined from beam experiments and their extrapolation to a 300 K measurement.

2.2. $He^+ + N_2$

The reaction rate constant for

$$He^+ + N_2 \rightarrow N^+ + N + He \tag{1a}$$

$$\rightarrow N_2^+ + He \tag{1b}$$

is very well measured by many groups, all in good agreement. As shown in Figure 2, the rate constant is independent of temperature from 300 to 900 K. It is also independent of N_2 vibrational temperature from 300 to 6000 K (Schmeltekopf et $al.$, 1968). The branching ratio is less well known, $k_{1a}/k_{1b} \sim 1.2$, and does depend somewhat on

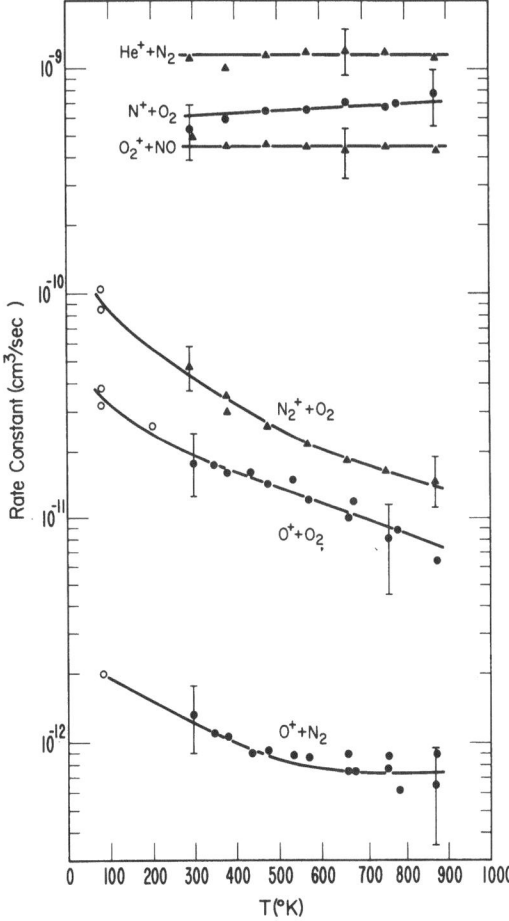

Fig. 2. Rate constants for several ionospheric reactions from 300 K (in some cases from 80 K) to 900 K from Lindinger et $al.$ (1974).

vibrational temperature. Reaction (1) is very fast, $k_1 \approx k_L = 2\pi e \sqrt{\alpha/\mu}$ the collision rate constant, as a consequence of the favorable Franck-Condon factor between the $N_2 X^1 \Sigma_g^+$ and $N_2^+ C^2 \Sigma_u^+$ states involved in the energy resonant charge-transfer which leads to a non-adiabatic electron jump.

2.3. O^+ REACTIONS

The reaction rate constants for

$$O^+ + N_2 \rightarrow NO^+ + N \tag{2}$$

and

$$O^+ + O_2 \rightarrow O_2^+ + O \tag{3}$$

are shown in Figure 2 from 80 to 900 K and in Figure 3 to a very much higher relative kinetic energy for reaction (2). It is easier experimentally to obtain k vs. ion kinetic energy than k vs. T and the available kinetic energy range is very wide whereas the few measurements of k vs. T that have so far been made fall far short of the highest F region temperatures of interest. The measurements should not necessarily agree since vibrational and rotational excitation may have an effect on rate constant. In

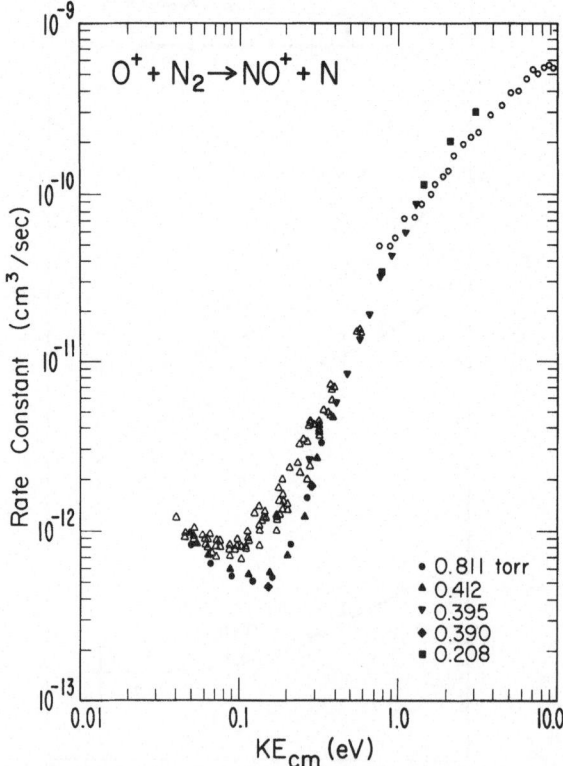

Fig. 3.   Rate constant for $O^+ + N_2 \rightarrow NO^+ + N$ as a function of center-of-mass kinetic energy. Solid symbols from McFarland *et al.* (unpublished); $\triangle$ from Johnson and Biondi (1973); $\bigcirc$ Rutherford and Vroom (1971).

the case of reactions (2) and (3) there is little difference between k vs. T and k vs. T_{KE} where $3/2kT_{KE} \equiv$ relative kinetic energy in the overlapping range.

2.4. N_2^+ REACTIONS

The rate constant for

$$N_2^+ + O_2 \rightarrow O_2^+ + N_2 \tag{4}$$

is shown vs. T in Figure 2 and vs. average ion kinetic energy in Figure 4. Reaction (4) is a charge-transfer with a very poor Franck-Condon factor for energy resonance so that a long range electron jump does not occur but rather the reaction occurs via an intermediate complex.

Note the characteristic behavior of rate constants with energy in Figure 2. For reactions in which $k \approx k_L$, the rate constant is relatively insensitive to temperature. In cases where $k \ll k_L$ the rate constant initially decreases as the temperature increases. This decrease has been attributed to the decrease in lifetime of the intermediate reaction complex with increased temperature and seems to be a very general phenomenon for 'slow' reactions. The complex lifetime decreases more rapidly with temperature (i.e., translation + vibration + rotation) than with translational energy alone and this might lead to a greater decrease of k with true temperature than with 'kinetic temperature' alone. In other words, the measured dependence of

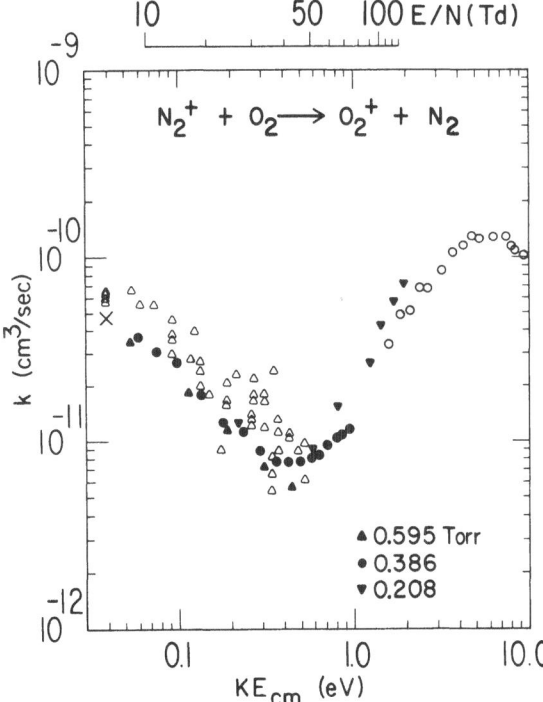

Fig. 4.   Rate constant for $N_2^+ + O_2 \rightarrow O_2^+ + N_2$ as a function of relative kinetic energy. Solid symbols McFarland *et al.* (1973); $\triangle$ Johnson *et al.* (1970); $\times$ Dunkin *et al.* (1968). $\bigcirc$ Rutherford and Vroom (1974).

rate constant on relative kinetic energy may not always give a reliable dependence on true temperature, since internal rotational and vibrational energy may have an effect on reaction rate constant.

The reaction

$$N_2^+ + O \rightarrow O^+ + N_2 \tag{5a}$$
$$\rightarrow NO^+ + N \tag{5b}$$

is more difficult to measure because of the chemical instability of O atoms. Flowing afterglow data exist on this reaction at 300 K and beam data down to ~0.4 eV. Very recently, measurements in the flow-drift tube have bridged this gap. Figure 5 shows k vs. T_{KE}. The fraction going into channel (5a) at 300 K is 7% and this increases slightly with energy. Reaction (5b), the dominant channel, is a case where it would be very desirable to know the reaction products, either $N(^2D)$ or $N(^4S)$ is energetically possible.

2.5. $N^+ + O_2$

The reaction of N^+ with O_2,

$$N^+ + O_2 \rightarrow NO^+ + O \tag{6a}$$
$$\rightarrow O_2^+ + N \tag{6b}$$

branches into two comparable channels. The rate constant is well established but

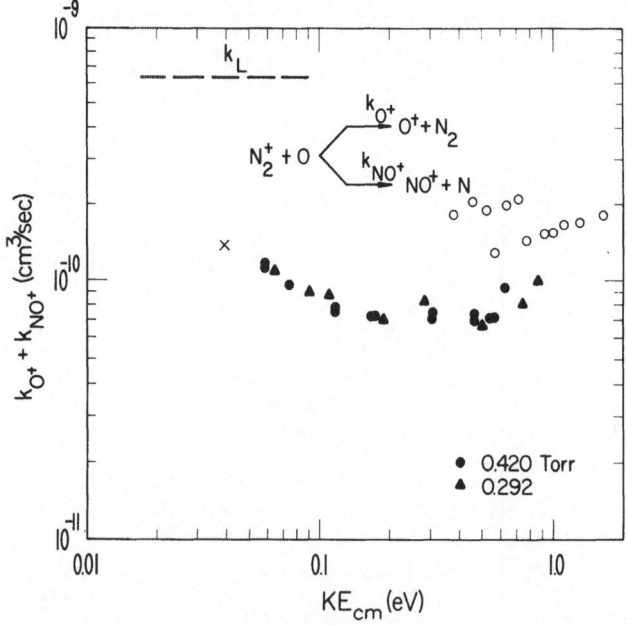

Fig. 5.   Rate constant for $N_2^+ + O$ as a function of relative kinetic energy. Solid symbols McFarland *et al.* (1974b); ○ Rutherford and Vroom (1974); × Fehsenfeld *et al.* (1970.) k_L is Langevin or orbiting collision rate constant.

measurements of the branching ratio are widely scattered. Channel (6b) could produce $N(^2D)$ and this is another case where product state determination is very important but has not yet been achieved.

2.6. $O^+ + NO$

The reaction

$$O^+ + NO \rightarrow NO^+ + O, \qquad (7)$$

is a very anomalous one. It is an exothermic charge-transfer which behaves as if it were endothermic, i.e. exhibits an apparent threshold (McFarland et al., 1974a). At 300 K, $k_7 < 8 \times 10^{-13}$ cm^3 s^{-1}, increasing to 10^{-11} cm^3 s^{-1} at 1 eV. This is a case where the Franck-Condon factor for charge-transfer is very low so that a non-adiabatic electron jump is forbidden and the adiabatic formation of NO_2^+ is endothermic as a consequence of the state correlation.

2.7. $NO^+ + O_3$

The exothermic reaction

$$NO^+ + O_3 \rightarrow NO_2^+ + O_2 \qquad (8)$$

is found not to occur at a measurable rate, $k_8 < 10^{-14}$ cm^3 s^{-1} at 300 K (Fehsenfeld et al., 1973). This is of importance since reaction (8) might have been a significant loss process for NO^+ and production of NO_2^+ and/or $O_2(^1\Delta_g)$ if k_8 had been large.

2.8. $S^+ + O_2$

The reaction

$$S^+ + O_2 \rightarrow SO^+ + O \qquad (9)$$

where $k_9 = 1.6 \times 10^{-11}$ cm^3 s^{-1} (Fehsenfeld and Ferguson, 1973a) is a strong atmospheric sulfur ion loss process, complicating the explanation of reported S^+.

3. *E* Region Metallic Ion Chemistry

There has been only a limited interest in ionospheric metal ion chemistry and very few laboratory studies. A laboratory data tabulation from 1972 (Ferguson, 1972) is still essentially up to date. A detailed review by Brown (1973) is also available.

The distribution of metal ions in the atmosphere is not as tidy a problem in chemistry as the distribution of the non-metallic ions for a number of reasons and this has led to a lack of research in these problems. The origin of the metals from meteor ablation is not quantitatively well defined and the metal ion distribution is controlled more by dynamic processes than chemical reactions in many cases. At present the loss processes for atmospheric metal ions (and atmospheric neutral metals) have not been well characterized. Some of the kinds of processes of importance are described below.

3.1. CHARGE TRANSFER

The ionospheric ions, NO^+, O_2^+, N_2^+, N^+ and O^+ all charge transfer rapidly to Na, Mg, Ca and Fe as determined in beam studies of Rutherford et al. (1971, 1972a, b) and Rutherford and Vroom (1972). An example is

$$NO^+ + Fe \rightarrow Fe^+ + NO, \qquad k_{10} = 9.1 \times 10^{-10}\ cm^3\ s^{-1} \tag{10}$$

where k_{10} is from beam data extrapolated to thermal energy. The large rate constants $k \sim k_L$ seem to be sufficiently general that one can reasonably assume that charge transfer will occur on essentially all collisions of these atmospheric ions with the lower ionization potential metals.

3.2. ION-ATOM INTERCHANGE

The atomic metal ions do not combine efficiently with electrons. As a consequence of their long lifetime against recombination, the metal ions are subject to large transport effects, such as concentration in dense sporadic E layers. Any reactions which convert atomic ions to molecular ions, thereby greatly enhancing the electron recombination rate, are therefore of very great significance. Because of the metals low ionization potential and therefore low energy content they cannot react exothermically with the abundant molecular species N_2 and O_2. The only possibility that has been considered is reaction with O_3 which has a very weak (1 eV) bond. The reactions of Ca^+, Mg^+, and Fe^+ have all been found to be fast, e.g.,

$$Mg^+ + O_3 \rightarrow MgO^+ + O_2, \qquad k_{11} = 2.3 \times 10^{-10}\ cm^3\ s^{-1}. \tag{11}$$

The ions Na^+ and K^+ have been found not to react efficiently with O_3, $k < 10^{-11}$ $cm^3\ s^{-1}$, from which it is supposed that these reactions are endothermic (Ferguson and Fehsenfeld, 1968).

The role of reaction (11) in converting the atomic ion Mg^+ to a molecular ion MgO^+, is largely mitigated in the E region however, because the reaction

$$MgO^+ + O \rightarrow Mg^+ + O_2, \qquad k_{12} \sim 1 \times 10^{-10}\ cm^3\ s^{-1} \tag{12}$$

is very fast (Ferguson and Fehsenfeld, 1968) and $[O] \gg [O_3]$. Presumably reactions similar to reaction (12) occur for other metal oxide ions but they have not been measured. The reaction

$$SiO^+ + O \rightarrow Si^+ + O_2, \qquad k_{13} \sim 2 \times 10^{-10}\ cm^3\ s^{-1} \tag{13}$$

is of interest (Si is not of course a metal) since both Si^+ and SiO^+ have apparently been observed in the ionosphere. The large value of k_{13} (Fehsenfeld, 1969) makes it extremely difficult to account for detectable SiO^+ concentrations in the E region that have been reported.

3.3. THREE-BODY REACTIONS

The ineffectiveness of binary loss processes for the atomic metal ions leads to the conclusion that they are probably lost by transport into the lower ionosphere where the pressure is large enough to make three-body reactions significant. All positive (and negative) ions have an affinity for all neutral molecules due to the electrostatic charge-induced-dipole attractive force so that all ions will cluster to all neutrals to some extent. Polar molecules, principally H_2O in the ionosphere and also SO_2 and HNO_3 in the lower atmosphere, have an even greater affinity for ions of course.

Reactions such as

$$Fe^+ + O_2 + M \rightarrow FeO_2^+ + M, \qquad k_{14} \sim 10^{-30} \; cm^6 \; s^{-1} \tag{14}$$

can be expected to occur rather non-selectively. Some rate constants have been determined (Ferguson and Fehsenfeld, 1968; Keller and Beyer, 1971) and reasonable estimates could be made for others. Three-body rate constants have the property of increasing with decreasing temperature, usually as fast or faster than T^{-1}.

Another kind of process which is very general has been labeled 'switching'. Exothermic solvent exchange or cluster switching reactions are found to be almost invariably fast, with rate constants comparable to the collision rate constant. For example one can be reasonably certain that the reaction

$$Fe^+ \cdot O_2 + H_2O \rightarrow Fe^+ \cdot H_2O + O_2 \tag{15}$$

would occur with a large rate constant, $k_{15} \sim 10^{-9} \; cm^3 \; s^{-1}$ with little temperature dependence. These kinds of reactions are of considerable importance in D region hydration as will be described below.

4. D Region Positive Ion Chemistry

The positive ion chemistry of the D region is shown schematically in Figure 6 which shows how the primary D region positive ions O_2^+ and NO^+ lead to water cluster ions $H^+(H_2O)_n$. There are problems with this scheme and it has not been found to be adequate to explain D region observations. I will restrict my comments to the presentation of a few unpublished new results and a brief discussion of the pressing experimental problems; for a more detailed discussion see Swider (1975).

Very recently it has been found by Fehsenfeld et al. (1975) that a reaction proposed by Heimerl et al. (1972)

$$NO^+(H_2O) + OH \rightarrow H_3O^+ + NO_2 \tag{16}$$

is too slow to be effective in H_3O^+ production, $k_{16} < 10^{-10} \; cm^3 \; s^{-1}$. It is also found that the reaction proposed by Burke (1970)

$$NO^+(H_2O) + H \rightarrow H_3O^+ + NO \tag{17}$$

is even slower than previously reported, $k_{17} < 7 \times 10^{-12} \; cm^3 \; s^{-1}$, which is too slow to be important.

Another reaction proposed by Heimerl and Vanderhoff (1974)

$$NO^+(H_2O) + HO_2 \rightarrow H_3O^+ + NO_3 \tag{18}$$

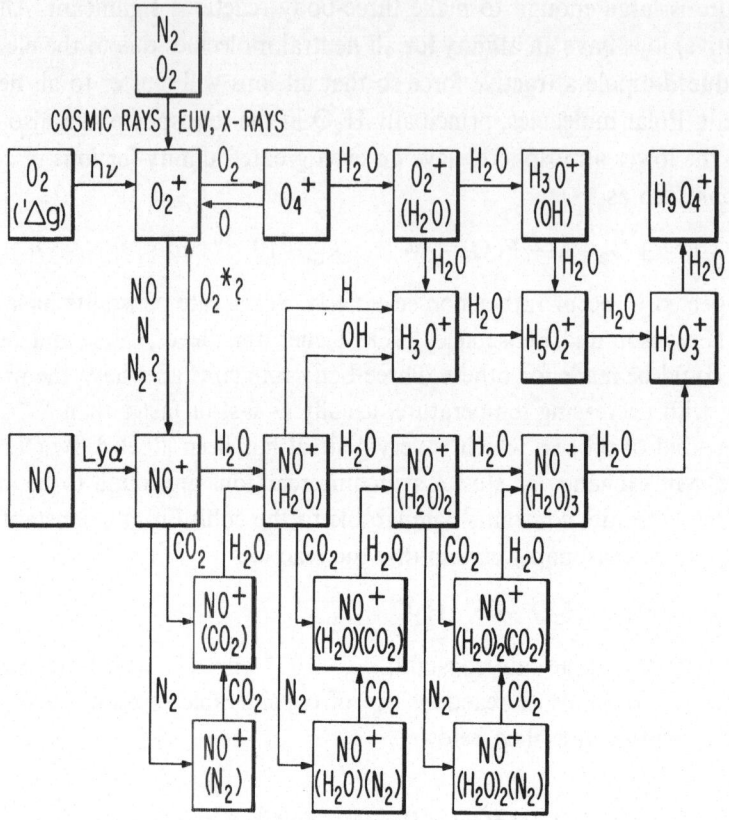

Fig. 6.   Schematic diagram of positive ion reactions in the D region.

remains unmeasured and is beyond present measurement capability. We estimate that reaction (18) cannot be significant with the maximum possible value of k_{18} using HO_2 model concentrations but it would nevertheless be desirable to have a measurement of k_{18}.

The most contentious D region laboratory question at present is the uncertain role of N_2 in promoting NO^+ hydration. Heimerl and Vanderhoff (1974) have made room temperature measurements of the association,

$$NO^+ + N_2 + M \rightleftarrows NO^+ \cdot N_2 + M \tag{19}$$

and Dunkin et al. (1971) made measurements of the equilibrium constant upper limit at 200 K. The results appear to be incompatible but they may not be since the measurements were carried out at different temperatures and conceivable (but not expected) values of ΔH (0.16 eV) and ΔS (16 e.u.) for reaction (19) would make the results compatible. Fehsenfeld has carried out further unpublished results in an attempt to resolve this dilemma. The reverse reaction (19) is exceedingly important because of the weak $NO^+ \cdot N_2$ bond and is extremely sensitive to temperature. Both k_{19}^- and K_{19} are very difficult to measure in the laboratory because of the fast breakup of

$NO^+ \cdot N_2$ and its fast loss by switching with almost any impurity in the system, e.g., NO, CO_2 or H_2O.

The present results are summarized in Table I. Only the equilibrium constant is necessary to establish the role of reaction (19) in the D region. It would be very helpful to have reliable data on reaction (19) at the appropriate D region temperatures and this is proving very difficult to come by.

Other reactions which should be measured include

$$NO^+ \cdot H_2O + N_2 \rightleftarrows NO^+ \cdot H_2O \cdot N_2 \qquad (20)$$

$$NO^+ \cdot H_2O + CO_2 \rightleftarrows NO^+ \cdot H_2O \cdot CO_2 \qquad (21)$$

etc., in order to establish the role of clustering and switching reactions in obtaining $NO^+(H_2O)_3$ which then reacts with H_2O in a binary reaction

$$NO^+(H_2O)_3 + H_2O \rightarrow H_3O^+(H_2O)_2 + HNO_2. \qquad (22)$$

Measurements of reactions like (20) and (21) are exceedingly difficult and are not currently being carried out.

TABLE I

$$NO^+ + N_2 + M \underset{k_r}{\overset{k_f}{\rightleftarrows}} NO^+ \cdot N_2 + M$$

T(K)	M	$k_f(cm^6 \, s^{-1})$	$k_r(cm^3 \, s^{-1})$	$K_{eq}(cm^3)$	$-\Delta G_T^\circ$ (kcal)	Ref.
200	He			$<1(-19)$	<0.5	[a]
300	NO	$3.4(-31)$	$<5(-11)$	$>6.8(-21)$	>-1.1	[b]
80	He	$\sim5(-30)$	$<2(-16)$	$>2(-12)$	>2.4	[c]
296	He			$<3.7(-20)$	<0	[c]

[a] Dunkin et al. (1971).
[b] Heimerl and Vanderhoff (1974).
[c] Fehsenfeld, unpublished results.

5. D Region Negative Ion Chemistry

The basic atmospheric negative ion reaction scheme known from the laboratory studies which have been carried out to date is outlined in Figure 7. The nature of the negative ion chemistry restricts it largely to high pressures, therefore low altitudes, essentially below about 80 km. The main initial negative ion O_2^- is formed in a three-body attachment. Then a series of reactions, involving the parallel sequences initiated by O_3^- and O_4^-, occur which leads ultimately to production of NO_3^-, the predicted 'terminal' negative ion of the D region. Atomic oxygen nips this process in the bud by associative-detachment

$$O_2^- + O \rightarrow O_3 + e \qquad (23)$$

which also limits significant negative ion concentrations to altitudes where the O concentration is low. Associative-detachment reactions which usually involve chem-

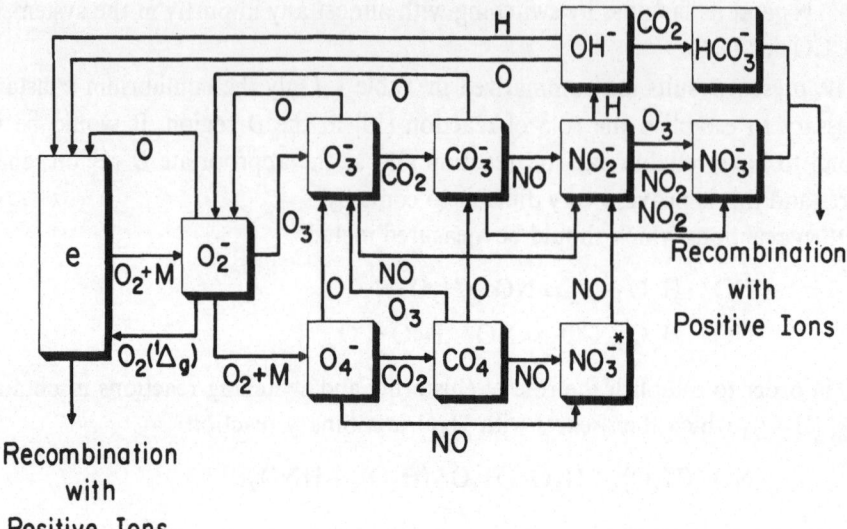

Fig. 7.   Schematic diagram of negative ion reactions in the D region, neglecting hydration.

ically unstable neutrals as in Equation (23), have mostly been measured in flowing afterglow systems. No energy dependences have been obtained for any of the ionospherically important associative-detachment reactions.

One of the chemical (and experimental) complications which arise in the negative chemistry of Figure 7 is the occurrence of two forms of NO_3^-. The form labeled NO_3^-* is found experimentally to have different reactivities and consequently a different energy than NO_3^-. It has been suggested that NO_3^-* is the form $O\!-\!O\!-\!N\!-\!O^-$, produced by switching reactions (exchange of neutrals 'clustered' to O_2^-)

$$O_2^- \cdot CO_2 + NO \rightarrow O_2^- \cdot NO + CO_2 \tag{24}$$

and

$$O_2^- \cdot O_2 + NO \rightarrow O_2^- \cdot NO + O_2 \tag{25}$$

while NO_3^- is the more stable form $O\!-\!N\!-\!O^-$, produced by transfer of a single oxygen atom

$$
\begin{array}{c}
\big| \\[-2pt]
O
\end{array}
$$

$$NO_2^- + O_3 \rightarrow NO_3^- + O_2 \tag{26}$$

or

$$NO_2^- + NO_2 \rightarrow NO_3^- + NO. \tag{27}$$

All ions, positive or negative, will hydrate in the D region if they are long lived enough. A question which has been of concern to a few people for many years is the role that hydration might play in altering the scheme of Figure 7. Only recently have measurements of gas phase hydrated negative ion reactions been obtained (Fehsenfeld and Ferguson, 1974). The first results indicate that hydration will have little effect. In the first place hydration, being a three-body reaction, will only occur to a significant extent for a few ions, NO_3^- of course which is long lived and perhaps CO_3^- and CO_4^- which are relatively long lived because of their small NO loss rates.

Secondly, hydration seems to have little effect. The charge-transfer of O_2^- with O_3 is little affected for example by either one or two waters of hydration, i.e.

$$O_2^-(H_2O)_n + O_3 \rightarrow O_3^-(H_2O)_m + O_2 + (n-m)H_2O \tag{28}$$

has a rate constant $\sim 3 \times 10^{-10}\ cm^3\ s^{-1}$ for $n = 0$, 1, and 2.

On the other hand, the electron detachment reactions of O_2^- by O and $O_2(^1\Delta_g)$ are completely quenched by one water of hydration, since the reactions

$$O_2^-(H_2O) + O \rightarrow O_3 + e + H_2O \tag{29}$$

and

$$O_2^-(H_2O) + O_2(^1\Delta_g) \rightarrow 2O_2 + H_2O \tag{30}$$

are endothermic.

The reaction of O_3^- with CO_2 is somewhat quenched by hydration, however the abundance of CO_2 makes the O_3^- lifetime too short for hydration to occur.

The D region negative ion observations to date are too few, too qualitative and too contradictory to allow more than qualitative comparison with laboratory derived models.

6. Sub-D Region Ion Chemistry

There has been little interest in ion chemistry in the atmosphere below the D region due to lack of ion composition measurements in this region. However, it might be of interest to mention a few of the predictions that can be made on the basis of present laboratory knowledge. In the case of the positive ion chemistry we know that the reactions

$$H_3O^+(H_2O)_n + NH_3 \rightarrow NH_4^+(H_2O)_n + H_2O \tag{31}$$

are fast and quite exothermic (Fehsenfeld and Ferguson, 1973b) so that solvated NH_4^+ ions might be expected to exist in the troposphere.

In the case of negative ion chemistry, it is known that reactions of the type

$$X^-(H_2O)_n + SO_2 \rightarrow X^-(H_2O)_{n-1}(SO_2) + H_2O$$

are fast and exothermic (Fehsenfeld and Ferguson, 1974). In a similar manner HNO_3 replaces both H_2O and SO_2 clustered to negative ions so that it might be expected that some negative ion clusters will include SO_2 and/or HNO_3 in the troposphere.

References

Brown, T. L.: 1973, *Chem. Reviews* **73**, 645.

Burke, R. R.: 1970, *J. Geophys. Res.* **75**, 1345.

Dunkin, D. B., Fehsenfeld, F. C., Schmeltekopf, A. L., and Ferguson, E. E.: 1968, *J. Chem. Phys.* **49**, 1365.

Dunkin, D. B., Fehsenfeld, F. C., Schmeltekopf, A. L., and Ferguson, E. E.: 1971, *J. Chem. Phys.* **54**, 3817.

Fehsenfeld, F. C.: 1969, *Can. J. Chem.* **47**, 1808.

Fehsenfeld, F. C. and Ferguson, E. E.: 1972, *J. Chem. Phys.* **56**, 3066.

Fehsenfeld, F. C. and Ferguson, E. E.: 1973a, *J. Chem.* **59**, 6272.

Fehsenfeld, F. C. and Ferguson, E. E.: 1973b, *J. Geophys. Res.* **78**, 1699.

Fehsenfeld, F. C. and Ferguson, E. E.: 1974, *J. Chem. Phys.* **61**, 3181.

Fehsenfeld, F. C., Ferguson, E. E., and Howard, C. J.: 1973, *J. Geophys. Res.* **78**, 327.

Fehsenfeld, F. C., Dunkin, D. B., and Ferguson, E. E.: 1970, *Planetary Space Sci.* **18**, 1267.

Fehsenfeld, F. C., Howard, C. J., Harrop, W. J., and Ferguson, E. E.: 1975, *J. Geophys. Res.*, in press.

Ferguson, E. E.: 1972, *Radio Sci.* **7**, 397.

Ferguson, E. E.: 1973, *Atomic Data and Nuclear Data Tables* **12**, 159.

Ferguson, E. E. and Fehsenfeld, F. C.: 1968, *J. Geophys. Res.* **73**, 6215.

Heimerl, J. M. and Vanderhoff, J. A.: 1974, *J. Chem. Phys.* **60**, 4362.

Heimerl, J. M., Vanderhoff, J. A., and Niles, F. E.: 1972, Report 1570, Aberdeen Proving Ground, Maryland.

Johnsen, R. and Biondi, M. A.: 1973, *J. Chem. Phys.* **59**, 3504.

Johnsen, R., Brown, H. L., Biondi, M. A.: 1970, *J. Chem. Phys.* **52**, 5080.

Keller, G. E. and Beyer, R. A.: 1971, *J. Geophys. Res.* **76**, 289.

Lindinger, W., Fehsenfeld, F. C., Schmeltekopf, A. L., and Ferguson, E. E.: 1974, *J. Geophys. Res.* **79**, 4753.

McFarland, M., Albritton, D. L., Fehsenfeld, F. C., Ferguson, E. E., and Schmeltekopf, A. L.: 1973, *J. Chem. Phys.* **59**, 6620.

McFarland, M., Albritton, D. L., Fehsenfeld, F. C., Schmeltekopf, A. L., and Ferguson, E. E.: 1974a, *J. Geophys. Res.* **79**, 2005.

McFarland, M., Albritton, D. L., Fehsenfeld, F. C., Ferguson, E. E., and Schmeltekopf, A. L.: 1974b, *J. Geophys. Res.* **79**, 2925.

McFarland, M., Albritton, D. L., and Tellinghuisen, J.: unpublished results.

Rutherford, J. A. and Vroom, D. A.: 1971, *J. Chem. Phys.* **55**, 5622.

Rutherford, J. A. and Vroom, D. A.: 1972, *J. Chem. Phys.* **57**, 3091.

Rutherford, J. A. and Vroom, D. A.: 1974, *J. Chem. Phys.* **61**, 2514.

Rutherford, J. A., Mathis, R. F., Turner, B. R., and Vroom, D. A.: 1971, *J. Chem. Phys.* **55**, 3785.

Rutherford, J. A., Turner, B. R., Mathis, R. F., and Vroom, D. A.: 1972a, *J. Chem. Phys.* **56**, 4654.

Rutherford, J. A., Mathis, R. F., Turner, B. R., and Vroom, D. A.: 1972b, *J. Chem. Phys.* **57**, 3087.

Schmeltekopf, A. L., Ferguson, E. E., and Fehsenfeld, F. C.: 1968, *J. Chem. Phys.* **48**, 2966.

Stebbings, R. F., Smith, A. C. H., and Ehrhardt, H.: 1964, *J. Geophys. Res.* **49**, 2349.

Swider, W.: 1975, this volume, p. 259.

OXYGEN CHEMISTRY

I. T. N. JONES

I.C.I. Limited, Corporate Research and Development Unit, Widnes, WA8 0QX, U.K.

1. Introduction

This review of oxygen chemistry attempts to highlight some of the recent progress in laboratory studies of the $O/O_2/O_3$ system. In addition to the ground states, $O(^3P)$, $O_2(^3\Sigma_g^-)$ and $O_3(^1A_1)$, reference will be made to the first two excited states of O_2 and O, viz. $O_2(^1\Delta_g)$, $O_2(^1\Sigma_g^+)$, $O(^1D)$ and $O(^1S)$, (see glossary), because of their importance in the atmosphere and intimate involvement in the oxygen photochemical system generally. This area of gas kinetics and photochemistry has benefited lately from an upsurge of interest and financial support for environmental reasons and particularly as a result of the suggestion (Johnston, 1971; Crutzen, 1970) that the injection of nitrogen oxides (NO_x) into the stratospheric O_3 layer from the exhausts of supersonic airplanes might lead to catalytic destruction of the Earth's O_3 shield. Clyne (1974) has proposed that oxides of chlorine and sulphur could have a similar effect. A state of controversy, not solely scientific, exists around this complex problem and those interested are referred to the relevant reports (AGARD, 1973; CIAP, 1972). It is sufficient to note here that, in the chemistry of the atmosphere, some minor atmospheric constituents have an importance far exceeding their abundance.

The precise position of the stratospheric O_3 layer depends on latitude and season. The O_3 concentration maximum occurs at about 25 km, although the maximum mixing ratio for O_3 is placed at slightly higher altitudes. The O_3 layer is approximately 10 km thick at half its peak concentration. It is this layer which, by absorbing solar radiation at wavelengths shorter than 290 nm, acts as an invaluable filter and protects living organisms at the Earth's surface from the harmful effects of UV rays. The O_3 in the atmosphere is maintained in a steady state by photochemically initiated reactions between O, O_2, O_3 and other trace constituents and the atmospheric winds. In the lower stratosphere, below 20 km, the atmospheric dynamics are much faster than the photochemical rates and hence the motions are predominant in determining the steady state O_3 distribution in both the vertical and horizontal directions. An important consequence of this effect is that the total global O_3 content is greater than that which could be expected from photochemical equilibrium conditions alone.

Information on the interaction of solar radiation with the gases that surround the Earth is of fundamental interest for understanding the natural atmosphere. The difficulty and expense of making direct measurements means that laboratory studies are essential in this area of great practical importance. Care must be exercised, however, in extrapolating to stratospheric conditions from the data obtained in the laboratory. In particular, in the upper atmosphere, pressures, temperatures and reactant concentrations are generally lower than those normally used in laboratories and the absence of the usual reactor walls may change any, possibly unappreciated, heterogeneous complications. Two important consequences of these differences are: (a) recombination reactions involving a third body which have a negative activation energy can become comparatively more important than more familiar two body processes, and (b) reactions with essentially no activation energy, as they are almost unaffected by temperature changes, remain rapid at the cooler temperatures, ~ 220 K, of the lower stratosphere.

2. Photochemistry

2.1. OZONE

In the UV and visible regions of the spectrum, between 200 and 660 nm, O_3 is the major naturally occurring atmospheric absorber. Its principal spectral feature is the intense Hartley band between 200 and 300 nm with a maximum at 255 nm which rapidly tails off into the much weaker Huggins band extending from 300 to 350 nm. Ozone also has a weak visible absorption, the Chappuis band, centered at 610 nm. Table I gives the long wavelength limits for the production of various electronic states of O and O_2 by O_3 photolysis:

$$O_3 + h\nu \rightarrow O_2 + O. \tag{1}$$

However, it must be emphasized that formation of all the energetically possible products at any given wavelength does not necessarily occur. At 254 nm, and probably for photolysis throughout the Hartley band, it is now generally agreed that the products of reaction (1) are $O_2(^1\Delta_g)$ and $O(^1D)$. Much evidence has been accumulated to show that O_3 photolysis in the visible spectral region leads to ground state products, $O_2(^3\Sigma_g^-)$ and $O(^3P)$ (see Wayne, 1973, for a review).

Photolytic production of $O(^1D)$ in reaction (1) is possible in a spin-conserved process at wavelengths shorter than about 310 nm. This reaction is of central importance to the whole of atmospheric chemistry since it is the major source of $O(^1D)$, which is, in turn, itself the initiator of many subsequent processes involving H_2O, CH_4 and N_2O. Hence the rate of $O(^1D)$ production needs to be known very reliably. As wavelengths increase in the 310 nm region, the O_3 absorption coefficients decrease (Griggs, 1968) and the solar flux incident on the atmosphere increases (Thekaekara, 1972). In addition to the problem of combining these rapidly changing quantities,

TABLE I

Long wavelength limits in nanometers for the production of various states of O and O_2 by photolysis of O_3

O_2 / O	$^3\Sigma_g^-$	$^1\Delta_g$	$^1\Sigma_g^+$	$^3\Sigma_u^+$	$^3\Sigma_u^-$
^3P	1180	611	463	230	170
^1D	411	310	266	167	150
^1S	234	196	179	129	108

early work (DeMore and Raper, 1966; Jones and Wayne, 1969) indicated that near 313 nm the mechanism of O_3 photolysis altered. With increasing wavelength the inferred $O(^1D)$ quantum yields fell off rapidly, reaching zero by 334 nm. Later work, however (Castellano and Schumacher, 1972; Simonaitis et al., 1973) reported much higher values, between 0.5 and 1.0, for the $O(^1D)$ quantum yield at 313 nm, although Castellano and Schumacher (1972) confirmed the zero $O(^1D)$ yield at 334 nm. To clarify this important issue, Lin and DeMore (1973) measured an extensive set of $O(^1D)$ quantum yields for reaction (1) in the 275 to 334 nm region. Their results showed a constant $O(^1D)$ quantum yield at wavelengths shorter than 300 nm, and then a fall off beginning at 305 nm, reaching zero at 320 nm. Lin and DeMore (1973) measured a $O(^1D)$ quantum yield at 313 nm of less than 0.1, in agreement with the values of Jones and Wayne (1970). This sharp fall in $O(^1D)$ production at $\lambda > 305$ nm in O_3 photolysis is reminiscent of a similar drop off in $O(^3P)$ quantum yield at $\lambda > 398$ nm in NO_2 photolysis (Jones and Bayes, 1973).

The basic atmospheric O_3 photochemistry theory was developed by Chapman (1930). The mechanism involves two photochemical and three chemical reactions:

$$O_3 + h\nu \quad \rightarrow O_2 + O \tag{1}$$
$$O_2 + h\nu \quad \rightarrow O + O \tag{2}$$
$$O + O_2 + M \rightarrow O_3 + M \tag{3}$$
$$O + O_3 \quad \rightarrow O_2 + O_2 \tag{4}$$
$$O + O + M \rightarrow O_2 + M. \tag{5}$$

Under the usual stratospheric conditions, O_3 photolysis is 10^3 times faster than oxygen photolysis, and reaction (3) is, to a similar degree, faster than reactions (4) and (5) for O removal.

Approximately 80% of the stratospheric O_3 produced by solar irradiation is not balanced by the Chapman scheme removal steps or by transport to lower altitudes. It is in this area presumably that reactions of O_3 with trace amounts of N, H, C, or halogen containing species is important, even in the unpolluted stratosphere.

2.2. OXYGEN

The principal absorption features of O_2 between 175 and 260 nm are bands of the $O_2(^3\Sigma_u^- \leftrightarrow ^3\Sigma_g^-)$ Schumann-Runge system between 175 and 210 nm and the weak $O_2(^3\Sigma_u^+ \leftrightarrow ^3\Sigma_g^-)$ Herzberg continuum from 175 to 260 nm. Data on the UV spectrum of O_2 have been reviewed by Krupenie (1972). Table II gives the long wavelength limits for the production of excited O in reaction (2). Quantum yield data for this process are not very extensive. At 185 nm, and presumably in the 175 to 242 nm region, the quantum yield for $O(^3P)$ formation is near two. At 147 nm, in the 133 to 175 nm region, the forbidden emission at 630 nm from the $O(^1D \rightarrow ^3P)$ system has been observed (Noxon, 1970) and an approximate calculation indicated that $O(^1D)$ and $O(^3P)$ are produced in equal amounts with a total O quantum yield of two. From 133 nm down to the ionization threshold at 103 nm O_2 absorption consists of a number of strong bands, with a notable narrow window at the Ly-α line (see glossary).

TABLE II

Long wavelength limits in nanometers for the
production of various states of O by photolysis
of O_2

O \ O	^3P	^1D	^1S
^3P	242.4	175.0	133.2
^1D	–	137.5	110.6
^1S	–	–	92.3

Flash photolysis of O_2 in this wavelength region results in emission at 558 nm from the $O(^1S \rightarrow {}^1D)$ system (Filseth and Welge, 1969).

In the atmosphere, production of O by O_2 photolysis occurs principally between 30 and 50 km. Oxygen atoms are effective quenchers of vibrationally excited N_2 (McNeal *et al.*, 1974) and the rate of this process keeps the vibrational temperature of N_2 low at these altitudes. Any singlet state atoms are rapidly deactivated to $O(^3P)$ by N_2 and O_2 (Heidner and Husain, 1973; Slanger *et al.*, 1972) and O_3 is then formed *via* reaction (3).

Several NO absorption bands occur in the same spectral region as the oxygen Schumann-Runge system. Very detailed analysis of the overlap between the two systems is necessary before the effective photolysis rate of NO in the stratosphere can be determined (Cieslik and Nicolet, 1973).

3. Reaction Rates

This review has benefited greatly from previous reaction data compilations. Particular thanks are due to the National Bureau of Standards team for their Chemical Kinetics Data Survey series prepared for CIAP (Garvin and Hampson, 1974). These documents should be referred to for more extensive lists of original references. Emphasis in this section has been placed on those areas of the $O/O_2/O_3$ system where there have been most new measurements. Rate data are given in units consistent with concentrations in molec cm^{-3}. Activation energies are in kcal mol^{-1} and the rate constant of reaction (n) is written as k_n.

3.1. $O(^3P)$ REACTIONS

3.1.1. $O + O_2 + M \rightarrow O_3 + M$

In the Chapman scheme this reaction is the only one reforming O_3. The rate, temperature dependence, and third body efficiency of this process are therefore of great importance in atmospheric oxygen chemistry. Garvin and Hampson (1974) list about a dozen determinations of k_3 over the past decade, most performed at 298 K by pyrolysis/discharge flow techniques. Huie *et al.* (1972) have used a flash photolytic/resonance fluorescence technique in a static system to evaluate k_3 in the 200 to 346 K region. They report $k_3 = 6.6 \times 10^{-35} \exp(1.0/RT)$ cm^6 molec^{-2} s^{-1} for M = Ar and quote relative third body efficiencies for Ar:He:N$_2$ = 1.0:0.9:1.6. The most extensive listing of third body efficiencies, all at 298 K, remains that of Kaufman and Kelso (1967).

3.1.2. $O + O_3 \rightarrow O_2 + O_2$

Early work on this reaction using discharge flow techniques was complicated by production in the discharge of excited species which also reacted with O$_3$. Ellis *et al.* (1971) have indicated that some of the O$_2$ produced is vibrationally excited. Mc-Crumb and Kaufman (1972) have measured the temperature dependence of k_4 in the range 269 to 409 K using thermal decomposition of O$_3$ to make O(^{3}P) and the chemiluminescence with NO to monitor atom concentrations. They quote $k_4 = 1.1 \times 10^{-11} \exp(-4.3/RT)$ cm^3 molec^{-1} s^{-1}. This activation energy is smaller than that determined from high temperature data. A subsequent study (Davis *et al.*, 1973a) using visible photolysis of O$_3$ to generate O(^{3}P) in the 220 to 353 K region obtained data for k_4 in reasonable agreement with those noted above for room temperature, although both the pre-exponential factor and activation energy obtained were slightly higher.

3.1.3. $O + O + M \rightarrow O_3 + M$

This reaction is slow enough to be ignored for most purposes in the lower atmosphere, although production of O$_2$($^1\Sigma_g^+$) in reaction (5) has been observed in the laboratory. A recent determination (Campbell and Gray, 1973) gives $k_5 = 4.8 \times 10^{-33}$ cm^6 molec^{-2} s^{-1} at 298 K for M = N$_2$.

3.1.4. $O + NO_2 \rightarrow NO + O_2$ (*Reaction (6)*)

This aeronomically important reaction is very fast and is now thought to have no measurable activation energy in the 230 to 339 K range. Measurements (Davis *et al.*, 1973b; Bemand *et al.*, 1973; Slanger *et al.*, 1973) by a variety of techniques all using resonance fluorescence to detect O(^{3}P) suggest a value for $k_6 = 9.3 \times 10^{-12}$ cm^3 molec^{-1} s^{-1}, somewhat higher than most previous determinations. An important consequence of these findings is that the rate of reaction (6) at stratospheric temperatures is twice as fast as previously believed.

3.1.5. $O + NO + M \rightarrow NO_2^* + M$ (*Reaction (7)*)

This reaction has been the subject of several classic studies in chemiluminescence. It has a small negative activation energy and consideration of several recent measurements (Stuhl and Niki, 1971; Slanger *et al.*, 1973; Becker *et al.*, 1973) suggests a value for $k_7 = 7 \times 10^{-32}$ cm^6 molec^{-2} s^{-1} at 300 K for M = O$_2$, He or Ar.

3.2. O(^{1}D) REACTIONS

In the atmosphere O(^{1}D) is produced primarily by UV photolysis of O$_3$, although some is formed by far UV photolysis of O$_2$ at higher altitudes. O(^{1}D) has a long radiative lifetime, ~ 145 s, and is chemically reactive. The temperature dependencies of $O(^1D)$ reactions are usually unknown, although they must be small. An extensive review of $O(^1D)$ chemistry has been prepared by Cvetanovic (1973); however, two reactions will be discussed further in this section.

3.2.1. $O(^1D) + O_2 \rightarrow O(^3P) + O_2(^1\Sigma_g^+$ or $^3\Sigma_g^-)$ (Reaction (8))

The absolute rate constant for this reaction, $k_8 = 7 \times 10^{-11}$ cm^3 $molec^{-1}$ s^{-1}, measured recently by Heidner and Husain (1973), is in agreement with earlier work of Noxon (1970). Energy transfer from $O(^1D)$ to O_2 is known to excite $O_2(^1\Sigma_g^+)$ but some vibrationally excited $O_2(^3\Sigma_g^-)$ is also formed. The efficiency of electronic energy transfer relative to the overall quenching of $O(^1D)$ in reaction (8) has been the subject of several studies (see Wayne, 1973). Values quoted for the efficiency of $O_2(^1\Sigma_g^+)$ production range from 0.3 to almost unity.

3.2.2. $O(^1D) + O_3 \rightarrow O_2 + (O_2$ or $2O)$ (Reaction (9))

This reaction is extremely fast (Gilpin et al., 1971; Heidner and Husain, 1973) with a rate constant $k_9 = 2.6 \times 10^{-10}$ cm^3 $molec^{-1}$ s^{-1}. $O_2(^1\Delta_g$ and $^1\Sigma_g^+)$ are not formed in reaction (9), but there is discussion (Webster and Bair, 1970; Davenport et al., 1972; Giachardi and Wayne, 1972) as to the relative importance of the various reaction branches and as to the mode of excitation, if any, of the O_2 formed.

3.3. $O_2(^1\Delta_g)$ REACTIONS

The long radiative lifetime of $O_2(^1\Delta_g)$, ~ 1 h, and its immunity (Wayne, 1973) to deactivation by atmospheric gases, reaction (10):

$$O_2(^1\Delta_g) + M \rightarrow O_2(^3\Sigma_g^-) + M \tag{10}$$

results in significant concentrations of $O_2(^1\Delta_g)$ in the atmosphere. In the 50 to 110 km region, $O_2(^1\Delta_g)$ concentrations are comparable to those of $O(^3P)$ and higher than those of all other potentially reactive species. The $O_2(^1\Delta_g \rightarrow {}^3\Sigma_g^-)$ emission at 1.27 μm is the strongest component of the airglow, although this wavelength band cannot be seen from the ground because of reabsorption in lower layers of the atmosphere.

3.3.1. $O_2(^1\Delta_g) + O_3 \rightarrow 2O_2 + O$ (Reaction (11))

Both $O_2(^1\Sigma_g^+$ and $^1\Delta_g)$ react with O_3 to produce $O(^3P)$. For $O_2(^1\Delta_g)$ this is the fastest atmospherically important reaction. At room temperatures, the results of several studies are in good agreement. Combination of measurements (Findlay and Snelling, 1971; Becker et al., 1972) in the range 283 to 360 K gives a value for $k_{11} = 5 \times 10^{-11}$ $\exp(-5.6/RT)$ cm^3 $molec^{-1}$ s^{-1}.

3.4. OZONE REACTIONS

Several of the more important reactions involving O_3 have been discussed in previous sections. The two most relevant processes that remain, viz. the reactions with NO_x are discussed below.

3.4.1. $NO + O_3 \rightarrow O_2 + (NO_2$ or $NO_2^*)$ (*Reaction* (12))

This reaction is chemiluminescent. By monitoring the light emission, Clyne *et al.* (1964) measured the overall rate constant, $k_{12} = 9.5 \times 10^{-13}$ exp$(-2.5/RT)$ cm^3 molec^{-1} s^{-1}; and also resolved k_{12} into its two component parts with different activation energies. Reaction (12) has assumed importance recently as the basis for measuring the very low, $\sim 10^8$ molec cm^{-3}, NO concentrations in the natural stratosphere (Ridley *et al.*, 1973; Loewenstein *et al.*, 1974). In conjunction with reaction (7), reaction (12) provides the means for the rapid interconversion of atmospheric NO and NO_2 with net destruction of O_3.

3.4.2. $NO_2 + O_3 \rightarrow NO_3 + O_2$ (*Reaction* (13))

Although this reaction is comparatively slow and is dominated by reaction (7) as a loss process for NO_2, in regions of low O concentrations it may be important as a source of NO_3. This radical has a strong absorption band in the visible and, depending on the photolysis products, may provide an alternative route into the O_3 destruction cycle. In the laboratory, reaction (13) is assumed to be followed by:

$$NO_2 + NO_3 + M \rightleftarrows N_2O_5 + M. \tag{14}$$

Recent work (Garvin and Hampson, 1974) suggests a smaller activation energy for reaction (13) than found previously and gives a value for $k_{13} = 1.2 \times 10^{-13}$ exp $(-4.9/RT)$ cm^3 molec^{-1} s^{-1} in the range 231 to 343 K.

4. Conclusion

This review has attempted to summarize some of the recent developments in laboratory studies of the oxygen system. Although laboratory kinetic data are not obtained solely for the benefit of atmospheric models, this particular field has received a great impetus from the various SST environmental programs. Emphasis has, therefore, been placed on the reactions of NO_x with the $O/O_2/O_3$ system, the role of O_3 photochemistry and low temperature rate data. However, typical jet engine exhaust has a higher hydrogen oxide/nitrogen oxide ratio than the natural stratosphere so that H chemistry must also be considered (Kaufman, 1975).

In stratospheric photochemical models, special care must be taken at sunrise and sunset when optical absorption paths are changing rapidly, since not all species may be in a photochemical steady state. At present there is no general agreement as to why stratospheric O_3 levels are as low as observed. Laboratory rate constants are probably sufficiently accurate to eliminate any of the current proposals. Other systems, e.g., those involving different trace constituents or aerosols, must therefore be important. Further speculation is of limited value as any conclusions hazarded at present will undoubtedly become dated within a short time.

References

AGARD: 1973, Conference on Atmospheric Pollution by Aircraft Engines, CP125.
Becker, K. H., Groth, W., and Schurath, U.: 1972, *Chem. Phys. Letters* **14**, 489.
Becker, K. H., Groth, W., and Thran, D.: 1973, *Symp. Combustion* **14**, 353.
Bemand, P. P., Clyne, M. A. A., and Watson, R. T.: 1973, *J. Chem. Soc. Faraday I* **69**, 1356.
Campbell, I. M. and Gray, C. N.: 1973, *Chem. Phys. Letters* **18**, 607.
Castellano, E. and Schumacher, H. J.: 1972, *Chem. Phys. Letters* **13**, 625.
Chapman, S.: 1930, *Mem. Roy. Meteorol. Soc.* **3**, 103.
CIAP: 1972, Second Conference Proceedings, November 1972.
Cieslik, S. and Nicolet, M.: 1973, *Planetary Space Sci.* **21**, 925.
Clyne, M. A. A.: 1974, *Nature* **249**, 796.
Clyne, M. A. A., Thrush, B. A., and Wayne, R. P.: 1964, *Trans. Faraday Soc.* **60**, 359.
Crutzen, P. J.: 1970, *Quart. J. Roy. Meteorol. Soc.* **96**, 320.
Cvetanovic, R. J.: 1973, in D. Garvin (ed.) NBSIR 73–206, p. 7.
Davenport, J., Schiff, H. I., and Welge, K. H.: 1972, *Disc. Farad. Soc.* **53**, 230.
Davis, D. D., Wong, W., and Lephardt, J.: 1973a, *Chem. Phys. Letters* **22**, 273.
Davis, D. D., Herron, J. T., and Huie, R. E.: 1973b, *J. Chem. Phys.* **58**, 530.
DeMore, W. B. and Raper, O. F.: 1966, *J. Chem. Phys.* **44**, 1780.
Ellis, D. M., McGarvey, J. J., and McGrath, W. D.: 1971, *Nature (Phys. Sci.)* **229**, 153.
Filseth, S. V. and Welge, K. H.: 1969, *J. Chem. Phys.* **51**, 839.
Findlay, F. D. and Snelling, D. R.: 1971, *J. Chem. Phys.* **54**, 2750.
Garvin, D. and Hampson, R. F.: 1974, Chemical Kinetics Data Survey, NBSIR 74–430.
Giachardi, D. J. and Wayne, R. P.: 1972, *Proc. Roy. Soc.* **A330**, 131.
Gilpin, R., Schiff, H. I., and Welge, K. H.: 1971, *J. Chem. Phys.* **55**, 1087.
Griggs, M.: 1968, *J. Chem. Phys.* **49**, 857.
Heidner, R. F. and Husain, D.: 1973, *Nature (Phys. Sci.)* **241**, 10.
Huie, R. E., Herron, J. T., and Davis, D. D.: 1972, *J. Phys. Chem.* **76**, 2653.
Johnston, H. S.: 1971, *Science* **173**, 517.
Jones, I. T. N. and Bayes, K. D.: 1973, *J. Chem. Phys.* **59**, 4836.
Jones, I. T. N. and Wayne, R. P.: 1969, *J. Chem. Phys.* **51**, 3617.
Jones, I. T. N. and Wayne, R. P.: 1970, *Proc. Roy. Soc.* **A319**, 273.
Kaufman, F.: 1975, this volume, p. 219.
Kaufman, F. and Kelso, J. R.: 1967, *J. Chem. Phys.* **46**, 4541.
Krupenie, P.: 1972, *J. Phys. Chem. Ref. Data* **1**, 423.
Lin, C. L. and DeMore, W. B.: 1973, *J. Photochem.* **2**, 161.
Loewenstein, M., Paddock, J. P., Poppoff, I. G., and Savage, H. F.: 1974, *Nature* **249**, 817.
McCrumb, J. L. and Kaufman, F.: 1972, *J. Chem. Phys.* **57**, 1270.
McNeal, R. J., Whitson, M. E., and Cook, G. R.: 1974, *J. Geophys. Res.* **79**, 1527.
Noxon, J. F.: 1970, *J. Chem. Phys.* **52**, 1852.
Ridley, B. A., Schiff, H. I., Shaw, A. W., Bates, L., Howlett, C., LeVaux, H., Megill, L. R., and
   Ashenfelter, T. E.: 1973, *Nature* **245**, 310.
Simonaitis, R., Braslavsky, S., Heicklen, J., and Nicolet, M.: 1973, *Chem. Phys. Letters* **19**, 601.
Slanger, T. G., Wood, B. J., and Black, G.: 1972, *Chem. Phys. Letters* **17**, 401.
Slanger, T. G., Wood, B. J., and Black, G.: 1973, *Int. J. Chem. Kinetics* **5**, 615.
Stuhl, F. and Niki, H.: 1971, *J. Chem. Phys.* **55**, 3943.
Thekaekara, M. P.: 1972, *Opt. Spectra*, March 1972.
Wayne, R. P.: 1973, in B. M. McCormac (ed.), *Physics and Chemistry of Upper Atmospheres*, D. Reidel
   Publ. Co., Dordrecht-Holland, p. 125.
Webster, H. and Bair, E. J.: 1970, *J. Chem. Phys.* **53**, 4532.

HYDROGEN CHEMISTRY: PERSPECTIVE ON EXPERIMENT AND THEORY

FREDERICK KAUFMAN

Department of Chemistry, University of Pittsburgh
Pittsburgh, Pa. 15260, U.S.A.

1. Introduction

In view of the rapidly accelerating pace of research in atmospheric kinetics, it is increasingly difficult to review a subject as large as the relevant 'hydrogen' chemistry. This subject may be discussed broadly in terms of the principal mechanisms which are thought to operate in planetary atmospheres, particularly in the terrestrial stratosphere and mesosphere. But such a discussion would necessarily be both superficial and repetitive. Alternately, one may be tempted to do a quick review of the vast store of kinetic information on the many reactions of H, OH, HO_2, H_2O_2, HCl, HNO_3, etc., in order to bring such information up to date. This is a laudable objective, but a note of caution must be sounded. Although it was very important that reviews of that type were first undertaken 10 to 15 yr ago (Kaufman, 1964) when experimental work on elementary reactions was barely getting under way, the situation is very different today, and the reviewing task is either a full-time occupation or an arduous labor of love. It is therefore particularly important that the users of reaction rate information, i.e. aeronomic modelers, not be confronted with too many alternate and often contradictory sources, because such alternate sources of input data cloud the comparison of modeling calculations. Modelers must therefore be encouraged to use the best, most up-to-date sources such as the Leeds evaluations (Baulch *et al.*, 1972) the NSRDS-NBS monographs (e.g., Johnston, 1968), the most recent NBS evaluations (e.g., Garvin and Hampson, 1974), review papers in the *Journal of Physical Chemical Reference Data* and in the *International Journal of Chemical Kinetics* such as those on OH (Wilson, 1972) and HO_2 reactions (Lloyd, 1974) and on stratospheric reactions (Hampson, 1973). The CODATA-3 Bulletin (Garvin, 1971) and the 'review of reviews' paper by Gevantman and Garvin (1973) are particularly valuable guides to and through the reviews thicket.

In this paper, I have chosen a somewhat different approach. Rather than do an extensive literature review, I have attempted to discuss briefly the arsenal of experimental methods available to the chemical kineticist (Section 2) and to indicate some of their advantages as well as shortcomings. Next, I review simple, theoretical approaches (Section 3) which are too often disregarded and which may help us choose or eliminate kinetic data, and finally I bring these considerations to bear on a few reactions of particular interest (Section 4).

B. M. McCormac (ed.), Atmospheres of Earth and the Planets, 219–232. All Rights Reserved.

2. Experimental Techniques

These have an enormous range, from classical global methods in which the overall change of a complex reaction mixture is analyzed in terms of assumed mechanisms, through flash photolysis and discharge flow methods in which the concentrations of reactive species are monitored and in which elementary steps may be successfully isolated for direct study, albeit with only slight state selection, to the crossed molecular beam method which unravels the detailed dynamics of a molecular encounter.

Among the classical methods we may distinguish thermal and photochemical ones. Among the former there are three categories: (a) The study of slow thermal reactions including the onset of explosion, usually in the temperature range 600 to 1000 K: (b) The study of flame processes, i.e., the monitoring of temperature and species concentrations as functions of distance in flames, and the comparison of such data with computer calculations, usually at 1000 to 2000 K; and (c) The study of endo-thermic reactions in shock tubes at temperatures above about 1500 K.

Category A has recently enjoyed a renaissance, in considerable measure due to the work of R. R. Baldwin's group, (e.g. Baldwin *et al.*, 1974), which started out with a careful characterization of the second explosion limit in H_2-O_2 mixtures (due to the competition of chain branching steps and HO_2 formation) and was followed by investigations of the partial inhibition of the H_2-O_2 reaction by small amounts of added hydrocarbons interpreted by the competitive reactions of H or OH with the hydrocarbons as well as with O_2 and H_2. On the assumption of complicated mecha-nisms and of reproducible behavior by various reactive species upon collision with the vessel surface, such data lead to rate constant ratios, and ultimately to absolute values of rate constants. Later expansion of this work included measurements and computer analysis of induction periods in H_2-O_2 explosions and of relative rates of the slow oxidation of mixed fuels, all giving rise to kinetic information. In spite of lingering doubts caused by the extreme complexity of the reaction systems, the results are amazingly consistent and the virtuosity of the analysis is admirable.

Category (b) was initiated by the work of Sugden and of Fenimore in the 1950's and is now mainly carried on by Dixon-Lewis and co-workers. It requires accurate con-centration and temperature vs. distance data in one-dimensional, premixed flames as well as binary diffusion coefficients of all species, and leads, via extensive computer analysis (Dixon-Lewis, 1968) to sets of rate constants consistent with flame experi-ments at various initial compositions and total pressures. As in (a) above, it is difficult to appraise the uniqueness of these computer fits and their sensitivity to simultaneous changes of several of the many parameters, but the analysis provides a valuable addi-tion to the more direct methods.

Category (c), shock tube studies, represents a vast area of research, predominantly at very high temperatures and for the study of dissociation-recombination processes not of primary interest here, and will not be discussed.

Classical photochemical methods mainly involve the measurement of quantum yields, i.e., fractional yields of reactant or product molecules destroyed or formed per quantum absorbed, as a consequence of the continuous (relatively weak) irradiation of a gas mixture, and therefore lead to rate constant ratios based on assumed mechanisms, usually invoking the steady-state approximation. Such indirect approaches have lately become less popular, mainly because of the complexity and uncertainty of the analysis and because of the availability of more direct methods.

The second group of experimental techniques, flash photolysis in stationary reactants or discharge excitation in flowing reactants, has become the mainstay of research on elementary atmospheric reactions and is carried on by 20 to 30 active groups around the world. The two basic configurations may be described and critically evaluated as follows: In the flash photolysis-resonance fluorescence apparatus of Braun, Davis, and co-workers (Braun and Lenzi, 1967; Davis et al., 1970), for example, repetitive vacuum UV flashes along the x-direction of a photolysis cell produce the desired atom or radical species, which are excited by a continuously operating resonance line source along the y-direction, and whose fluorescence is observed along the z-direction. If necessary, the decay of the monitored species is coherently summed over many photolytic flashes on a multichannel analyzer until statistical errors become sufficiently small. The cell temperature may be controllable from 150 to 600 K. The original gas mixture is prepared in such a manner that the active species are generated reasonably uniformly in space and that the other reactant is present in concentrations which assure a manageable decay rate. Where applicable, this method has several outstanding advantages over all others such as the absence of surface effects, high detection sensitivity, easy attainment of large signal-to-noise ratio, applicability to experiments at moderately high pressures, and absence of interference by subsequent reactions because of low concentration of reactive species (10^8 to 10^{10} cm^{-3}). The variant of the method which employs resonance line absorption rather than fluorescence is both less sensitive and less accurate, but has the advantage of a more quantitative estimate of the absolute concentration of absorbers and of the absence of interference effects due to quenching of the resonance fluorescence. It has been used extensively by Husain and co-workers in the measurement of rate constants for the reactions of many ground-state and metastable atomic species with a variety of other reactants.

The limitations of the method include the photolytic generation of the species to be studied which is sometimes difficult (e.g., for N-atoms) and which may restrict the choice of other reactants because of their absorption; and lack of flexibility due to the requirement that a single gas mixture be prepared which is to contain all the reactant and precursor molecules and which must be unreactive before the flash.

The absence of these limitations makes the discharge-flow technique particularly attractive. To describe it briefly, time variation is here traded for space variation under steady-state (preferably one-dimensional) flow condition. The atom/radical generation region is thereby removed from the reaction zone, usually a cylindrical tube, and a variety of production mechanisms such as glow discharge, high temperature thermal dissociation, photolysis, or chemical reaction of other precursors can be utilized. Two or more gas streams carrying different reactive species can therefore be brought together immediately upstream of the reaction zone. A large variety of detectors may be used. These included chemiluminescent emission, electron spin resonance, resonance absorption or fluorescence, catalytic probe calorimetry, mass spectrometry, and chemical analysis. Any one of three reaction geometries is commonly employed: (a) Fixed position of radical source, of reactant addition inlet(s) (single or multiple), and of detector. When a single addition inlet is used, kinetic information is then obtained only by varying the concentration of added reactant under otherwise constant conditions, a mode of operation which has been used mainly in ion-molecule reaction studies (Ferguson *et al.*, 1969). With multiple inlets, a constant reactant flow may be switched from port to port, but such kinetic data are both less reliable and less extensive than those in (b) and (c) below. (b) Fixed source, fixed reactant addition inlet, and movable detector. This is feasible for detection by chemiluminescence, light absorption, or probe calorimetry, but not for resonance fluorescence or mass spectrometry where the detector is complex or delicately adjusted. However, this is the optimum geometry for steady-state plug flow experiments, because it is equivalent to monitoring the desired species concentration in real time when one-dimensional plug flow is applicable, i.e. $d/dt = \bar{v}\, d/dx$ where $\bar{v}$ is the average flow velocity. (c) Fixed source and detector, movable reactant injector. This geometry provides the advantages of simple operation, optimization of detector performance, and easy temperature control of the reaction zone. It permits simple data reduction if and only if the radical decay processes are pseudo-first-order and surface loss processes are either unaffected by the added reactant or simply and reversibly related to its concentration downstream of the movable inlet (Westenberg and deHaas, 1967, 1969; Anderson *et al.*, 1974). These provisos become distinct disadvantages when higher-order radical reactions or complex surface effects are present, because the measured concentration at the fixed detector represents the integrated effect of such non-linear rate terms and thus becomes a complicated function of the desired rate constant and of other parameters.

Common to the successful interpretation of all flow experiments is the need for the characterization of transport effects such as radial and axial diffusion and viscous pressure drop, and for the inhibition of surface removal of the monitored species (Kaufman, 1961; Poirier and Carr, 1971). It is these considerations which limit the applicability of the discharge-flow method: At pressures above about 10 torr, the one-dimensional plug flow approximation becomes increasingly poor, because radial diffusion is no longer able to counteract the effects of the laminar flow velocity distribution; and at low pressure, surface reactions, which always compete with the processes under investigation, are often irreproducible, poorly controlled, and too fast.

Little can be said here about the rapidly growing field of molecular beam studies and chemical dynamics. The detailed insight into energy requirement and energy disposal in reactive encounters which these advanced methods provide is not usually required in atmospheric research. Moreover, the method is relatively weakest at establishing total reactive cross sections with good accuracy. It is important, however, to point out what marvelous progress has recently been achieved in this field through improvements in high vacuum techniques, in universal detectors, in beam intensity through the introduction of nozzle beams, and in computer data handling. Thus, reactions of H-, O-, and F-atoms are now being studied successfully, as are those of large organic molecules, and it is only a matter of a short time before beam sources of OH, O_3, and perhaps even HO_2 will be developed. The interested reader may wish to peruse the recent Faraday Discussion No. 55, 1973, to gauge the present status of this rapidly accelerating field.

3. Reaction Rate Theory in Atmospheric Science

It is clear that this brief discussion can deal only with the simple, gross features of rate theory. It will restrict itself to neutral reactions and then mainly to those which take place adiabatically, i.e., on a single electronic hypersurface. Two problems will be touched: the two-body encounter, $A + B \rightarrow C + D$, and the recombination-dissociation process $A + B \rightleftarrows AB$. These differ in an essential way, since the former normally comes about in a single encounter whereas the latter is necessarily made up of a large set of energy transfer (inelastic but nonreactive) collisions with other molecules, in addition to a single bondforming (or breaking) step.

3.1. BINARY REACTIONS, $A + B \rightarrow C + D$

It must first be realized that the accurate, *ab initio* solution of the full problem of reaction dynamics is incredibly complex and totally out of reach of even the best of theorists armed with the fastest computers except in the simplest cases such as $H + H_2$ or perhaps $F + H_2$. In the Born-Oppenheimer approximation the problem may be broken down into two parts, the calculation of the potential energy surface and that of the dynamics on that surface. For the latter, it seems that the quantum-mechanical scattering calculations may often be substituted by classical trajectory calculations with random initial conditions and statistical averaging of results, a relatively simple procedure on a fast computer. The potential energy surface problem, however, is still in poor shape. Simple semi-empirical approximations such as LEPS and particularly BEBO often estimate barrier heights with fair accuracy, but one can not be confident of success in every case. Important atmospheric reactions are usually so fast, moreover, and temperatures so low that inordinate accuracy would be required for a really useful prediction. One tends to fall back, therefore, on a combination of experimental data, transition-state theory or its quasi-thermodynamical analog (Benson, 1968), chemical analogy, and intuition.

Yet, this is not meant to imply the rejection of all simple theoretical arguments nor is it an invitation to accept any and all experimental result uncritically. Before trying to reduce this problem to a set of useful guidelines, let us briefly examine the foundation of transition-state theory and its expectation of success. That theory (Laidler, 1969) postulates the existence of an intermediate, activated, half-reacted state of well-defined energy and geometry, in equilibrium with reactants, whose rate of crossing from reactants to products, along a single, internal reaction coordinate, is given by a universal frequency, independent of all other degrees of freedom of the pseudo-molecule. This disposes of the dynamics of the problem and leaves only the energetics and statistical mechanics of the critical, intermediate state to be determined. Now, for reactions with a sizable energy barrier and with a clearly predictable reaction coordinate and transition state geometry, this simple theory is quite successful, i.e., it will predict rate constants to better than an order of magnitude accuracy. Unfortunately, these prerequisites are not met by many important atmospheric reactions which frequently have small energy barriers and uncertain or variable transition state geometries. In such reactions, the notion of a transition state may lose its usefulness and meaning altogether. If the theory is used nevertheless, to calculate the rate constant, errors of several types will arise: (a) The assumption of too restricted a transition state configuration will lead to an underestimate of the rate constant by discriminating too much against other geometries for successful encounters; (b) In the use of harmonic oscillator partition functions for weak bending vibrations of the transition state, the rate constant may be overestimated and an equivalent collisional solid angle $\geqslant 4\Pi$ introduced which is obviously meaningless (Mayer *et al.*, 1966). (c) If the dynamics of the reaction are such that excitation of specific degrees of freedom of the reactants is particularly efficient in bringing about reaction, for example if vibrational energy is required to overcome the energy barrier, the overall rate constant will be smaller than that calculated by transition state theory, i.e. will correspond to a transmission coefficient, $\varkappa$, less than unity. (d) Quantum effects (tunnelling) need to be corrected for (very imperfectly) in the case of H-atom reactions at low temperatures. They will tend to increase the rate constant.

It is particularly important to distinguish among the several energy quantities which characterize such rate processes and to analyze correctly the experimental temperature dependence of the rate constant. The magnitude of experimental errors is such that one is rarely justified to use more than two paramenters to represent laboratory data over a fairly small temperature range, and normally one presents them in the form of the Arrhenius expression, $k = A \exp(-E/RT)$. The temptation to introduce an additional fixed power of T, e.g. $A'T^{1/2} \exp(-E'/RT)$, should be resisted, as it adds neither to data analysis nor to theoretical insight. It must be remembered, however, that neither A nor E have clear, physical significance and that they both depend on T in a complicated manner. In simple transition state theory, $k = (kT/h) \times \times Q_{AB}^{\dagger}/Q_A Q_B \exp(-\Delta E_0/RT)$, where $Q_{AB}^{\dagger}$ is the total partition function of the transition state (exclusive of the degree of freedom corresponding to the reaction coordinate), Q_A and Q_B are the total partition functions of the reactants, and ΔE_0 is the energy difference between $AB^{\dagger}$ and reactants at $0\,K$. Translational and rotational partition functions are proportional to $T^{1/2}$ and contribute $RT/2$, each, to E, but the vibrational degrees contribute from O to RT, each, depending on frequency and temperature (or to $RT/2$ if the vibration becomes an internal rotation) and the total effect of the Q's is normally one of rising n with increasing T in a $T^n \exp(-\Delta E_0/RT)$ representation of the rate constant, or a rising $|E|$ in the Arrhenius expression. This is due, essentially, to the fact that in a two-body process there is a change from two molecules to one in the formation of the transition state with an attendant replacement of translational and rotational degrees of freedom (of constant heat capacity) by vibrational ones (of increasing heat capacity with increasing T).

This natural increase of $|E|$ over a large temperature range throws doubt on the wisdom of trying to fit experimental data from 300 to 2000 K to a single Arrhenius expression, just as it would be unwise to expect ΔH and ΔS of a recombination reaction to remain constant over such a large temperature range. The effect is frequently encountered experimentally, i.e., simple reactions (e.g., OH + CO, OH + OH) are found to have substantially lower Arrhenius activation energies near 300 K than near 2000 K.

Regarding non-adiabatic, curve-crossing processes, a brief cautionary note: Apart from the difficult problem of the crossing probability it is important to make a distinction between direct and compound encounters, i.e., between those in which the collision lasts $\sim 10^{-13}$ s and those in which an intermediate complex of much longer lifetime is formed. For the latter, the curve-crossing probability may be greatly increased, because the critical curve-crossing geometry is traversed many times (Tully, 1974).

An equally brief word about excited state reactions. Such processes are controlled by electronic state correlation. When that correlation is fully allowed, and if there are no energy barriers, rate constants may be very large, indeed, because the higher energy of the excited state brings about a large variety of encounters along attractive potential energy surfaces. It should therefore not be surprising that $O(^1D)$ reacts with most hydride molecules at gas-kinetic rates when we remember that ground-state $OH(^2\Pi)$ correlates with ground-state $H(^2S)$ plus $O(^1D)$. The higher energy $O(^1S)$ has no such correlation and reacts very much more slowly.

Lastly, a brief check list to facilitate an enlightened estimate of a rate constant: (1) Check electronic states of reactants and products for correlation (at least spin, in simple cases also orbital angular momentum); (2) If adiabatic, estimate a likely transition state geometry and barrier energy by analogy or simple calculation (BEBO), and (3) Calculate the rate constant in transition state (or thermochemical kinetics) approximation within the restrictions mentioned above.

3.2. RECOMBINATION-DISSOCIATION REACTIONS

This is obviously not the place to present a thorough discussion of unimolecular reactions, and the reader is referred to two fine, recent monographs (Robinson and Holbrook, 1972; Forst, 1973). Let us keep in mind the crux of this problem, viz., the all-important competition between vibrational energy transfer and spontaneous dissociation of energy-rich molecules which control the rate of the overall process. The lifetime for spontaneous dissociation is a sensitive function of molecular complexity and this has three important consequences: It shifts the 'fall-off' pressure range, where the recombination is in transition from its high pressure (second-order) to its low pressure (third-order) limit, from very high pressures (hundreds of atmospheres) for atom recombination to very low pressure (<1 torr) for complex radical recombination; it broadens the fall-off pressure range over which this transition takes place; and it changes the temperature dependence from that of a simple, binary, bond-forming encounter at the high pressure limit (where subsequent energy transfer steps are fast and do not control the rate) to that of the lifetime of the energy-rich molecule multiplied by two simple collisional rate constants at the low pressure limit (where the strong, negative temperature dependence of that lifetime controls the T behavior of the process for polyatomic species). The realization that the recombination of OH and NO_2 to form NHO_3, for example, may be in the fall-off regime in the lower stratosphere and that its temperature dependence may change from near zero in the troposphere to moderate negative in the mid-stratosphere has come as a surprise to aeronomers.

The parameters of the detailed models needed in the quantum-statistical (RRKM) theory of such reaction rates are somewhat arbitrary, since they are dependent on the choice of a transition state in the recombination step. Normally, this transition state is a loosened, extended version of the final product molecule, and its geometry and vibration frequencies may be estimated reasonably well. In that sense, the check list for recombination reactions corresponding to that for binary steps of the preceding section is simpler, at least when the recombination is electronically adiabatic.

4. Brief Review of Selected Reactions

In the spirit of the above evaluations of experiment and theory, the last section will review the present state of affairs for a number of important OH- and HO_2-reactions. It is subdivided simply along lines of 'proprietary' interest into one subsection dealing with OH-reactions recently investigated in our laboratory, and another one dealing with a few critical HO_2-reactions studied elsewhere.

4.1. RECENT MEASUREMENTS OF OH-REACTIONS FROM OUR LABORATORY

The steady-state discharge-flow method using the fixed source, fixed detector, movable reactant injector configuration has been successfully applied to the study of five important OH-reactions over temperature ranges of about 230 to 440 K. The principal advance over earlier flow tube investigations has been the use of resonance fluorescence for the measurement of OH concentrations which has increased detection sensitivity by two to three orders of magnitudes over electron spin resonance and UV absorption, to a level of about 10^9 cm^{-3} for the minimum detectable concentration (S/N = 1). Two flow tube systems were used, of which the newer one incorporated the following improvements: (a) modular construction of the flow tube for easy replacement in whole or in part with several sections including additional ports making up a fixed total length; (b) stainless steel construction of upstream mixing chambers and downstream fluorescence cell; (c) multiple resonance fluorescence cell for simultaneous monitoring of up to three reactive species; and (d) vacuum UV monochromator (McPherson, Model 218) and discharge resonance lamps for the measurement of atomic and molecular species. This last improvement has not yet been utilized in the applications described below.

A general remark may be made regarding the assessment of experimental errors. Random errors as measured by one single standard deviation in the least squares analysis of sets of kinetic experiments under similar conditions were usually near 5%. Systematic errors were estimated and were thought to bring the overall accuracy to the range of about $\pm 15\%$ to 20%. Much interesting information on the surface removal of OH as a first-order process or as a heterogeneous reaction with NO and NO_2 was also obtained and is described in some of the publications. Rate constant expressions for the five reactions will next be presented, compared with other results, and briefly discussed.

4.1.1. $OH + O_3 \rightarrow HO_2 + O_2$

The Arrhenius expression, $k = 1.3 \times 10^{-12} \exp(-1900 \text{ cal mole}^{-1}/RT) \text{ cm}^3 \text{ s}^{-1}$ over the temperature range 220 to 450 K has remained the only result of a direct measurement (Anderson and Kaufman, 1973). Two recent estimates by indirect photochemical methods have also been reported. DeMore (1973) photolyzed mixtures of O_2, H_2O and CO until they reached a steady-state O_3 concentration and determined the $OH + O_3$ rate constant by the competition of O_3 with CO, whose rate constant for reaction with OH is well known, or alternately by the dependence of the O_3 yield in the presence of large excess CO, where rate constant ratios of HO_2 reactions are also required. Based on a 10-step reaction mechanism a range of 7 to $8 \times 10^{-14} \text{ cm}^3 \text{ s}^{-1}$ was obtained at 300 K, compared to $(5.5 \pm 1.5) \times 10^{-14}$ by the direct method. Simonaitis and Heicklen (1973) photolyzed O_3-H_2O mixtures in the presence and absence of O_2 and arrive at a range of 0.3 to 1.5×10^{-14} on the basis of the decrease in the quantum yield for O_3 destruction when O_2 is present, using a 15-step mechanism, uncertain chain length and surface termination, and a paucity of experimental data. Such a result is an order of magnitude estimate whose error limits can not be ascertained and which shows the limitations and pitfalls of some bulk photochemical methods.

A cursory application of the above-mentioned theoretical guidelines suggests that (a) correlation rules present no problem, i.e., that the reaction is adiabatic; (b) the energy barrier should be low (in uncertain analogy to the simpler $OH + O$ and $H + O_3$ reactions); and (c) a pre-exponential factor near $10^{-12} \text{ cm}^3 \text{ s}^{-1}$ seems qualitatively reasonable. A rigorous application of Eyring theory would seem to be of limited usefulness.

4.1.2. $OH + NO_2 + M \rightarrow HNO_3 + M$

For the details of this and the equivalent $OH + NO$ study, the reader is referred to the recent publication (Anderson et al., 1974). Third-order rate constants of 1.0, 1.0, and $2.3 \times 10^{-30} \text{ cm}^6 \text{ s}^{-1}$ at 295 K for M=He, Ar, and N_2 and a temperature dependence of $T^{-2.5}$ or $\exp(+1800 \text{ cal mole}^{-1}/RT)$ were found. Although other studies are too numerous to be discussed here, it is gratifying to note the good agreement, with the study of Westenberg and deHaas (1972a), and with the discharge-flow, far-IR-laser-magnetic-resonance absorption study of Howard and Evenson (1974).

RRKM theory was applied to this well-behaved simple recombination reaction by Tsang (1973) with very reasonable results except that the calculations predict a moderate ($\sim 40\%$) decrease in the low pressure, third-order rate constant in the 1 to 10 torr pressure range, which has not been observed experimentally. The reaction is, of course, of very great importance in the NO_x—HO_x—O_3 photochemistry of the lower stratosphere where it ties up part of the NO_x in the catalytically inactive and more readily removable form of HNO_3. It must be stressed that both the pressure and the temperature dependence of the recombination rate constant are in their transition range in the lower stratosphere so that modeling calculations should neither use simple two- or three-body rate expressions nor a single temperature dependence. The above $T^{-2.5}$ or $\exp[+1800 \text{ cal mole}^{-1}/RT]$ expression, for example, applies at altitudes above about 30 to 35 km, but the T exponent becomes more positive at lower altitudes, reaching an as yet undetermined value, probably between -1 and O, at ground pressure.

4.1.3. $OH + NO + M \rightarrow HNO_2 + M$

This reaction is kinetically very similar to the preceding one although atmospherically much less important because of the lower stability of HNO_2. On the basis of simple theory one would expect its low pressure, third-order rate constant to be smaller and its temperature dependence to be slightly less negative than that of its HNO_3 counterpart. This is experimentally borne out (Anderson et al., 1974), $k = 0.33$, 0.34, and $0.58 \times 10^{-30} \text{ cm}^6 \text{ s}^{-1}$ at 295 K for M = He, Ar, and N_2, and a $T^{-2.4}$ or $\exp(+1700 \text{ cal mole}^{-1}/RT)$ dependence was reported, in good agreement with several other investigations which used static or flow methods. The theory would also predict the transition regime to lie at somewhat higher pressures and to approach a somewhat larger high pressure, second-order rate constant than for HNO_3. Experimental data on these points are sparse and conflicting.

4.1.4. $OH + CH_4 \rightarrow H_2O + CH_3$

This reaction which represents a major initial step in the stratospheric oxidation of CH_4 (and formation of water) was studied at 290 to 440 K under conditions where secondary radical-radical reactions do not affect the decay of OH because of its low initial concentration due to the great sensitivity of the resonance fluorescence detection method (Margitan et al., 1974). An Arrhenius expression of $3.8 \times 10^{-12} \exp(-3660 \text{ cal mole}^{-1}/RT)$ was reported, in good agreement with earlier flash photolysis work by Greiner (1970), but not with Wilson's review (1972) which proposes a higher activation energy of 5000 cal mole^{-1} in order to fit a single Arrhenius expression over a wide temperature range. Although the magnitude of the energy barrier of this reaction can not be predicted quantitatively, the larger pre-exponential factor is quite reasonable, because of the reaction path degeneracy, $l = 4$, i.e., equivalent reaction probability at each of the four H-atoms of CH_4.

4.1.5. $OH + HCl \rightarrow H_2O + Cl$

Increasing concern over the possibility of catalytic O_3 destruction by Cl and ClO_x reactions has led to a study of this reaction over the temperature range 224 to 460 K in the flow tube resonance fluorescence system (Zahniser *et al.*, 1974). An Arrhenius expression of $2.0 \times 10^{-12} \exp(-620 \text{ cal mole}^{-1}/RT)$ was obtained, in good agreement with static flash-photolysis results of Smith and Zellner (1974), but not with an estimate by Takacs and Glass (1973) which combined their own room temperature value (in excellent agreement with ours) with high temperature results of Wilson *et al.* (1969). The rate constant expression seems to be entirely consistent with simple theory, both in terms of the small Arrhenius activation energy and the magnitude of the pre-exponential factor.

4.2. Brief Discussion of some HO_2 Reactions

To conclude this paper, the present state of the kinetics of two important radical-radical reactions involving HO_2 will be summarized. These show very well the diversity of experimental methods which can be brought to bear on such problems, but they also show the need for direct measurements.

4.2.1. $OH + HO_2 \rightarrow H_2O + O_2$

This reaction is a major loss process for odd H in the stratosphere and thereby plays an important role in establishing the total HO_x concentration, which, in turn, determines the magnitude of the catalytic destruction of O_3 by HO_x. Its rate constant was first estimated by this author to be $\geqslant 10^{-11} \text{ cm}^3 \text{ s}^{-1}$ (Kaufman, 1964) on the basis of Foner and Hudson's (1962) experimental mass spectrometer investigation of HO_2. A direct study of this reaction, using UV absorption of HO_2 (Hochanadel *et al.*, 1972) following the flash photolysis of H_2O in mixtures with O_2 and excess He, Ar, or H_2 led to the very high value of $2 \times 10^{-10} \text{ cm}^3 \text{ s}^{-1}$ which is widely used in modeling calculations, but is undoubtedly incorrect as further discussed below. Computer analysis of detailed composition and temperature profile measurements for several fuel-rich, near stoichiometric and fuel-lean flames has led Dixon-Lewis *et al.* (1974) to propose a value of $(3 \pm 1.5) \times 10^{-11} \text{ cm}^3 \text{ s}^{-1}$, independent of temperature in the 1300 to 1600 K range. Although these calculations are complicated and involve 5 to 10 elementary reactions, a value as large as 2×10^{-10} seems to be excluded.

Experimental objections to the flash photolysis result are based on the serious concentration gradients which must have been present in the 1 or 2 cm diameter cell. At water concentrations of 0.03 atm ($7.4 \times 10^{17} \text{ cm}^{-3}$) whose absorption cross section is between 2 and $5 \times 10^{-18} \text{ cm}^2$ at 1600 to 1750 Å, the corresponding attenuation for a 1 cm path length ranges from 4 to 40-fold, i.e., H and OH were generated at much higher concentrations near the cell wall than at its center. For the radical-radical reactions $HO_2 + OH$ and $HO_2 + HO_2$, such inhomogeneity effects are squared in the rate expression, will lead to an underestimate of radical concentrations (overestimate of the absorption coefficient of HO_2) and to a large overestimate of the rate constants. All of these errors are apparent: The absorption coefficient is 50 to 60% too large

(Paukert and Johnston, 1972), the rather well established rate constant for the $HO_2 + HO_2$ reaction, 3.0×10^{-12} (Foner and Hudson, 1962) and 3.6×10^{-12} (Paukert and Johnston, 1972), is reported to be 9.5×10^{-12} cm^3 s^{-1}, and the $HO_2 + OH$ is overestimated still more severely.

It is, of course, obvious on simple theoretical grounds that a diatomic plus triatomic atom transfer reaction can not have a hardsphere, gas-kinetic rate constant. For valid comparison one may look at the pre-exponential factors of the $O + H_2O_2$ and $OH + H_2O_2$ reactions which are, approximately, 3.6×10^{-11} and 1.7×10^{-11} cm^3 s^{-1}. When we keep in mind that these reactions have a reaction path degeneracy of 2, the pre-exponential factor is reduced to 1 to 2×10^{-11} per H-atom. On this basis, the rate constant for $OH + HO_2$ is estimated to be $(2 \pm 1) \times 10^{-11}$ cm^3 s^{-1} near 300 K and to have a small, positive temperature dependence.

4.2.2. $H + HO_2 \xrightarrow{a} H_2 + O_2$
$$\xrightarrow{b} 2OH$$
$$\xrightarrow{c} H_2O + O$$

This set of reactions which is important in the mesosphere as a loss process for odd H, as a source for H_2, and thereby as a major link in the hydrogen escape, has been studied by most of the methods summarized earlier with contradictory results. The work is reviewed extensively by Lloyd (1974). Four discharge-flow studies near 300 K have attempted to measure the relative magnitude of k_a, k_b, and k_c. For the fraction $k_a/(k_b + k_c)$ values of 0.51 (Clyne and Thrush, 1963); 0.77 (Dodonov et al., 1969); 0.75 (Bennett and Blackmore, 1971); and 1.63 (Westenberg and deHaas, 1972b) have been reported, but all experimental studies are open to serious criticism such as neglect of the very rapid surface recombination loss of OH under steady-state conditions where [OH] is 10^{12} to 10^{13} cm^{-3}. In the absence of special surface treatment (Anderson et al., 1974) first-order surface recombination rate constants of ~ 50 s^{-1} have usually been measured in one inch diameter flow tubes which amounts to a much larger loss rate than the gas phase $OH + OH \rightarrow H_2O + O$ reaction. Moreover, the ultimate products of the surface reaction are unknown. Other potential sources of error include the neglect of all radical reactions subsequent to reactions (a), (b), and (c) (Dodonov et al., 1969), and excessively high pressures for plug flow interpretation (Bennett and Blackmore, 1971).

Dixon-Lewis and co-workers (Day et al., 1973) have reported very much lower ratios of ~ 0.15 for $k_a/(k_b + k_c)$ in hydrogen-rich H_2—O_2—N_2 flames at 1000 to 1100 K, a result which is supported by Baldwin's group, and has led Dixon-Lewis to propose an activation energy for reactions (b) and (c) which is 1.5 to 2 kcal mole^{-1} higher than that for reaction (a). In the most recent study (Dixon-Lewis et al., 1974), $k_a/(k_b + k_c) = 3.6 \times 10^{-2} \exp(+2200 \text{ cal mole}^{-1}/RT)$, a result which must be considered very uncertain. In order to arrive at estimates for the absolute magnitude of these rate constants, one must make use of computer studies of the H_2—O_2 reaction near 770 K (Baldwin et al., 1967, 1974) which leads to values of 7.8×10^{19} cm^{-3} s, and more recently 1.5×10^{20} and 2.0×10^{20} cm^{-3} s for the ratio $k_b^2/k_1^2 k_2$ where k_1 and k_2 are the reasonably well-known rate constants for the reactions $H + O_2 \rightarrow OH + O$ and $HO_2 + HO_2 \rightarrow H_2O_2 + O_2$. Dixon-Lewis thus reports $(k_b + k_c) = 5 \times 10^{-10} \exp(-2200$ kcal $mole^{-1}/RT)$ and $k_a = 1.2 \times 10^{-11}$ cm^3 s^{-1} as well as $k_b \approx 10 k_c$. Although the magnitude of the rate constants and the k_b/k_c ratio are reasonable, the pre-exponential factor of k_b is excessively high as is its activation energy. The three processes might be visualized as follows: Reaction (a), attack of H on the H-end of HO_2, clearly a 'direct' reaction, i.e., not involving a long-lived complex, but one which may be expected to have a small activation energy and a pre-exponential factor in the 10^{-11} range.

Reaction (b) is very different, since it may be expected to lead to a relatively long-lived, energetic H_2O_2 intermediate, with a lifetime in the 10^{-10} to 10^{-11} s range, which invariably dissociates into 2 OH, i.e., a typical 'chemical activation' process in its low pressure limit where one would expect little or no energy barrier and a moderately large pre-exponential factor, perhaps near 1×10^{-10} but not as large as 5×10^{-10} cm^3 s^{-1}.

Reaction (c) would probably be direct, since its transition state does not correspond to a vibrationally excited, stable molecule, but little can be said about its rate constant parameters except that it is unlikely to be as fast as (a) or (b). It is disturbing, of course, that the expected temperature dependence of the rate constant ratio k_a/k_b runs contrary to the present experimental results, but the latter involve a large number of assumptions. Experimental work involving the direct measurement of several of the reactive species is clearly needed. Until definitive measurements are reported, aeronomers are advised to do their calculations for a wide range of $H + HO_2$ rate constants and of their ratios.

Acknowledgments

This work was supported by the National Aeronautics and Space Administration under Grant No. NGR 39-011-161 and by the Defense Atomic Support Agency monitored by the U.S. Army Research Office-Durham under Grant DA-AR0-D-31-124-72-G89.

References

Anderson, J. G. and Kaufman, F.: 1973, *Chem. Phys. Letters.* **19**, 483.

Anderson, J. G., Margitan, J. J., and Kaufman, F.: 1974, *J. Chem. Phys.* **60**, 3310.

Baldwin, R. R., Jackson, D., Walker, R. W., and Webster, S. J.: 1967, *Trans. Faraday Soc.* **63**, 1665, 1676.

Baldwin, R. R., Fuller, M. E., Hillman, J. S., Jackson, D., and Walker, R. W.: 1974, *J. C. S. Faraday I*, **70**, 635.

Baulch, D. L., Drysdale, D. D., Horne, D. G., and Lloyd, A. C.: 1972, 1973, *Evaluated Kinetic Data for High Temperature Reactions*, vols. 1 and 2, Butterworth and Co., London, Chemical Rubber Co., Cleveland, Ohio, U. S. A.

Bennett, J. E. and Blackmore, D. R.: 1971, *13th Symposium (International) on Combustion*, The Combustion Institute, p. 57.

Benson, S. W.: 1968, *Thermochemical Kinetics*, Wiley and Sons, New York.

Braun, W. and Lenzi, M.: 1967, *Faraday Discussions* **44**, 252.

Clyne, M. A. A. and Thrush, B. A.: 1963, *Proc. R. Soc. London* **A275**, 559.

Davis, D. D., Braun, W., and Bass, A. M.: 1970, *Intl. J. Chem. Kin.* **2**, 101.

Day, M. J., Thompson, K., and Dixon-Lewis, G.: 1973, *14th Symposium (International) on Combustion*, The Combustion Institute, p. 47.

DeMore, W. B.: 1973, *Science* **180**, 735.

Dixon-Lewis, G.: 1968, *Proc. R. Soc. Lond.* **A307**, 111.

Dixon-Lewis, G., Greenberg, J. B., and Goldsworthy, F. A.: 1974, Paper, *15th Combustion Symposium*, Tokyo.

Dodonov, A. F., Lavroskaya, G. K., and Talroze, V. L.: 1969, *Kin. i. Kat.* **10**, 701.

Ferguson, E. E., Fehsenfeld, F. C., and Schmeltekopf, A. L.: 1969, in *Advances in Atomic and Molecular Physics*, vol. 5, Academic Press Inc., New York, p. 1.

Foner, S. N. and Hudson, R. L.: 1962, *J. Chem. Phys.* **36**, 2681.

Forst, W.: 1973, *Theory of Unimolecular Reactions*, Academic Press, New York.

Garvin, D.: 1971, CODATA 3 Bulletin, NBS, Washington, DC 20234, U.S.A.

Garvin, D. and Hampson, R. F. (eds.): 1974, *Chemical Kinetics Data Survey VII*, Data for Modelling of the Stratosphere, NBSIR 74–430.

Gevantman, L. H. and Garvin, D.: 1973, *Intl. J. Chem. Kin.* **5**, 213.

Greiner, N. R.: 1970, *J. Chem. Phys.* **53**, 1070.

Hampson, R. F., Editor: 1973, *Phys. Chem. Ref. Data* **2**, 267.

Hochanadel, C. J., Ghormley, J. A., and Ogren, P. T.: 1972, *J. Chem. Phys.* **56**, 4426.

Howard, C. J. and Evenson, K. M.: 1974, *J. Chem. Phys.* **61**, 1943.

Johnston, H. S.: 1968, *Gas Phase Reaction Kinetics of Neutral Oxygen Species*, NSRDS-NBS 20.

Kaufman, F.: 1961, in G. Porter (ed.), *Progress in Reaction Kinetics*, vol. 1., Pergamon, New York, p. 1.

Kaufman, F.: 1964, *Ann. Geophys.* **20**, 106.

Laidler, K. J.: 1969, *Theories of Chemical Reaction Rates*, McGraw-Hill, New York.

Lloyd, A. C.: 1974, *Intl. J. Chem. Kin.* **6**, 1969.

Margitan, J. J., Kaufman, F., and Anderson, J. G.: 1974, *J. Geophys. Res. Letters* **1**, 80.

Mayer, S. W., Schieler, L., and Johnston, H. S.: 1966, *J. Chem. Phys.* **45**, 385.

Morley, C. and Smith, I. W. M.: 1972, *J. Chem. Soc. Faraday II* **68**, 1016.

Paukert, T. T. and Johnston, H. S.: 1972, *J. Chem. Phys.* **56**, 2824.

Poirier, R. V. and Carr, R. W.: 1971, *J. Phys. Chem.* **75**, 1593.

Robinson, P. J. and Holbrook, K. A.: 1972, *Unimolecular Reactions*, Wiley-Interscience, London.

Simonaitis, R. and Heicklen, J.: 1973, *J. Photochem.* **2**, 309.

Smith, I. W. M. and Zellner, R.: 1974, *J. Chem. Soc. Faraday II*, in press.

Takacs, G. A. and Glass, G. P.: 1973, *J. Phys. Chem.* **77**, 1948.

Tsang, W.: 1973, *Intl. J. Chem. Kin.* **5**, 947.

Tully, J. C.: 1974, *J. Chem. Phys.* **61**, 61.

Westenberg, A. A. and deHaas, N.: 1967, *J. Chem. Phys.* **46**, 490.

Westenberg, A. A. and deHaas, N.: 1969, *J. Chem. Phys.* **50**, 707.

Westenberg, A. A. and deHaas, N.: 1972a, *J. Chem. Phys.* **57**, 5375.

Westenberg, A. A. and deHaas, N.: 1972b, *J. Phys. Chem.* **76**, 1586.

Wilson, W. E., Jr.: 1972, *J. Phys. Chem. Ref. Data* **1**, 535.

Wilson, W. E., Jr., O'Donovan, J. T., and Fristrom, R. M.: 1969, *12th Symposium (International) on Combustion*, The Combustion Institute, p. 929.

Zahniser, M. S., Kaufman, F., and Anderson, J. G.: 1974, *Chem. Phys. Letters* **27**, 507.

PART V

ATMOSPHERIC AND IONOSPHERIC MODELS

NEUTRAL ATMOSPHERE MODELING

G. KOCKARTS

Institut d'Aéronomie Spatiale de Belgique, 3 Avenue Circulaire,
B – 1180 Bruxelles, Belgium

1. Introduction

In this introductory paper, no attempt will be made to describe any particular model or to compare different models. Emphasis will be given to the physical phenomena occurring in the terrestrial upper atmosphere.

A perfect atmospheric model is actually a set of numbers which is intended to represent the physical properties of the atmosphere in the past, in the present and in the future. This is a very ambitious goal which is far from being reached. There are, however, essentially three types of atmospheric models used for our partial understanding of the upper atmosphere.

The theoretical models result from the solution of some equations with simplifying hypotheses. These models always require a knowledge of boundary conditions which should come out of experimental data. Most of these models are discussed by Blum *et al.* (1972).

The semi-empirical models are usually based on a certain amount of observational data combined with some theoretical background such as the diffusive equilibrium hypothesis. Among the most widely used models are those of Jacchia (1965, 1971) which are based on satellite drag data. These semi-empirical models were initially guided by the theoretical approach developed by Nicolet (1961a, b).

The purely experimental models are a direct representation of physical parameters measured *in situ* by satellites. Since these models are by definition limited in time and in space, they lead very often to semi-empirical models by extrapolation and interpolation processes. As an example, the OGO-6 model (Hedin *et al.*, 1974) gives a worldwide distribution of temperature and concentrations by a spherical harmonic analysis of measured data. The ESRO-4 data described by von Zahn (1975) will probably give the next experimental model of the upper atmosphere.

It is clear, however, that all these models can make a valuable contribution to the knowledge of the upper atmosphere, not only through their internal results, but also by their mutual interaction.

2. Application of the Conservation Equations

The general conservation equations (Chapman and Cowling, 1970; Landau and Lifshitz, 1959) are usually simplified for atmospheric applications but, unfortunately, the simplifying assumptions are not always clearly stated in the literature. This situation can lead to some confusion when different results are compared.

B. M. McCormac (ed.), Atmospheres of Earth and the Planets, 235–243. All Rights Reserved.
Copyright © 1975 by D. Reidel Publishing Company, Dordrecht-Holland.

For each atmospheric species with concentration n_i, the general continuity equation is

$$(\partial n_i/\partial t) + \nabla \cdot n_i(\mathbf{v}_0 + \mathbf{V}_i) = P_i - L_i \tag{1}$$

where P_i and L_i are, respectively, the production and loss rates of the constituent $i \cdot \mathbf{V}_i$ is the diffusion velocity and $\mathbf{v}_0$ is the mass average velocity. In a gas mixture, the diffusion velocity is characterized by the relation

$$\Sigma n_i m_i \mathbf{V}_i = 0 \tag{2}$$

where m_i is the mass of the particle i. Using this relation, the total density $(\varrho = \Sigma n_i m_i)$ continuity equation is obtained from Equation (1) as

$$(\partial \varrho/\partial t) + \nabla \cdot (\varrho \mathbf{v}_0) = 0. \tag{3}$$

The term $\Sigma_i(P_i - L_i)$ vanishes, since the total mass of the system is conserved. It should be noted that the use of Equation (3) for a single constituent is incorrect unless the diffusion velocity $\mathbf{V}_i$ and the production and loss terms are zero. Extensive use has been made of Equation (1) to solve the problem of atomic and molecular oxygen distributions in the lower thermosphere.

Although the molecular diffusion velocity is given by gas kinetic theory (Chapman and Cowling, 1970), the mass average velocity $\mathbf{v}_0$, however, requires the solution of the momentum equation. Considering the viscosity μ as constant and neglecting the gradient of $\nabla \cdot \mathbf{v}_0$, the momentum equation for a gas mixture, can be written

$$\varrho\left[(\partial \mathbf{v}_0/\partial t) + \mathbf{v}_0 \cdot \nabla \mathbf{v}_0\right] + \nabla_p - \mu \nabla^2 \mathbf{v}_0 - \sum_i n_i \mathbf{X}_i = 0, \tag{4}$$

where the total pressure is the hydrostatic pressure and $\mathbf{X}_i$ is any external force acting on the i-type particles. In a rotating frame of reference

$$\mathbf{X}_i = [\mathbf{g} - 2\omega x \mathbf{v}_0 - \omega x(\omega x \mathbf{r})]\, m_i + \mathbf{F}_i, \tag{5}$$

where the first three terms, respectively, represent the gravitational force, the Coriolis force and the centrifugal force at a geocentric distance $\mathbf{r}$ when the angular velocity is ω. The external forces $\mathbf{F}_i$ are the frictional forces such as ion drag. One sees that it is through these frictional forces that the ionosphere can interact with the neutral atmosphere. Applications of the momentum equation to the determination of winds in the upper atmosphere are described by Kohl (1975).

Any solution of the momentum equation requires a knowledge of the thermal structure given by the energy equation. When P and L are respectively, the heat production and loss, the energy equation is given by

$$\varrho\left[(\partial c_v T/\partial t) + \mathbf{v}_0 \cdot \nabla(c_v T)\right] + p\nabla \cdot \mathbf{v}_0 + \nabla \cdot \mathbf{E} - \sum_i n_i \mathbf{F}_i \mathbf{V}_i = P - L, \tag{6}$$

where T is the temperature and c_v is the specific heat at constant volume for the gas mixture. The heat flow $\mathbf{E}$ is given by

$$\mathbf{E} = -\lambda \nabla T + T \sum_i c_{pi} \varrho_i \mathbf{V}_i, \tag{7}$$

where λ is the heat conductivity of the gas mixture and c_{pi} is the specific heat at constant pressure for the i-type particle.

With the addition of an expression for the diffusion velocity, the system of coupled Equations (3), (4), and (6) is complete and can in principle be solved. Up to now, simplifying assumptions have always been used. They are discussed by Banks and Kockarts (1973).

Solutions of the energy equation are usually based on the assumption that the atmospheric constituents are in diffusive equilibrium, i.e. $\mathbf{V}_i = 0$. The diffusion terms are, therefore, omitted in Equations (6) and (7), with the consequence that the diffusion time should be small compared to the heat conduction time. On the other hand, such an assumption, mainly introduced to simplify the mathematical problem, is not necessarily correct. The diffusion time τ_D is usually defined as

$$\tau_D = H^2/D \tag{8}$$

where H is the atmospheric scale height and D is the molecular diffusion coefficient. The heat conduction time τ_H can be written

$$\tau_H = \varrho c_v H^2/\lambda, \tag{9}$$

where all the symbols have been previously defined.

The comparison between τ_D and τ_H is given in Figure 1 which is computed using the parameters of the mean COSPAR international reference atmosphere (Champion and Schweinfurth, 1972). The expression for the heat conductivity is taken from Banks and

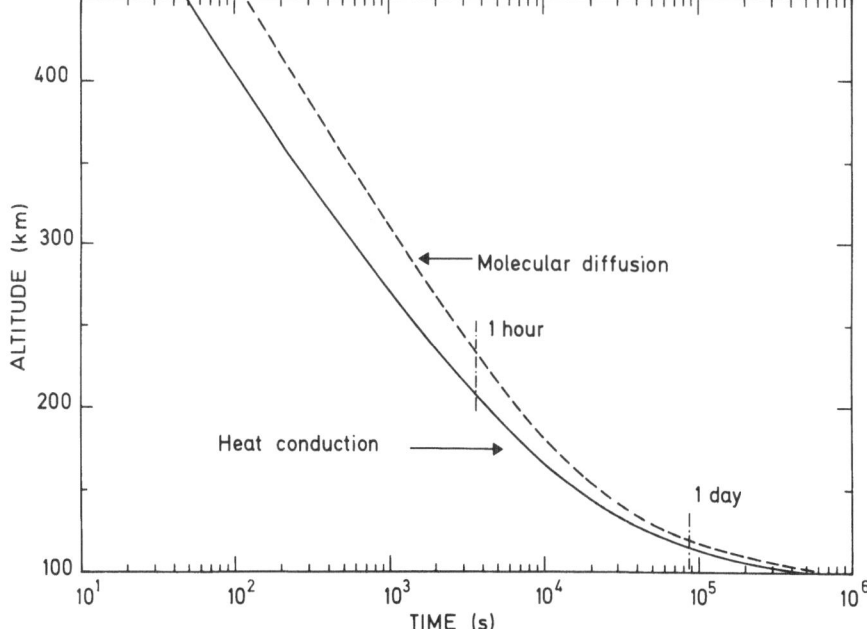

Fig. 1. Comparison between diffusion time and heat conduction time in the mean COSPAR international reference atmosphere.

Kockarts (1973). It can be seen from Figure 1 that the heat conduction time and the molecular diffusion time are of the same order of magnitude over the whole thermospheric height range. This means that the diffusion terms are not necessarily negligible in the energy equation, but further studies are required to clarify this point. Since the diffusion time is of the order of one day around 120 km, variations of the thermal structure will add to the difficulty for reaching diffusive equilibrium in that height region. Possible deviations from diffusive equilibrium have also been suggested (Stubbe, 1972) as a consequence of horizontal winds. There is, however, no general agreement for the magnitude of such an effect (Blum, 1974; Stubbe, 1974; Rishbeth et al., 1974).

Up to now, almost all theoretical models were based on fixed boundary conditions at 120 km altitude. As a result of the variations observed in this height region, it seems better now to adopt boundary conditions at lower heights. In this case the transport effects of O and O_2 become more important and the energy equation must be solved simultaneously with the continuity and momentum equations. This procedure has been used by Chandra and Sinha (1974) in a one-dimensional case. These authors stressed, however, the importance of downward heat fluxes by eddy transport. Such a flux is to be added to the molecular conduction given in Equation (7). Hunten (1974) pointed out that the introduction of eddy conduction may result in an overestimate of the downward heat transfer since dissipative phenomena such as gravity waves may also contribute to the heating. This shows that lowering the boundary level results not only in more complicated mathematics, but introduces also complex physical problems which are not yet completely understood.

Even if the mathematical techniques were available to solve the three-dimensional time-dependent conservation equations, one has still to determine the different production and loss terms as well as the external forces acting on the system.

3. Energy Sources and Sinks

The terms involving the mass average velocity in the energy equation are sometimes considered as compressional heat sources during the night and expansive heat sinks during the day. Since the total contribution is zero over a 24 h period, these terms do not constitute a real global source or sink for the system. They must, however, be included in the energy equation, since they strongly influence the diurnal temperature distribution. When the momentum and energy equations are not solved simultaneously it is difficult to estimate the effect of the compression and expansion terms. It has been shown (Kockarts, 1973) how such an effect can be simulated in a one-dimensional calculation by introducing a very small heat source during the night and a compensating sink during the day. Although the absolute magnitude of these sources or sinks is negligible compared with the total UV absorbed energy, their effect is important since they play a role in a height region above 200 km where very little UV radiation is absorbed.

The major heat sources of the thermosphere are solar UV radiation, Joule dissipation from ion-neutral interactions and atmospheric waves such as gravity waves and tidal oscillations. Although the physical mechanisms are fairly well understood, there is still some uncertainty with regard to the amount of energy deposited in the upper atmosphere.

Most of the theoretical models are based on solar UV heating alone. For practical purposes the solar spectrum is often divided into two parts: the Schuman-Runge continuum below 1750 Å and the wavelength region below Ly-β (1026 Å) down to 80 Å. Longer and shorter wavelengths are not considered, since they are not significantly absorbed above 100 km altitude. The amount of energy available in the Schumann-Runge continuum is of the order of 15 erg cm^{-2} s^{-1} according to the tabulation presented by Ackerman (1971). Between 1026 Å and 80 Å the UV flux is more variable with solar activity and for medium solar activity the measured flux is of the order of 2.3 erg cm^{-2} s^{-1} (Hinteregger, 1970). The validity of Hinteregger's fluxes has been questioned several times in relation with the neutral and ionospheric heat balance. Recently, Prasad and Furman (1974) concluded that the arguments advanced for doubling the solar UV fluxes below 1300 Å are not compelling, since too many uncertain parameters are involved in such studies. The effects of changing solar UV fluxes from 1.7 to 4.5 erg cm^{-2} s^{-1} are indicated by Banks and Kockarts (1973) and by Kockarts (1973). New measured solar UV fluxes are now available below 1000 Å (Schmidtke *et al.*, 1974). Figure 2 shows a comparison between the fluxes measured onboard AEROS-A (Schmidtke *et al.*, 1974) on March 2, 1973, and the values given by Hinteregger (1970). With the exception of an Si xii, an Fe xv line, and the 176–155 Å wavelength interval, Hinteregger's values are always smaller than the fluxes given by Schmidtke *et al.* (1974). The ratio of these measured values is presented in the lower part of Figure 2. The total flux below 1000 Å measured onboard AEROS-A is 3.8 erg cm^{-2} s^{-1}. More data are, however, required before a definite conclusion can be reached with regard to the absolute UV fluxes. Furthermore, long term variations of the UV radiation emitted by the whole sun are not yet available, despite their fundamental importance in aeronomic studies.

Another important mechanism is the Joule heating resulting from ionospheric currents (Cole, 1971, 1972). Since Joule heating is proportional to the electron concentration and to the square of the electric field, its effect is predominant in the auroral region during disturbed conditions (see Wickwar, 1975). Joule dissipation is actually a linking mechanism between the neutral and the ionized atmospheres, as can be seen from the momentum and energy equations. Neutral winds related to the convection electric fields have been measured in the E region using the incoherent radar facility at Chatanika, Alaska. At nighttime, during storm conditions, neutral wind velocities up to 200 m s^{-1} are not uncommon (Brekke *et al.*, 1974). Horizontal winds as high as 1000 m s^{-1} have been deduced at 140 km (Chang *et al.*, 1974) from geomagnetic variations recorded at College, Alaska. As a result of the involved physical mechanism, Joule heating is strongly dependent on the latitude. A global estimation of the amount of heat available through this process requires simultaneous knowledge of the world-wide electric field and the electron concentration distributions. Using a global

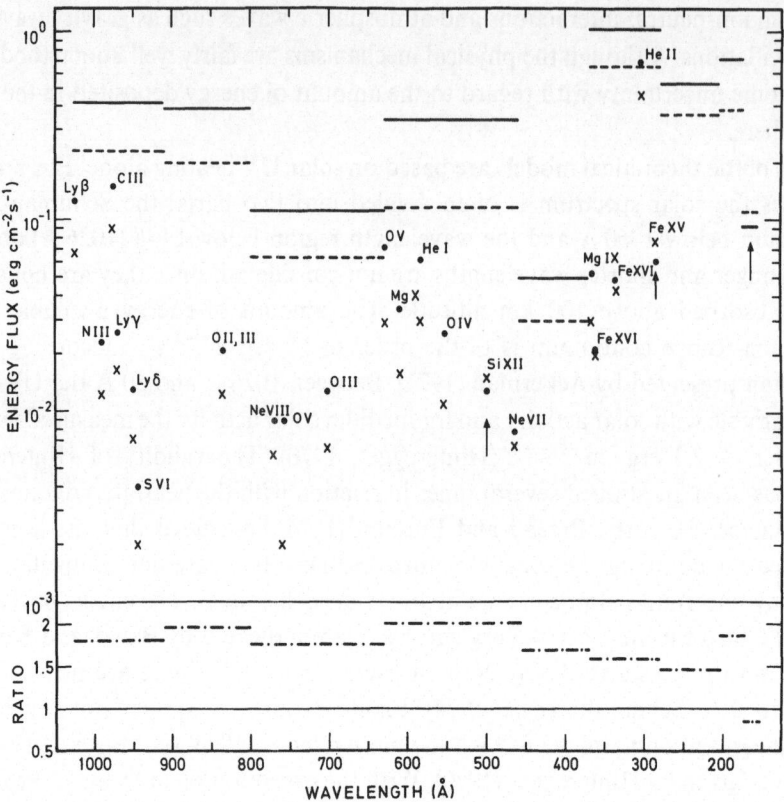

Fig. 2.   Solar UV fluxes available at the top of the atmosphere. Full lines and points are the values obtained by Schmidtke *et al.* (1974). Dashed lines and crosses are Hinteregger's (1970) values. The lower part of the figure gives the ratio between the two sets of data.

ionospheric model and electric field data for high and middle latitudes, Ching and Chiu (1973) computed the Joule dissipation in a given neutral model and concluded that it was similar in magnitude and height profile to the global solar UV absorption. Although Joule heating is mainly concentrated at high latitudes, a redistribution of the heat in the whole thermosphere occurs. A quantitative analysis of this transport phenomenon can only be made with a three-dimensional model. Furthermore, the Joule heating mechanism indicates that the complete self-consistent construction of a neutral atmosphere model would require a simultaneous computation of the ionospheric structure, i.e., the three neutral conservation equations should be coupled to the corresponding ionic and electronic equations.

The potential significance of atmospheric waves as contributors to the thermospheric heat budget has been pointed out by Hines (1965, 1973) who estimated an average energy heat input of the order of 0.1 erg cm^{-2} s^{-1} above 120 km altitude. The variations of the actual supply of energy are, however, not yet known and it is, therefore, very difficult to introduce this effect in theoretical models. Furthermore, wave dissipation can result from medium scale waves generated in the lower atmosphere, or from high altitude large scale wave generated in the auroral regions and traveling over large horizontal distances toward the equator. This means that atmospheric waves can introduce energy into the thermosphere from below and from above. Klostermeyer (1973) has shown that gravity waves generated in the auroral regions during geomagnetic disturbances can lead to temperature increases which are in agreement with the empirical relation deduced from satellite drag (Jacchia *et al.*, 1967).

The energy sources previously described, however can, not explain all the observational facts. Using observed temperatures, densities and winds, Barlier *et al.* (1974) have shown that a definite north-south asymmetry exists in the thermosphere. As an example, Figure 3 gives the ratio of the observed densities to Jacchia's 1971 model. It is seen that, even at the equinoxes, there is an asymmetry when the solar illumination is identical for both hemispheres. The total density is greater in the south. A similar

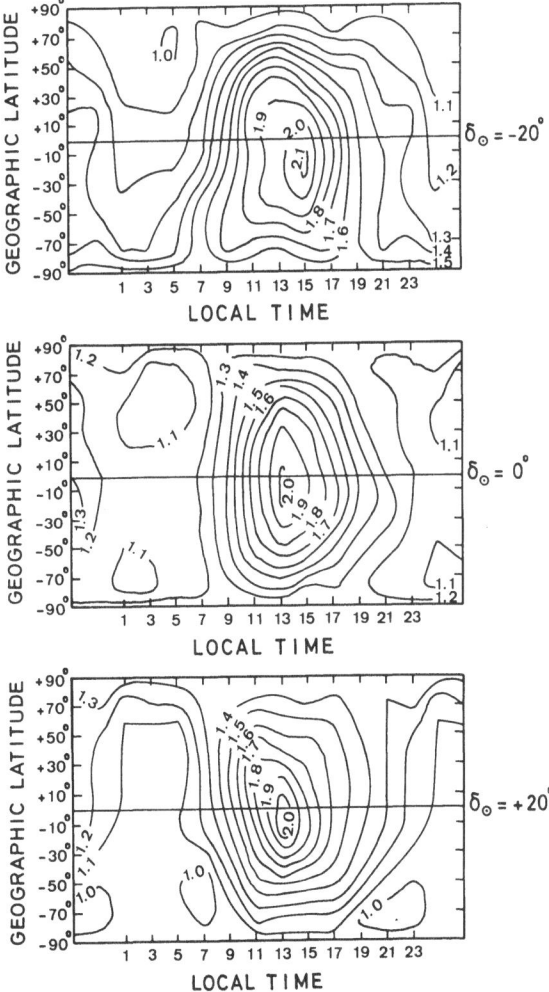

Fig. 3.   Ratio at 280 km between observed densities and the nighttime minimum density of Jacchia (1971) as a function of local solar time and latitude. The solar declination is respectively $-20°$, $0°$ and $+20°$.

asymmetry has also been observed in the global temperature distribution. From the large volume of data accumulated during the last solar cycle maximum, Barlier *et al.* (1974) conclude that the temperature is, on the average, higher in the southern thermosphere. These authors suggest that this fact could be related to the geomagnetic field asymmetry and to the tidal wave dissipation which is linked to the asymmetrical worldwide O_3 distribution in the stratosphere and mesosphere.

Since the construction of the first atmospheric models, heat sources other than the solar UV radiation have been discovered. Up to now, no thermospheric model has been published taking into account all these sources. The former problem of the lack of sufficient sources of energy is now perhaps replaced by the problem of how to lose sufficient energy to avoid high temperatures which are not observed. Besides the downward heat conduction mechanism, the IR emission of O at 63 μm is the sole loss process usually introduced in thermospheric models. It has been shown (Kockarts and Peetermans, 1970) that, below 150 km, radiative transfer strongly reduces the 63 μm volume emission rate and, therefore, decreases the cooling rate resulting from O. At 100 km altitude the 63 μm cooling rate is negligible. Molecular heat conduction can then lead to large temperature gradients in the 100 to 120 km region. Such a situation can, however, be avoided if other loss processes are introduced.

Some experimental results seem to indicate that other IR emissions could play a role in the upper atmosphere heat budget. Very high IR emission in the 3 to 8 μm spectral region has been observed at altitudes above 150 km (Markov, 1969) but no complete physical explanation has been given. Recently the IR emission of O_3 at 9.6 μm and of CO_2 at 15 μm has been observed in the mesosphere and lower thermosphere (Stair *et al.*, 1974). According to these data, the 15 μm CO_2 emission at 120 km and 100 km corresponds, respectively, to 1 and 10 erg cm^{-2} s^{-1}. It is, therefore, not excluded that CO_2 could be the cooling agent in the 100 to 120 km height range.

4. Conclusion

It appears from the above discussion that there is a need for more investigations designed to develop the numerical techniques used for the solution of the three-dimensional conservation equations. Furthermore, longterm observations are still needed. A better understanding of the time and space evolution of all heat sources and sinks which are the fundamental input parameters of any atmospheric model is necessary.

References

Ackerman, M.: 1971, in G. Fiocco (ed.), *Mesospheric Models and Related Experiments*, D. Reidel Publishing Company, Dordrecht-Holland, p. 149.
Banks, P. M. and Kockarts, G.: 1973, *Aeronomy*, Part B, Academic Press, New York.
Barlier, F., Bauer, P., Jaeck, C., Thuillier, G., and Kockarts, G.: 1974, *J. Geophys. Res.* **79**, 5273.
Barlier, F., Falin, J. L., Ill, M., and Jaeck, C.: 1973, *Space Res.* **13**, 349.
Blum, P. W.: 1974, *Planetary Space Sci.* **22**, 183.
Blum, P., Harris, I., and Priester, W.: 1972, in *Cospar International Reference Atmosphere 1972*, Akademie-Verlag, Berlin, p. 399.

Brekke, A., Doupnik, J. R., and Banks, P. M.: 1974, *J. Geophys. Res.* **79**, 2448.

Champion, K. S. W. and Schweinfurth, R. A.: 1972, in *Cospar International Reference Atmosphere 1972*, Akademie-Verlag, Berlin, p. 3.

Chandra, S. and Sinha, A. K.: 1974, *J. Geophys. Res.* **79**, 1916.

Chang, S.-C., Wu, S. T., and Smith, R. E.: 1974, *J. Atmospheric Terrestr. Phys.* **36**, 889.

Chapman, S. and Cowling, T. G.: 1970, *The Mathematical Theory of Non-Uniform Gases*, 3rd ed., Cambridge Univ. Press, London.

Ching, B. K. and Chiu, Y. T.: 1973, *Planetary Space Sci.* **21**, 1633.

Cole, K. D.: 1971, *Planetary Space Sci.* **19**, 59.

Cole, K. D.: 1972, *Planetary Space Sci.* **20**, 2205.

Hedin, A. E., Mayr, H. G., Reber, C. A., Spencer, N. W., and Carignan, G. R.: 1974, *J. Geophys. Res.* **79**, 215.

Hines, C. O.: 1965, *J. Geophys. Res.* **70**, 177.

Hines, C. O.: 1973, *Planetary Space Sci.* **21**, 2238.

Hinteregger, H. E.: 1970, *Ann. Geophys.* **26**, 547.

Hunten, D. M.: 1974, *J. Geophys. Res.* **79**, 2533.

Jacchia, L. G.: 1965, *Smithsonian Contr. Astrophys.* **8**, 215.

Jacchia, L. G.: 1971, *Smithsonian Inst. Astrophys. Obs. Spec. Rep.* **332**.

Jacchia, L. G., Slowey, J., and Verniani, F.: 1967, *J. Geophys. Res.* **72**, 1423.

Klostermeyer, J.: 1973, *J. Atmospheric Terrestr. Phys.* **35**, 2267.

Kockarts, G.: 1973, in B. M. McCormac (ed.), *Physics and Chemistry of Upper Atmospheres*, D. Reidel Publishing Company, Dordrecht-Holland, p. 54.

Kockarts, G. and Peetermans, W.: 1970, *Planetary Space Sci.* **18**, 271.

Kohl, H.: 1975, this volume, p. 87.

Landau, L. D. and Lifshitz, E. M.: 1959, *Fluid Mechanics*, Pergamon Press, Oxford.

Markov, M. N.: 1969, *Appl. Opt.* **8**, 887.

Nicolet, M.: 1961a, *Smithsonian Inst. Astrophys. Obs. Spec. Rep.* **75**.

Nicolet, M.: 1961b, *Planetary Space Sci.* **5**, 1.

Prasad, S. S. and Furman, D. R.: 1974, *J. Geophys. Res.* **79**, 2463.

Rishbeth, H., Moffett, R. J., and Bailey, G. J.: 1974, *Planetary Space Sci.* **22**, 189.

Schmidtke, G., Rawer, K., Fischer, W., and Rebstock, C.: 1974, Paper Cospar, Sao Paolo.

Stair, A. T., Jr., Ulwick, J. C., Baker, D. J., Wyatt, C. L., and Baker, K. D.: 1974, *Geophys. Res. Letters*, **1**, 117.

Stubbe, P.: 1972, *Planetary Space Sci.* **20**, 209.

Stubbe, P.: 1974, *Planetary Space Sci.* **22**, 186.

von Zahn, U.: 1975, this volume, p. 133.

Wickwar, V. B.: 1975, this volume, p. 111.

MODELS OF THE IONOSPHERE

JOHN S. NISBET

Ionosphere Research Laboratory, The Pennsylvania State University, University Park, Pa. 16802, U.S.A.

Abstract. The purposes of model ionospheres are reviewed and used to determine the types of models required. The problems associated with obtaining the data required for empirical and theoretical models are considered and used to develop recommendations for approaches to ionospheric modeling. The components of a model ionosphere are discussed and some of the key problems in each area reviewed.

1. Introduction

Model ionospheres have as their prime purpose the representation of the properties of the real ionosphere as accurately as possible for a given set of conditions. The most accurate representation of the real ionosphere over a range of conditions should be obtained by using the most complete theory possible. Realistic values of neutral atmospheric parameters, solar fluxes, etc., must be provided in a manner that will allow the real ionosphere to be reproduced, which is frequently the most difficult problem in the development of a model ionosphere.

Three main types of model ionosphere would appear to be useful: static models, dynamic models and 'forecasting' models.

Static models are of importance for calculations about the importance of a mechanism such as ion drag on the neutral atmosphere, the refraction of a radio signal, or the altitude at which a radio astronomy satellite must be located to ensure that the plasma frequency exceeds the measuring frequency. What is required are average profiles with some statistical parameters which will allow the distribution about these average conditions to be determined. The quality of a model of this type is dependent on how closely the average behavior agrees with the model, how well the geophysical indices are chosen to minimize the fluctuations about the mean, and how well the geophysical indices themselves can be predicted for the operating conditions.

Dynamic models can be designed to reproduce typical time varying events in such a way that the sequence of events and gradients in parameters is typical of those likely to be encountered in practice. Simple examples of this type of model are those designed to reproduce sunrise, sunset or eclipse behavior, or the response of the ionosphere to a gravity wave in the neutral atmosphere. More complicated models could be envisioned to reproduce the entire evolution of a magnetic storm.

Ionospheric forecasting models are required to predict in time, or extrapolate in distance, observed ionospheric characteristics. Such a model would be of considerable use in estimating such things as navigation satellite errors in real time or predicting optimum frequencies for HF communication channels.

2. Approaches and Logistics

The similarities between a model atmosphere, such as CIRA (1965) or CIRA (1972), and a model ionosphere would make it appear that a similar approach to that which has been so successful in the past would be the best course to follow. Indeed the projected International Reference Ionosphere (I.R.I.) will consist of a set of tables based on empirical data (Rawer, 1974).

The differences are however very real because of the much larger number of parameters at any one location and because of the much greater effect of the Earth's magnetic field on the ions which does not allow longitude and time to be freely interchanged. To gain some idea of the size of the problem, Table I lists the range of variables that could be used to interpolate for most engineering purposes.

TABLE I

Variables for an ionospheric model

Variable	Number of values	Variable	Number of values
Altitude	100	Season	6
Latitude	17	Magnetic activity	3
Longitude	36	Solar activity	5
Time	24		

It is possible to cut the number of variables down considerably as has been done by most empirical modelers such as Nisbet (1964), Damon and Hartranft (1970), Bent *et al.* (1972), Ching and Chiu (1973), and others by fitting empirical models to the altitude distributions and modeling these parameters. It would seem that about 10 parameters can give a minimal representation of the region cutting down the data requirements considerably. Even these data requirements are large and it is necessary to see to what extent measurements are available to provide the required information for a strictly empirical model (Table II).

The small ratio of the number of bits available to the number of bits required is mainly due to regions of the world in which no measurements have been made and on

TABLE II

Data requirements for an empirical model

Region	Number of parameters	Bits required	Approximate number of independent bits available
D region	2	10^7	600
E region	2	10^7	1×10^5
F_1 region	2	10^7	1×10^5
F_2 region	2	10^7	2.8×10^5
F_2 above peak	2	10^7	9×10^4
TOTAL		5×10^7	5.8×10^5

the limited number of conditions for which topside soundings are available. The ten parameter model would give a rather primitive picture of the ionosphere and would not be useful in a predictive mode even after any conceivable development.

With a theoretically based model the problem becomes quite different.

TABLE III

Data requirements for a theoretical model

Parameter	Bits required	Approximate number of independent bits available
Thermosphere	512	374
Mesosphere	200	100
Solar flux	256	192
Cross sections	1024	768
Winds	256	192
Electric fields	512	384
Reaction rates	400	300
Particle fluxes	400	200
Magnetic field	256	256
TOTAL	3816	2766

This leaves only about 1000 bits of information undetermined as opposed to almost 5×10^7 bits of information for the empirical model and the sources for obtaining this data are much more comprehensive. Part of the approximately 6×10^5 bits of ionospheric data can be used to make intelligent choices to fill in missing data requirements and the remainder to provide an independent check on the model predictions.

It seems, therefore, that from a purely practical viewpoint much better model ionospheres for engineering purposes can be obtained through theoretical models than by attempts at strictly empirical modeling. For scientific purposes the theoretical models are required because only in this way can internal consistency be obtained between the densities of the neutral and ionized species, the electric fields and wind systems, the ionizing radiation and the temperatures. Each new measurement of electron or ion density or temperature, EUV or particle flux, electric field or airglow intensity can be evaluated for information content by comparing it with the predictions of the model. The process of constructing the model distinguishes between those aspects of the ionosphere where basic measurements and understanding are missing and those that are well understood, those parameters of which measurements are critical and those which are repetitious and inconsequential.

3. Components of a Model Ionosphere

Figure 1 shows the components of a theoretical model ionosphere. The input control parameters serve to specify the range of spatial and temporal conditions, the range of solar and magnetic activity for which it is desired to make the calculations. Data

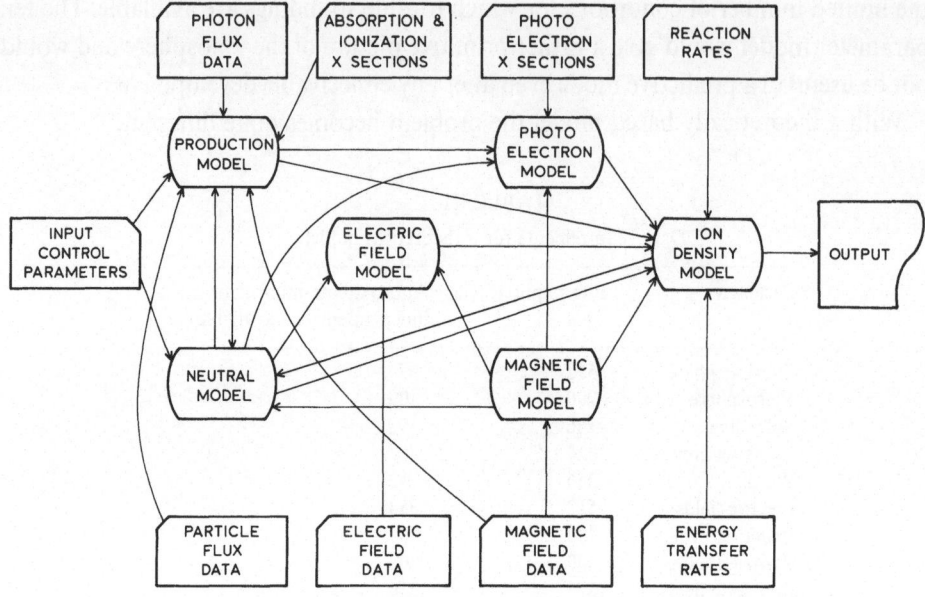

Fig. 1.   Components of a theoretical model ionosphere.

sources required are the photon and particle fluxes, cross sections for all of the important reactions, and data on the magnetic and electric fields.

Most important of all components of the model is that of the neutral atmosphere. Ions are always a minor constituent, and production loss and transport depend directly on the neutral densities. The requirements placed on a neutral model for these purposes are much more rigorous than are needed for a model to be used to study orbital decay; for example, in addition to total density the relative densities of several constituents, some of them very minor, the temperature and the horizontal gradients in the pressure which control the wind system must be accurately modeled.

The production model uses the neutral model together with the photon and particle fluxes to calculate the production rates of the positive ions and photoelectrons including, if necessary, their initial energy. The magnetic field model would seem to have few problems, but the electric field model is very complicated and important particularly at high latitudes and interacts very directly with the wind system. The photoelectron model is important in the F region because of nonlocal heating which controls the electron and ion temperaure and consequently the scale heights in the upper ionosphere.

The ion model contains most of those things normally associated with an ionospheric model, the system of solving the series of complex continuity equations for the various ionic species depending on production loss and transport.

4. Problems and Predictability

A good model ionosphere should behave as closely as possible like the real ionosphere and the statistical variations about that mean model should be as small as possible. In this section attention will be concentrated on those aspects of the modeling process related to the predictability of the variations.

Several studies have been made of the planetary variations of some of the more notable ionospheric characteristics such as $f_0 F_2$ and $f_0 E$ by King (1966), Noonkester (1966), Potapova (1966), and others. As an example Figure 2 taken from Potapova (1966) shows the coefficient of variation of the maximum electron density about a seasonal average for 1 yr and longitude region. Such plots give a good idea of the variability about a mean model because the ability to predict monthly averages greatly exceeds the ability to predict fluctuations about those averages. Variability of the peak density of the E and F layers is much smaller than the fluctuations in electron densities at fixed heights because of vertical motions as can be seen from Figure 3 taken from Swartz et al. (1974a) which shows the variability of the ionosphere for the magnetically quiet period on April 11 and 12, 1972. The degree to which ionospheric variation can be predicted depends ultimately on the degree to which the variability of the model components is understood.

4.1. PRODUCTION MODEL

To a certain extent the EUV and X-ray fluxes would seem to be capable of being handled in a real time model providing sufficient satellite measurements could be measured accurately in real time. This is not true, however, for particle precipitation

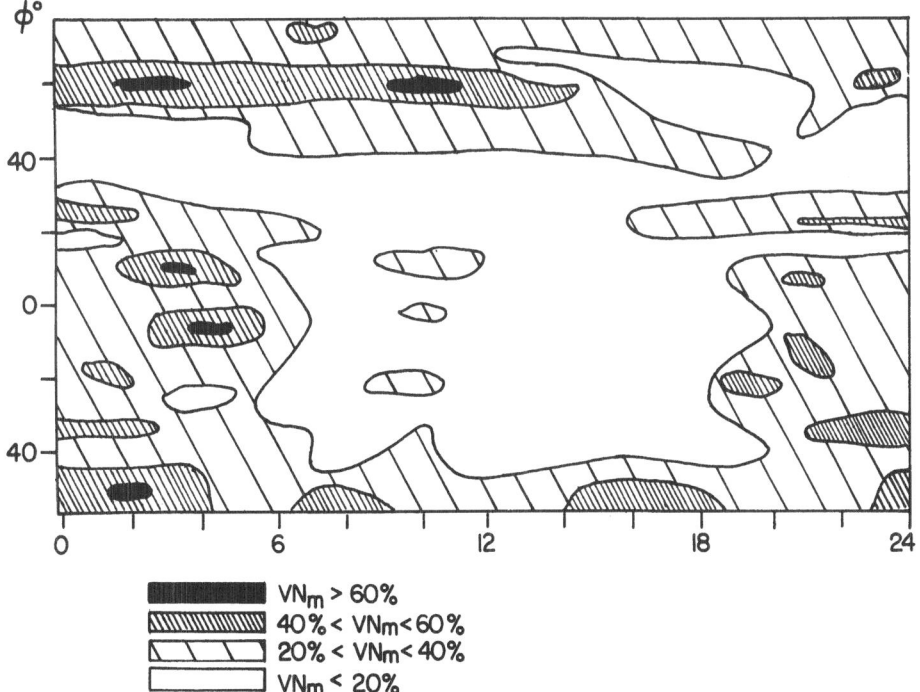

Fig. 2. Coefficient of variation in $N_m F_2$ for June 1958 Eastern Sector (from Potapova, 1966).

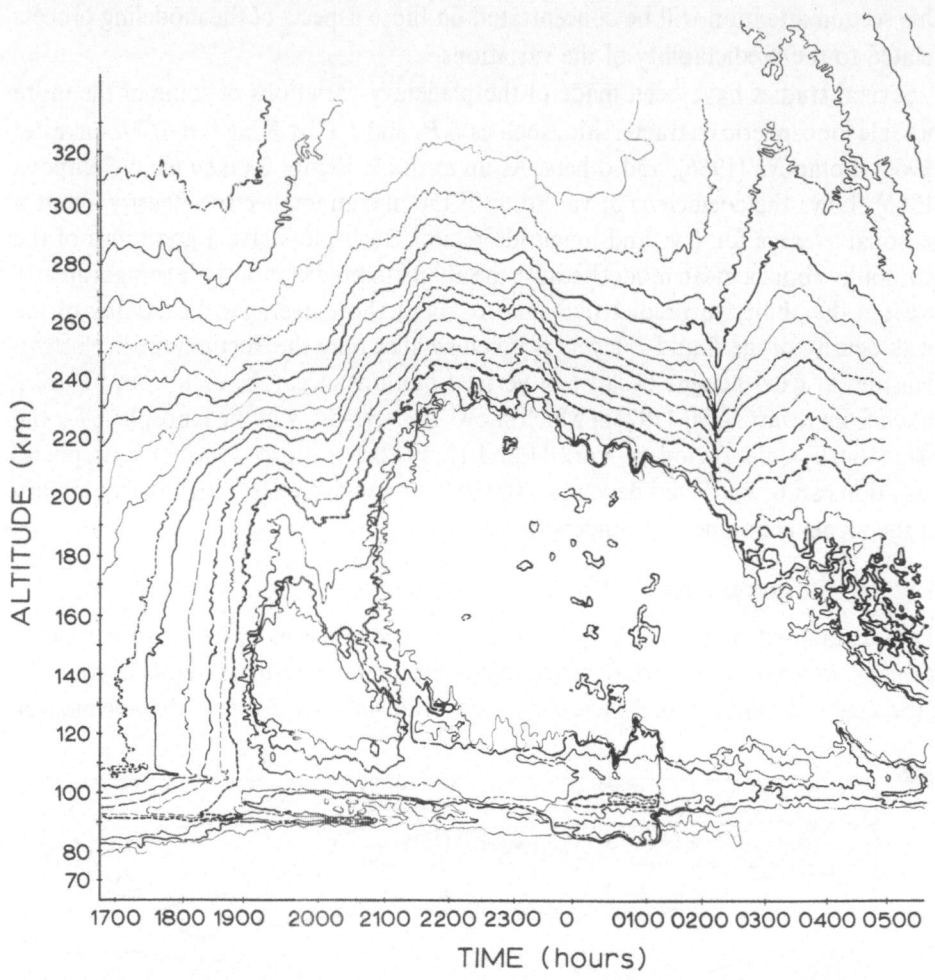

Fig. 3.   Electron density at Arecibo April 11–12, 1972 (from Swartz et al., 1974a).

where the geographical distribution varies in a systematic way with the evolution of
a storm but not in any way that can be simply related to a parameter such as the Ap,
Kp, or AE indices. Studies have shown that the diameter of the auroral oval is related
to the southward component of the interplanetary magnetic field. The understanding
of the morphology of the oval is vital to the study of the precipitation patterns that
control not only the production but also the electric fields, neutral temperatures, and
winds in the high latitude ionosphere. Calculations of ionization rates are also more
complicated for auroral particles as discussed by Banks et al. (1974).

   In the intercomparison of data taken at different times a major problem is at-
tempting to characterize the very complex EUV spectrum by a suitable character
figure. Ground based indices such as the Zurich sunspot number, the Ca II index and
the 2800 MHz solar flux are well correlated on a solar cycle basis with such variables

as the thermospheric temperatures and neutral and ionic densities. The solar rotation or 27 day component is not nearly so well correlated and very large differences occur in the relative variation of individual lines and in the overall correlation from one solar rotation to the next. Woodgate *et al.* (1973) have discussed the variations of the solar EUV flux observed on OSO 6 and show the correlation with the 2800 MHz solar index and the Zürich sunspot number. In Figure 4 the intensity of the 304 Å line and the 2800 MHz solar index is plotted as a function of the average temperature measured from 6300 Å airglow spectra in the altitude region around 275 km within the latitude region ±10° of the equator by Blamont and Luton (1972). In this region the correlation of the temperatures with magnetic activity is small. It is apparent that the EUV intensities are much better correlated with the exospheric temperatures than is the 2800 MHz solar index. It would seem that indices based on satellite EUV measure-

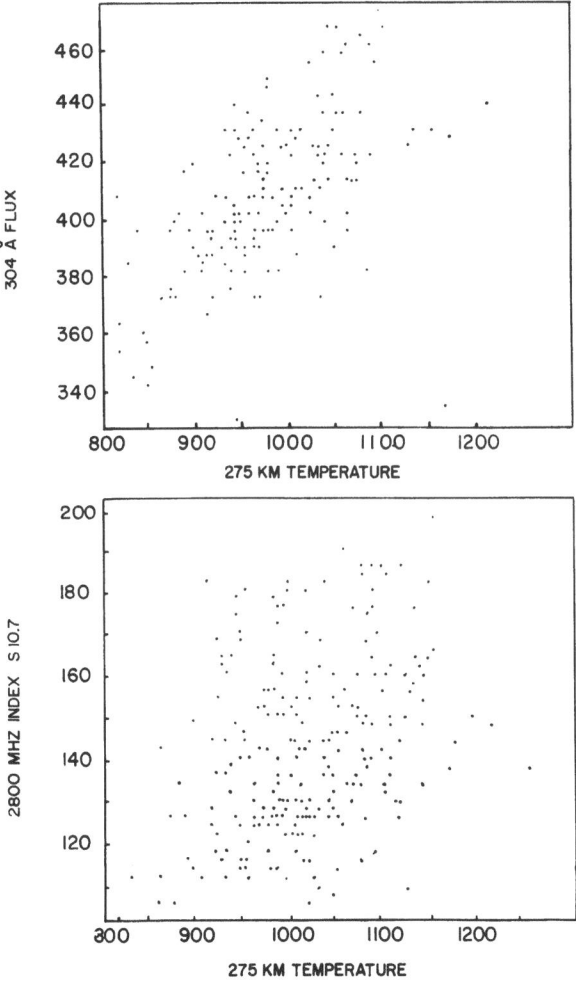

Fig. 4.   Relation of 304 Å intensity and 2800 MHz index with 275 km temperature.

ments must be developed that are well correlated with the exospheric temperature and the ionization rates in the different ionospheric regions.

4.2. Neutral model

The study of the thermospheric response to magnetic storms (Mayr and Volland, 1973; Blamont and Luton, 1972) is an extremely complicated problem that will be very challenging in the years to come. From a modeling point of view it is extremely difficult, involving, as it does, heat inputs from particles and the currents they produce, winds induced by pressure gradients caused by the temperature changes as well as those produced by ion-neutral collisions, and energy transported by gravity waves as well as by the wind system. The problem is so complicated that it would seem that only by realistic theoretical neutral atmospheric and ionospheric modeling can the interrelation of the mechanisms be understood.

Much remains to be learned about the behavior of the thermosphere even under quiet conditions. The F region is extremely sensitive to the ratio of O to O_2 because the former controls the production rate and the latter influences the loss rate. Johnson (1973) has discussed the seasonal behavior of this ratio and Mayr and Volland (1972) have developed a first order model to explain the behavior. Stubbe (1973) has used his theoretical models to relate the densities of O, O_2 and N_2 to measurable F region parameters. Figure 5 shows the seasonal variations in the O densities at 0° and 60° showing that while at low latitudes the variation is primarily semiannual, at high latitudes the effect is predominantly annual. This gives, however, only a picture of the

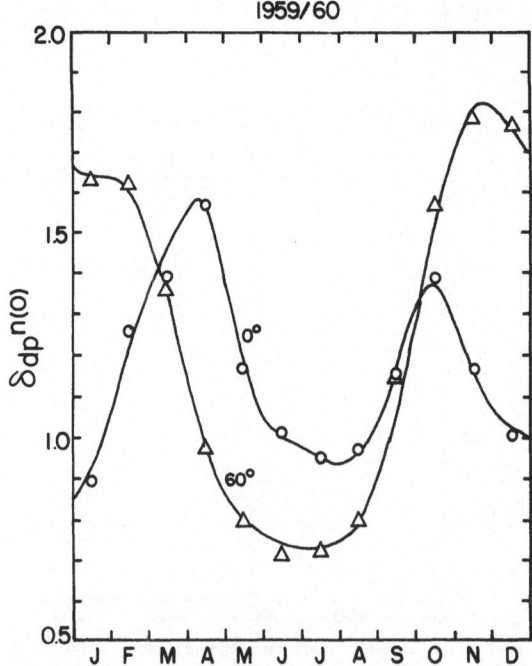

Fig. 5.   Atomic oxygen parameter as a function of season for 0° and 60° latitude.

main seasonal changes. Large local perturbations occur in the neutral density ratios with small scales in both time and distance as has been shown by von Zahn (1974) from ESRO 4 measurements and Becker (1974) from ionosonde measurements.

The effects of the thermospheric wind system have been frequently analyzed since the original paper of King and Kohl (1965). It has recently been realized that much of the difference between ionospheric behavior at different longitudes but the same latitude is due to the effect of the Earth's declination (Kohl *et al.*, 1969). Baran (1973) has used electron densities from a group of stations to develop models for the zonal and meridional winds. Figure 6 shows the differences in behavior between three stations of differing declination but similar latitudes.

Figure 7 shows the comparison of the Baran (1973) winds calculated for 51°N with theoretical winds calculated from the MK III three-dimensional neutral atmospheric

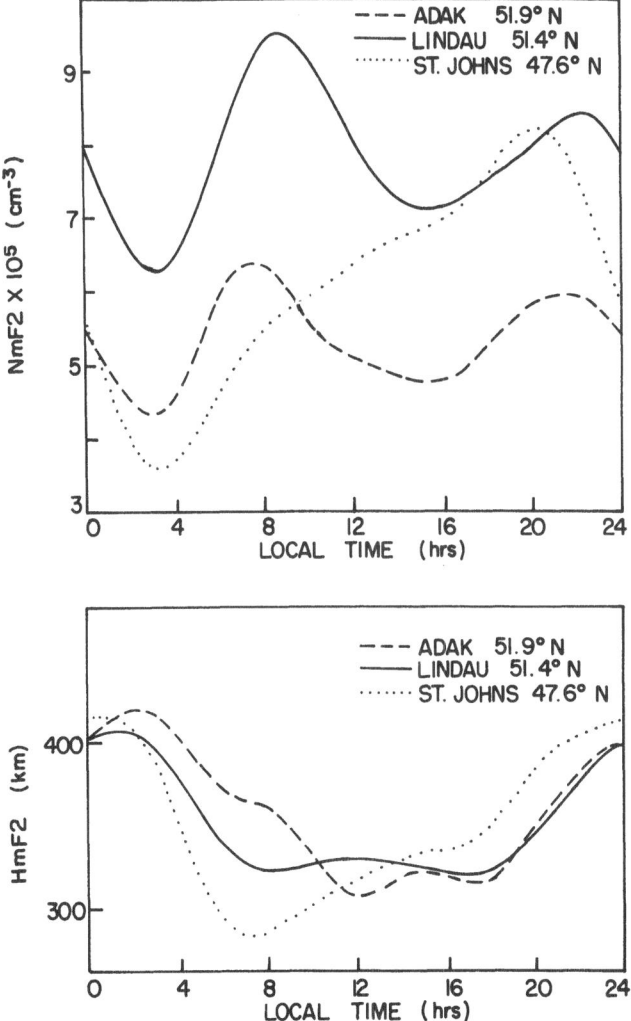

Fig. 6.   Variations in N_mF_2 and H_mF_2 for three stations at different longitudes but at similar latitudes.

model of Vest (1973). It is apparent that fluctuations in the zonal wind velocities will produce changes which are anticorrelated at St. Johns and Adak because of the difference in the sign of the declination.

The uses of ionospheric models in studying neutral atmospheric variability is not confined to the thermosphere. Changes in ionospheric absorption in the D region such

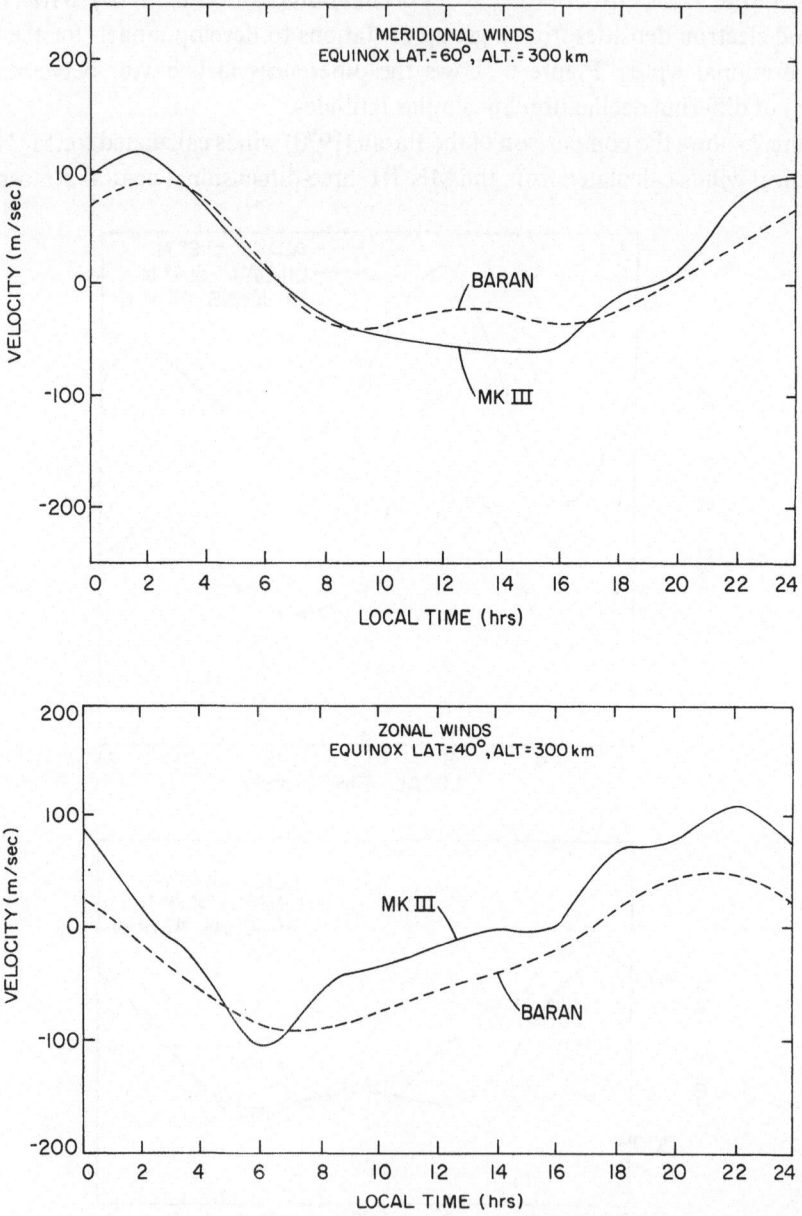

Fig. 7. Comparison of zonal and meridional winds calculated by Baran (1973) with the MK III model of Vest (1973).

as the winter anomaly, provide a sensitive remote method of studying the geographical variation of a minor neutral constituent. If, as has been suggested by Chesworth (1974), these changes are due to ice crystals it would appear that modeling studies could provide important information on the relation of the complex temperature changes which occur during stratospheric warmings (Labitzke, 1974) and variations in the mesosphere and D region.

4.3. ELECTRIC FIELD MODEL

Considerable progress has been made in recent years in measuring the electric field systems in the upper atmosphere. Data have been obtained using artificial ion clouds (Haerendel *et al.*, 1967), balloon probes (Mozer and Serlin, 1969), satellite probes (Heppner, 1972), and incoherent scatter sounding. Incoherent scatter measurements have been made at Jicamarca by Woodman (1970), at Arecibo by Behnke and Harper (1973), at Millstone Hill by Carpenter and Bowhill (1971) and Evans (1971), at Nançay by Amayenc and Vasseur (1972), and at Chatanika by Doupnik *et al.* (1972). At low latitudes Anderson *et al.* (1973) have had considerable success in modeling the combined effects of the neutral winds and electric field system, as have Behnke and Kohl (1974) at Arecibo. Kirchhoff and Carpenter (1974a), based on measurements at Millstone, demonstrated the importance of electric fields of plasmaspheric origin at night well within the plasmapause. Figure 8 shows measurements made at Millstone in

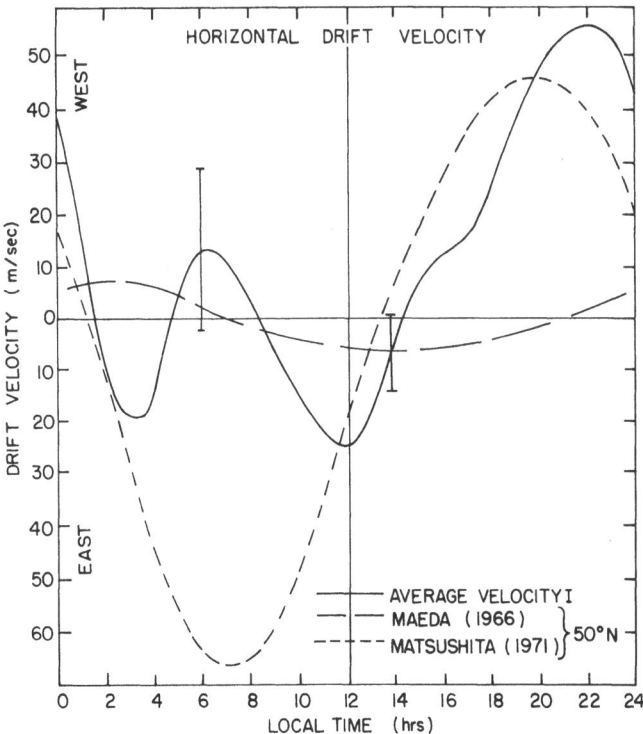

Fig. 8.   Drift velocities measured at Millstone (solid curve) compared to theoretical models.

which it can be seen that the nighttime fields of plasmaspheric origin are much larger than the daytime fields of E region origin. At high latitudes the magnetospheric convection field causes rapid $E \times B$ drift of the ionospheric plasma greatly complicating the solution of any type of continuity equation. The situation is further complicated by the effect of the ion motion on the thermospheric wind system.

4.4. THE PHOTOELECTRON MODEL

Considerable progress has been made in the reconciliation of the photoelectron densities and fluxes calculated by Monte Carlo methods, modified diffusion, and a two stream method (Cicerone *et al.*, 1973; Cicerone, 1974). All the methods now give essentially identical results though each has particular advantages for certain types of calculation (Swartz, 1974). Work is continuing on the effects of interhemispheric transport of the photoelectrons (Swartz *et al.*, 1974b).

4.5. ION MODEL

At low and middle latitudes the ion models have to be able to handle the major ionic densities dynamically under the influence of production loss and transport controlled by diffusion, neutral winds, and electric fields. Heating by photoelectrons must include both local and non-local effects and the effects of interhemispheric coupling must be included. Considerable progress has been made in these areas by Anderson (1973), Bailey *et al.* (1969), Mayr *et al.* (1972), Ruster and King (1973), Sterling *et al.* (1969), Strobel and McElroy (1970), Stubbe (1973), Torr and Torr (1969), and others. It appears at the present time that all of the important mechanisms in the F region can be handled as far as the average low and midlatitude ionosphere is concerned. Such major improvements as are necessary involve combining the features of the various models into one model, in the data sources, and associated electric field and neutral models.

At high latitudes, starting in the region of the midlatitude trough, the situation is completely different. In addition to the severe problems with the electric fields, magnetic field and wind system models, there is the need to develop models including the polar wind flow of H^+ and He^+ escaping from the high latitude ionosphere (Banks and Holzer, 1969; Lemaire and Scherer, 1973). Plasma motions and field fluctuations make it difficult to consider a static model and the degree of coupling of the particle precipitation, neutral wind, and electric field is much greater than at low latitudes.

In the E region the major problems lie in explaining the 'sporadic' behavior as far as the ion models themselves are concerned (Whitehead, 1971; Atkinson, 1973). As with the F region models the neutral atmosphere with particular emphasis on winds, meteoric residues, and nitric oxide, electric fields and ionization sources particularly at night requires major effort. With the D region the problems lie not in the programs to compute the ion models but in the understanding of the chemistry and physics they express and the minor constituents that take part in the reactions.

5. Conclusions

Of the three types of model ionosphere discussed the static model is by far the most simple to develop. At the low and middle latitudes sufficient understanding of the basic physical and chemical processes exists and it would appear that theoretical models can be developed from first principles that are more accurate than empirical models. Much work remains to be done on the neutral atmospheric and electric field models in particular and on the variations of the individual parameters that control the day to day variations in the ion densities. At high latitudes a start has been made on the problem but it is exceedingly complex and a considerable amount of work remains on the models of the particle inputs, the electric field and current systems and associated heat inputs and neutral winds, and the He^+ and H^+ escape along field lines. For these regions a considerable amount of theoretical understanding is required to even order empirical data in a meaningful manner and the only hope of understanding the complex interactions taking place lies eventually in complex neutral and ion models.

As far as the dynamic models are concerned some progress has been made at low and middle latitudes but much more work is required particularly on the neutral atmosphere and the electric field system. For the high latitude ionosphere probably only dynamic models are really applicable and the problems discussed above are even more difficult.

For the weather forecasting models very much remains to be done. The ionosphere is not well correlated at distances of a few thousand kilometers and only through a thorough understanding of the changing wind systems, electric field systems, neutral densities and gravity waves can good predictive models be developed.

References

Amayenc, P. and Vasseur, G.: 1972, *J. Atmospheric Terrestr. Phys.* **34**, 351.
Anderson, D. N.: 1973, *Planetary Space Sci.* **21**, 421.
Anderson, D. N., Matsushita, S., and Tarpley, J. D.: 1973, *J. Atmospheric Terrestr. Phys.* **35**, 753.
Atkinson, K. R.: 1973, *J. Atmospheric Terrestr. Phys.* **35**, 469.
Bailey, G. J., Moffett, R. J., and Rishbeth, H.: 1969, *J. Atmospheric Terrestr. Phys.* **31**, 253.
Banks, P. M. and Holzer, T. E.: 1969, *J. Geophys. Res.* **74**, 6713.
Banks, P. M., Chappell, C. R., and Nagy, A. F.: 1974, *J. Geophys. Res.* **79**, 1459.
Baran, D.: 1973, Presentation at AGU Fall Meeting, San Fransisco, December 1973.
Bauer, S. J.: 1972, Penn. State University Ionosphere Research Report, PSU-IRL-SCI-401, 131.
Becker, W.: 1974, *Kleinheubacher Berichte* **17**, 157.
Behnke, R. A. and Harper, R. M.: 1973, *J. Geophys. Res.* **78**, 8222.
Behnke, R. A. and Kohl, H.: 1974, *J. Atmospheric Terrestr. Phys.* **36**, 325.
Bent, R. B., Llewellyn, S. K., and Wallock, M. K.: 1972, SAMSO T. R. 72-239.
Blamont, J. E. and Luton, J. M.: 1972, *J. Geophys. Res.* **77**, 3534.
Carpenter, L. A. and Bowhill, S. A.: 1971, *Radio Sci.* **6**, 203.
Chesworth, E. T.: 1974, Ph.D. Thesis, Penn State University.
Ching, B. K. and Chiu, U. T.: 1973, *J. Atmospheric Terrestr. Phys.* **35**, 1615.
Cicerone, R. J.: 1974, *Rev. Geophys. Space Phys.* **12**, 259.
Cicerone, R. J., Swartz, W. E., Stolarski, R. S., Nagy, A. F., and Nisbet, J. S.: 1973, *J. Geophys. Res.* **76**, 8299.
CIRA: 1965, North-Holland Publishing Co., Amsterdam.

258   JOHN S. NISBET

CIRA: 1972, Akademie-Verlag, Berlin.
Damon, T. D. and Hartranft, F. R.: 1970, *Aerospace Environmental Support Center Technical Memorandum*, 70-3.
Doupnik, J. R., Banks, P. M., Baron, M. J., Rino, C. L., and Petriceks, J.: 1972, *J. Geophys. Res.* **77**, 4271.
Evans, J. V.: 1971, *Radio Sci.* **6**, 609.
Haerendel, G., Lust, R., and Rieger, E.: 1967, *Planetary Space Sci.* **15**, 1.
Heppner, J. P.: 1972, *Planetary Space Sci.* **20**, 1475.
Johnson, F. S.: 1973, *Rev. Geophys. Space Phys.* **11**, 741.
King, G. A. M.: 1966, *J. Atmospheric Terrestr. Phys.* **28**, 531.
King, J. W. and Kohl, H.: 1965, *Nature* **206**, 699.
Kirchhoff, V. W. and Carpenter, L. A.: 1974a, *J. Atmospheric Terrestr. Phys.* **36**, to be published.
Kirchhoff, V. W. and Carpenter, L. A.: 1974b, private communication.
Kohl, H., King, J. W., and Eccles, D.: 1969, *J. Atmospheric Terrestr. Phys.* **31**, 1011.
Labitzke, K.: 1974, paper presented at COSPAR.
Lemaire, J. and Scherer, M.: 1973, *Rev. Geophys. Space Phys.* **11**, 427.
Mayr, H. G. and Volland, H.: 1972, *J. Geophys. Res.* **77**, 6774.
Mayr, H. G. and Volland, H.: 1973, *J. Geophys. Res.* **78**, 2251.
Mayr, H. G., Fontheim, E. G., Brace, L. H., Brinton, H. C., and Taylor, H. A.: 1972, *J. Atmospheric Terrestr. Phys.* **34**, 1659.
Mozer, F. S. and Serlin, R.: 1969, *J. Geophys. Res.* **74**, 4739.
Nisbet, J. S.: 1964, H. R. B. Singer Technical Note.
Noonkester, V. R.: 1966, *J. Geophys. Res.* **71**, 4192.
Potapova, H. I.: 1966, *Geomag. Aeron.* **6**, 800.
Rawer, K.: 1974, COSPAR Working Group Annual Review.
Rino, C. L.: 1972, *Radio Sci.* **7**, 1049.
Ruster, R. and King, J. W.: 1973, *J. Atmospheric Terrestr. Phys.* **35**, 1317.
Sterling, T., Hanson, W., Moffett, R. J., and Baxter, R. G.: 1969, *Radio Sci.* **4**, 1005.
Strobel, D. F. and McElroy, M. B.: 1970, *Planetary Space Sci.* **18**, 1181.
Stubbe, P.: 1973, Penn State University Ionosphere Research Scientific Report, PSU-IRL-SCI-**418**.
Swartz, W. E.: 1974, private communication.
Swartz, W. E., Ionnidis, G. A., Shen, J. S., Brice, N. M., and Rowe, J. F.: 1974a, *Radio Sci.*, to be published.
Swartz, W. E., Bailey, G. J., and Moffett, R. F.: 1974b, *J. Geophys. Res.*, submitted.
Torr, M. R. and Rorr, D. G.: 1969, CSIR Research Report 271, Pretoria, South Africa.
Vest, R.: 1973, Penn State University Ionosphere Research Scientific Report, PSU-IRL-SCI-**412**.
Whitehead, J. D.: 1971, *J. Geophys. Res.* **76**, 3127.
Woodgate, B. E., Knight, D. E., Uribe, R., Sheather, P., Bowles, J., and Nettelship, R.: 1973, *Proc. Roy. Soc.* **332A**, 291.
Woodman, R. F.: 1970, *J. Geophys. Res.* **75**, 6249.
von Zahn, U.: 1974, this volume, p. 133.

THE *D* AND *E* REGIONS

WILLIAM SWIDER

Air Force Cambridge Research Laboratories, L. G. Hanscom Field,
Bedford, Mass. 01730, U.S.A.

1. Introduction

The peaks of the daytime *E* and *D* regions are considered to obey a simple $q = \alpha_{\text{eff}} [e]^2$ law, where q is the ionization production rate, α_{eff} is the effective electron loss coefficient, and $[e]$ is the electron concentration. Beyond this, the similarity of these adjacent ionospheric regions is inconsequential.

    E region chemistry is fairly simple, the region being dominated by the presence of O_2^+ and NO^+ ions. The conversion of major precursor ions like N_2^+ and O^+ into NO^+ and O_2^+ ions is rapid enough at these altitudes (about 90 to 150 km) to prevent these ions from becoming prominant species. However, meteor ions can occasionally be very important, especially when dynamical processes are active. Transport effects and the presence of meteor ions can complicate determinations of α_{eff}.

    The chemistry of the *D* region is far more complex than any other portion of the ionosphere because it is the only section of the ionosphere where three-body processes are important. The formation of negative ions in the ionosphere originates with the processes $e + O_2 + O_2(N_2) \rightarrow O_2^- + O_2(N_2)$ and only at *D* region altitudes are neutral concentrations large enough to provide for significant O_2^- formation. Radiative processes like $e + O_2 \rightarrow O_2^- + h\nu$ are too slow.

    D region dynamical processes on the other hand are generally unimportant in regards to the movement of charged particles. However, the transport of minor neutral species, especially NO, is important to both this region and the *E* region.

2. *E* Region

The *E* region has been more or less successfully modeled by Keneshea *et al.* (1970). Their work has been extended into the *F* 1 region by Torr *et al.* (1972). Swider (1972a) has provided considerable discussion concerning many of the details required for accurate modeling of the *E* region. A historical account of the development of *E* region research has recently been given by Bates (1973).

2.1. *E* REGION RECOMBINATION COEFFICIENT

The electron recombination coefficient of the *E* region may be written as (Swider, 1972a)

$$\alpha_{\text{MR}} = \frac{\alpha(NO^+)[NO^+] + \alpha(O_2^+)[O_2^+] + \alpha(M^+)[M^+]}{[NO^+] + [O_2^+] + [M^+]}$$

B. M. McCormac (ed.), Atmospheres of Earth and the Planets, 259–268. All Rights Reserved.

in order to accommodate the sometimes presence of atomic metallic ion concentrations, $[M^+]$. These ions are normally absent, resulting in

$$\alpha_{MR}=\alpha_M=\frac{\alpha(NO^+)\,[NO^+]+\alpha(O_2^+)\,[O_2^+]}{[NO^+]+[O_2^+]}$$

which, since $\alpha(NO^+)\approx 2\alpha(O_2^+)\approx 4\times 10^{-7}(300/T)$ cm^3 s^{-1} (Biondi, 1969), results in a mean recombination coefficient of about 3×10^{-7} $(300/T)$ cm^3 s^{-1} for the midday E region where $[NO^+]\approx[O_2^+]$.

The metallic ions deposited by the continual ablation of meteors are most likely to be significant at night when only scattered HLy-α and HLy-β are present to weakly form the nighttime E region (Swider, 1965; Ogawa and Tohmatsu, 1966; Keneshea et al., 1970; Tohmatsu and Wakai, 1970). A meteor ion layer can lead to misleading estimates of the nighttime ionization source (Swider, 1972a), as shown in Figure 1. Electron precipitation may occassionally be important at mid-latitudes (Nicolet and Swider, 1963). However, such events may be of more significance to the D region in regards to explaining the 'winter anomaly' (Manson and Merry, 1970). The E region model of Keneshea et al. (1970) suffices even for high latitudes under quiet conditions (Figure 2).

Attempts to determine α_{eff} from bulk analysis of the E region and relate it to α_M have failed for a variety of reasons in addition to sometimes strong layering of meteor ions. E region eclipse studies in the earlier days invariably led to α_{eff} values of 0.5 to

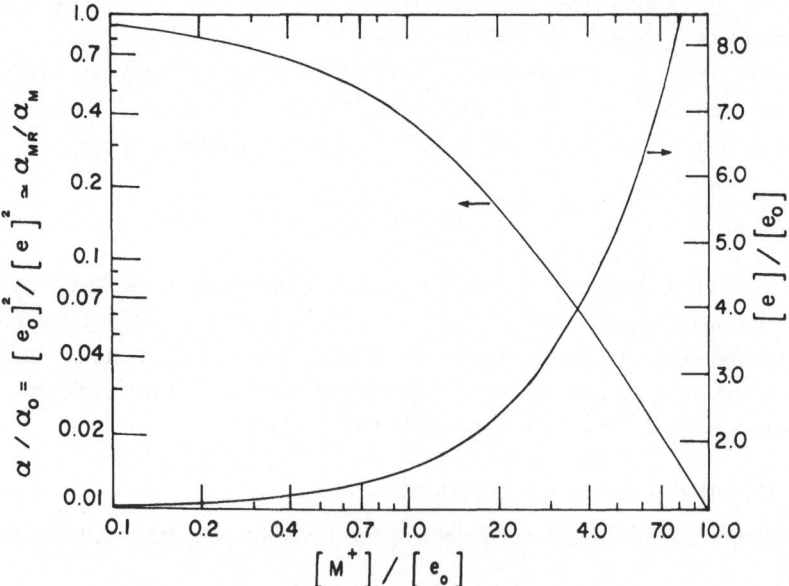

Fig. 1.   Relative electron concentrations and recombination coefficients as a function of the ratio of metallic ions $[M^+]$ to the original electron concentration $[e_0]$ in the absence of these ions. The mathematical development is given by Swider (1969). The ionization production rate is constant. If the recombination rate is assumed constant then the increasing (apparent) production rate, q, is given by $q=q_0[e^2]/[e_0^2]$.

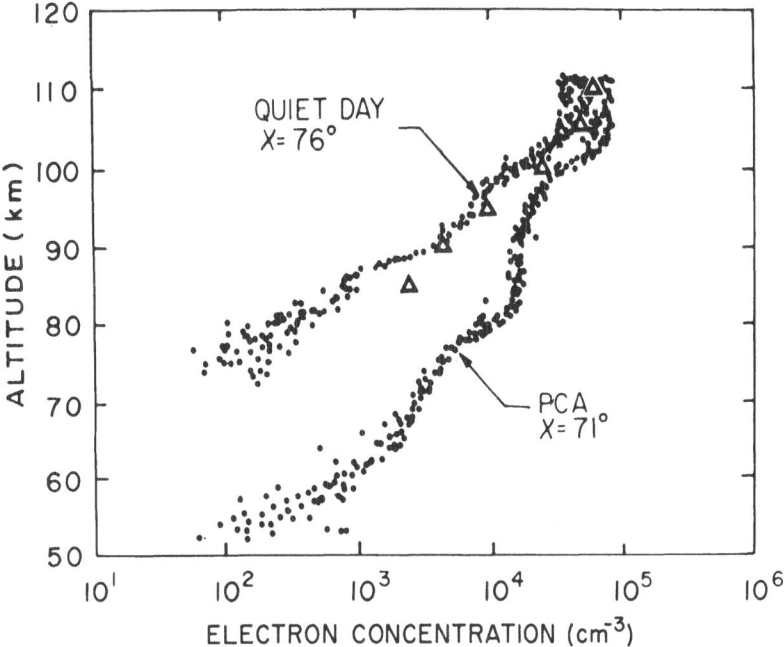

Fig. 2. The quiet day electron concentration of Grieder and Burt (1973) at Thule, Greenland for March 22, 1971 as compared to the E region model (Δ) of Keneshea *et al.* (1970) for the same solar zenith angle, 76°. Data (Grieder and Burt, 1973) obtained during a PCA event on April 6, 1971 are also shown.

2.0×10^{-8} cm^3 s^{-1} (Ratcliffe, 1965a), the importance of X-rays at totality being recognized in the mid-fifties (e.g., Hunaerts and Nicolet, 1955) and implying that α_{eff} could be as great as 10^{-7} cm^3 s^{-1} (Ratcliffe, 1956b). Eclipse studies have proved to be a difficult way of determining α_{eff}.

Measurements of the relaxation time of the ionosphere, Δt, also failed (Swider, 1972a) since the negative values sometimes attained could hardly be correlated with $\alpha_{eff} = 1/(2\Delta t \, [e])$ since α_{eff} must be a positive number if it represents α_M. Palluconi (1963) and other workers he cites have attempted to clarify the problem by use of a modified form of the E region electron continuity equation. However, inclusion of a simple divergence term has not led to physically meaningful results even if there is a consistent pattern which suggests effects linked to the Sq current system or some other sort of transport effect (Swider, 1972a; Butcher, 1974).

2.2. E REGION ION TRANSPORT

Mid-latitude sporadic-E is generally believed to result from the buildup of ions by the action of winds and wind shears in conjunction with the presence of the Earth's magnetic field. These sporadic-E layers apparently are quite common at night and sunset, and at Arecibo are most pronounced during winter months (Rowe, 1974). Keneshea and MacLeod (1970) have successfully modeled a sporadic E region experimental profile using the specific wind profiles measured at the same time as the ion composition data of Narcisi *et al.* (cited by Keneshea *et al.*, 1970). Similar studies have been undertaken by Chen and Harris (1971). The sporadic-E layers are often but not always composed of meteor ions. Meteor showers have been observed to strongly perturb the nighttime E region (Rowe, 1973; see also Swider, 1965, for older references).

2.3. NITRIC OXIDE

The minor neutral gas NO is important in E region chemistry. It is a major factor in the existence of NO^+ ions because the process $O_2^+ + NO \rightarrow NO^+ + O_2$ is generally competitive with electrons as an O_2^+ loss mechanism. In fact, at night, although the prime source of ionization is the ionization of O_2 by scattered HLy-β radiation (Swider, 1965; Ogawa and Tohmatsu, 1966; Keneshea et al., 1970), the dominant ion is NO^+ since the electron concentration is low. Ionization of nitric oxide by the direct HLy-α flux is a major twilight ion source (Swider, 1965; Swider and Keneshea, 1968) and ionization of NO by scattered HLy-α is important to the lower E region at night.

The distribution of NO is not yet well understood. Typical profiles are shown in Figure 3. The profile adopted by Keneshea et al. (1970) essentially followed Barth's (1966) results except at 85 km where a lower value was needed. Meira's (1971) data are higher than Barth's (1966) but may reflect different atmospheric conditions at mid-latitudes. Tisone's (1973) values are for about 20° north latitude. The Swider and Narcisi (1970, 1974a) results were inferred from auroral ion composition data and suggest (also Barth and Rusch, 1971) that NO is enhanced at auroral latitudes.

3. D Region

The daytime quiet D region has not been successfully modeled to date. On the other

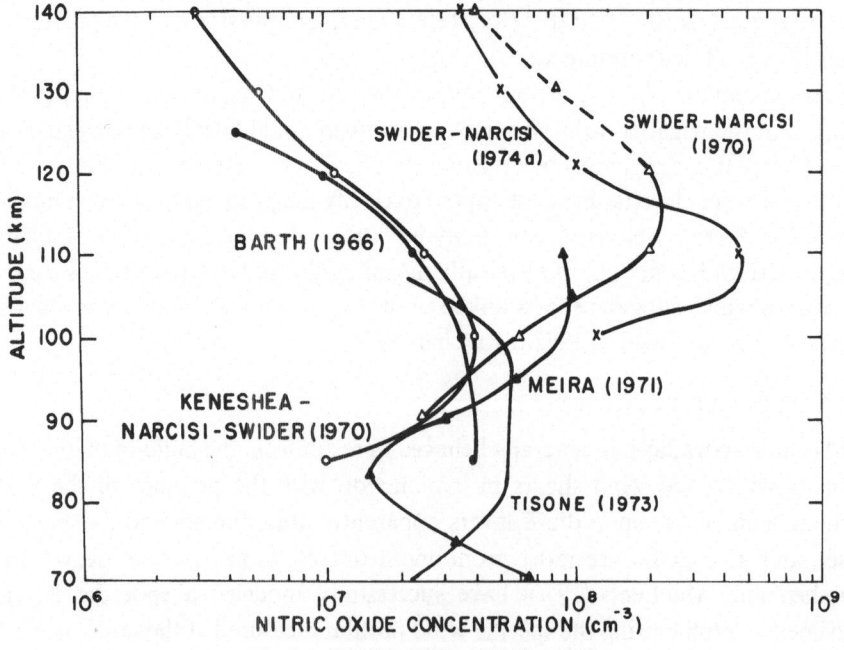

Fig. 3. E region NO profiles (see text).

hand, recent studies of *D* region observations during the November 2–5, 1969, PCA event have yielded fairly straightforward interpretations of the positive ion chemistry. The negative ion chemistry is somewhat uncertain but appears to be inconsequential at night for the majority of the *D* region because electrons are irretrievably lost below about 75 km by the action of three-body processes producing O_2^- ions (Swider *et al.*, 1971). See Mitra (1968), Sechrist (1972), and Thomas (1971, 1974) for major *D* region reviews.

3.1. THE QUIET *D* REGION

At present there is no understanding of how the initially formed ions in the daytime *D* region, presumably NO^+ ions, become converted into oxonium ions, $H_3O^+ \cdot$ $\cdot(H_2O)_n$. Figure 4, from the paper of Swider (1972b), illustrates that Meira's (1971) NO profile must be reduced by at least a factor of five if the ionization of NO is the initial quiet *D* region process. In fact, a factor of ten decrease in Meira's (1971) mixing ratio of 5×10^{-8} would appear to be appropriate (see also Strobel, 1972).

3.2. THE DISTURBED *D*-REGION

Although the chemistry of the *D* region is extremely complex, studies of PCA events or SPE's can be especially worthwhile because the ionization level is much higher than that for normal conditions and varies only very slowly with time. Swider and Narcisi (1974b) have shown that the modeling of the nighttime *D* region during a

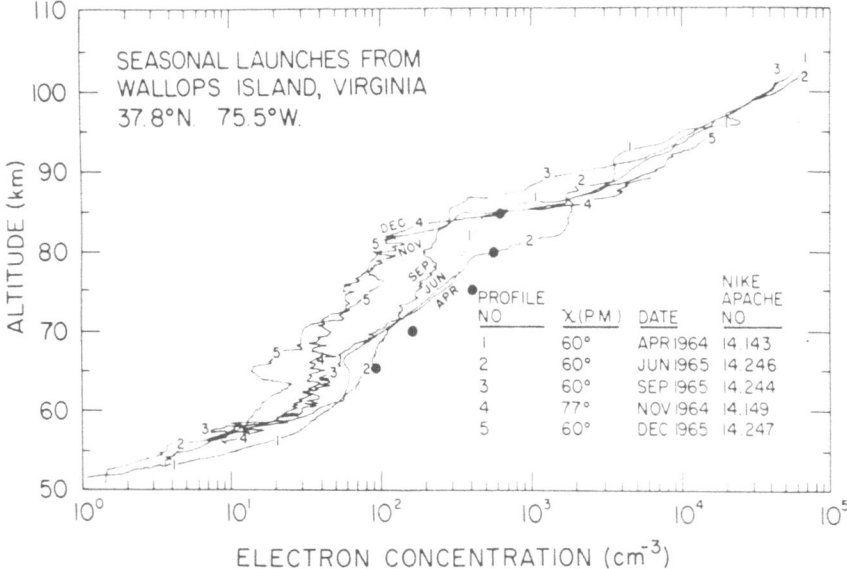

Fig. 4.   The observational data of Mechtly and Smith (1968) are compared to equilibrium calculations by Swider (1972a) with $[NO] = 10^{-8}[M]$. Dissociative recombination rates of 6×10^{-7} cm^3 s^{-1} for 85 km and 4×10^{-6} cm^3 s^{-1} for the other altitudes were adopted since NO^+ is the dominant ion at 85 km and oxonium ions dominate at the other altitudes (Narcisi and Roth, 1970).

PCA event is quite feasible. Figure 5 represents the ion composition at night for an ionization level of about 300 ion-pairs $cm^{-3} s^{-1}$ for all altitudes shown. The results are compatible with the observational data in regards to the major ions O_2^+, NO^+ and $H_5O_2^+$ (Swider and Narcisi, 1974b). However, the chemistry of certain of the minor ions is doubtful since some of the rate coefficients must be estimated. A single rate coefficient was adopted for all ion-ion recombinations. This appeared to be permissible because ion-ion recombination is a minor loss process for positive ions over this altitude region; oxonium ions recombining with electrons at a rate at least ten times faster than as with negative ions.

Analysis of a previous PCA event (Swider *et al.*, 1971) yielded an approximate formula for the electron concentration at night

$$[e] = \{(L(A)^2 + 4\alpha_D q)^{1/2} - L(A)\}/2\alpha_D$$

where $L(A) = k_1 [O_2]^2 + k_2 [O_2] [N_2]$, is the electron loss rate and O_2^- formation rate (s^{-1}), and $\alpha_D \simeq 4 \times 10^{-7} cm^3 s^{-1}$, with q the ionization production rate. This equation, uncertain by a factor of 2 near 80 km, attains greater accuracy with decreasing altitude as it collapses to the simpler form,

$$[e] \simeq q/L(A).$$

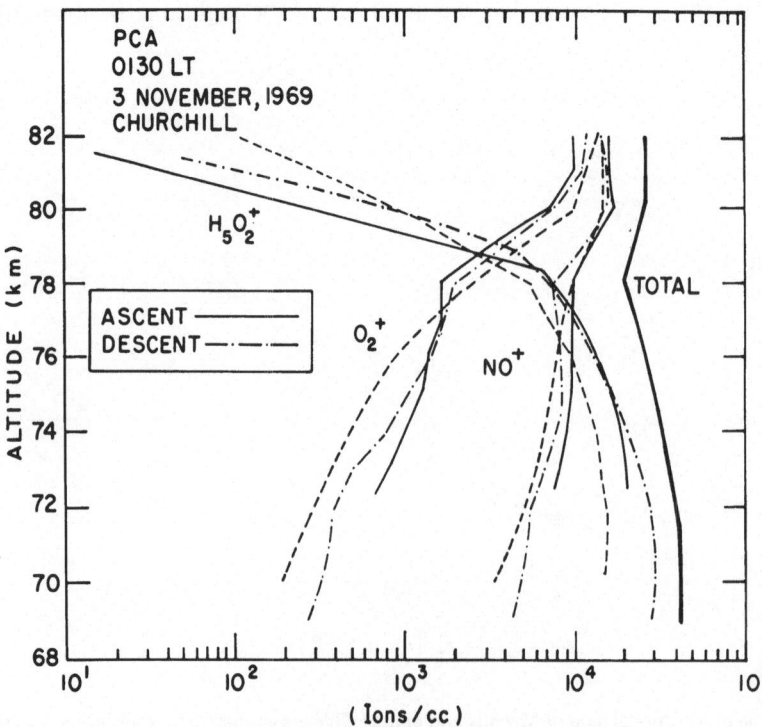

Fig. 5. Nighttime major positive ions in a PCA event (Narcisi *et al.*, 1972) as compared to model calculations by Swider and Narcisi (1974b).

The effective recombination coefficient thus becomes

$$\alpha_{\text{eff}} = q/[\text{e}]^2 = L(\text{A})^2/q$$

being directly proportional to the FOURTH POWER of the total neutral concentration and inversely proportional to the ionization rate.

The ratio of the total negative ion population to that of the electrons is now given by (Swider *et al.*, 1971)

$$\lambda = \frac{\sum_i [n_i^-]}{[\text{e}]} \simeq \frac{\sum_i [n_i^-]}{q/L(\text{A})} \cong \frac{(q/\alpha_i)^{1/2}}{q/L(\text{A})} = \frac{L(\text{A})}{(q\alpha_i)^{1/2}}$$

with α_i representing the mean ion-ion recombination coefficient which appears to be about $1 \times 10^{-7} \text{ cm}^3 \text{ s}^{-1}$ within a factor of 2.

Electron precipitation other than in auroras may influence the D region also, in a less dramatic way than in a PCA event. Earlier in the text we suggested it was generally not important at mid-latitudes. There is a zone just equatorward of the auroral zone where D region electron precipitation is most common (Whalen *et al.*, 1971). Weak electron precipitation events are most significant at night since solar ionization effects are absent except for scattered HLy-α radiation. Gough and Collin (1973) report that at South Uist ($L=3.5$) precipitating electrons are the dominant nighttime D region ionization source $15 \pm 11\%$ of the time during solar minimum and about $35 \pm 20\%$ during solar maximum.

Besides electron precipitation and sporadic meteor effects, cosmic X-rays may also contribute to D region ionization, particularly at night. Svennesson *et al.* (1972) have studied VLF signal phases along several paths and found effects related to the stellar X-ray sources Sco XR-1, Cen XR-2, Cen XR-4 and possibly Tau XR-1. Baird and Francey (1972) have argued that restriction of electron precipitation effects to $L \gtrsim 2$ would explain many of the conflicts in VLF data, noting, as compatible, the fact that the X-ray effects discussed by Svennesson *et al.* (1972) all concern VLF paths with midpoints at $L < 1.7$. A rough picture thus emerges in which for the nighttime D region cosmic X-ray sources can be important over the region $0 \leqslant L \lesssim 2$, with electron precipitation effects becoming more pronounced with increasing geomagnetic latitude, peaking near $70°$. Meteor ion effects, coupled to the meteor influx and transport effects, compound the problem.

3.3. DAYTIME NEGATIVE ION POPULATION

The negative ion chemistry of the daytime D region must be understood in detail because electrons may be detached from negative ions through solar processes. The most direct way is photodetachment. In this case the photodetachment cross section of the negative ion must be reasonably well known as a function of wavelength. Except for a few ions, even the threshold for such processes are unknown.

A less direct process is associative detachment in the sense that primary agents like O and $O_2(^1\Delta)$ are generated in conjunction with the daytime O_3 chemistry. These species react with O_2^- to free the electron. They also react more indirectly by reducing complex negative ions to simpler ions more susceptible to photo- or associative detachment processes.

The role of photodetachment and O is illustrated in Figure 6. The sudden jump in the electron concentration near $\chi = 98°$ is a detachment effect primarily as a result of the reaction $O_2^- + h\nu \rightarrow O_2 + e$. Detailed computations show that at $\chi = 97.45°$, the rate at which electrons are being produced by this process is 2.39×10^2 cm^{-3} s^{-1} with $O_2^- + O \rightarrow O_3 + e$, yielding 1.26×10^1 electrons cm^{-3} as the second most productive electron source, not including the ionization production rate, 1.56×10^3 cm^3 s^{-1}. At $\chi = 89.78°$, the yields of these processes are about equal, 8.47×10^2 cm^{-3} s^{-1} and 8.72×10^2 cm^{-3} s^{-1}, respectively. According to the rate coefficients used (see Swider and Keneshea, 1972) the yields of these reactions will be equal when $[O] = 1.1 \times 10^9$ cm^{-3}. The initial rise in electrons thus appears to be a result of photodetachment, associative detachment becoming important near about $\chi = 93°$ when sufficient O is present. However, O also acts to reduce the more complicated ions to simpler species, e.g., $O + CO_3^- \rightarrow O_2^- + CO_2$, which are more readily subject to detachment processes.

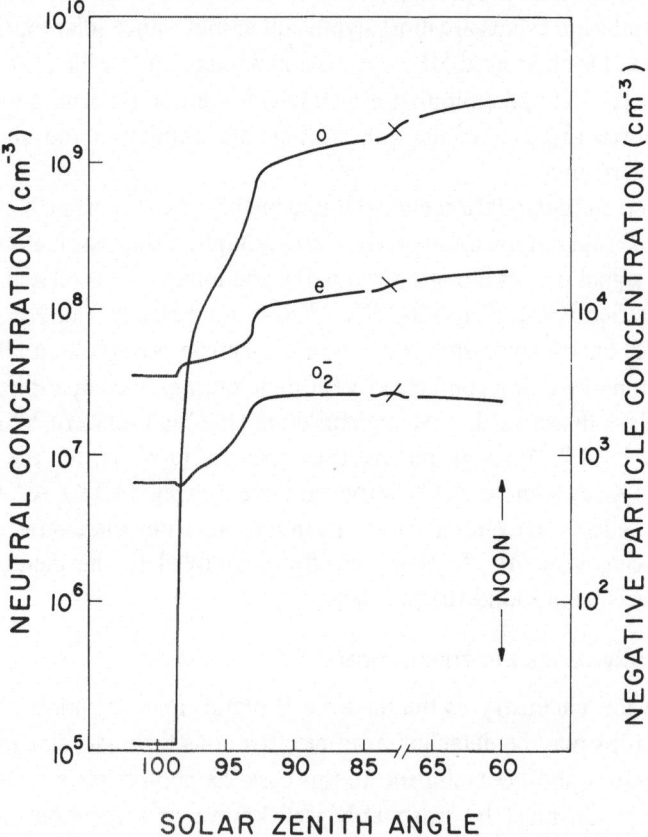

Fig. 6.   Computed (Swider and Keneshea, 1972) [O], [O$_2$], and [e] concentrations at sunrise (70 km) for a PCA event.

The initial rise of O is a result of the photodissociation of O_3 by visible light. Photodissociation of O_3 in the UV becomes more important for $\chi \leqslant 94°$. At sunset, the finite lifetime of O, about 20 min at 70 km, will contribute to [e] being somewhat larger, for the same χ, at sunset than at sunrise (see also Adams and Megill, 1967), contributing to a sort of hysteresis loop for [e] vs. χ.

Minor neutral gases may also exert a profound effect upon the negative ion chemistry. Thus, inclusion of NO_2 (Swider and Keneshea, 1972) in daytime PCA D region calculation results in reduced 30 MHz absorption. This effect is caused by the more rapid formation of negative ions less readily destroyed by detachment processes for the same chemistry considered leading to fewer electrons and hence less absorption at 30 MHz.

4. Conclusions

It appears that the physics of the mid-latitude E region is generally well understood. Some details regarding transport processes, particularly in regards to the NO distribution, require further analysis. The positive ion chemistry of the disturbed D region is reasonably known. More work is required in order to properly interpret the quiet D region. The negative ion population distributions are unclear for either D region condition. Laboratory work is needed for further clarification. At night the negative ion chemistry appears to be irrelevant in regards to the electron chemistry.

Acknowledgments

This work has been supported in part by the Defense Nuclear Agency under Subtask S99QAXHD028, Work Unit 23 entitled 'Composite PCA 69 Study'.

References

Adams, G. W. and Megill, L. R.: 1967, *Planetary Space Sci.* **15**, 1111.

Baird, G. A. and Francey, R. J.: 1972, *J. Geophys. Res.* **77**, 1966.

Barth, C. A.: 1966, *Ann. Géophys.* **22**, 198.

Barth, C. A. and Rusch, D. W.: 1971, IUGG paper, Moscow.

Bates, D. R.: 1973, *J. Atmospheric Terrestr. Phys.* **35**, 1935.

Biondi, M. A.: 1969, *Can. J. Chem.* **47**, 1911.

Butcher, E. C.: 1974, *J. Atmospheric Terrestr. Phys.* **36**, 177.

Chen, W. M. and Harris, R. D.: 1971, *J. Atmospheric Terrestr. Phys.* **33**, 1193.

Gough, M. P. and Collin, H. L.: 1973, *J. Atmospheric Terrest. Phys.* **35**, 835.

Grieder, W. F. and Burt, D. A.: 1973, in M. J. Rycroft and S. K. Runcorn (eds.), *Space Research XIII*, Akademie-Verlag, Berlin, p. 575.

Hunaerts, J. and Nicolet, M.: 1955, *J. Geophys. Res.* **60**, 537.

Keneshea, T. J. and MacLeod, M. A.: 1970, *J. Atmospheric Sci.* **27**, 981.

Keneshea, T. J., Narcisi, R. S., and Swider, W.: 1970, *J. Geophys. Res.* **75**, 845.

Manson, A. H. and Merry, M. W. J.: 1970, *J. Atmospheric Terrestr. Phys.* **32**, 1169.

Mechtly, E. A. and Smith, L. G.: 1968, *J. Atmospheric Terrestr. Phys.* **30**, 1555.

Meira, L. G.: 1971, *J. Geophys. Res.* **76**, 202.

Mitra, A. P.: 1968, *J. Atmospheric Terrestr. Phys.* **30**, 1065.

Narcisi, R. S. and Roth, W.: 1970, *Adv. Electronics Elect. Phys.* **29**, 79.

Narcisi, R. S., Philbrick, C. R., Thomas, D. M., Bailey, A. D., Wlodyka, L. E., Baker, D., Federico, G.,

Wlodyka, R., and Gardner, M. E.: 1972, in J. C. Ulwick (ed.), *Proc. COSPAR Symp. Solar Particle Event of November 1969*, AFCRL-72-0474, p. 421.

Nicolet, M. and Swider, W.: 1963, *Planetary Space Sci.* **11**, 1459.

Ogawa, T. and Tohmatsu, T.: 1966, *Rep. Ionosphere Space Res. (Japan)* **20**, 395.

Palluconi, F. D.: 1963, Ionosphere Res. Lab. Sci. Report No. 198, Penna. State University.

Ratcliffe, J. A.: 1956a, *J. Atmospheric Terrestr. Phys.* **6**, 1.

Ratcliffe, J. A.: 1956b, *J. Atmospheric Terrestr. Phys.* **6**, 306.

Rowe, J. F.: 1973, *J. Geophys. Res.* **78**, 6811.

Rowe, J. F.: 1974, *J. Atmospheric Terrestr. Phys.* **36**, 225.

Sechrist, C. F., Jr.: 1972, *J. Atmospheric Terrestr. Phys.* **34**, 1565.

Strobel, D. F.: 1972, *J. Geophys. Res.* **77**, 1337.

Svennesson, J., Reder, F., and Crouchley, J.: 1972, *J. Atmospheric Terrestr. Phys.* **34**, 49.

Swider, W.: 1965, *J. Geophys. Res.* **70**, 4859.

Swider, W.: 1969, *Planetary Space Sci.* **17**, 1233.

Swider, W.: 1972a, *J. Atmospheric Terrestr. Phys.* **34**, 1615.

Swider, W.: 1972b, *J. Geophys. Res.* **77**, 2000.

Swider, W. and Keneshea, T. J.: 1968, *Space Research* **8**, 370, North-Holland, Amsterdam.

Swider, W. and Keneshea, T. J.: 1972, in J. C. Ulwick (ed.), *Proc. COSPAR Symp. Solar Particle Event of November 1969*, AFCRL-72-0474, p. 589.

Swider, W. and Narcisi, R. S.: 1970, *Planetary Space Sci.* **18**, 379.

Swider, W. and Narcisi, R. S.: 1974a, *J. Geophys. Res.* **79**, 2849.

Swider, W. and Narcisi, R. S.: 1974b, *J. Geophys. Res.*, in press.

Swider, W., Narcisi, R. S., Keneshea, T. J., and Ulwick, J. C.: 1971, *J. Geophys. Res.* **76**, 4691.

Thomas, L.: 1971, *J. Atmospheric Terrestr. Phys.* **33**, 157.

Thomas, L.: 1974, *Radio Sci.* **9**, 121.

Tisone, G. C.: 1973, *J. Geophys. Res.* **78**, 746.

Tohmatsu, T. and Wakai, N.: 1970, *Ann. Geophys.* **26**, 209.

Torr, D. G., Torr, M. R., and Laurie, D. P.: 1972, *J. Geophys. Res.* **77**, 203.

Whalen, J. A., Buchau, J., and Wagner, R. A.: 1971, *J. Atmospheric Terrestr. Phys.* **33**, 661.

THE *F* REGION

PETER STUBBE

Max-Planck-Institut für Aeronomie, 3411 LINDAU/Harz, West Germany

Abstract. The basics of *F* region modeling, i.e., the pertinent transport equations and physical processes, are briefly discussed, and the present problems in *F* region modeling are listed.

1. Introduction

The *F* region of the ionosphere covers the height range of approximately 140 to 1000 km. These limits, however, are somewhat arbitrary, in particular the upper limit which separates the topside *F* region from the protonosphere, i.e. the region where H^+ ions are the predominant charged particles. The major ions in the *F* region are O_2^+ and NO^+ in the lower portion (the so-called *F*1 region), O^+ around the *F* region maximum (the so-called *F*2 region), and H^+ in the upper portion. Under certain conditions He^+ may also be an important ion in the upper *F* region. A typical ion density distribution is shown in Figure 1.

Since there are practically no negative ions in the *F* region, the electron density is simply given by the sum of the ion densities, due to the quasi-neutrality condition.

Modeling the *F* region means to theoretically reproduce the properties of the electron and ion gases. The most basic properties are the particle densities, macroscopic velocities and temperatures, corresponding to the first three moments of the velocity distribution functions. What it takes, therefore, is to solve the transport equations for the first three moments, i.e., the continuity, momentum and energy equations.

For the discussion of these equations we introduce a Cartesian coordinate system with the *x*-axis pointing towards the south, the *y*-axis towards the east, and the *z*-axis towards the zenith.

2. Transport Equations

2.1. CONTINUITY EQUATION

The continuity equation expresses the conservation of mass. It reads for constituent *j*

$$\frac{\partial n_j}{\partial t} = P_j - L_j - \text{div}(n_j \mathbf{v}_j), \tag{1}$$

where n_j is the particle density, t the time, P_j and L_j are the production and loss rates, respectively, and $\mathbf{v}_j$ is the macroscopic velocity. It is a useful, and in most cases valid, assumption to consider the *F* region as being horizontally stratified. This is equivalent to saying that the horizontal derivatives in the div-term are negligible in comparison with the vertical derivative. With this assumption the continuity equation simplifies

B. M. McCormac (ed.), Atmospheres of Earth and the Planets, 269–280. All Rights Reserved.

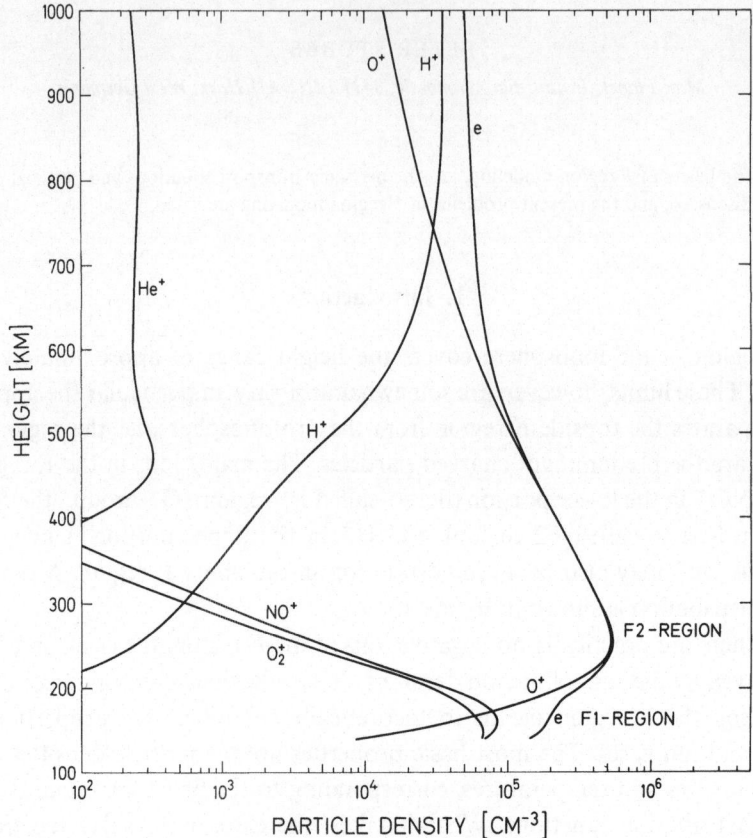

Fig. 1.   Typical ion density distribution in the *F* region. Theoretical results for Wallops Island ($\varphi = 38°$ N), March 2, 1966, noon (Stubbe, 1973), being in good agreement with mass spectrometer results (Brinton *et al.*, 1969).

to

$$\frac{\partial n_j}{\partial t} = P_j - L_j - \frac{\partial}{\partial z}(n_j v_{jz}). \tag{1a}$$

Equation (1a) holds true at middle and high latitudes, except for the description of small-scale horizontal structures. Equation (1a) cannot be applied, however, in the vicinity of the equator where horizontal transport must be taken into account, due to the geometry of the geomagnetic field.

2.2. Momentum Equation

The momentum equation, expressing the conservation of momentum, is given by

$$\frac{d\mathbf{v}_j}{dt} = -\frac{1}{\varrho_j}\,\mathrm{grad}\,p_j + \frac{\eta_j}{\varrho_j}\left(\Delta\mathbf{v}_j + \tfrac{1}{3}\,\mathrm{grad}\,\mathrm{div}\,\mathbf{v}_j\right) + \mathbf{a}_j, \tag{2}$$

where ϱ_j is the mass density, η_j the viscosity coefficient, and $p_j = n_j k T_j$ the partial pressure. The terms, from left to right, have the following meaning: inertial term, acceleration due to a pressure gradient force, acceleration due to viscosity, and acceleration due to an external force. Inserting typical values for v_j and the other parameters, one finds that the inertial and viscosity terms are negligible at F region altitudes, so that the velocity v_j is determined by a balance between the pressure gradient force and the external force;

$$O = -\frac{1}{\varrho_j} \operatorname{grad} p_j + \mathbf{a}_j. \tag{2a}$$

The dominant contributions to the external force are the gravity force, the electrostatic force, the Lorentz force and the frictional force with the neutral gas:

$$\mathbf{a}_j = \mathbf{g} + \frac{e}{m_j}(\mathbf{E} + \mathbf{v}_j \times \mathbf{B}) - \frac{\mu_{jn}}{m_j} v_{jn}(\mathbf{v}_j - \mathbf{v}_n), \tag{2b}$$

where e is the elementary charge (positive for ions, negative for electrons), $\mathbf{E}$ the electric field strength, m_j the particle mass, $\mu_{jn} = m_j m_n/(m_j + m_n)$ the reduced mass, and v_{jn} the momentum transfer collision frequency of one particle of species j with the neutral particles. The momentum transfer collision frequency appearing in the frictional force is defined in such a way that, if applied to rigid spheres, it yields the number of real collisions. This remark is necessary since in some papers another definition for the collision frequency is used, resulting, of course, in another expression for the frictional force (Stubbe, 1968). A typical value for v_{jn} is

$$v_{jn} \approx 10^{-9} \, n_n (\text{cm}^{-3}) \, [\text{s}^{-1}]$$

although in a more detailed consideration one has to allow for v_{jn} to be different for the different ion and neutral species.

2.3. Energy Equation

The conservation of energy is expressed by the equation

$$\frac{\partial T_j}{\partial t} = -\mathbf{v}_j \operatorname{grad} T_j + \frac{L}{n_j C_{vj}} \{ Q_{pj} + Q_{Tj} + \operatorname{div}(\varkappa_j \operatorname{grad} T_j) -$$
$$- p_j \operatorname{div} \mathbf{v}_j + \eta_j [\tfrac{4}{3}(\operatorname{div} \mathbf{v}_j)^2 - (\operatorname{curl} \mathbf{v}_j)^2] \} \tag{3}$$

where $L = 6.02 \, 10^{23}$, C_{vj} is the specific heat per mole at constant volume, and $\varkappa_j$ the heat conductivity. The right hand terms correspond, from left to right, to convection, heat production due to UV absorption, heat transfer between gases at different temperatures, heat conduction, conversion of work into heat or vice versa, and viscosity heating. An order of magnitude estimate shows that all the velocity dependent terms may be omitted. Again we may neglect horizontal derivatives and thus obtain for the energy equation of constituent j

$$\frac{\partial T_j}{\partial t} = \frac{L}{n_j C_{vj}} \left\{ Q_{pj} + Q_{Tj} + \frac{\partial}{\partial z}\left(\varkappa_j \frac{\partial T_j}{\partial z}\right) \right\}. \tag{3a}$$

There is a strong coupling between the three transport equations. The particle density depends on the velocity via the div-term in the continuity equation. The velocity depends on the particle density and the temperature via the pressure gradient force. The temperature depends on the particle density since a given heat supply is more efficient if it is shared among a smaller number of particles. Thus, F region modeling requires a simultaneous solution of the three transport equations for the electron gas and the O_2^+, NO^+, O^+ and H^+ gases.

In the following, we will briefly discuss the processes and mechanisms that determine the F region properties.

3. Production of Ionization

The major electron-ion source in the F region is provided by photoionization of neutral particles by the EUV part of the solar spectrum, i.e., between about 50 Å and the wavelength corresponding to the ionization energy. For the main thermospheric constituents, N_2, O, and O_2, the ionization wavelengths are 796, 911, and 1027 Å, respectively.

The number of electron-ion pairs produced per unit volume and time at height z is given by

$$P_j(z) = n_k(z) \int_0^{\lambda_k} I(z, \lambda)\, \sigma_k^{(i)}(\lambda)\, \eta_k(\lambda)\, d\lambda, \tag{4}$$

where $I(z, \lambda)$ is the solar photon flux (number of photons per unit area, time and wavelength), $\sigma_k^{(i)}(\lambda)$ the ionization cross section of neutral spesies k from which ion species j originates, $\eta_k(\lambda)$ the ionization yield (number of electron-ion pairs per ionizing photon), and λ_k the ionization wavelength (wavelength corresponding to ionization energy). The introduction of an ionization yield is necessary since the primary photoelectron may have enough energy to cause further ionization. This, however, is possible only if the photon energy is more than twice the ionization energy. Since the photoelectrons lose their energy through many different channels and not just by ionization, a description of the ionization yield is highly complicated and can best be attained by means of the Monte-Carlo method.

The photon flux at height z is related to the photon flux outside the atmosphere through

$$I(z) = I_\infty\, e^{-\tau(z)}, \tag{5}$$

where

$$\tau(z) = \sec\chi \sum_1 \sigma_1^{(a)} \int_z^\infty n_1\, dz \tag{5a}$$

is the so-called optical depth. The subscript 1 denotes the individual neutral species, $\sigma_1^{(a)}$ is the absorption cross section, and χ the solar zenith angle. Since the number density of the ionizable neutral gas decreases with altitude, while the solar UV flux increases with altitude, the electron-ion production rate $P_j(z)$ possesses a maximum. The height of the maximum is the lower, the smaller the absorption cross section and the smaller the solar zenith angle.

Up-to-date values for the absorption and ionization cross sections can be found in the *DNA Reaction Rate Handbook* (1972), while values for the solar photon flux are given by Donally and Pope (1973).

4. Electron and Ion Chemistry

We may distinguish two types of chemical reactions: Firstly, reactions between ions and neutral particles, leading to a redistribution of ionization. Secondly, reactions between ions and electrons, leading to a loss of ionization. Attachment reactions between electrons and neutral particles are without any practical importance in the F region.

4.1. ION-NEUTRAL REACTIONS

The primary ions produced by means of photoionization are N_2^+, O_2^+ and O^+. The other ions present in the F region are produced through chemical reactions. The main F region ion-neutral reactions are listed below. Also given are the reaction rates according to the *DNA Reaction Rate Handbook* (1972):

$$O^+ + O_2 \rightarrow O_2^+ + O \qquad k_1 = 1.6 \times 10^{-11} \text{ cm}^3 \text{ s}^{-1} \qquad \text{(R1)}$$
$$O^+ + N_2 \rightarrow NO^+ + O \qquad k_2 = 6 \times 10^{-13} \text{ cm}^3 \text{ s}^{-1} \qquad \text{(R2)}$$

$$O^+ + H \rightarrow H^+ + O \qquad k_3 = \tfrac{9}{8}k_4 \qquad \text{(R3)}$$
$$H^+ + O \rightarrow O^+ + H \qquad k_4 = 4 \times 10^{-10} \text{ cm}^3 \text{ s}^{-1} \qquad \text{(R4)}$$

$$N_2^+ + O_2 \rightarrow O_2^+ + N_2 \qquad k_5 = 5 \times 10^{-11} \text{ cm}^3 \text{ s}^{-1} \qquad \text{(R5)}$$
$$N_2^+ + O \rightarrow NO^+ + N \qquad k_6 = 1.4 \times 10^{-10} \text{ cm}^3 \text{ s}^{-1}. \qquad \text{(R6)}$$

There is new evidence that k_1 may be lower than given above (Ferguson, 1975). Reactions (R1) and (R2) act as loss reactions around the $F2$ peak since the ions O_2^+ and NO^+, which are produced at the expense of O^+, are rapidly destroyed by recombination with electrons. Reactions (R3) and (R4) determine the abundance of H^+ in the topside ionosphere. Around 500 km, H^+ is close to chemical equilibrium, so that the number density of H^+ is approximately given by

$$n(H^+) \approx \tfrac{9}{8} \frac{n(H)}{n(O)} n(O^+). \qquad (6)$$

Reactions (R5) and (R6) are responsible for N_2^+ being a minor ion in the $F1$ region although it is produced at a photoionization rate comparable to that of O_2^+ and O^+. Reactions (R2) and (R6) provide the production mechanisms for NO^+. Direct production of NO^+ through photoionization of NO is important in the lower ionosphere, but negligible in the F region.

Usually, the ion-neutral reaction rates are taken to be temperature independent. The justification for this lies in the fact that the ion-neutral reaction rate as a function of temperature typically exhibits a shallow minimum at temperatures representative of F region conditions (Ferguson, 1975). In the case of reaction (R2), however, the increase of k_2 beyond the minimum is very steep, so that the assumption of a constant k_2 may not be justified for high thermospheric temperatures. On the other hand, it remains to be seen whether the measurements of k_2 as a function of the ion kinetic energy may be simply converted into a $k_2(T)$ dependence.

4.2. ION-ELECTRON REACTIONS

Ions and electrons can recombine in two ways, either by emission of a photon or by dissociation of the resulting neutral molecule. Since radiative recombination has a very small cross section, there is practically no O^+ recombination. This is the reason for O^+ being the predominant ion over a wide height range and for the existence of an $F2$ layer. The dissociative recombination processes are (*DNA Reaction Rate Handbook*, 1972)

$$N_2^+ + e^- \rightarrow N + N \qquad k_7 = 2.7 \times 10^{-7} \, (T/300)^{-0.2} \qquad (R7)$$
$$O_2^+ + e^- \rightarrow O + O \qquad k_8 = 2.1 \times 10^{-7} \, (T/300)^{-0.7} \qquad (R8)$$
$$NO^+ + e^- \rightarrow N + O \qquad k_9 = 4.0 \times 10^{-7} \, (T/300)^{-0.5}. \qquad (R9)$$

The given temperature dependence of the recombination rates relates to a situation where the electron, ion and neutral temperatures are the same. Measurements are also available for a variation of the electron temperature alone, while the neutral and ion temperatures are kept at 300 K. The general temperature dependence, however, is not known yet.

5. Electron and Ion Motions

At $F2$ region altitudes the production rate of O^+ is proportional to $n(O)$ because $I(z)$ is practically indentical with I_∞. On the other hand, loss of O^+ through reactions (R1) and (R2) is proportional to $n(O_2)$ and $n(N_2)$. Since O_2 and N_2 decay more rapidly with altitude than O, the production/loss ratio increases with altitude, leading to a monotonic increase of $n(O^+)$ with altitude. Thus, the existence of an $F2$ region maximum cannot be understood in terms of photochemistry alone. This emphasizes the need to introduce an electron-ion transport.

According to the momentum Equation (2a), a plasma motion may have three causes. Firstly, if the pressure gradient force and the gravity force do not balance each other, the plasma will be accelerated until the frictional force balances the sum of the pressure gradient and gravity force. This is diffusion. Due to the smallness of the electron mass, the frictional force is much smaller for the electron gas than for the ion gases. Consequently, the electron gas will initially diffuse faster than the ion gases, giving rise to charge separation and, thus, to a polarization field which henceforth forces the electron and ion gases to move together. This process is called ambipolar diffusion. Secondly, the plasma can be moved by the neutral gas. Since the plasma is confined to the magnetic field lines when the collision frequency is much smaller than the gyrofrequency, a horizontal neutral wind is able to give the plasma a vertical velocity component, unless the dip angle is either O or 90°. Thirdly, a plasma motion can be caused by an electric field. Again, the presence of the geomagnetic field is of decisive importance. Without a magnetic field, electrons and ions would be moved in opposite directions, until a polarization field terminates further motion. In a magnetic field, however, electrons and ions have the same macroscopic velocity perpendicular to both the electric and magnetic field. From the coupled ion and electron equations, the vertical velocity of ion constituent *j* follows to be

$$v_{jz} = -D_a \left[\frac{1}{n_j(T_e + T_j)} \frac{\partial}{\partial z}(n_j T_j) + \frac{1}{n_e(T_e + T_j)} \frac{\partial}{\partial z}(n_e T_e) + \frac{m_j g}{k(T_e + T_j)} \right] +$$

$$+ v_{nx} \sin I \cos I + \frac{E_y}{B} \cos I, \tag{7}$$

where

$$D_a = \frac{k(T_e + T_j)}{v_{jn}\mu_{jn}} \sin^2 I$$

is the so-called ambipolar diffusion coefficient and *I* the dip angle which, by definition, is positive in the northern and negative in the southern hemisphere. The right-hand terms in Equation (7) describe, from the left to right, ambipolar diffusion, wind induced drift, and electric field induced drift. We note that an equatorward neutral wind and an eastward electric field cause an upward plasma motion.

We have seen that ion production and loss tend to establish an O^+ density – height – distribution which shows a monotonic increase with altitude. On the other hand, diffusion tends to establish a distribution following the barometric law. As a result, diffusion in the *F*2 region leads to a downward plasma motion into an area of higher O^+ losses. It is this downward motion which gives rise to the formation of an *F*2 layer with a distinct maximum.

Neutral winds play an important role for the diurnal variation of the *F*2 layer. The meridional wind component is directed towards the pole at day and towards the equator at night. Thus, the daytime *F*2 layer is lowered by winds, while the nighttime *F*2 layer is raised. This mechanism is essential for the maintenance of the nighttime *F*2 layer (Kohl *et al.*, 1968). Electric field induced plasma drifts are far less important at mid-latitudes, but they have a significant effect on the properties of the equatorial *F* region, accounting for the so-called equatorial anomaly (Hanson and Moffett, 1966), i.e. the existence of an electron density trough at the magnetic equator with crests at about 15 to 20° latitude.

6. Electron and Ion Temperatures

The major heat supply to the electron-ion gas stems from the incident solar EUV radiation. The conversion mechanism from radiation energy to heat energy consists of a complex chain of processes which may be roughly described as follows: By photoionization of a neutral particle, an electron is released with a kinetic energy greatly in excess of the thermal energy of the ambient electron gas. These photo-electrons lose their excess energy by elastic collisions with neutral particles, ions, and ambient electrons and by inelastic collisions, mainly with neutral particles. Since the energy transferred in an elastic collision depends on the mass ratio of the colliding particles, the energy given to an ambient electron is by three to four orders of magnitude larger than the energy given to an ion or a neutral particle. This is the reason why the electron temperature generally exceeds the ion or neutral temperature. In a self-consistent F region model, the heat supply to the electron gas has to be calculated subject to the solar UV intensity and the conditions of the neutral atmosphere representative of the respective time and geographic location. This amounts to solving the photoelectron continuity equation in the variables height and energy, which, in itself, is a tremendous task. Therefore, in F region modeling one usually works with the simple 'heating efficiency concept', i.e. one relates the electron heating rate to the electron ionization rate by

$$Q_{Pe} = \varepsilon P_e, \tag{8}$$

where ε is the so-called heating efficiency. The shortcoming of relation (8) is that it disregards nonlocal heat deposition of the photoelectrons which may become important above about 300 km. An analytic expression for ε as a function of the electron density and the O_2, N_2, and O densities is given by Swartz and Nisbet (1972).

The main energy loss of the thermal electron gas occurs through inelastic collisions with neutral particles by exciting internal degrees of freedom and electronic states (Stubbe and Varnum, 1972) and through elastic collisions with ions (Banks, 1966).

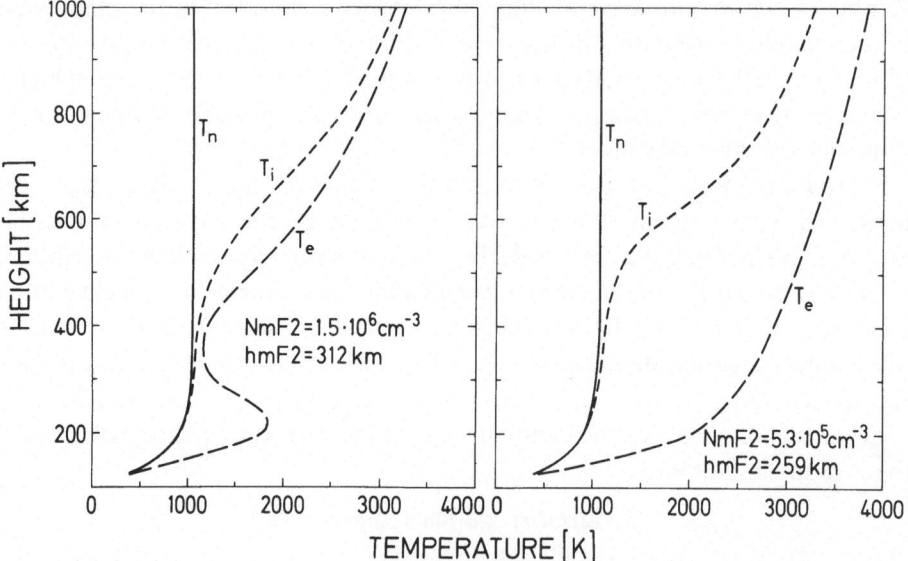

Fig. 2.   Typical electron and ion temperature profiles. Theoretical results, obtained with a heating efficiency artificially increased by a factor of 3 in order to reach agreement with measured temperatures. Left section: High electron density case. Right section: Low electron density case.

The first process predominates below about 250 km and forces the electron temperature to approach the neutral temperature towards lower altitudes. The second process is important around the $F2$ maximum and may lead to an electron temperature minimum in the vicinity of the $F2$ maximum if the electron-ion density is sufficiently high. Above about 300 km heat conduction becomes increasingly important, causing the electron temperature to monotonically increase with height. This is because photoelectrons escaping the ionosphere deposit energy in the protonosphere and thereby give rise to a heat flux from the protonosphere into the ionosphere. Typical electron temperature profiles are shown in Figure 2.

Also shown are the corresponding ion temperature profiles. The ion temperature is determined by heat exchange with the electron gas, which acts as a heat source, and by heat exchange with the neutral gas, which acts as a heat sink. Since the electron to neutral density ratio increases with altitude, the ion temperature follows the neutral temperature in the lower F region and the electron temperature in the upper F region. Strictly speaking, the individual ion gases possess their individual temperatures, but the differences are only of the order 100 deg in the upper F region so that the use of a common ion temperature is justified (Banks, 1967).

7. Problems in Midlatitude *F* Region Modelling

The F region is probably the best understood area in aeronomy. Nonetheless, there is a good number of problems which, so far, have made F region modeling not always fully satisfactory.

7.1. THERMOSPHERIC MODEL PROBLEM

The F region properties strongly depend on the thermospheric properties like composition, density, and temperature. Existing thermospheric models (e.g. Hedin *et al.*, 1974) can, of course, only describe the average thermospheric behavior. Day to day variations are not included in these models, and variations during magnetic storms are necessarily represented in a highly idealized way. Correspondingly, theoretical F region modeling is restricted to a description of the average F region behavior.

7.2. NEUTRAL WIND PROBLEM

Neutral winds strongly affect the F region. A calculation of the wind velocity requires the knowledge of the horizontal pressure gradient. Since this is a differential quantity, its accuracy is highly limited. Consequently, the calculated wind velocities may be subject to considerable errors.

7.3. BOUNDARY VALUE PROBLEM

In order to solve the continuity and energy equations, the particle and energy fluxes through the upper boundary have to be known. Only in rare cases are measurements available. Otherwise, one has to make more or less crude assumptions about these quantities. Two difficulties in F region modeling are clearly related to the boundary value problem. The first is that existing models cannot explain nighttime increases in the electron density which are sometimes observed. The second is the old maintenance problem of the nighttime F region. Although the introduction of winds into F region theory has considerably eased this problem, there is still a remainder which requires the existence of a downward particle flux at night, especially during the first hours of the night (see Figure 3).

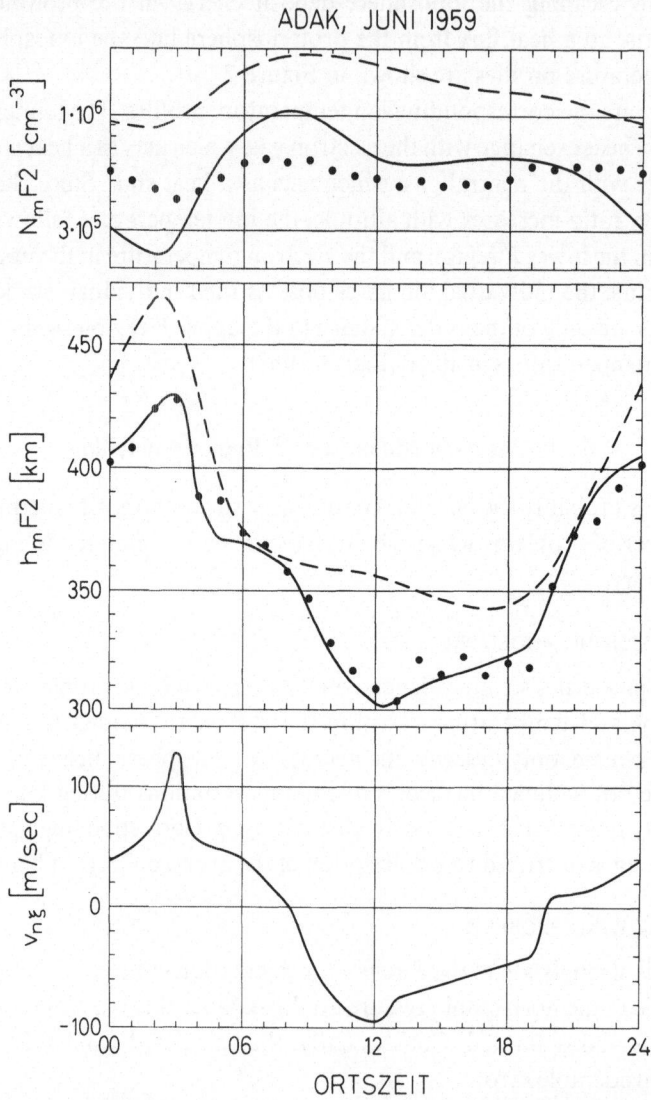

ADAK, JUNI 1959

ORTSZEIT

Fig. 3. Comparison of calculated and observed F region results for Adak, June 1969. Upper section: Experimental (dots) and theoretical (solid curve) N_mF2 values as a function of local time (the dashed curve is of no concern here). The 'Summer Morning Problem' consists in the much too steep early morning increase of the theoretical N_mF2 curve. Also seen is the 'Boundary Value Problem' in the evening hours. Middle section: h_mF2, height of $F2$ layer maximum. The agreement between the observed and theoretical values is enforced by properly specifying, at h_mF2, the wind velocity in the magnetic N-S direction, $v_{n\xi}$. Lower section: $v_{n\xi}$ at h_mF2 as required to get agreement in h_mF2.

7.4. EUV PROBLEM

The solar EUV spectrum is known to be subject to considerable changes within a solar cycle. It is not fully known, however, to which extent the intensities in the different wavelength intervals change, and, even more important, how a sensible ground-based monitoring of the UV spectrum can be achieved. Therefore, the calculated photoionization rates at any particular day are not very accurate.

7.5. ION CHEMISTRY PROBLEM

Some of the chemical reaction rates are not known well enough. This is particularly obvious for the N_2^+ reaction scheme. Measured N_2^+ densities are typically a factor of 10 greater than calculated N_2^+ densities (Stubbe, 1973). This can mean either that the given reaction scheme is not complete for N_2^+, or that at least one of the reaction rates k_5, k_6, k_9 is strongly overestimated. The recombination rates are not known with respect to their full temperature dependence, as outlined before. The reaction rate k_2 poses a problem with respect to its dependence on the kinetic and vibrational temperature (Schmeltekopf *et al.*, 1968; Ferguson, 1975).

7.6. ELECTRON TEMPERATURE PROBLEM

Using the EUV intensities, heating efficiencies and heat transfer rates published in literature, one obtains electron temperatures in the bottomside F region which are typically 30 to 40% lower than observed temperatures. This is a very large deviation considering that the electron temperature cannot drop below the neutral temperature. It would take a factor of 3 increase of the heating efficiency in order to reach reasonable agreement between theoretical and experimental electron temperatures (Stubbe, 1973).

7.7. SUMMER MORNING PROBLEM

This is a rather special problem which is described in Figure 3.

Attempts to solve the problem have been mainly centered on modifying the neutral wind velocity. Figure 3 shows why these attempts could not be fully successful: Although the wind velocity is specified here in such a way that the observed and theoretical h_mF2 values closely agree, the forenoon increase in the theoretical N_mF2 curve is much too steep. Thus, it is not likely that the explanation to the problem is to be found in terms of a transport process.

Conclusion

In the interest of conciseness, the peculiarities of equatorial F region modeling could not be discussed in the foregoing. The reader is referred to the works by Sterling *et al.* (1969) and Anderson (1973). Additional information on midlatitude F region modeling can be found in Rüster (1971), Strobel and McElroy (1970), Stubbe (1970), and Tanaka and Hirao (1973).

References

Anderson, D. N.: 1973, *Planetary Space Sci.* **21**, 409

Banks, P.: 1966, *Planetary Space Sci.* **14**, 1085.

Banks, P.: 1967, *Planetary Space Sci.* **15**, 77.

Brinton, H. C., Pharo, M. W., Mayr, H. G., and Taylor, H. A.: 1969, *J. Geophys. Res.* **74**, 2941.

DNA Reaction Rate Handbook: 1972, DoD Nuclear Information Center, General Electric, Santa Barbara, California 93102.

Donally, R. F. and Pope, J. H.: 1973, NOAA Technical Report ERL 276-SEL25.

Ferguson, E. E.: 1975, this volume, p.197.

Hanson, W. B. and Moffett, R. J.: 1966, *J. Geophys. Res.* **71**, 5559

Hedin, A. E., Mayr, H. G., Reber, C. A., and Spencer, N. W.: 1974, *J. Geophys. Res.* **79**, 215.

Kohl, H., King, J. W., and Eccles, D.: 1968, *J. Atmospheric Terrestr. Phys.* **30**, 1733.

Rüster, R.: 1971, *J. Atmospheric Terrestr. Phys.* **33**, 137.

Schmeltekopf, A. L., Ferguson, E. E., and Fehsenfeld, F. C.: 1968, *J. Chem. Phys.* **48**, 2966.

Sterling, D. L., Hanson, W. B., Moffett, R. J., and Baxter, R. G.: 1969, *Radio Sci.* **4**, 1005.

Strobel, D. F. and McElroy, M. B.: 1970, *Planetary Space Sci.* **18**, 1181.

Stubbe, P.: 1968, *J. Atmospheric Terrestr. Phys.* **30**, 1965.

Stubbe, P.: 1970, *J. Atmospheric Terrestr. Phys.* **32**, 865.

Stubbe, P.: 1973, Pennsylvania State University Report PSU-IRL-SCI 418.

Stubbe, P. and Varnum, W. S.: 1972, *Planetary Space Sci.* **20**, 1121.

Swartz, W. E. and Nisbet, J. S.: 1972, *J. Geophys. Res.* **77**, 6259.

Tanaka, T. and Hirao, K.: 1973, *J. Atmospheric Terrestr. Phys.* **35**, 1443.

PART VI

OPTICAL OBSERVATIONS

THE GLOBAL PATTERN OF 6300 Å ATOMIC OXYGEN
EMISSION AS SEEN FROM THE ISIS-2 SPACECRAFT

G. G. SHEPHERD

Centre for Research in Experimental Space Science, York University, 4700 Keele Street Downsview, Ontario,
M 3J-1P3, Canada

1. Introduction

The 6300 Å line of O is a useful detector of low energy ionospheric processes. The 100 eV magnetosheath-type electrons incident in the dayside cleft produce red auroras, and red-enhanced auroras appear as polar cap arcs and at high latitudes on the nightside. During storms, thermal excitation inside the plasmapause generates SAR Arcs. Outside the plasmapause conjugate photoelectrons produce significant increases of 6300 Å emission by electron impact. Dissociative recombination of O_2^+ produces 6300 Å emission, dramatically evident at the equator. All of these mechanisms are identifiable in the 'instantaneous' global maps produced by the Red Line Photometer (Shepherd *et al.*, 1973a) on the ISIS-2 spacecraft. The Auroral Scanning Photometer (Anger *et al.*, 1973) also yields data at 5577 Å and 3914 Å. Parameters measured in the circular orbit at 1400 km include (Shepherd *et al.*, 1973b) electron density and temperature, ion composition and temperature, and energy spectra of electrons and protons; while the electron density profile down to the ionosphere is traced by the topside sounder.

2. The Red Auroral Oval

It seems curious that the awareness of dayside aurora has developed so recently. This may be attributed to the low visual response at 6300 Å, and the small amount of land mass, all of it lightly inhabited, lying near 78° IN Lat and high geographic latitude. But from all-sky camera photographs (Feldstein, 1966; Akasofu, 1972; Lassen, 1972), and aircraft flights (Heikkila *et al.*, 1972; Whalen and Pike, 1973) the auroral oval became familiar to us.

The oval pattern of 6300 Å emission obtained by ISIS-2 during a quiet period was shown by Shepherd and Thirkettle (1973). To plot auroral data we have found it convenient and useful to arbitrarily choose 1 kR to define the auroral boundary (since albedo in the polar cap will normally double the intensity, this is in fact closer to 500 R). The 1 kR boundary for the Shepherd and Thirkettle (1973) data is shown in Figure 1, plotted along with Feldstein's oval. The agreement is remarkably good, except at noon where the red aurora lies a little poleward of the oval. Although the oval does not close under these conditions, it does close at lower intensities, or at 1 kR during slightly disturbed periods. (For the nightside we have used 700 R here.)

On the nightside, the discrete forms plotted by Feldstein (1966) lie with the 6300 Å emission in what Eather and Mende (1971) have called the soft auroral zone, at about

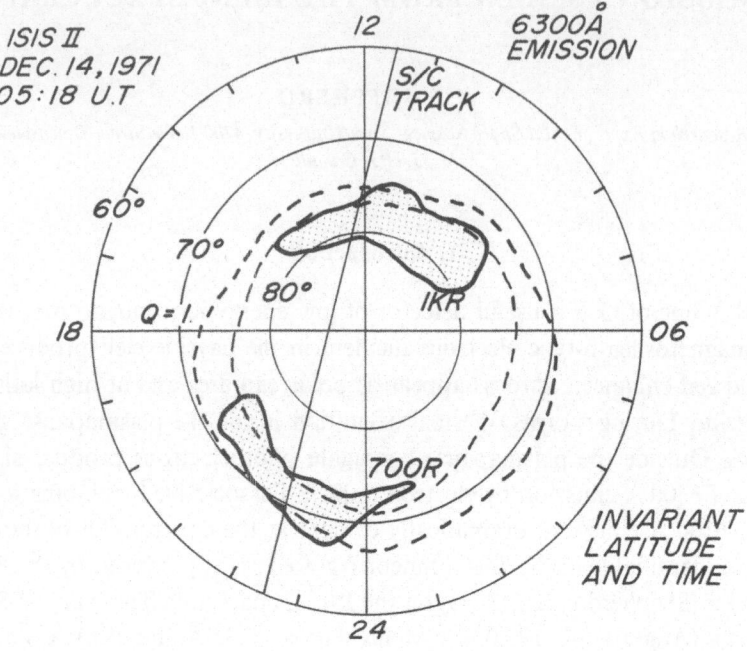

Fig. 1.   A plot of 6300 Å data from an ISIS 2 pass at 05 : 18 UT, December 14, 1971. The 1 kR contour on the dayside and the 700 R contour on the nightside are plotted in invariant coordinates along with Feldstein's $Q=1$ oval.

70° latitude and where the 6300 Å emission occurs. During slightly disturbed times the diffuse aurora (Lui and Anger, 1973) brightens at about 65° IN Lat, and discrete forms appear there. Figure 2 is a shading map taken when the two zones are clearly evident, one day before the pass of Figure 1. From Anger's (1974) data the ratio $I(6300)/I(5577)$ is seen to be about three in the poleward region and unity in the equatorward region. Under these conditions ($Q=4$, say) Feldstein's oval broadens to include both the discrete (soft) and the diffuse regions. So the red auroral oval is, almost exactly, the Feldstein oval but the more circular belt of hard precipitation lies outside it.

3. Global Storms

Optical aurora is always visible somewhere, no matter how quiet conditions are, but the pattern during disturbed times is dramatically different. The global pattern of 6300 Å emission on December 18, 1971 is shown in Figure 3. On the dayside the bright 6300 Å emission extended from the invariant pole right down to the limit of sunlight. Only a small region, on the nightside of the pole did not exhibit the intense emission. It is questionable whether present magnetospheric models are adequate to understand this kind of pattern.

A more major storm occurred on August 4 to 5, 1972, and onwards. The ISIS-2 spin axis was perpendicular to the orbit plane, near dawn-dusk and only limited

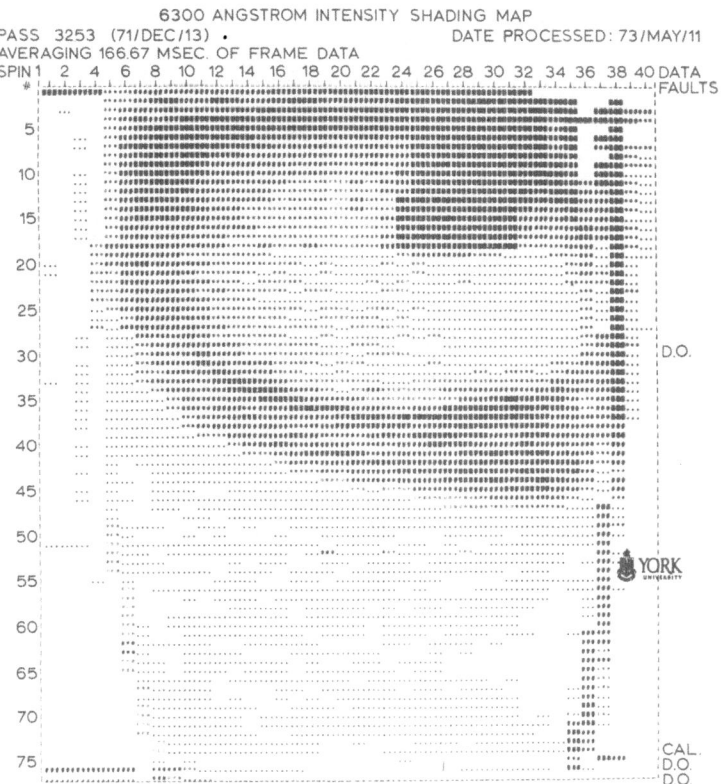

Fig. 2. A 'shading map' for the 6300 Å O emission, obtained from a pass at 04:41 UT, December 13, 1971. One line of print corresponds to one spacecraft rotation and so the picture is built up from successive spins. Noon is at the top of the picture (with a rectangular box pattern arising from baffle-scattered sunlight) and midnight is at the bottom. The curved Earth limbs can be seen and an almost continuous auroral oval with two clearly separated regions on the nightside.

data were obtained. Figure 4 shows a slide obtained solely from limb data, plotting the peak intensity at the limb versus geographic latitude. The remarkable feature of these data, taken from two orbits at the onset of the storm, is a sheet of auroral intensity, with a marked intensity gradient, beginning at 45° IN Lat and increasing up to about 70°, the region of discrete auroras. A similar, though less marked effect occurred in the December 18, 1971, storm and it may be more common than that, but normally at a level that is difficult to distinguish from airglow.

4. Mid-Latitude Airglow and the Plasmapause

In early passes of the first northern hemisphere winter of ISIS-2, October 1971, a consistent fringe of 6300 Å intensity was observed equatorward of the aurora. Comparison with the topside sounder and Langmuir probe data showed that this equatorward fringe boundary was at the plasmapause. The explanation offered by Shepherd

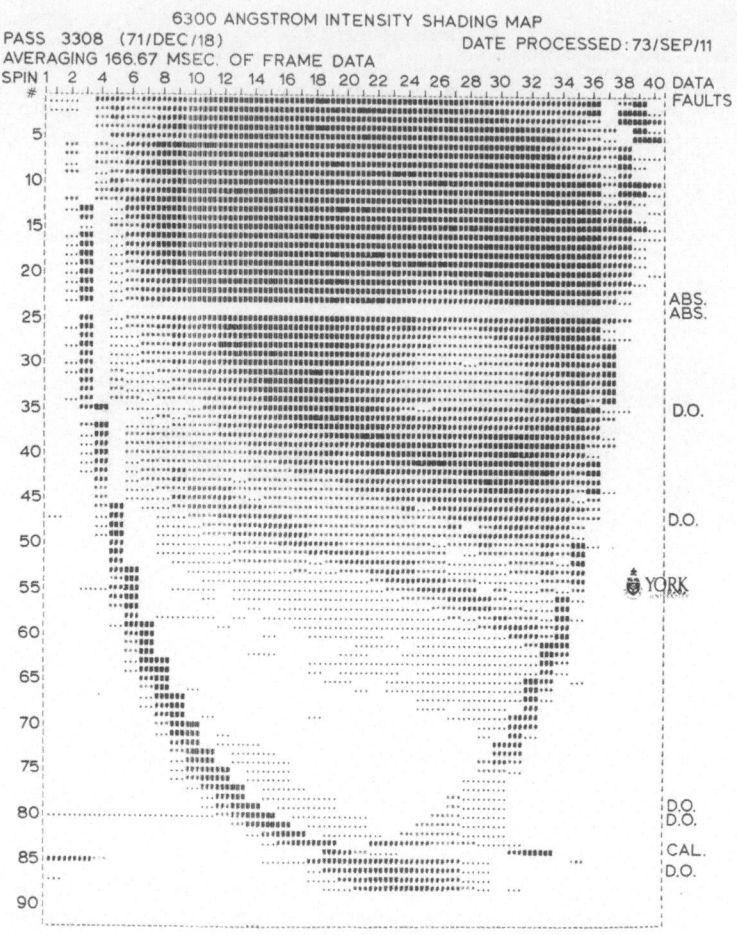

Fig. 3. A shading map showing the 6300 Å emission during the December 16–21, 1971 storm, at 04:02 UT, December 18. The viewing geometry is similar to that of the pass of December 13, shown in Figure 2. The area of coverage by 6300 Å emission exceeding 1 kR is remarkable, particularly on the dayside. A SAR-arc is visible equatorward of the aurora.

et al. (1973c) was that conjugate photoelectrons were incident on both sides of the plasmapause, but producing significant 6300 Å emission only outside the plasma-pause, since inside it they were attenuated by the higher total electron content along the field lines. For this paper I have done a limb analysis of one of the Shepherd *et al.* (1973c) passes to confirm that it is not a spurious effect in the downlooking data. The result is shown in Figure 5, along with an approximate latitudinal profile of electron density provided by L. H. Brace. The inverse relation between the two is very striking.

5. Equatorial 6300 Å Emission

Tropical arcs are a spectacular source of 6300 Å emission (Reed *et al.*, 1973) and

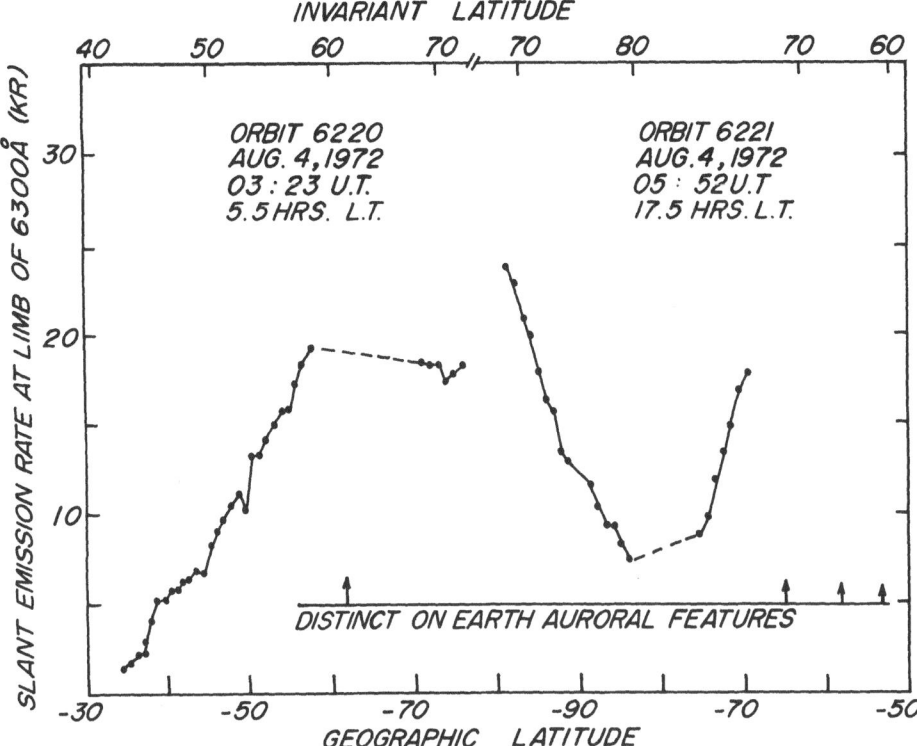

Fig. 4. Plots of 6300 Å limb intensity vs. geographic latitude for two orbits near the height of the August 4–5, 1972 storm, indicating a sheet of auroral intensity. The location of some distinct auroral features is also shown.

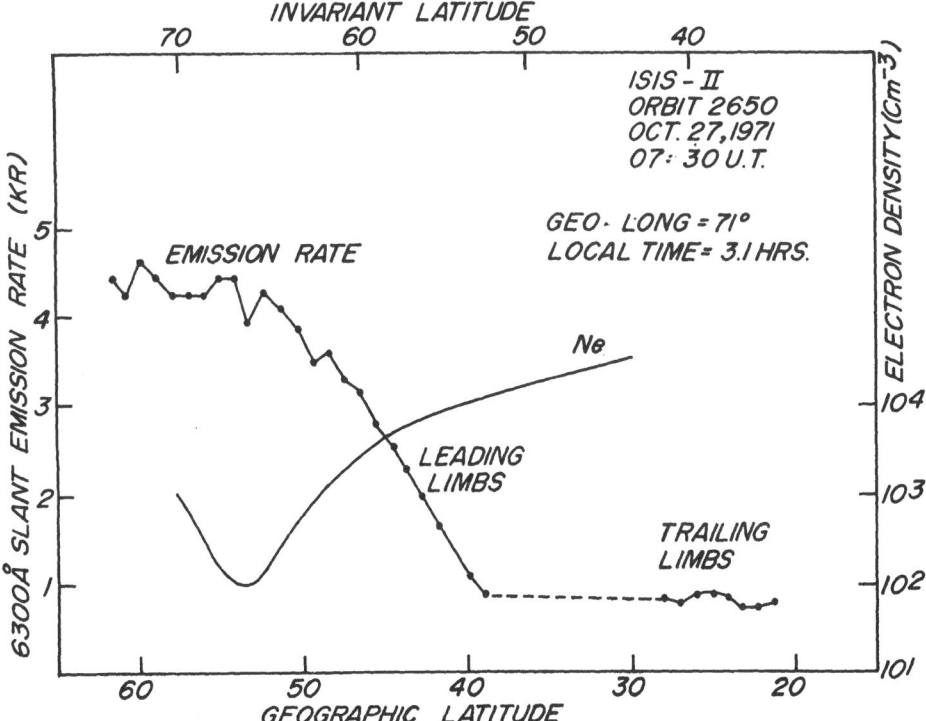

Fig. 5. A plot of 6300 Å limb intensity vs. geographic latitude compared with the electron density at 1400 km, indicating a plasmapause effect on the emission rate.

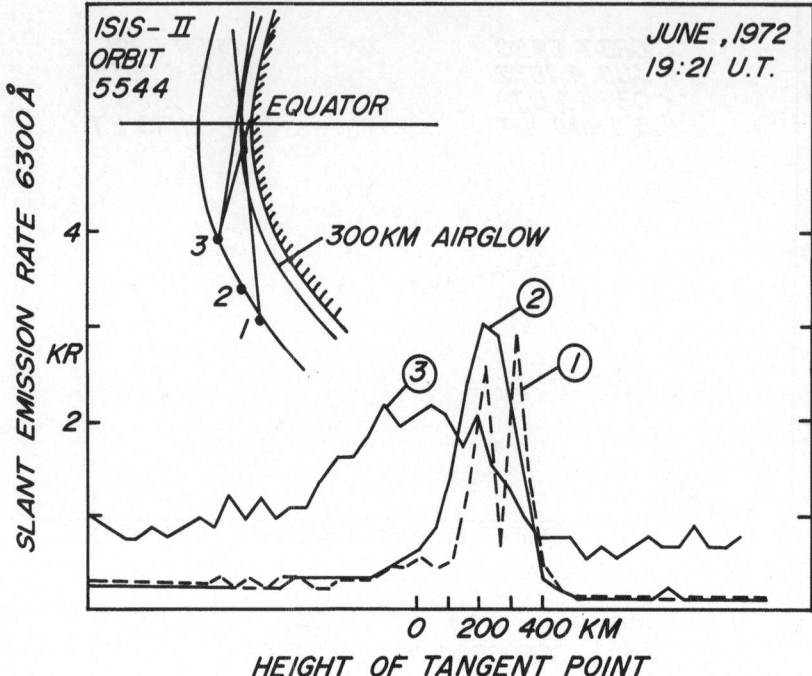

Fig. 6. Data plots (linear in time) of three selected limb scans across an equatorial arc. The inset shows the viewing geometry.

should be very interesting to study with the ISIS-2 spacecraft. So far only a very limited attempt has been made. A spectacular arc of about 1 kR intensity was detected over Africa, in cartwheel orientation. The result of three limb scans is shown in Figure 6. The location of the spacecraft while trace 1 was obtained is shown in the inset; a double layer is evident. With the spacecraft in position 2, the feature appears single. In position 3 the arc is no longer at the limb but appears as an on-Earth feature. The inset also shows a possible model, with an airglow layer at about 300 km and the localized arc rather lower at about 200 km.

References

Akasofu, S.-I.: 1972, *J. Geophys. Res.* **77**, 2303.
Anger, C. D.: 1974, private communication.
Anger, C. D., Fancott, T., McNally, J., and Kerr, H. S.: 1973, *Appl. Opt.* **12**, 1753.
Eather, R. H. and Mende, S. B.: 1971, *J. Geophys. Res.* **76**, 1746.
Feldstein, Y. L.: 1966, *Planetary Space Sci.* **14**, 121.
Heikkila, W. J., Winningham, J. D., Eather, R. H., and Akasofu, S.-I.: 1972, *J. Geophys. Res.* **77**, 4100.
Lassen, K.: 1972, *Geofys. Publ.* **29**, 87.
Lui, A. T. Y. and Anger, C. D.: 1973, *Planetary Space Sci.* **21**, 799.
Reed, E. I., Fowler, W. B., and Blamont, J. E.: 1973, *J. Geophys. Res.* **78**, 5658.
Shepherd, G. G. and Thirkettle, F. W.: 1973, *Science* **180**, 737.
Shepherd, G. G., Fancott, T., McNally, J., and Kerr, H. S.: 1973a, *Appl. Opt.* **12**, 1767.
Shepherd, G. G., Anger, C. D., Brace, L. H., Burrows, J. R., Heikkila, W. J., Hoffman, J., Maier, E. J., and Whitteker, J. H.: 1973b, *Planetary Space Sci.* **21**, 819.
Shepherd, G. G., Brace, L. H., and Whitteker, J. H.: 1973c, *J. Geophys. Res.* **78**, 4689.
Whalen, J. A. and Pike, C. P.: 1973, *J. Geophys. Res.* **78**, 3848.

OGO-6 OBSERVATIONS OF 5577 Å

T. M. DONAHUE*

Dept. of Physics, The University of Pittsburgh, Pittsburgh, Pa. 15260, U.S.A.

1. Introduction

There will be very little material discussed in this lecture that has not already found its way into the literature. Therefore, I shall content myself with the briefest meaningful review of the data obtained by the horizon scanning 5577 Å airglow photometer flown aboard OGO 6 and the interpretations we have put upon them. Most of my fresh remarks will have to do with the limitations in these interpretations and will amount to a list of caveats and comments about the meaning of the results we have obtained.

2. The Experiment

OGO 6 was launched June 5, 1969, and flew for about 1 yr of useful life in a polar orbit of 82° inclination. Apogee was at about 1100 km and perigee at about 400 km. Our photometer had a field of view that allowed us to scan in a vertical plane 20° from the orbital plane. The field of view was 7.5' vertical × 4.2° horizontal and was scanned in 128 steps each of 7.5' during a time of 18.4 s. The field scanned was from 10° to 26° below the horizontal at the spacecraft. The field of view was so narrow that it spanned only about 5 km in the vertical along the line of sight when that line passed 100 km above the Earth's surface at closest approach. Unfortunately we could observe the nightglow layer at 100 km only when the satellite was below 700 km. Thus we lost an appreciable amount of data on each orbit. There was an internal calibration system. Where comparisons were made with ground-based photometers after launch we found that we generally agreed very well with them (i.e., within about 25% at worst.) The photometer output to telemetry was in 256 steps of 0.02 V each. During most of the lifetime of the satellite the sensitivity in the green channel was about 400 kR V^{-1} so the lowest detectable slant emission rate was 10 ± 5 R and the upper limit to any observation after background or dark current subtraction was 5 R. For low slant emission rates the empirical relationship between integrated vertical emission rate and maximum slant rate was 22 R kR^{-1}. Hence the minimum detectable equivalent integrated vertical emission rate was only 0.22 ± 0.11 R. Since our slant rates typically ranged from about 1 kR to about 6 kR, the vertical rates implied ran from about 22 R to about 105 R. Although there were some unusual tropical observations when the contribution from the 100 km layer was as low as 0.20 kR (4.5 R vertical) and there were mid-latitude observations close to 10 kR (180 R vertical). These results should be kept in mind when comparing our deductions with those of other observers.

* Present Address: Dept. of Atmospheric and Oceanic Science, University of Michigan, Ann Arbor, Michigan.

B. M. McCormac (ed.), Atmospheres of Earth and the Planets, 289–307. All Rights Reserved.

According to our calibration, about 60 R would be a typical value for the chemical or Chapman airglow vertical emission rates. Seasonal maxima in the latitude range 40 to 50° would be about 170 R. Given the generally low brightness found for the Chapman layer in the tropics (20 to 40 R) we have found that the ionospheric region contribution often exceeding 40 R can be a major portion of the tropical green airglow in the region of the Appleton anomaly.

3. The F_1 Region Contribution

Figure 1 shows the slant emission rate observed in one vertical scan near local midnight on November 3, 1969 when the line of sight of the photometer while viewing the upper maximum at about 220 km above the Earth passed over Huancayo. Since the ionosonde at Huancayo was taking data at the time we obtained a profile for the electron density vs. attitude above that station at 4.56 UT November 3, 1969. This is an example of several such coincidental overlaps we were able to discover.

The field of view of the photometer is so limited in the vertical and the scan rate so high that each very small emitting volume element in the atmosphere is observed a number of times (usually 20 to 30) in successive scans. This fact enables us to unfold from the sequence of scans in the horizontal plane the local emission rate without resorting to a uniform layer assumption. The procedure is tedious and described in detail in the paper by Thomas and Donahue (1972). The result was that we obtained

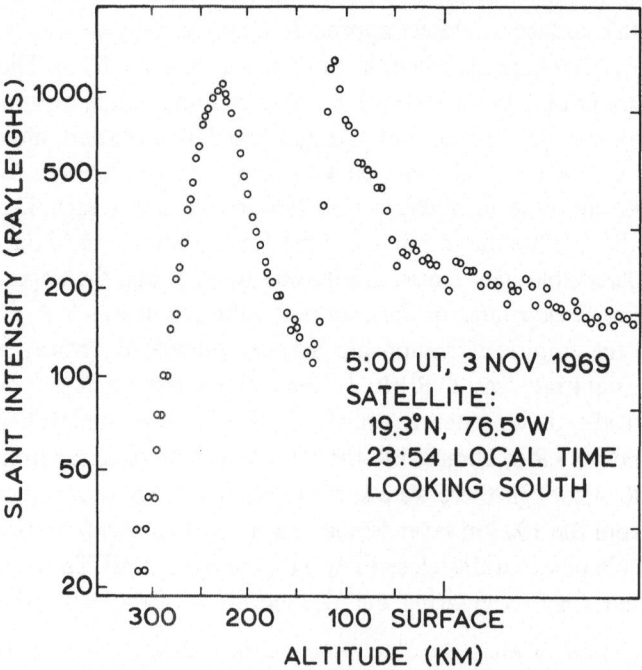

Fig. 1.   Green line (5577 Å) photometer data points plotted as a function of the distance of closest approach to the Earth's surface on the optic axis.

maps giving emission rates for the 5577 Å line as functions of latitude and altitude over the range of latitudes from which useful F region data were obtained in specific orbits. These rates are reliable we believe to within $\pm 5\%$.

Assuming that the emission is all produced by the reaction

$$O_2^+ + e \rightarrow O(^3P) + O(^1S) \tag{1}$$

$$O(^1S) \rightarrow O(^1D) + hv \tag{2}$$

it is a simple matter to show that throughout the altitude region in question quenching of $O(^1S)$ is not likely to be important and that for all practical purposes the local 5577 Å emission rate, E, is given by

$$E = 0.94 f\alpha_3 n_e [O_2^+], \tag{3}$$

where 94% of all atoms excited to $O(^1S)$ emit the green line and $f\alpha_3$ is the rate constant for production of $O(^1S)$ in dissociative recombination. Since the O_2^+ at night is presumably formed exclusively as a result of the charge exchange reaction

$$O^+ + O_2 \rightarrow O_2^+ + O \, (k_1) \tag{4}$$

and we are confronted with a steady state, we have

$$E = 0.94 f k_1 [O_2] [O^+]. \tag{5}$$

The relation between n_e and $[O^+]$ is not simple below 250 km because of the reaction

$$O^+ + N_2 \rightarrow NO^+ + N \tag{6}$$

and the relative importance of NO^+ and O_2^+ densities compared to those of O^+, so that a bit of complexity has to be introduced into the relation between E and n_e below that altitude (Van Zandt and Peterson, 1968), but for the most part of the altitude range covered

$$n_e \cong [O^+] \tag{7}$$

and

$$E = 0.94 f k_1 [O_2] n_e \tag{8}$$

or

$$n_e = E/0.94 f k_1 [O_2]. \tag{9}$$

In our comparisons with ground ionosonde stations or incoherent scatter stations we found that we had to reduce the quantity

$$f k_1 [O_2] \tag{10}$$

by a factor of 2.1 from its nominal value to obtain agreement. The sort of agreement between observed E and that predicted from the sounder data then achieved is shown in Figure 2.

The nominal values referred to are the O_2 densities taken from Jacchia (1971) for the appropriate exospheric temperature a value for f of 0.1 (Zipf, 1970) and

$$k_1 = 2.0 \times 10^{-11} \sqrt{300/T} \ cm^3 \ s^{-1} \tag{11}$$

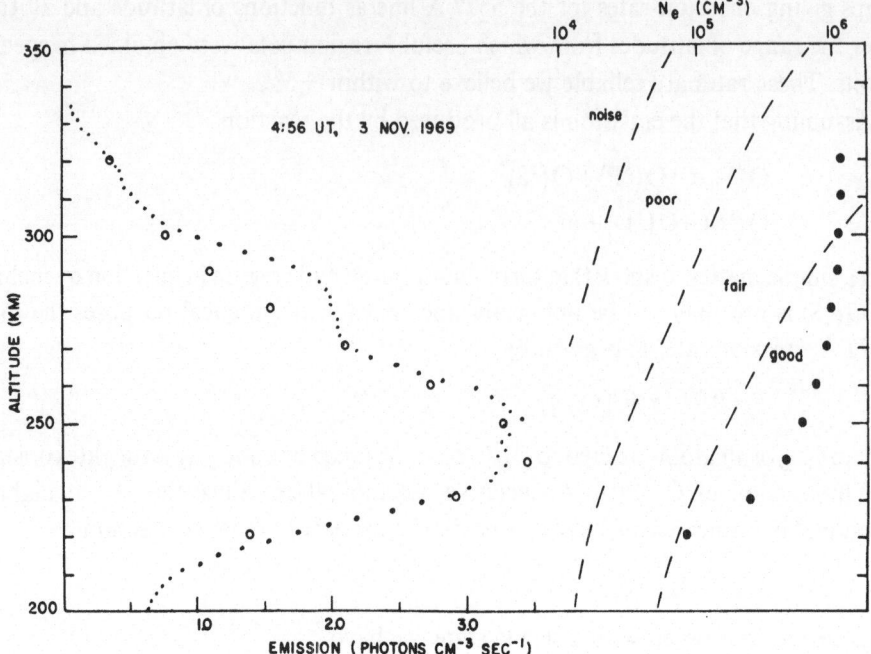

Fig. 2.   Electron density profile observed at Huancayo (solid circles), the 5577 Å emission rates inferred
from this profile (open circles) compared with the measured rates (small dots).

(Ferguson, 1969) or 1.2×10^{-11} at $T = 900$ K. I have long argued that ionospheric data seem to call for a value of only about 5.8×10^{-12} cm^3 s^{-1} at 900 K for k_1 (Donahue, 1966, 1968, 1972). This correction would make up the entire factor of 2.1, leaving f at 10%.

We have found that our calibration gives emission rates about 34% higher than one ground station (Fritz Peak), over which we passed. If Fritz Peak is right and we are wrong, the factor $f k_1 [O_2]$ would be too high by 2.8. Since the Jacchia (1971) O_2 densities seem to be in satisfactory accord with current direct measurement we would conclude that $f k_1$ is probably smaller by a factor of 0.48 to 0.36 than the product of the Zipf-Ferguson values as far as the ionosphere is concerned. More direct application of this technique that is in general agreement with our interpretation was the rocket flight of Hays and Sharp (1973) in which they could measure ion densities as well as 5577 Å emission rates directly.

The horizon scanning photometer can thus be considered a bottom side ionosonde operating from a satellite above the F_2 peak and capable of giving maps of electron density like those in Figure 3 as soon as $f k_1 [O_2]$ is pinned down definitively.

4. The Chapman Airglow Layer

A procedure similar to that discussed above can be used to unfold the local volume emission rates in the other nightglow layer located near 95 km. We are currently

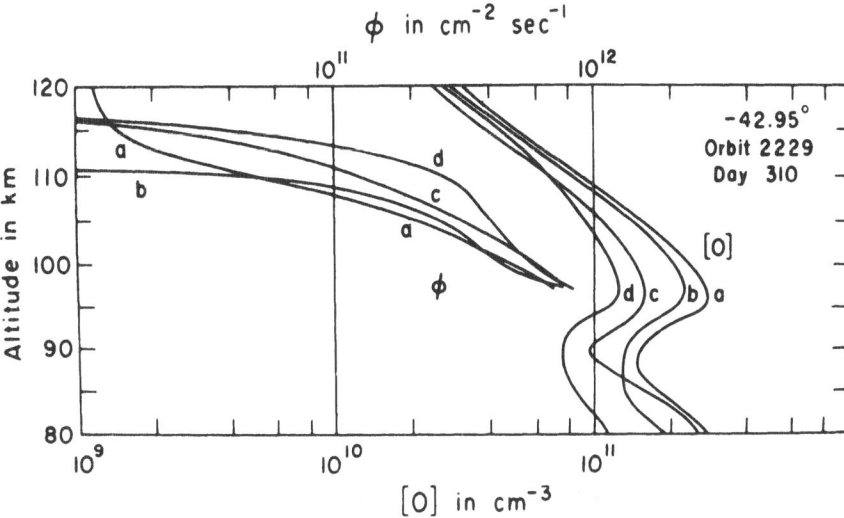

Fig. 3.   Atomic oxygen density and downward flux inferred from airglow emission rates for the four
assumptions concerning k_3, k_4 and k_5 discussed in the text.

trying to use this method in its full blown complexity to obtain accurate emission rates below 90 km, but thus far we have only published results (Donahue *et al.*, 1973, referred to henceforth as DGTa) based on a uniform layer assumption. This assumption has little effect on emission rates deduced at high altitude (i.e., above 110 km) but can seriously distort the true picture below 95 km. We also have a problem subtracting background if we use this assumption when the data look like those of Figure 1, but not at mid-and high-latitudes where the emission rates drop to dark current values for all lines of sight passing more than about 120 km above the Earth. It is only in a latitude band about $\pm 20°$ around the dip equator that we suffer serious uncertainties in our deduced emission rates above 105 km. The F region background is certainly too small to bother us around the 95 km maximum.

We have analyzed these low altitude nightglow data in an attempt to deduce something about the variation of O with altitude, latitude and season from them. To do this we have assumed that the excitation mechanism is that of Chapman (1931).

$$O(^3P)+O(^3P)+O(^3P)\rightarrow O_2+O(^1S) \quad (k_3) \tag{12}$$
$$O(^1S)\rightarrow O(^1D)+h\nu(5577\ \text{Å}) \tag{13}$$

in competition with

$$O(^1S)+O(^3P)\rightarrow 2O(^3P) \quad (k_4) \tag{14}$$

and

$$O(^1S)+O_2\rightarrow O(^3P)+O_2 \quad (k_5). \tag{15}$$

In this case

$$[O] = \left\{ \frac{1 + \tau_R k_4 [O] + \tau_R k_5 [O_2]}{k_3} \right\}^{1/3} [E]^{1/3} \tag{16}$$

where τ_R is the radiative lifetime of the ^1S state.

We have faith in our profiles for E, at least above 90 km. Obviously, for our deduction of [O] we are at the mercy of the people who do or do not measure the rate constants as a function of T, and of our knowledge of the temperature profile. I cannot protest often enough, vigorously enough or with greater futility, that we are not to be accused of advocating low or high values of [O] from these data. $k_3(T)$, $k_4(T)$ and $T(z)$ will need to be pinned down with good accuracy before we can go from E to [O] with confidence. We do have an advantage over upward looking measurements from sounding rockets in that the Van Rhijn effect allows us to sound the altitude range 115 to 120 km. And we do have more faith in our calibration than we do in those of photometers that habitually report hundreds of Rayleighs for the 5577 Å airglow.

In our analyses we have tried a variety of assumptions. We have performed a complete treatment of all of our data assuming

$$k_3 = 4.8 \times 10^{-33} \ cm^6 \ s^{-1} \ (\text{Felder and Young, 1972}) \tag{17}$$

independent of T. We have also tried the assumption

$$k_3(T) \ 2.7 \times 10^{-34} \ \exp(790/T) \ cm^6 \ s^{-1} \tag{18}$$

(or sometimes)

$$k_3(T) \ 3.4 \times 10^{-34} \ \exp(790/T) \ cm^6 \ s^{-1}. \tag{19}$$

The temperature dependence selected here is the one appropriate to the process

$$O + O + M \rightarrow O_2 + M, \tag{20}$$

and we have no rationale to justify it when M is $O(^3P)$ on the left and $O(^1S)$ on the right, except that, for certain reasons, we wanted to try an assumption that would reduce the rate at which [O] decreased between 97 km and 115 km. We have tried

$$k_4 = 7.5 \times 10^{-12} \ cm^3 \ s^{-1} \ \text{independent of } T \tag{21}$$

as measured at 300 K by Felder and Young (1972) and also

$$k_4 = 1.2 \times 10^{-11} \ e^{-227/T} \ cm^3 \ s^{-1} \ (\text{Slanger and Black, 1973}) \tag{22}$$

$$k_5 = 4.3 \times 10^{-12} \ e^{-853/T} \ cm^3 \ s^{-1} \ (\text{Slanger and Black, 1973}). \tag{23}$$

The temperature dependent rate constants reduce the values of [O] deduced at 97 km, multiplying them by a factor of 0.47 but leave the values deduced near 120 km virtually unaltered. The kind of altitude profiles deduced for [O] are shown in Figures 3 and 4 where the labels distinguish among various assumptions concerning the temperature

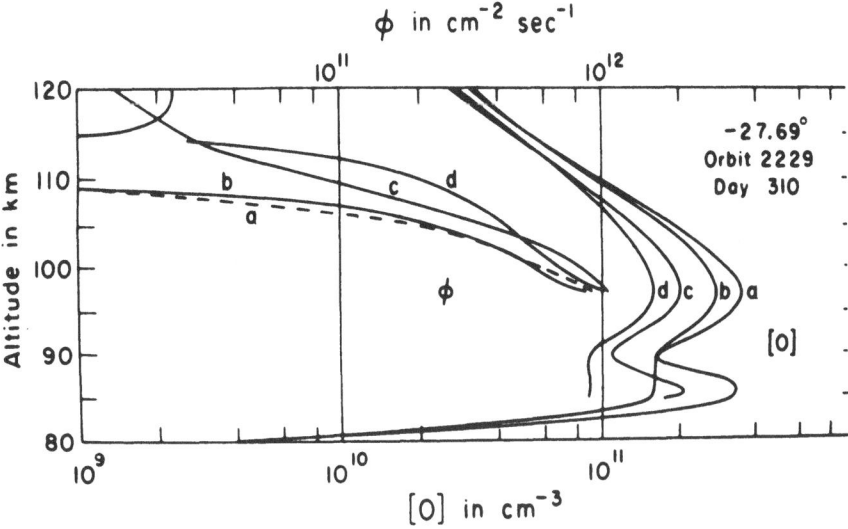

Fig. 4. Atomic O density and downward flux inferred from airglow emission rates for the four assumptions concerning k_3, k_4 and k_5 discussed in the text.

dependence of rate constants:

(a) k_3 constant, k_4 and k_5 constant
(b) k_3 constant, k_4 and k_5 variable
(c) k_3 variable, k_4 and k_5 constant, and
(d) k_3 variable, k_4 and k_5 variable.

The ratio of case (c) maximum densities to case (a) is 0.57.

These'curves demonstrate several important characteristics of the density profiles that we have deduced using the information concerning rate constants that we have at our disposal now:

(1) Above about 110 km quenching is not important – note that only a change in assumption concerning the temperature dependence of k_3 makes any difference in the value of [O] obtained and even that is not very significant. Hence in this region, where T is close to 300 K

$$[O] \cong (E/k_3)^{1/3}. \tag{24}$$

To increase [O] by a factor of 3 at 120 km will call for a 30 fold decrease in k_3.

(2) At 120 km the value of [O] tends to range from 2 to 6×10 cm^{-3}.

(3) At 97 km the value of [O], deduced subject to all of the uncertainties in the values of k_3, k_4 and k_5 at the temperature prevailing there runs between about 1 and 3×10^{11} cm^{-3} if k_3 decreases with temperature as supposed in Equation (18) or about 1.7 and 5×10^{11} cm^{-3} if it changes little with T and the Young-Felder value is correct.

(4) Under all assumptions taken there is a very steep gradient in the densities deduced especially between about 105 and 115 km. For this reason we have tried the effect of the temperature variations in k_3 and have tried to devise other modes for exciting the 5577 Å airglow. These have modified the old suggestions of Barth and

Hildebrandt (1961) in an attempt to reduce the amount of O needed near 95 km compared to that needed near 115 km to produce the radiation we observe. No *ad hoc* assumption has succeeded in curing our problem.

The nature of this problem caused by the steep gradients is clear when the downward flux of O calculated from the standard diffusion equation

$$\phi = -D\left\{\frac{\partial[O]}{\partial z} + \frac{[O]}{H(O)} + \frac{[O]}{T}\frac{\partial T}{\partial z}\right\}$$

$$-K\left\{\frac{\partial[O]}{\partial z} + \frac{[O]}{H_{av}} + \frac{[O]}{T}\frac{\partial T}{\partial t}\right\} \tag{25}$$

is calculated. (Here D is the molecular diffusion coefficient for O in air, K the eddy diffusion coefficient, $H(O)$ the oxygen scale height and H_{av} the average scale height.) In a steady state with photolytic production the downward flux of O atoms at 115 km should be about 22% of the maximum at 97 km. The fluxes calculated and shown in Figures 3 and 4 are based on the assumption of an altitude independent K that produces an effect on the flow even beyond the turbopause. The flows resulting go toward zero much more rapidly than they should if the flux and the production were more or less in a steady state. In fact for case (a) the flux even reverses direction. The possibility that we are dealing with a situation far removed from a steady state as Keneshea and Zimmerman (1970) insist runs up against the strange observational fact that we cannot find very much difference among typical [O] vertical profiles obtained shortly after sunset, near midnight and just before sunrise (DGTa).

One possible solution to the defect observed is to abandon the assumption that K is constant with altitude so that mixing can help to move the O downward in the neighborhood of 115 km. The result, as one might fear, is to move the turbopause up toward 115 or 120 km, and that runs us into trouble with measurements of the [A]/[N$_2$] ratio (Donahue *et al.*, 1974, DGTb). The effect obviously depends on the true value of the solar flux in the Schumann-Runge region. If the flux were as small as the Parkinson and Reeves (1969) values the average downward O flow would only be about 2.25×10^{11} cm^{-2} s^{-1}, the value of K at 97 km only about 5×10^5 cm^2 s^{-1}, and its maximum value at the turbopause about 3×10^6 cm^2 s^{-1} in order to keep the flux in balance with production. In this case the problems with the [A]/[N$_2$] ratio would not be very serious. However, if Widing *et al.* (1970) or Detwiler *et al.* (1961) are correct with their solar flux we have a serious difficulty.

Of course there are other remedies that can be tried. Perhaps the temperature profile we have taken from the Jacchia (1971) model is not correct. Hunten (1974) has privately suggested to me that we try this assumption. The temperature gradient might be steeper than the model near 110 km and tend to send more oxygen downward. To demonstrate what is the effect of the various terms in Equation (25) on the flux and how much the temperature gradient would need to change from model values I have plotted the contribution of each term to the flux in Figure 5. This calculation is for the density profile obtained at $-27.69°$ on orbit 2229, day 310 (Figure 3) and discussed

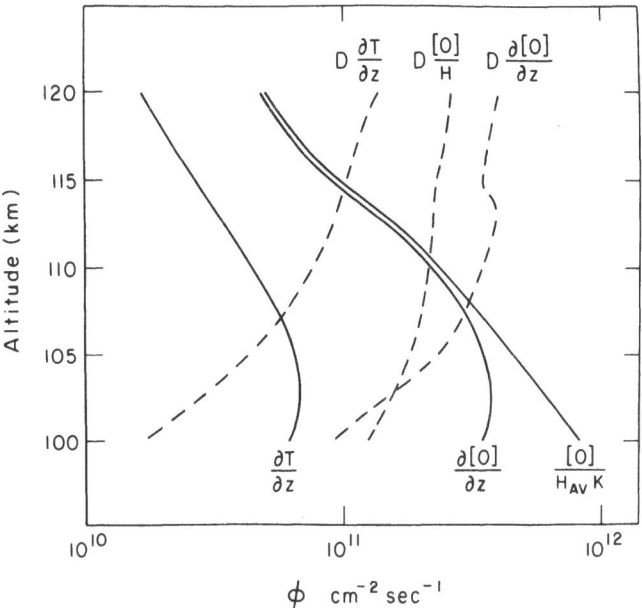

Fig. 5.   Absolute contribution to the O flux from the 6 terms in equation 25 for the case of Figure 3a. Solid curves are the eddy coefficient contributions (only the $D(\partial O]/\partial z)$ term is positive).

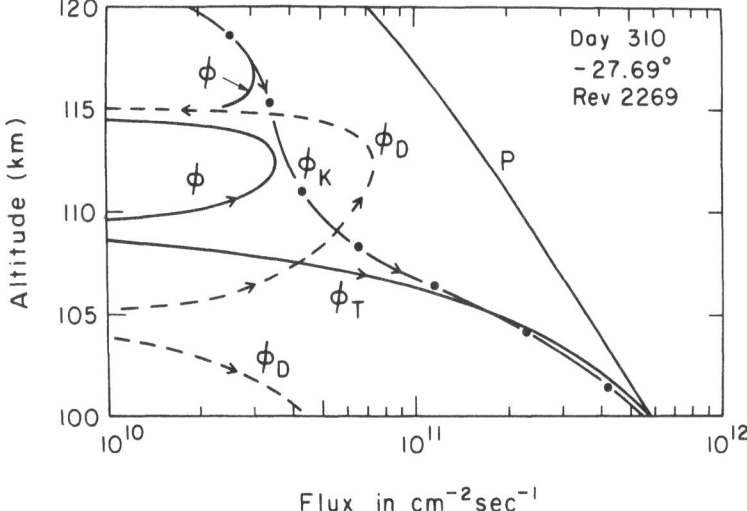

Fig. 6.   Contribution of molecular diffusion ϕ_D, eddy diffusion ϕ_K, to the O flux, the total flux ϕ or ϕ_T and the O production rate P. Arrows indicate direction of flow.

in connection with Figure 6 in DGTa. The value of K is taken to be 1.6×10^6 cm^2 s^{-1} and constant with altitude. In Figure 6 the total flux from molecular diffusion, eddy diffusion, their sum, and the production rate are plotted. The contribution of $-(D+K)$ ([O]/T) $\partial T/\partial z$ to the flux is the smallest of the three terms in both the D and K parts of the flux. Nevertheless, I have proceeded to determine by how much T and $\partial T/\partial t$

would have to be changed in order to make ϕ and P equal – always assuming K to be constant with z. The changes required are large but, perhaps, not forbiddingly so. The worst discrepancy, as a look at Figures 4 and 6 will show, is at 109 km where a gradient of 21.5 K km^{-1} instead of 5.5 K km^{-1} would be needed. A temperature profile that would be suitable to produce the missing downward flux and its gradient is compared

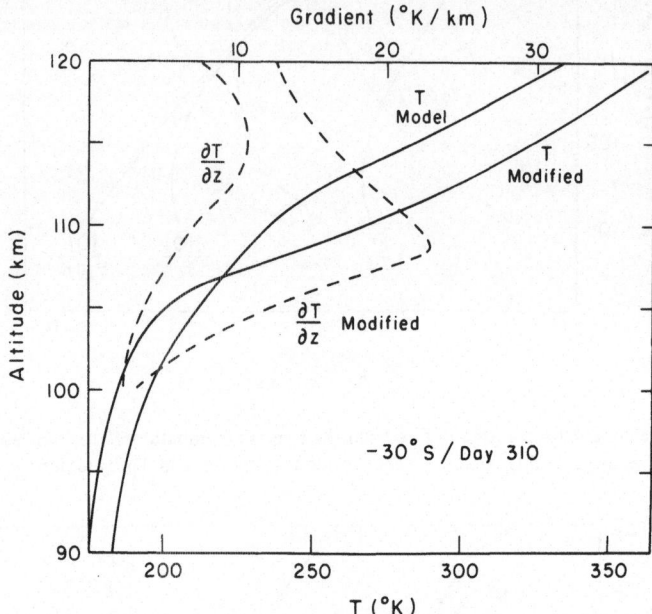

Fig. 7.   Temperature and gradient as modified to balance downward O flux and P compared to the model (Jacchia, 1971).

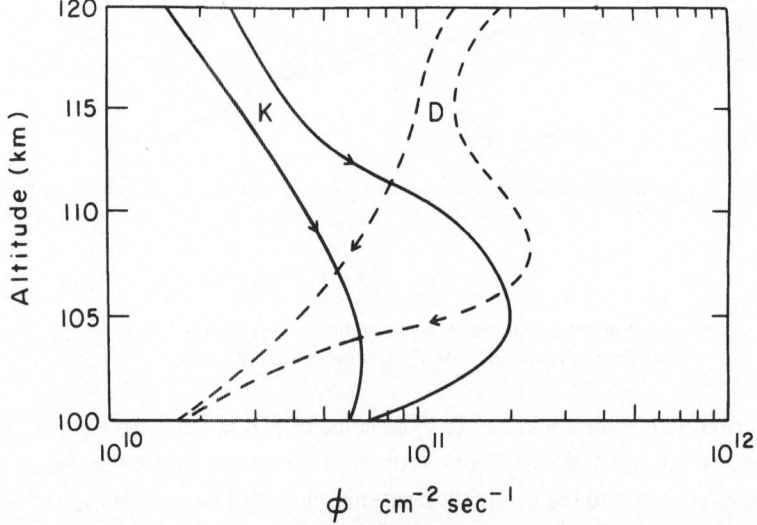

Fig. 8.   The contribution of the temperature gradient terms to the O flux in Figure 5 and the norm this contribution would take if ϕ were to equal P in Figure 5 and the temperature gradient were as in Figure 7.

with the model used in Figure 7. The effect on the contribution to the thermal gradient terms on the flux is shown in Figure 8.

When it is considered that the assumption of constant K is probably not realistic and that a modest increase in eddy coefficient between 100 and 110 km would relax the conditions on $\partial T/\partial z$ it begins to appear that the problem posed by the large gradients in our O profiles between 105 and 115 km may have an explanation that does not require pushing the turbopause to unacceptably large altitudes. At any rate I am now prepared to propose that our data suggest that the temperature profile must be considered with care in evaluating the O fluxes implied by our data. In many cases our profiles make most sense if temperature gradients near 107 km are greater than most models give. I would suggest that some increase in the eddy diffusion coefficient be allowed between 100 and 110 km. However, we would have a problem in keeping the flux and production in balance if the effect of eddy mixing were to vanish where the molecular diffusion and eddy diffusion contributions to the flux become equal (Keneshea and Zimmerman, 1970).

Our data showed very large latitudinal and seasonal variations. These are illustrated in Figure 9 where the maximum slant emission rates are plotted against time for each of 12 latitude bands 10° wide from $+60°$ to $-60°$. These slant rates can be converted roughly to vertical contributions from the Chapman layer by multiplying slant rates below 2×10^3, Rayleighs by 22×10^{-3} and by 18×10^{-3} when slant rates are above 2 kR (see Figure 5 of DGTa). They may be converted to densities with the help of Figures 11 and 12 from DGTa.

In Figures 10 and 11 we try to indicate the nature of the latitude variations and how they change with some combination of longitude and UT by connecting density values on consecutive orbits with 'iso-density' contour lines. In Figure 10 the deep tropical trough that sometimes develops (in August and January most spectacularly) is apparent. The other common feature of a maximum between 40° and 55° latitude in the local winter hemisphere, a minimum in the tropics and a weak maximum near 30° in the local summer hemisphere can be discerned. In the maps for days 310 and 311 (Figure 11) note how drastically the pattern can change during a 24 h period over the same part of the Earth. There is a solemn warning here for all sounding rocket observers not to generalize from the results of one or even a few flights.

From these data we can compute the global average densities (at least for the local times of our observations). These averages showed a very large semidiurnal effect at 97 km. A large maximum occurs in May, minima in July and January, and a secondary but quite decided maximum in October (Figure 16 of DGTa). The ratio of maximum to minimum is somewhat greater than 2/1. Since the global photolysis rate is constant, this density variation must imply a variation in the effectiveness of global downward transport that is inverse to the density effect. Much of the semiannual effect is produced by the development of the deep tropical troughs, but we have no way of telling yet whether this is because most of the variation in K occurs in the tropics or whether, at least in part, there are strong meridional winds at 100 km level transporting O from the hemisphere where it is being most copiously produced to the opposite hemisphere

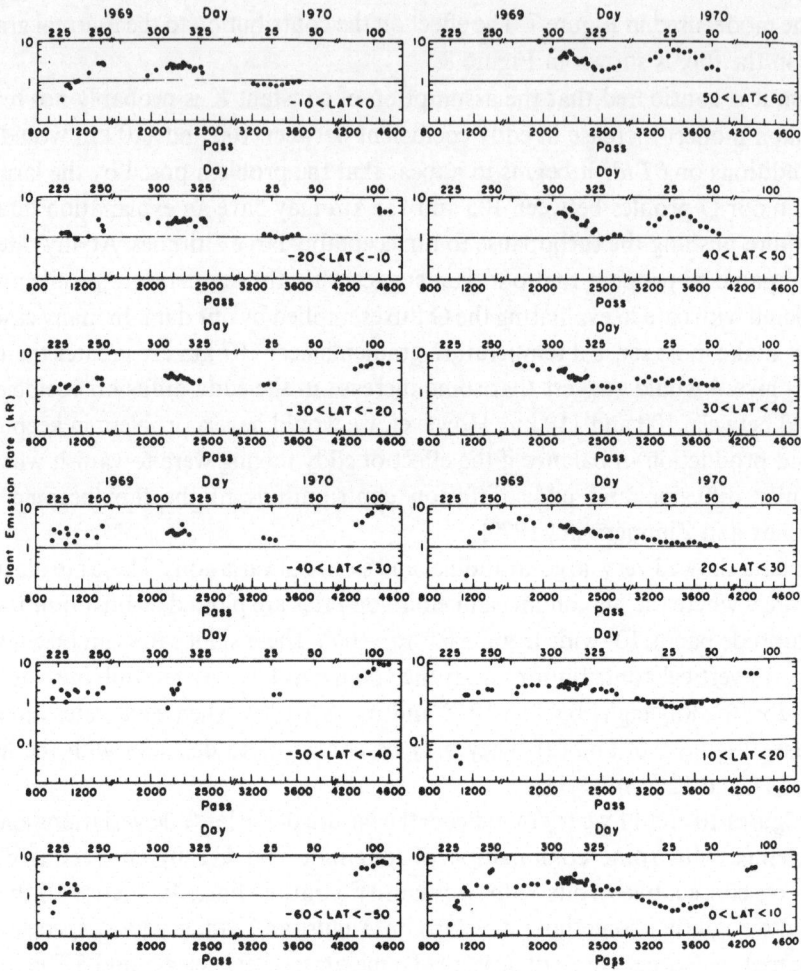

Fig. 9.   Average 5577 Å maximum slant intensities during 24 h periods for all observations within 10°
wide latitude strips.

and that they blow strongest through the tropics (DGTb). If this were the dominant
effect there would need be some mechanism to cause a global variation in K so that
downward mixing of O has maxima in July and January and a very deep minimum in
April. By global I mean the effect would have to be in phase throughout the world
and would be accompanied by the meridional transport pattern just described. The
magnitude of the winds is calculated in (DGTb). It could attain velocities of 40 to 50
m s^{-1} across the equator at 97 km. Again a caveat: our satellite data are restricted in
local time.

5. Variations in the Altitude of the 5577 Airglow

Something of a controversy has developed concerning the possibility that the altitude
of the green airglow maximum undergoes a sizable variation – staying up close to
100 km throughout most of the year but dropping as low as 85 km in the middle of

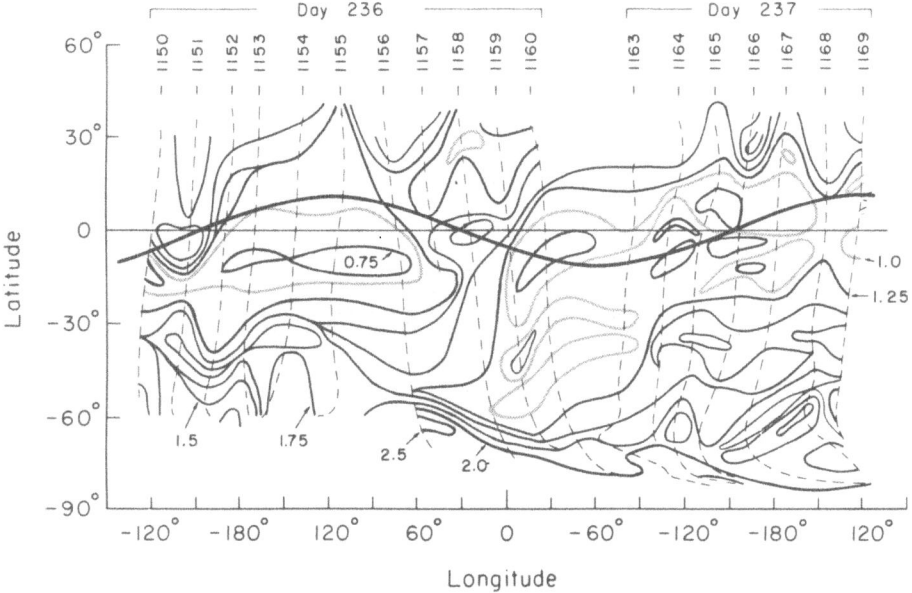

Fig. 10.   Contours of equal maximum O density, 1969. The dip equator is the heavy black line. The labels are in 10^{11} atoms cm^{-3}. The k_3 is assumed constant. Orbit numbers are indicated.

August (Offermann and Drescher, 1973). Because of quenching, such a variation in the O distribution would contribute to a variation in airglow emission rates also. The evidence cited for the variation is from sounding rocket flights and depends strongly on two observations, one conducted at Ft. Churchill (Dandekar, 1972) and the other at an unspecified latitude by Tarasova (1963). From Figure 9 it is apparent that the nature of the seasonal variation in the green line brightness is greatly a function of latitude. Those of us who have flown sounding rockets in the auroral zone are very much aware that weak auroras, inhomogeneous in space and time, can trick an up-ward looking photometer on a sounding rocket as it moves laterally and vertically into mistaking changes in brightness overhead for vertical gradients when the varia-tions are really caused by auroral forms moving across the field of view or changing rapidly in brightness.

Our own observations of the altitude of the peak slant emission rates do not show any evidence for systematic variations in the altitude of the emitting layer. Although our problems with satellite altitude have prevented us from achieving complete coverage of the period of interest in the northern hemisphere, we do cover a large part of it and also have data from the southern hemisphere 6 mo away in time. What we have found from satellite attitude and altitude data alone is quite a spread in apparent altitude on any orbit. If we plot histograms that give the frequency of observation of altitudes for the slant maxima within 2 km wide bins the maximum frequency is always in the 94 to 96 km bin and the spread can be characterized as 95 ± 4 km on a statistical basis. We have no way of knowing whether anomalies that occur as high as

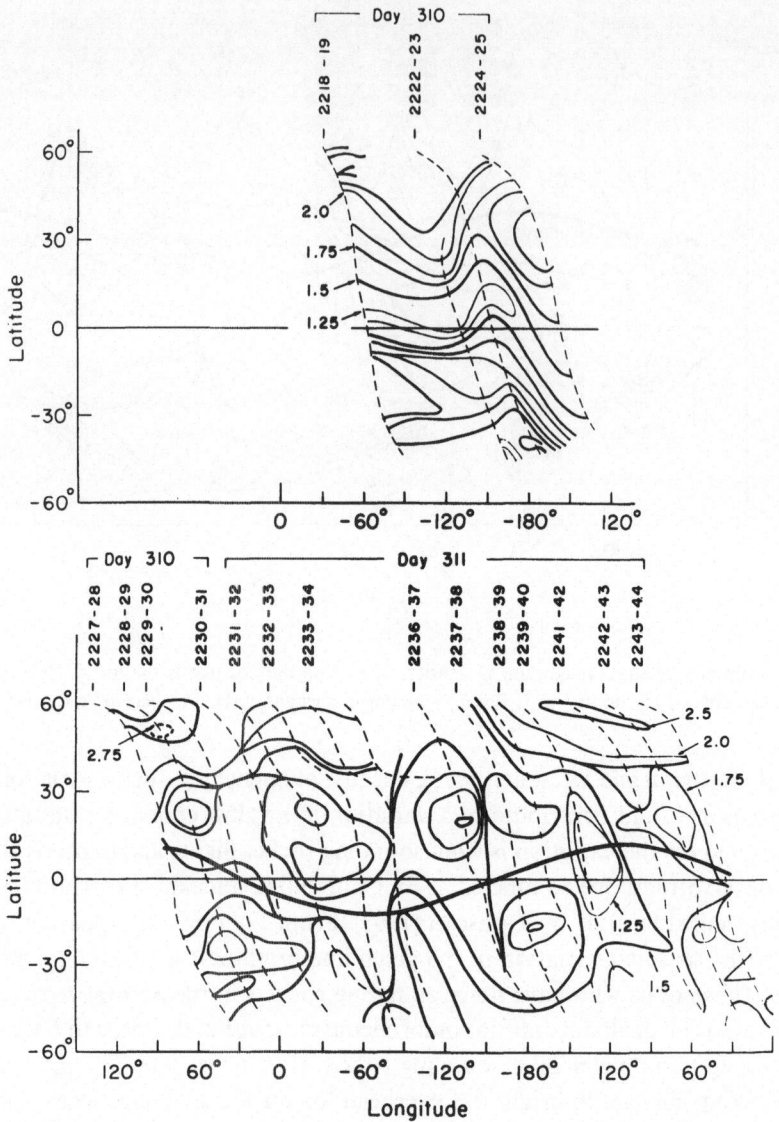

Fig. 11.   Contours of equal maximum density, 1969. The dip equator is the heavy black line. The labels are
in 10^{11} atoms cm^{-3}. The k_3 is assumed constant. Orbit numbers are indicated.

120 km or as low as 75 km represent real altitude variations or inaccurate satellite
orientation information. We do know that we never see the centroid shift away from
95 km, or the standard deviation change appreciably. These figures based on a mass
of data – 550 observations in 24 h and about 65 per day per 10° of available latitude on
which to base our statistics.

We have also tried to find evidence for any correlation between maximum emission
rate and the apparent altitude. The results are plotted in Figures 12 and 13. All data for
all latitudes observed on 3 passes are shown in each figure. Orbits traced during days

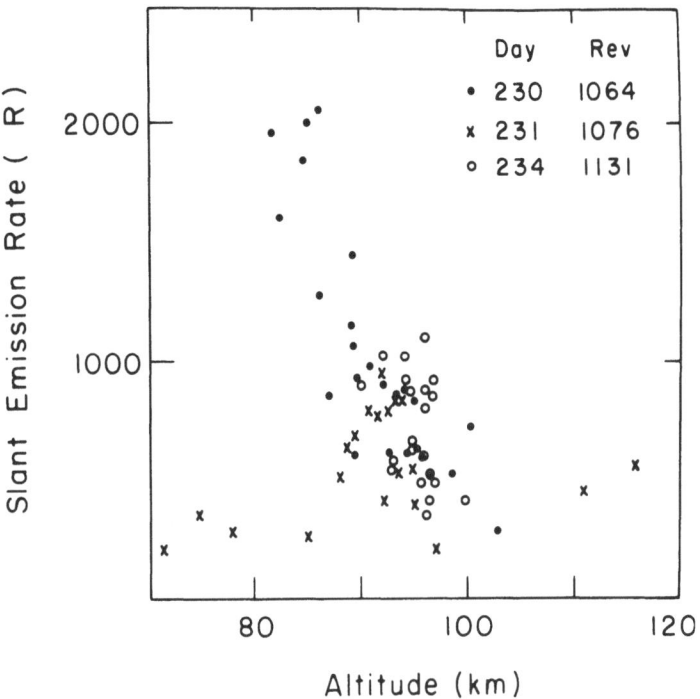

Fig. 12. Slant intensity plotted against apparent altitude for each observation on the orbits indicated.

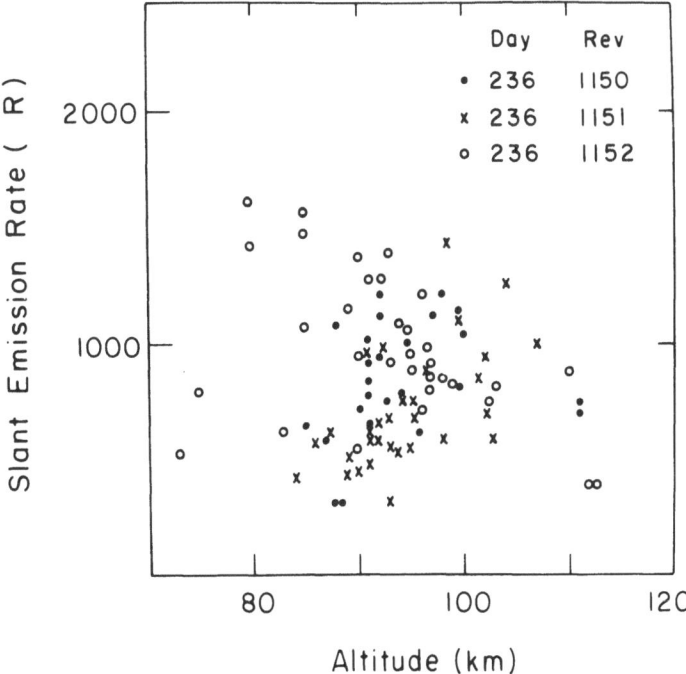

Fig. 13. Slant intensity plotted against apparent altitude for each observation on the orbits indicated.

230, 231, and 234 are covered in Figures 13 and three successive orbits on day 236 shown in Figure 13. Except for day 230, orbit 1064, where a correlation opposite in sense to the one sought can easily be claimed, the data as a whole show no correlation whatever between apparent altitude and maximum brightness.

We have found that the peak slant emission rate lies about 2 km below the peak local volume emission rate. We have also found that the altitude of the maximum varies by only ± 2 km from the inflection near 50 km caused presumably by O_3 absorption, Rayleigh scattering, and some other extinction process. So far we have not been able to account for the attenuation profile with models using O_3 and air alone. Some other attenuator near 25 km seems to be at work. It was this effect that led us to claim that the peak emission of 5577 Å occurs at 97 ± 2 km in DGTa instead of 97 ± 4 km. Until we thoroughly understand the connection between O at 97 km and our attenuator between 50 and 20 km it would probably be safer to take the altitude as 97 ± 4 km. We have a strong suspicion, however, that the real random variation is not as great as 8 km, and we are certain that a systematic variation in altitude from 97 km for the times and places where we have data is not anything like ± 4 km. We also are confident that there is no evidence from these satellite data for any systematic correlation of airglow brightness and the altitude of the layer, real or apparent.

6. The NO + O Continuum

One final remark. It has to do with the air afterglow continuum

$$NO + O \rightarrow NO_2 + hv \quad (k_6). \tag{26}$$

In recent years the rate constant for this reaction has been chased down to low pressure (Becker *et al.*, 1972; Cody and Kaufman, 1974) and shown to be asymptotic to a value

$$k_6(5890 \text{ Å}) = (1.3 \pm 0.5) \times 10^{-21} \text{ cm}^3 \text{ s}^{-1} \text{ Å}^{-1} \tag{27}$$

below 0.1 mtorr. This is a value much smaller than will be found in most airglow literature. Nevertheless, it is still large enough to produce significant radiance within the 24 Å pass band of the Na filter on our photometer if there is as much NO and O at 110 km and higher as the received wisdom has it. In Figure 14 we show a series of observations in the Na channel – again very typical – of the slant emission rate in Rayleighs (Division by 24 converts the scale to R Å^{-1}). Now, some of the radiation observed might be Na airglow, but the observations certainly put an upper limit on the product [O] [NO]. The error bars in the figure take into account the finite voltage steps telemetered from the photometer output – amounting to 12.5 R in the case of the Na channel. We also show the O profiles obtained in adjacent 5577 Å scans for k_3 constant and $k_3(T)$ plus the product of [NO] [O] using Meira's (1971) NO densities. Along with the observed radiance we have plotted that predicted from the product of our densities and those of Meira and from Jacchia (1971) O densities and the NO profile of Meira. The former is almost always consistent with the upper limit for [O] [NO]. The latter is usually a factor of 5 too high. Of course a valid con-

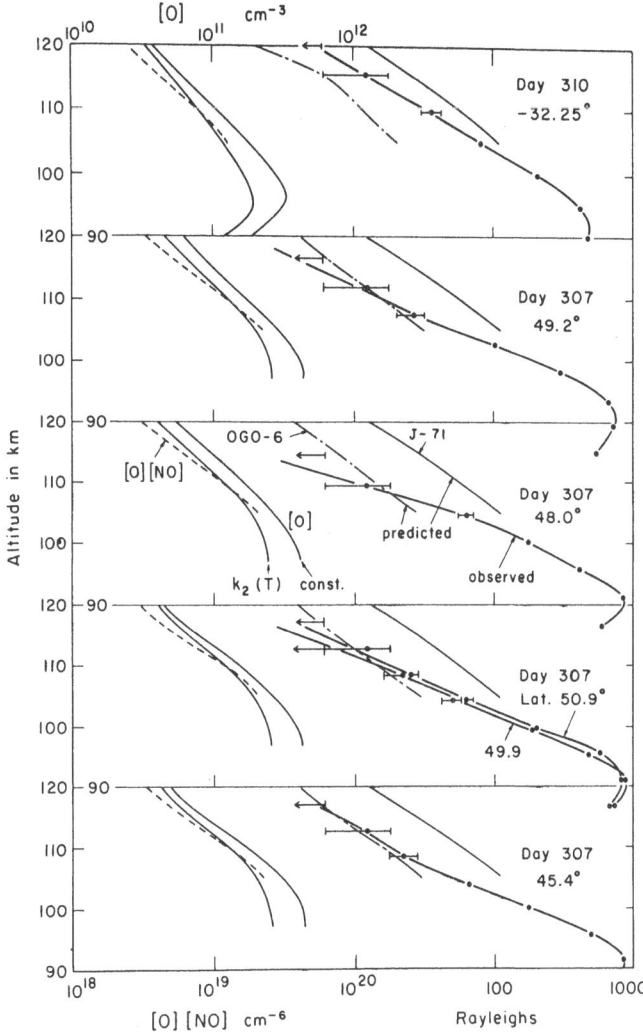

Fig. 14.   Measured slant emission rate in yellow channel (curve with dots and error bars). Contribution expected from NO—O reaction for Jacchia 71 [O] and Meira [NO] marked J-71. Contribution to the signal expected from the reaction for OGO 6 [O] and Meira [O] marked predicted, the O densities deduced from companion 5577 Å observations marked $k(T)$ and constant and the product of [O] and [NO] for the $k_3(T)$ O profile.

clusion would be that the Jacchia [O] values are correct and Meira's [NO] values too high by at least 5. This interpretation would help in the resolution of the famous D region ion and electron density problems (Donahue, 1972). But if the Meira NO values are valid above 110 km there is support for the values of O densities above 110 km deduced from our 5577 Å analysis.

7. Conclusions

(1) Rate constants with activation energies appropriate to the temperature range 150 to 500° are desperately required for k_3, k_4 and k_5.

(2) Observation of the 5577 Å and 6300 Å nightglow from orbiters is the simplest method of obtaining synoptic data for electron densities below (5577 Å) and above (6300 Å) the F_2 peak. However, the product (10) must be tied down firmly before this technique can be exploited to its fullest potential.

(3) Until objective (1) is fulfilled absolute atomic O densities should not be attributed to the OGO-6 green line photometer results or any other green line data. All authors, referees and editors should be alert to violations on this interdiction. Nevertheless, all should be aware of the need to change k_3 by a devastating factor in order to boost the value of [O] deduced above 115 km by a large amount. The evidence from [NO] [O] air afterglow should also be held in mind.

(4) There is a large semi-annual variation in the average effective transport properties of the lower thermosphere and probably some meridional circulation pattern that transports material in the lower thermosphere from the summer to the winter hemisphere. This last conclusion is hardly to be considered novel (Johnson and Gottlieb, 1973; Reber and Hays, 1973).

(5) The O profiles near 110 km suggest a maximum in eddy diffusion at about that altitude accompanied by generally steeper temperature profiles than those found in current models.

(6) There probably is no systematic variation in the altitude of the O maximum by more than 2 km.

(7) Observation of the 5577 Å nightglow at high spatial resolution should be a feature of all aeronomical orbiting satellites on this and our neighboring planets.

Acknowledgments

The construction and calibration of the photometer was basically a triumph of the Service d'Aeronomie, J. E. Blamont the leader. Support for the work reported was provided by NASA under Grants NA5-11077, NGR-39-011-155 and NSF Grants GA-27638 and GA-37744 moderated by the Aeronomy Program Office of the Atmospheric Sciences.

References

Becker, K. H., Groth, W., and Thran, D.: 1972, *Chem. Phys. Letters* 15, 215.
Barth, C. and Hildebrandt, A. F.: 1961, *J. Geophys. Res.* 66, 958.
Chapman, S.: 1931, *Proc. Roy. Soc., Ser. A.* 132, 353.
Cody, R. J. and Kaufman, F.: 1974, private communication.
Dandekar, B. S.: 1972, *Planetary Space Sci.* 20, 1781.
Detwiler, C. R., Garrett, D. L., Purcell, J. D., and Tonsey, R.: 1961, *Ann. Geophys.* 17, 263.
Donahue, T. M.: 1966, *Planetary Space Sci.* 14, 33.
Donahue, T. M.: 1968, *Science* 159, 489.
Donahue, T. M.: 1972, *Radio Sci.* 7, 73.
Donahue, T. M., Guenther, B., and Thomas, R. J.: 1973, *J. Geophys. Res.* 78, 6662.
Donahue, T. M., Guenther, B., and Thomas, R. J.: 1974, *J. Geophys. Res.* 79, 1959.
Felder, W. and Young, R. A.: 1972, *J. Chem. Phys.* 56, 6028.
Ferguson, E. E.: 1969, *Ann. Geophys.* 25, 819.
Hays, P. B. and Sharp, W. E.: 1973, *J. Geophys. Res.* 78, 1153.
Hunten, D. M.: 1974, private communication.

Jacchia, L. G.: 1971, *Spec. Rep. 332*, Smithsonian Astrophys. Obs., Cambridge, Mass.

Johnson, F. S. and Gottlieb, B.: 1973, *Planetary Space Sci.* **21**, 1001.

Keneshea, T. M. and Zimmerman, S. P.: 1970, *J. Atmospheric Sci.* **27**, 831.

Meira, L. A., Jr.: 1971, *J. Geophys. Res.* **76**, 202.

Offermann, D. and Drescher, A.: 1973, *J. Geophys. Res.* **78**, 6690.

Parkinson, W. H. and Reeves, E. M.: 1969, *Solar Phys.* **10**, 342.

Reber, C. A. and Hays, P. B.: 1973, *J. Geophys. Res.* **78**, 2977.

Slanger, T. G. and Black, G.: 1973, *Planetary Space Sci.* **21**, 1757.

Tarasova, T. M.: 1963, *Space Res.* **3**, 162.

Thomas, R. J. and Donahue, T. M.: 1972, *J. Geophys. Res.* **77**, 3557.

Van Zandt, T. E. and Peterson, V. L.: 1968, *Ann. Geophys.* **24**, 747.

Widing, K. G., Purcell, J. D., and Sandlin, G. D.: 1970, *Solar Phys.* **12**, 52.

Zipf, E. C.: 1970, *Bull. Amer. Phys. Soc.* **15**, 418.

AURORAL ULTRAVIOLET EMISSIONS

JEAN-CLAUDE GÉRARD*

Institut d'Astrophysique, Université de Liège, 4200 – Ougrée – Belgium

1. Introduction

The identification of most of the important features of the UV spectrum has progressed simultaneously with the development of sensitive rocket-borne and satellite-borne UV spectrometers. It may be considered, at the present time, that our knowledge of the auroral spectrum above 1200 Å is satisfactory as far as identification of important transitions is concerned. Unfortunately, the spectral range covered by such measurements has often been very limited, thus preventing accurate comparisons to be made of intensities of features spread over the whole spectrum. As far as theory is concerned, a great deal of effort has been devoted to the prediction of UV emission intensities (Green and Barth, 1965; Stolarski and Green, 1967). The calculations were generally performed for a range of electron energies and comparison was made with available observational results. Ambiguities remain in the interpretation to be given to the discrepancies which may be due either to the use of inappropriate cross sections or excitation mechanisms, or to large differences between real physical conditions existing at the time of the observation. For the first time, Sharp and Rees (1972) combined satellite-borne and airborne equipment to observe simultaneously the same spot of aurora in the range 1200 to 4000 Å. Below 1200 Å, the presence of strong emissions has been detected photometrically by Landensperger (1971) and Paresce *et al.* (1972) but the spectral distribution is still unknown. Most of the excitation of UV emission is direct e.g., resulting from impact of primary and secondary electrons on ground state species. This makes interpretation of intensities often easier than in the case of visible emissions. However, radiation trapping complicates analysis of O I $\lambda 1304$ Å and N I $\lambda 1200$ Å altitude distribution. This paper reviews the present knowledge on excitation processes and related intensity of the UV auroral emissions. It will also report on new results concerning $N_2 A^3 \Sigma_u^+$ deactivation as determined by the V-K bands altitude profile. Finally, UV photometric data obtained recently with the ESRO TD-1 satellite concerning early evening polar cap aurora will be reported.

2. Excitation of Ultraviolet Emissions

The experimental results collected so far concerning the intensities of the UV spectrum are listed in Table I. They have been normalized to an IBC II aurora ($I(3914)=$ $=10$ kR). For comparison the intensites calculated by Sharp and Rees (1972) and

* Aspirant of the Belgian National Foundation for Scientific Research (FNRS).

TABLE I

Auroral UV intensities (kR) normalized to $I(3914) = 10$ kR

Feature	(1)	(2)	(3)	(4)	(5)	(6)	(7)	(8)	(9)
N_2 2P						18	16	20	23
N_2 V-K, $v=0$						5		5.3	6
$v=1$						3		4.7	4
N_2 LBH	8			25[a]	15[b]		21[a]	15	19 to 27
NO $\gamma(1-0)$								1.2	
OI 1304 Å	6	6.5	1.9	21	11.5		24	7	10
1356 Å	<1				2.1		2.6	1.3	1.6
2972 Å								0.5	0.4
NI 1200 Å	0.6	0.6	0.3				<4	1.6	
1240 Å							0.5		
1493 Å	<0.5						1.3	0.7	

(1) Miller *et al.* (1968); (2) Peek (1970), flight 1; (3) Peek (1970), flight 2; (4) Opal *et al.* (1970); (5) Barth and Schaffner (1970); (6) Sharp (1971); (7) Theobald and Peek (1969); (8) Sharp and Rees (1972), observations; (9) Sharp and Rees (1972) and Strickland and Rees (1974), theory.

[a] Corrected for O_2 absorption

[b] Deduced from $I(1384 \text{ Å}) = 1.4$ kR

Strickland and Rees (1974) for a high altitude nadir observation are also displayed in the last column. These results may be discussed together with the excitation mechanism of the various transitions.

2.1. OXYGEN LINES

The main O emissions in the UV spectrum are $\lambda 1304 \text{Å}$ ($^3P-^3S$) and $\lambda 1356 \text{Å}$ ($^3P-^5S$). Of minor importance are $\lambda 2972 \text{ Å}$ ($^3P-^1S$), $\lambda 1152 \text{ Å}$ ($^1D-^1D°$), $\lambda 1218 \text{ Å}$ ($^1S-^1P°$) and $\lambda 1027 \text{ Å}$ ($^3P-^3D°$).

The radiation transfer equations for $\lambda 1304 \text{ Å}$ multiplet have been solved by Donahue and Strickland (1970) and Strickland and Rees (1974). Adopting cross sections recently measured (by Stone and Zipf (1974) for the direct excitation of $O(^3S)$ and $O(^5S)$, by Mumma and Zipf (1971) for dissociative excitation), Strickland and Rees (1974) find dissociative excitation of O_2 to contribute negligibly to the total intensity. The predicted intensites at 1356 Å and 1304 Å for electron impact on O are, respectively, 3 and 8 times larger than those observed by rocket. These discrepancies are likely to be attributed to invalid cross sections. The calculated nadir intensity ratio $I(1304)/I(1356) \simeq 6$ is however in good agreement with the observations of Sharp and Rees. Intensity prediction of features at 1027, 1152 and 1218 Å have been made by Stolarski and Green (1967), but none of them has been detected due to their low intensity.

2.2. NITROGEN LINES

Strong nitrogen lines are observed at 1200 Å ($^4S°-^4P$), 1493 Å ($^2D°-^2P$), 1743 Å ($^2P°-^2P$) and 3466 Å ($^4S°-^2P°$). Cross sections for the dissociative excitation of N_2 giving rise to the first three emissions have been measured by Ajello (1970) and

Mumma and Zipf (1971). Prasad and Green (1971), basing their discussion on theoretical estimates of these cross sections, concluded that dissociative excitation of the molecule prevails over direct impact on the atom due to the low abundance in atmospheric N. Laboratory work has confirmed these views (Stone and Zipf, 1973) but comparison between observed and calculated intensites at 1200 Å is complicated by radiative imprisonment by N, even if present in weak concentration (Strickland, 1971). The intensity ratio predicted by Prasad and Green ($I(1493)/I(3914)=4$) is in satisfactory agreement with Miller *et al.*'s (1968) observations 0.25 when considering the important absorption of the low altitude emission by O_2.

2.3. N_2 SECOND POSITIVE SYSTEM (2P): $C^3\Pi_u \rightarrow B^3\Pi_g$

Recent experimental determination of the N_2 2P cross section has been obtained by Shemansky and Broadfoot (1971a), Cartwright *et al.* (1971) and Imami and Borst (1974). The Einstein transition probabilities of this system were obtained by Shemansky and Broadfoot (1971b), whereas Franck-Condon factors relating $C^3\Pi_u$ to the $N_2X^1\Sigma_g$ ground state were calculated by Albritton *et al.* (1975). Experimental and theoretical determination of the system intensity indicate that $I(2P)/I(3914) \simeq 2$ (cf. Table I) and that the ratio remains independent of altitude (Sharp, 1971; Sharp and Hays, 1974).

2.4. N_2 VEGARD-KAPLAN SYSTEM (V-K): $A^3\Sigma_u^+ \rightarrow X^1\Sigma_g^+$

There has been controversy during recent years as to the importance of various contributions to the population of $A^3\Sigma_u^+$. The A state is populated by two main contributions:

(1) direct impact electron impact X→A, and

(2) cascade from $B^3\Pi$, itself partly populated through C→B cascades.

Consideration of Franck-Condon factors (Albritton *et al.*, 1975) readily shows that direct excitation is most efficient for levels $v' > 7$. On the other hand, the importance of the cascade contribution decreases steadily with increasing vibrational numbers. Several recent works have calculated the relative importance of the two contributions to population rates. Three areas of disagreement may be noted:

(i) Cartwright *et al.* (1971, 1973) indicate that, besides direct excitation and C→B cascades, the lower levels of the B state are populated by cascades from the $W^3\Delta_u$. However, this contribution does not increase the population rate of A levels by more than 10%. On the other hand, Shemansky and Broadfoot (1973) find no laboratory evidence of the presence of this transition.

(ii) According to Cartwright *et al.* the A→B reverse first positive system depopulates the levels $v' > 7$ rapidly. Consequently, the intensity of V−K emissions arising from these levels should be very weak in the aurora. Here again, opposing views exist as to the efficiency of this depopulation. Only observation of transitions arising from high vibrational levels may solve this problem.

(iii) The electron impact cross section for excitation of $B^3\Pi$ used by Shemansky *et al.* (1971) is about twice as high as that determined by Cartwright *et al.* (1971). In the case of the $A^3\Sigma_u^+$ cross section, the disagreement is still more important between Shemansky *et al.*'s (1971) results on one hand and Cartwright *et al.* (1973) and Borst and Chang (1973) on the other hand. The importance for the prediction of population rates of this discrepancy is certainly larger than the role played by $A \to B$ and $W \to B$ transitions.

Recently, Borst and Chang (1974) have calculated relative population rates for various values of the spectral index n of the electron energy distribution between 6 and 50 eV. They find that population rates depend only slightly on the flux distribution over a large range of n values. It should be pointed out that observed spectra (Sharp, 1971) do not indicate the presence of high vibrational transitions with the intensity calculated by combining Shemansky and Broadfoot's (1971a) cross section and Shemansky's (1969) transition probabilities.

2.5. N_2 LYMAN-BIRGE-HOPFIELD SYSTEM (LBH): $a^1\Pi \to X^1\Sigma_g^+$

Excitation cross section of the LBH system has been measured by Ajello (1970), Borst (1972), and Freund (1971). Table I indicates that theory predicts an intensity ratio $I(\mathrm{LBH})/I(3914)$ between 2 and 3, whereas satellite observations give a ratio of 1.5. Rocket measurements yield ratios ranging between 0.33 and 2.5. Opal *et al.* (1970) observed a fall of this ratio below 110 km. To determine whether such differences are real, further experimental work is needed (see also Section 4).

It must be noted that the intense short wavelength part of this system is absorbed in the O_2 Shumann-Runge continuum (1350 to 1750 Å) below 120 km. Consequently, the fraction of LBH intensity absorbed by O_2 when observing the aurora from above may depend on the altitude of maximum energy deposition. This fact, coupled with calibration errors, may explain the observed differences.

2.6. NO γ SYSTEM: $A^2\Sigma \to X^2\Pi$

Discovery of the presence of the NO γ system has been claimed by Duysinx and Monfils (1972) and Sharp and Rees (1972). In both cases, this identification is based on the presence of an emission feature near 2150 Å, tentatively identified as the $(1-0)$ γ band. Various problems are however raised by this identification: (i) Other bands of the $v' = 1$ progression should be observed. However, the Einstein coefficients calculated by Nicholls (1964) do not agree with the intensity ratios measured in laboratory works (Poland and Broida, 1971), which are in satisfactory agreement with a $q_{1,v''}v^3$ dependence. The strongest of these bands may be found among the unidentified features of Sharp and Rees' spectrum but they cannot be found at all in Duysinx and Monfils' spectrum whose sensitivity and resolution should however allow them to be seen; (ii) Comparing the intensity to be expected from direct electron impact on NO molecules, on the one hand, and energy transfer from $N_2(A^3\Sigma_u^+)$, on the other hand, Sharp and Rees concluded that the latter dominates by more than one order of magnitude. In this case, the vibrational population ratio $(v = 0:1)$ should be close to 6 (Callear and Wood, 1971), in disagreement with observation. Consequently, the attribution cannot be considered as definite until other bands have been unambiguously identified with the proper relative intensity.

In any case, an emission feature is undoubtedly present at 2150 Å, with an intensity comparable to the $(0-6)$ V-K band.

3. Quenching of $N_2 A\ ^3\Sigma_u^+$ State

Spectra of the V-K bands in the spectral range 2600 to 3500 Å have been obtained by Sharp (1971). The altitude profile of this emission has been analyzed in terms of production and quenching mechanisms. He concluded that $N_2(A)$ states is quenched by O with a deactivation coefficient $d_O = 9 \times 10^{-11}$ cm^3 s^{-1}, a value exceeding laboratory results by two orders of magnitude. From non-linear dependence of $I(2P)/I(V-K)$ vs. O density, he concluded that O_2 deactivates the A state at a rate $d_{O_2} = 4.4\ d_O$. This deduction is based on the assumption that the population rate of the A state is proportional to the intensity of the 2P bands. The same experimental material has been reanalyzed by Shemansky et al. (1971) who confirmed the value of d_O but attributed low altitude deactivation to nitric oxide with a rate $d_{NO} = 8.5 \times$ $\times 10^{-11}$ cm^3 s^{-1} and concluded that quenching by O_2 is insignificant. This result is in agreement with laboratory works indicating

$$d_{NO}/d_{O_2} \simeq 18 \text{ and } d_{O_2} \simeq 3.8 \times 10^{-12} \text{ cm}^3 \text{ s}^{-1}.$$

Emission profiles of V-K and 1N bands deduced from spectra obtained by Duysinx and Monfils (1972) have been reported by Duysinx (1973). We may use the ratio of the $(0-0)$ 1N to $(0, 6)$ V-K intensities to derive a value of the $N_2(A^3\Sigma_u^+)$ deactivation rate. Indeed, Sharp's measurements show that the production rate ratio $\eta(2P)/$ $\eta(1N)$ remains independent of altitude as expected on theoretical grounds. In this case, assuming $A^3\Sigma_u^+$ is quenched by O at rate d_O:

$$I(1N, 0-0) = K\left(1 + \frac{d_O n(O)}{A_O}\right) I(V-K, 0-6) \tag{1}$$

K being a proportionality constant and $A_O = 0.52$ s^{-1}, the radiative lifetime of the $v' = 0$ level. Since individual spectra were scanned in a short time (1 s), groups of 20 scans were added to improve signal/noise. The $\lambda 3914$ Å intensity was measured by a photometer pointing in the same direction (perpendicular to the spin axis) with an identical field of view and averaged over the period of 20 spectral scans. We have plotted in Figure 1 the $I(0-0, 1N)/I(0-6, V-K)$ intensity ratio, observed by Duysinx, vs. O density. In order to compare our result with that of Sharp's, his model atmosphere A has been adopted. Since a linear relationship is obtained, a linear least squares fit is made, which allows a value $d_O = 4 \times 10^{-10}$ cm^3 s^{-1} to be deduced. This value is 4 times higher than that derived from Sharp's measurements. The effect of quenching by O_2 or NO in these data would be indicated by the presence of experimental points above the straight line for large values of $n(O)$. On the contrary, for high $n(O)$ (data not plotted in Figure 1), the experimental points tend to lie below

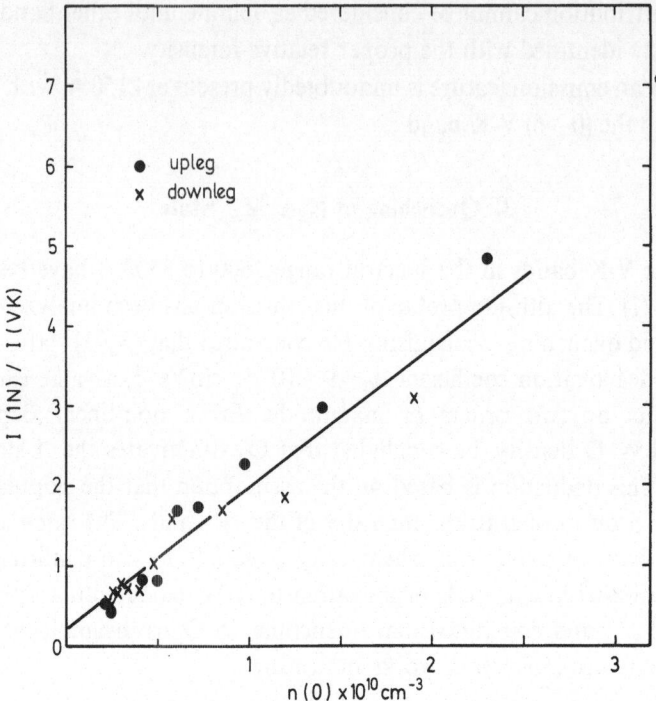

Fig. 1.   Intensity ratio as a function of O number density from Sharp's model atmosphere A. The line
is the least-squares fit to the experimental results.

this line. This may be due to non validity of relation (1) near the lower border of
the aurora.

4. Ultraviolet Satellite Photometry of Polar Cap Aurora

Measurements of auroral UV emission have been made recently using the ESRO
TD-1A satellite. This spacecraft was launched in March 1971 into Sun-synchronous
orbit inclined at 98° on the equator (Tilgner, 1971). The S2/68 experiment consists
in a UV stabilized telescope giving star spectra in the spectral range 1350 to 3200 Å.
When extended sources are observed, the instrument response is similar to a four-
channel photometer whose sensitivity is illustrated in Figure 2. Particular attention
was given to the calibration of the experiment and its absolute response is known
to be accurate within 20%. The presence of a dayglow emission in Channel A1 in
the vicinity of the dip equator has been attributed to Mg II resonance scattering at
2800 Å (Boksenberg and Gérard, 1973; Gérard and Monfils, 1974). During winter
periods, astronomical observations are stopped and the satellite spins about its solar
axis. Auroral and airglow observations have been made in December 1973 and
January 1974 in these conditions. The aurorally excited emission features observed
in the four channels are V-K bands, LBH bands, and atomic lines. Table II lists the
spectral range, peak sensitivity, and spectral response of the 4 channels to these

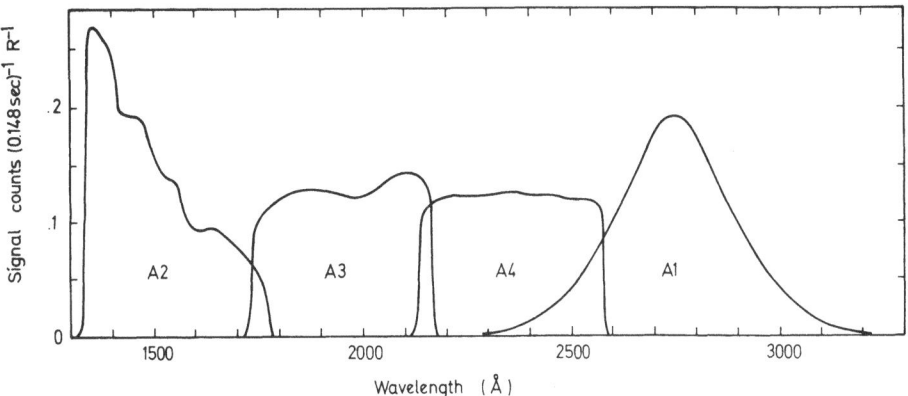

Fig. 2. Sensitivity of the four UV channels onboard satellite TD-1.

emissions. In the case of molecular emissions, the intensities refer to the whole system intensity. Roughly, A_2 and A_3 contain LBH and A_4 and A_1 respond to V-K. The spectral intensity distribution adopted in these calculations is given by Vallance Jones (1971). LBH relative intensities are in agreement with Ajello's laboratory observations whereas A state population rates are deduced from Broadfoot and Hunten (1964).

Recent measurements indicate that these rates are not correct but the influence on the channel response is very weak. A part of the A_2 signal is due to 1356 Å line of O: it has been taken as 10% of the unabsorbed LBH intensity (see Table I). Figure 3 shows an example of a high latitude (76° N) scan plotted in 3 channels. The rotation period of the satellite about the solar axis is 190 s and the observations were made parallel to the twilight plane. The two horizon crossings can be clearly seen: the northern one contains auroral radiation whereas the southern one is indicated by some airglow emission in channel A_1. Three regions of strong auroral emission are

TABLE II

Sensitivity of UV channels onboard TD1 satellite

Channel	Spectral range	Identification	Sensitivity (count/kR)	Typical intensity (kR)	Arc intensity (kR)
A_1	2300–3200 Å	V-K	32.5	12	19
		2P	4.5		
A_2	1350–1800 Å	LBH	85.5	7	11.5
		(with O_2 absorption)			
		[O] λ 1356 Å	265	1	4
A_3	1700–2200 Å	LBH	32	10	37
		(unabsorbed)			
A_4	2100–2600 Å	V-K	13.4	10.5	25
		LBH	1.6		

readily seen in addition to the northern horizon: two are situated in the polar cap and the third one may be associated with the auroral oval. As discussed in another paper (Gérard and Monfils, 1974) channels A_1, A_3 and A_4 are sensitive to particle precipitation causing scintillation in the P.M. window while A_2 is not. Several passes

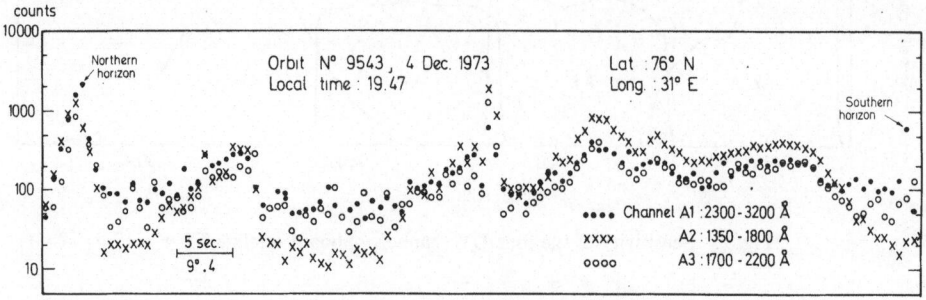

Fig. 3.   High latitude UV scan, parallel to the twilight plane.

have been analyzed so far: interesting conclusions may be drawn about UV intensity ratios. Table II lists intensities observed in a strong high latitude arc and in the lower latitude aurora. Figure 3 and other examples show that A_2/A_3 intensity ratio is constant, as expected, since both channels mainly respond to LBH bands. Noticeable exceptions occur on two occasions.

(1) When observing aurora near the horizon, absorption by O_2 along the line of sight decreases the A_2/A_3 ratio.

(2) In the center of the arcs A_2/A_3 drops down indicating a hardening of the precipitation which lowers the altitude of maximum LBH production. Consequently, the short wavelength part of LBH (A_2) is absorbed below 120 km $(\sigma_{abs.} \simeq 10^{-17} \text{ cm}^2)$ whereas the high wavelength part (A_3) is not affected $(\sigma_{abs.} \simeq 10^{-23} \text{ cm}^2)$.

This indicates that occasionally about 70% of the LBH radiation below 1700 Å is absorbed by O_2. The LBH/V-K system intensity ratio varies between 2 and 0.6. The dispersion reflects the increasing effect of quenching of $A^3\Sigma_u^+$ when the spectrum of the primary particles is hardening, larger LBH/V-K ratios being associated with lower A_2/A_3 ratios. These values are in agreement with results reported in Table I.

5. Conclusions

The composition of the UV auroral spectrum is reasonably well understood but several questions concerning relative intensities are still to be solved. They mainly concern:

(i) the spectral distribution of auroral radiation below 1000 Å;

(ii) the prediction of intensities of optically thick emissions (O I 1304 Å, N I 1200 Å);

(iii) the determination of the absolute O I $^3P - {}^5S$, $N_2 X \rightarrow C$ and $X \rightarrow A$ excitation cross sections by electron impact; and

(iv) the role played by the $N_2 A \rightarrow B$ and $W \rightarrow B$ transitions in the population of the A state.

Moreover, no observational data on the proton excited UV aurora have been obtained. Comparison between predicted and observed intensities will be of great interest.

References

Ajello, J. M.: 1970, *J. Chem. Phys.* **53**, 1156.
Albritton, D. L., Schmeltekopf, A. L., and Zare: 1975, *Diatomic Intensity Factors*, Harper and Row Publishers, in preparation.
Barth, C. A. and Schaffner, S.: 1970, *J. Geophys. Res.* **75**, 4299.
Boksenberg, A. and Gérard, J. C.: 1973, *J. Geophys. Res.* **78**, 4641.
Borst, W. L.: 1972, *Phys. Rev.* **A5**, 648.
Borst, W. L. and Chang, S. L.: 1973, *J. Chem. Phys.* **59**, 5830.
Borst, W. L. and Chang, S. L.: 1974, private communication.
Broadfoot, A. L. and Hunten, D. M.: 1964, *Can. J. Phys.* **42**, 1212.
Callear, A. B. and Wood, P. M.: 1971, *Trans. Faraday Soc.* **67**, 272.
Cartwright, D. C., Trajmar, S., and Williams, W.: 1971, *J. Geophys. Res.* **76**, 8368.
Cartwright, D. C., Trajmar, S., and Williams, J.: 1973, *J. Geophys. Res.* **78**, 2365.
Donahue, T. M. and Strickland, D. J.: 1970, *Planetary Space Sci.* **18**, 691.
Duysinx, R.: 1973, paper, ESRO Symposium; Spatind.
Duysinx, R. and Monfils, A.: 1972, *Ann. Geophys.* **28**, 109.
Freund, R. S.: 1971, *J. Chem. Phys.* **54**, 1407.
Gérard, J. C. and Monfils, A.: 1974, *J. Geophys. Res.* **79**, 2544.
Green, A. E. S. and Barth, C. A.: 1965, *J. Geophys. Res.* **70**, 1083.
Imami, M. and Borst, W. L.: 1974, *J. Chem. Phys.* **61**, 1115.
Landensperger, W.: 1971, *Space Res.* **11**, 1195, Akademie Verlag.
Miller, R. E., Fastie, W. G., and Isler, R. C.: 1968, *J. Geophys. Res.* **73**, 3353.
Mumma, M. J. and Zipf, E. C.: 1971, *J. Chem. Phys.* **55**, 1661.
Nicholls, R. W.: 1964, *Ann. Geophys.* **20**, 144.
Opal, C. B., Moos, H. W., and Fastie, W. G.: 1970, *J. Geophys. Res.* **75**, 788.
Paresce, F., Lampton, M. and Holberg, I.: 1972, *J. Geophys. Res.* **77**, 4773.
Peek, H. M.: 1970, *J. Geophys. Res.* **75**, 6209.
Poland, H. M. and Broida, H. P.: 1971, *J. Quant. Spectr. Radiative Transfer* **11**, 1868.
Prasad, S. S. and Green, A. E. S.: 1971, *J. Geophys. Res.* **76**, 2419.
Sharp, W. E.: 1971, *J. Geophys. Res.* **76**, 987.
Sharp, W. E. and Hays, P. B.: 1974, *J. Geophys. Res.* **79**, 4319.
Sharp, W. E. and Rees, M. H.: 1972, *J. Geophys. Res.* **77**, 1810.
Shemansky, D. E.: 1969, *J. Chem. Phys.* **51**, 689.
Shemansky, D. E. and Broadfoot, A. L.: 1971a, *J. Quant. Spectr. Radiative Transfer* **11**, 1401.
Shemansky, D. E. and Broadfoot, A. L.: 1971b, *J. Quant. Spectr. Radiative Transfer* **11**, 1385.
Shemansky, D. E. and Broadfoot, A. L.: 1973, *J. Geophys. Res.* **78**, 2357.
Shemansky, D. E., Zipf, E. C., and Donahue, T. M.: 1971, *Planetary Space Sci.* **19**, 1669.
Stolarski, R. S. and Green, A. E. S.: 1967, *J. Geophys. Res.* **72**, 3967.
Stone, E. J. and Zipf, E. C.: 1973, *J. Chem. Phys.* **58**, 4278.
Stone, E. J. and Zipf, E. C.: 1974, *J. Chem. Phys.* **60**, 4237.
Strickland, D. J.: 1968, Ph.D. Thesis, University of Pittsburgh.
Strickland, D. J. and Rees, M. H.: 1974, *Planetary Space Sci.* **22**, 465.
Theobald, K. and Peek, H. M.: 1969, Los Alamos Scientific Lab. Internal Report, LA. 3929.
Tilgner, B.: 1971, *ELDO/ESRO Sci. Techn. Rev.* **3**, 567.
Vallance Jones, A.: 1971, *Space Sci. Rev.* **11**, 776.

OBSERVED INTENSITY RATIOS OF THE
N₂ 1PG BANDS IN AURORA*

R. L. GATTINGER and A. VALLANCE JONES

Astrophysics Branch, National Research Council of Canada, Ottawa, K1A OR8, Canada

1. Introduction

In evaluating the relative importance of direct electron impact vs. cascading in the excitation of the N_2 first positive (1PG) system it is necessary to measure accurately the relative intensities of bands from as many different v' levels as possible to establish the relative populations. Gattinger and Vallance Jones (1974) have obtained measurements of the vibrational development of the system in medium intensity aurora, and have compared their results with various theoretical distributions. Vallance Jones and Gattinger (1972) made similar but less comprehensive measurements on bright aurora and arrived at ratios for $I(1,0):I(2,1):I(3,2)$ which were quite different from those determined in the later paper (Table I).

The possibility exists that the vibrational development is very dependent upon the auroral brightness, but it must be borne in mind that there is a large uncertainty involved in the measured 1,0 band brightness since it occurs very near the upper wavelength limit of response of the photomultiplier tube employed (RCA C31000F) which had an extended red multialkali photocathode surface. It is also possible that in the preparation of the synthetic comparison spectra the effects of blending of various auroral features especially in the 8600 to 8800 Å region introduced further inaccuracies into the results.

In an attempt to reduce the uncertainty inherent in the results the auroral observations were repeated in February 1974 at Ft. Churchill using an improved photomultiplier (RCA 31034A) having a lower dark count and a Ga-As photocathode which possessed a higher quantum efficiency over the pertinent spectral region. Operation at about twice the resolution (4 Å) in the 8700 Å region as compared with the

TABLE I

N₂ 1PG vibrational development

Band	Wavelength Å	Relative intensity		
		1972	1974	Current
1,0	8912	1.16	1.66	1.45
2,1	8723	1.00	1.00	1.00
3,2	8542	0.38	0.23	0.23
		Very bright	Med. int.	Med. int.

B. M. McCormac (ed.), Atmospheres of Earth and the Planets, 319–322. All Rights Reserved.

previous system (about 8 Å) resulted in further reducing the uncertainty in the blending ratios to the point where this aspect was no longer a problem as will be shown later. However, the 1,0 band is still in the region of sharp cutoff of response of the C31034A photomultiplier tube so that this component of the uncertainty in band intensity ratios still exists, although with the careful calibration techniques employed it is hoped that the uncertainty has been sufficiently reduced.

2. Results

The higher resolution spectrum (Figure 1) will be discussed first so that the relative intensities of the blending features in the 8700 Å region may be determined. Six scans on medium intensity aurora were averaged to produce the observed spectrum which is compared with the composite synthetic spectrum along with its components. By using an auroral intensity monitor it was possible to ensure that spectral scans were selected during which the auroral intensity fluctuations were not large. Contained in Table II are the relative intensities of the predominant emission features which were used to construct the synthetic spectrum.

   The lower resolution spectrum (Figure 2) was obtained by averaging sixteen scans, and is also representative of medium intensity aurora. The resolution was reduced to improve the signal-to-noise ratio for the N_2 1PG 1,0 band because of the rapidly decreasing response of the photomultiplier tube in this wavelength region. The syn-

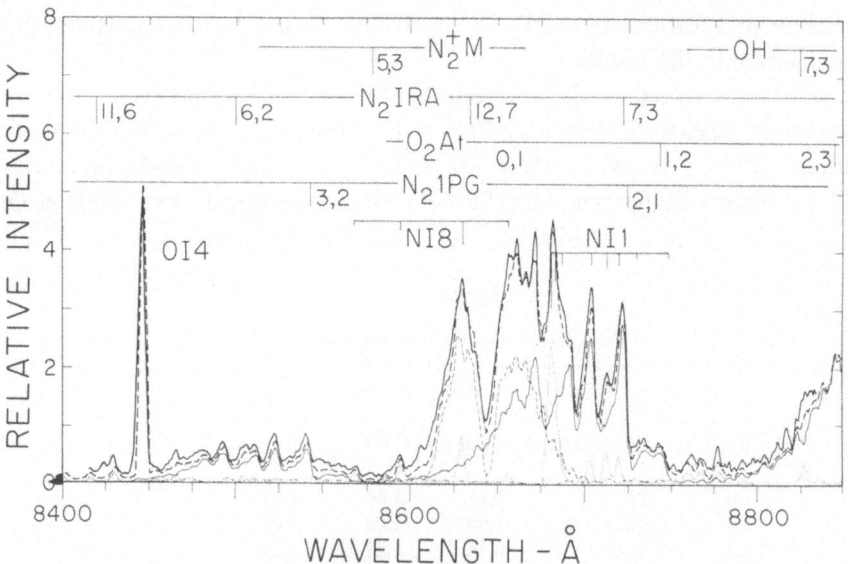

Fig. 1.   Comparison between observed and synthetic auroral spectra in the 8400–8850 Å region at a resolution of 4 Å. Observed spectrum – heavy solid line; total synthetic spectrum – heavy dashed line; N_2 1PG component – light solid line; O_2 At component – light dashed line; Atomic line component – dot, dash; OH night airglow component – dots; N_2^+ M component – dot, dot, dash; N_2 IRA component – dot, dash, dash.

TABLE II

Relative intensities of predominant emission features in 8400–9000 Å region

N_2 1PG		O_2 At		Atomic multiplets		
Band	Intensity	Band	Intensity	Ident.	Major λ(Å)	Total I
1,0	120	0,1	58	N_I 1	8680	11.6
2,1	82	1,2	1.6	N_I 8	8629	3.2
3,2	19	2,3	1.4	O_I 4	8446	11.0

thetic comparison spectrum in the region of the N_2 1PG 2,1 band was prepared using the blending ratios determined in Figure 1; the observed N_2 1PG 1,0 band intensity (Table II) was matched by the synthetic spectrum using the spectral information available in Figure 2.

3. Discussion

According to the results summarized in Table I it appears that the vibrational development of at least the $v' = 1$, 2, and 3 levels of the N_2 1PG is dependent upon the

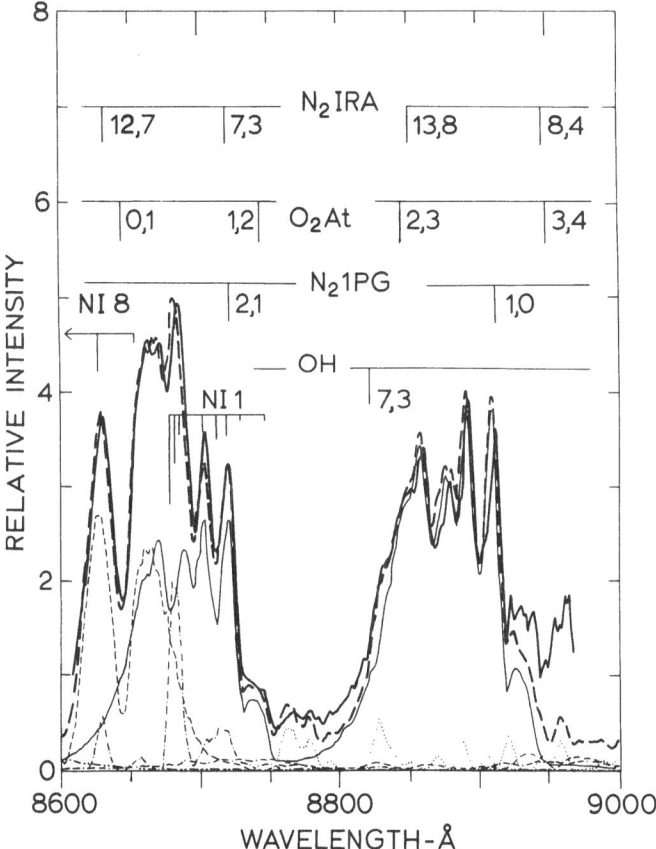

Fig. 2. Comparison between observed and synthetic auroral spectra in the 8600–9000 Å region at a resolution of 8 Å. See Figure 1 for coding of curves.

auroral intensity. This is based on the belief that blending of auroral emission features has been properly accounted for, and that the relative calibration for system response with wavelength over the spectral region is sufficiently accurate.

If an inverse relationship between auroral intensity and altitude exists (Chamberlain, 1961; Boyd *et al.*, 1971; Gattinger and Vallance Jones, 1972), then the trend in the change in vibrational development is as Cartwright *et al.* (1973) predict. They suggest that the population of the three lowest N_2 1PG v' levels is due largely to cascading with at least one of the cascading mechanisms being subject to quenching at lower altitudes.

Acknowledgment

The authors wish to thank Mr D. Tyler for his assistance in obtaining the experimental results.

References

Boyd, J. S., Belon, A. E., and Romick, G. J.: 1971, *J. Geophys. Res.* **76**, 7694.
Chamberlain, J. W.: 1961, in *Physics of the Aurora and Airglow*, Academic Press Inc., New York, p. 129.
Cartwright, D. C., Trajmar, S., and Williams, W.: 1973, *J. Geophys. Res.* **78**, 2365.
Gattinger, R. L. and Vallance Jones, A.: 1972, *Ann. Geophys.* **28**, 91.
Gattinger, R. L. and Vallance Jones, A.: 1974, *Can. J. Phys.* **52**, 2343.
Vallance Jones, A. and Gattinger, R. L.: 1972, *Can. J. Phys.* **50**, 1833.

PROCESSES AND EMISSIONS ASSOCIATED WITH
ELECTRON PRECIPITATION

M. H. REES*

National Science Foundation, Washington, D.C. 20550, U.S.A.

1. Introduction

The interaction of energetic auroral electrons with atmospheric gases provides a variety of challenging problems in several areas of physics and chemistry. Electron precipitation profoundly affects the state of ionization of the atmosphere, changes the neutral composition, and produces electromagnetic radiation over a wide range of wavelengths. At this time, we believe that the major processes have been identified, and models have been constructed that tie together several observable parameters: particle fluxes, ion densities and temperatures, optical emissions, compositional changes, and others.

The physical processes related to the energy loss suffered by electrons as they pass through the atmosphere are reviewed, indicating expected changes in various atmospheric and ionospheric parameters, and evaluating the resulting optical emission rates. The major uncertainties, present limitations in our model computations, and areas of research most needed and most promising are identified.

Energy spectra and angular distributions of auroral electrons have been and are being measured *in situ* by satellite-borne instruments and rocket payloads. A wide variety of population distributions is observed; some events are dominated by electrons with energies of 100 eV or less, others have large fluxes of electrons in the keV range. In the magnetospheric cleft region, the precipitation is predominantly 'soft', consisting of electrons with energies of a few hundred eV or less. The 'common' aurora is produced by a wide range of energy spectra. At the same time, angular distributions vary from monodirectional to nearly isotropic. In describing the various processes I shall employ a specific model electron spectrum to quantify the arguments, and I shall arrive at measurable parameters. The type of analysis described here is currently being applied to ISIS-2 satellite observations and to data obtained from a rocket flight in Alaska.

2. Auroral Processes and Emissions

We assume that a differential auroral electron number flux, described by the equation

$$N(E)\, \mathrm{d}E = 3.6 \times 10^4\, E \exp(-E/600)\, \mathrm{d}E\ \mathrm{cm}^{-2}\, \mathrm{s}^{-1}\, \mathrm{eV}^{-1},$$

bombards the atmosphere; an initially isotropic angular distribution is selected. This

* On leave from the University of Colorado, Boulder, Colo., U.S.A.

B. M. McCormac (ed.), Atmospheres of Earth and the Planets, 323–333. All Rights Reserved.
Copyright © 1975 by D. Reidel Publishing Company, Dordrecht-Holland.

corresponds to a total electron flux of 1.3×10^{10} cm^{-2} s^{-1} and an energy flux of 25 erg cm^{-2} s^{-1}.

The important physical processes associated with energetic electron bombardment and their effects are indicated in Figure 1, where measurable quantities are designated by the more heavily outlined boxes. The primary electron flux, unaffected by atmospheric scattering, is shown in Figure 2, labeled ∞. In practice, a low energy component of secondaries, tertiaries, etc., is always found, even at high altitudes where an appreciable back scattered component adds to local production. The high energy tail of the electron energy distribution in our model aurora is so small that both the production rate of bremsstrahlung X-rays associated with such an event and the ionization due to X-rays are negligible. Fundamental to all auroral processes is the altitude profile of the energy deposition rate. This quantity can be derived from appropriate laboratory data (Rees, 1963) or from energy degradation computations that yield the differential electron flux at all altitudes in the atmosphere (Berger *et al.*, 1970, 1974; Rees and Maeda, 1973; Banks *et al.*, 1974). Figure 3 shows the energy deposition rate; we note that in our model aurora the maximum occurs at 140 km. The processes that absorb most of the primary electron energy are ionization, dissociative ionization, and dissociation, where product ions, molecules, and atoms may be left in excited states. The reactions for each species, the rate coefficients, and computational procedures used in our model are given in the papers by Jones and Rees (1973) and Rees and Jones (1973). Ionization and dissociative ionization produce secondary electrons. Recent computations along this line have used the laboratory experiments by Opal *et al.* (1971) to obtain the production rate of secondary electrons in nitrogen and oxygen. An electron detector onboard a rocket or satellite can measure only the combined

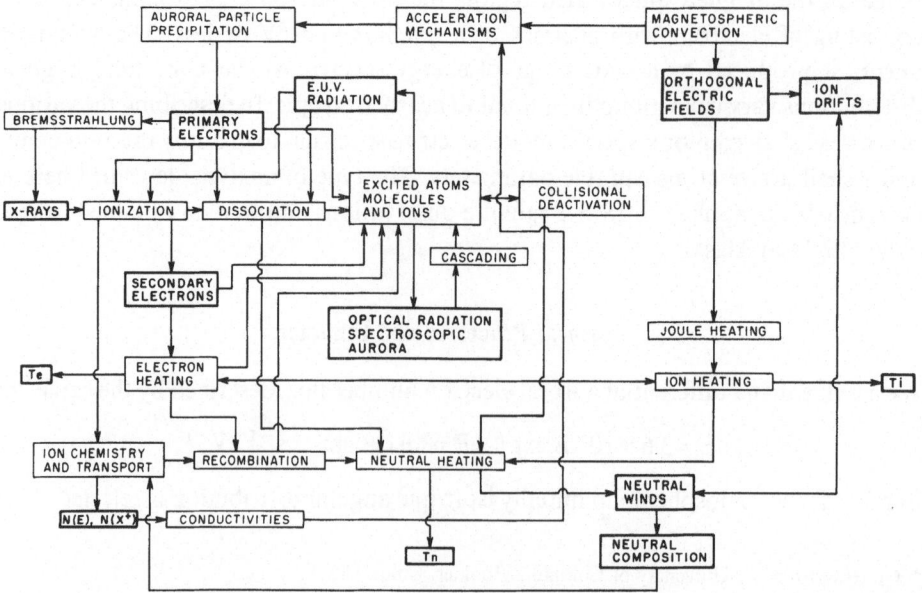

Fig. 1.   Flow chart of auroral processes and effects due to electron bombardment.

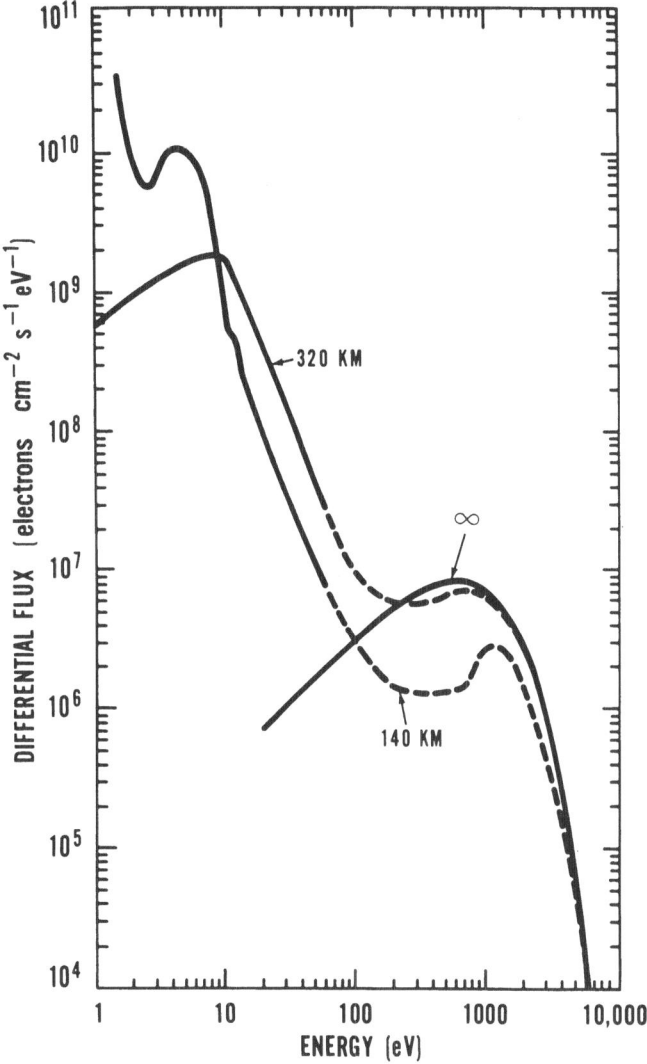

Fig. 2.   Differential electron fluxes at 140 km and 320 km associated with the primary flux labeled ∞. The dashed portions of the curves are inferred from previous computations of spectra and from energy conservation requirements.

spectrum of degraded primaries, secondaries, etc. Recently, computations by Rees and Maeda (1973) and Banks *et al.* (1974) have succeeded in modeling the gross features of observed spectra reasonably well. Comparison with the observations is difficult since satellite-borne detectors obtain a sample at only a single altitude along a field line, and rockets seldom can be fired along the field line to provide corresponding altitude profiles of electron spectra. The differential electron fluxes associated with our model aurora at 140 km, the altitude of maximum energy deposition, and at 320 km are shown in Figure 2. The large flux of low energy electrons accounts for the bulk of the optical excitation in auroras. To compute excitation rates of auroral

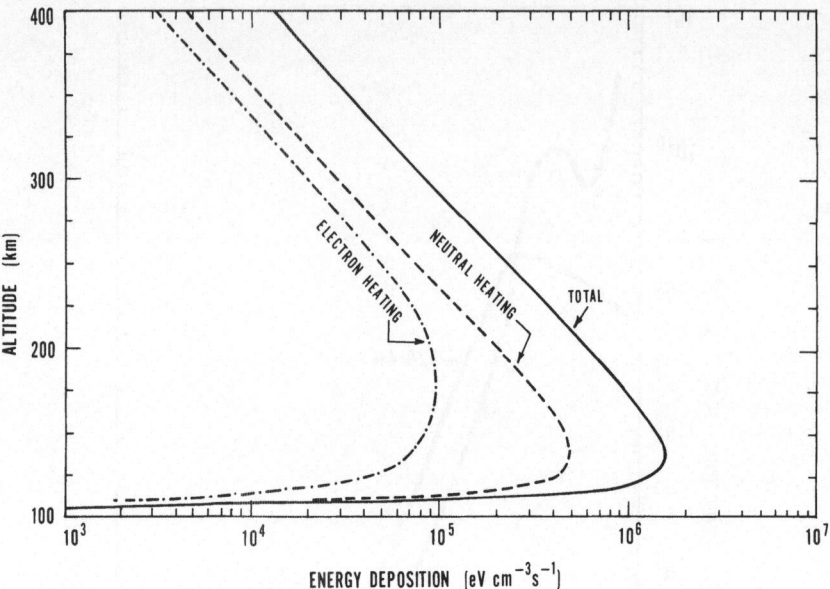

Fig. 3.   Altitude profiles of the total energy deposition rate, neutral heating, and electron heating associated with the model aurora under investigation.

radiations, the spectral distribution of the electron flux together with the energy
dependent cross sections must be known.

The primary electron flux ionizes atmospheric gases, and molecular species may be
dissociated. While ion production rate is not a measurable quantity, the concentrations of various species of ions have been measured. The ionic composition is close to a
steady state after a 20 min interval of invariant electron bombardment, and we show
the results for our model aurora in Figure 4. NO^+ is the dominant species below 160
km, but O^+ becomes the most abundant above this level. Generally, observations of
the NO^+/O_2^+ ratio made by rocket-borne mass spectrometers yield larger values than
those predicted by the model. Jones and Rees (1973) found the ratio obtained from the
model can be considerably larger under non-steady state conditions. There could be
processes that convert O_2^+ to NO^+ in addition to the ones considered in our model
computations, or the appropriate rate coefficients may be larger than the values we
have used. We also notice in Figure 4 that nitrogen ions and excited O^+ are minor
species.

A noteworthy consequence of auroral chemistry is the buildup of NO to levels
considerably higher than those present in the unperturbed atmosphere. The two
processes thought to be responsible are

$$O_2^+ + N_2 \rightarrow NO^+ + NO$$

which has a reported rate coefficient of 5×10^{-16} cm^3 s^{-1} (Ferguson, 1967), and

$$N + O_2 \rightarrow NO + O$$

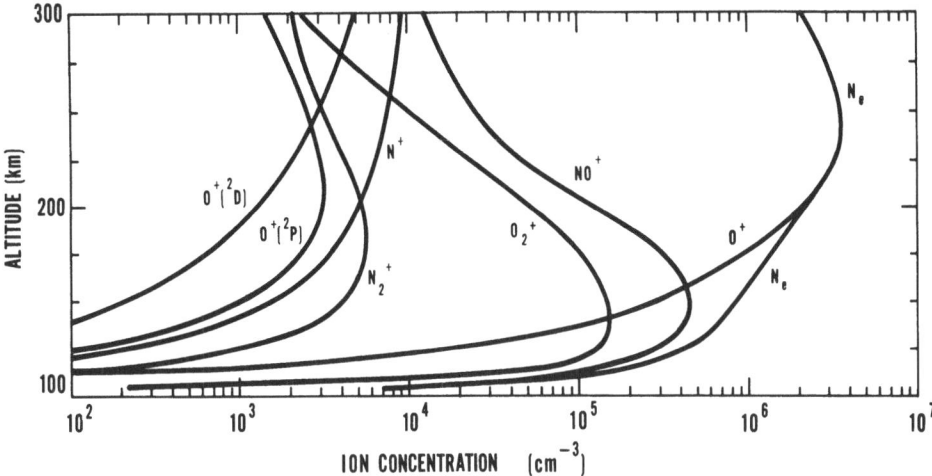

Fig. 4.   Altitude profiles of major ion concentrations appropriate to the model aurora.

which is slow for N in the ground state, ^{4}S, but which has a large rate coefficient for N(^{2}D) and N(^{2}P). The rate coefficient, 5×10^{-12} cm^3 s^{-1}, given by Lin and Kaufman (1971), was used in our computations. Nitric oxide disappears by ion-atom interchange with O_2^+ and recombination with N to form N and O. In the case of the latter process a steady state is reached only after a matter of hours. Thus, auroral bombardment lasting for several hours, even if intermittent, can result in the substantial build-up of NO observed by Zipf (1970). The enhancement predicted in our model is shown in Figure 5.

Returning to Figure 1, we see that a host of processes can lead to the formation of excited atoms, molecules and ions: electron impact, dissociative excitation, recombination, cascading, and quenching reactions. These processes are discussed in detail by Rees and Jones (1973). Excited species that have not been collisionally deactivated will produce radiation. Extreme UV radiation may be trapped, causing additional ionization or dissociation, while the visible radiation is the aurora, and a part of the IR radiation may add to local heating. Altitude profiles of the volume emission rate of four frequently observed auroral features are shown in Figure 6 for the model under investigation; quenching has been included. The column emission rates are also given in the Figure. The N_2^+ 1NG (0, 1) feature at 4278 Å and the radiations from the ground-state configuration of O I at 5577 Å and 6300 Å may be used to infer the magnitude and spectral character of precipitating electron fluxes (Rees and Luckey, 1974). The radiation from O_2 $^1\Delta_g$ at 1.27 μm has been included to show that this auroral emission feature is also enhanced; however, since·the radiative lifetime of the excited state is of the order of 1 h, and negligible quenching occurs, a substantial buildup of the excited state develops over protracted periods of auroral bombardment. Likewise, a buildup of the observed emission rate occurs, as shown by Gattinger and Vallance Jones (1973). Numerous radiations from O, N_2, N and O_2 and their ions are present in the auroral spectrum, excited by the processes listed above and by energy transfer collisions. There have been reports of auroral radiations from some minor constituents, NO, O_3, CO_2, NO_2, and OH.

Atoms formed by molecular dissociation and by electron-ion recombination may be left with excess kinetic energy; collisional quenching may also create hot atoms and molecules. We find that about one-third of the energy deposited by auroral electron

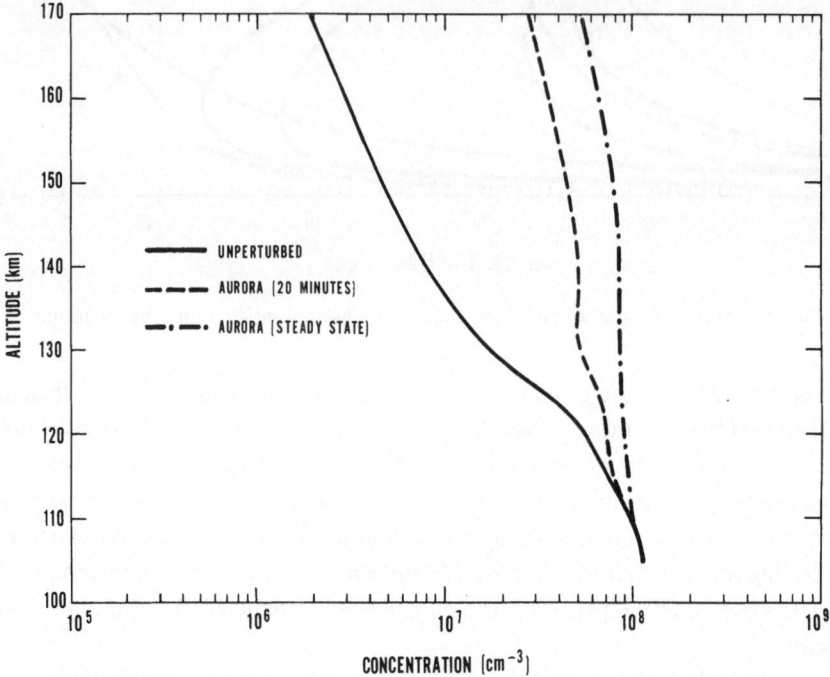

UNPERTURBED

AURORA (20 MINUTES)

AURORA (STEADY STATE)

Fig. 5. Altitude profile of the concentration of NO for unperturbed and auroral conditions of the model (20 min). The steady state concentration, reached in a few hours, is also shown.

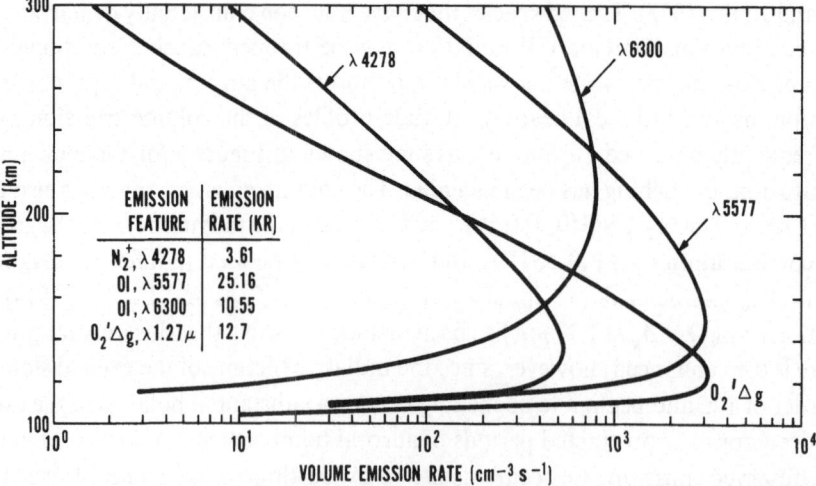

EMISSION FEATURE	EMISSION RATE (KR)
N_2^+, $\lambda 4278$	3.61
OI, $\lambda 5577$	25.16
OI, $\lambda 6300$	10.55
$O_2{}'\Delta g$, $\lambda 1.27 \mu$	12.7

Fig. 6. Altitude profiles and column emission rates of four auroral spectroscopic features predicted by our model.

fluxes is expended in heating of the neutral gas. The altitude profile of neutral heating appropriate to our model is shown in Figure 3. Hays *et al.* (1973) showed that local heating produces an upwelling of the neutral gas, a vertical wind, that causes the atmosphere to become better mixed, and, therefore, changes the composition at all altitudes above the level of heat input. The vertical wind is rather modest, a few meters per second in the example examined here; the altitude profile of the vertical wind is given in Figure 7. The changes in composition resulting from this vertical wind after one hour are shown in Figure 8. The increased mixing increases the ratio of molecular to atomic species. Using the fact that diffusive separation dominates above 300 km, the profiles may be extended to higher altitudes.

Although the effects of electric fields are not included in this lecture, we note, referring to Figure 1, that substantial heating of the neutral gas occurs through joule energy dissipation. In fact, joule heating is probably the most important source of neutral atmospheric heating in aurora (Cole, 1971; Hays *et al.*, 1973). However, in the model computations presented here we investigate the effects of particle heating alone. In addition ion drag will influence the neutral winds. Changes in the composition of the neutral atmosphere will influence the relative ion production rates, the ion-neutral chemistry, as well as all subsequent processes that depend on these parameters.

While the principal effect of neutral heating is to change the composition of the neutral atmosphere, there will also be a small increase in the neutral gas temperature

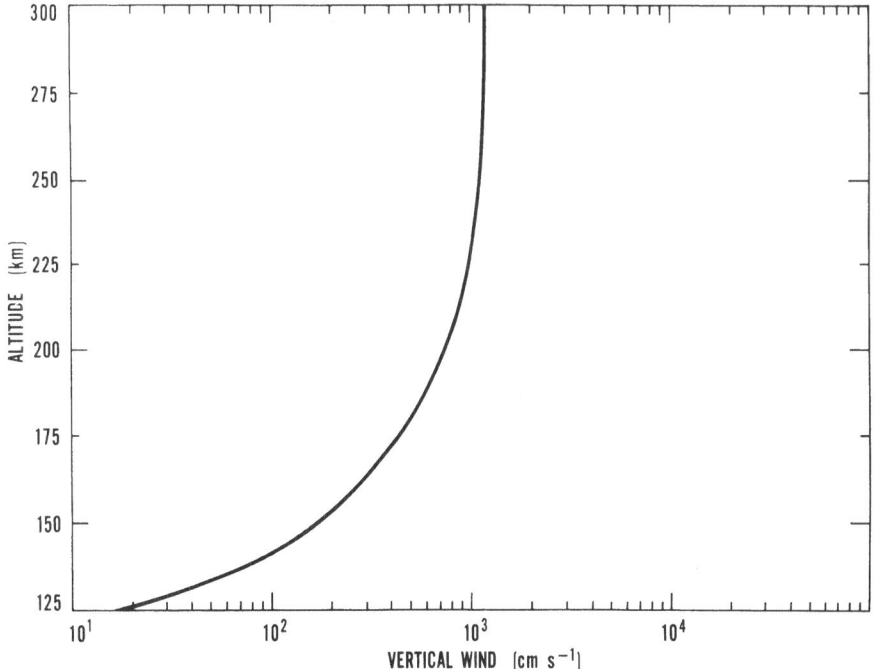

Fig. 7.   Vertical wind resulting from heating of the neutral gas associated with the model aurora.

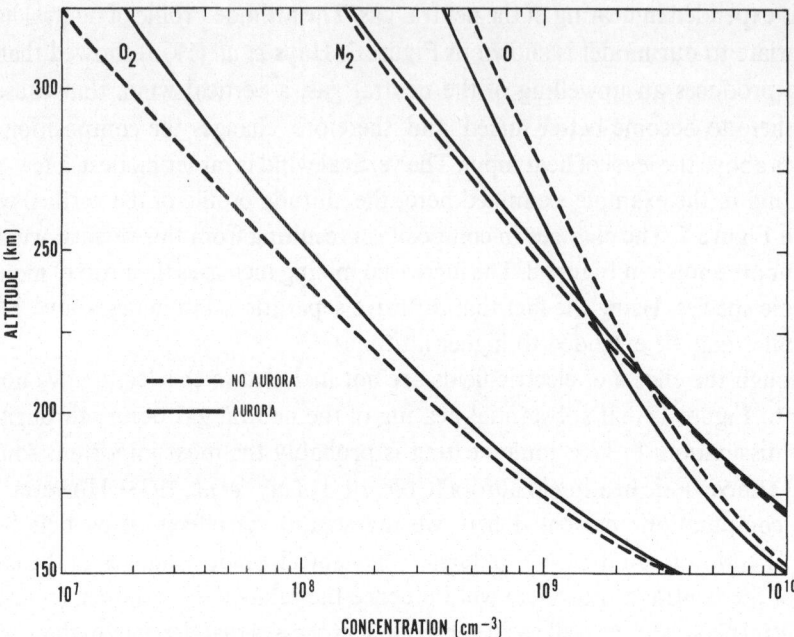

Fig. 8.   Compositional change of the neutral atmosphere after 1 h of auroral heating.

(Hays *et al.*, 1973). After 1 h of heating, this increase, which is inconsequential below 150 km, is predicted to be only about 60 K above 300 km.

About 8% of the energy flux in this model aurora goes into heating of the ambient electron gas by electron-electron collisions. The altitude profile of electron heating is shown in Figure 3; this profile indicates that most of the electron heating occurs at higher altitudes, where the degree of ionization is larger. At low altitudes excitation of electronic states in atoms and molecules, vibrational and rotational excitation in molecules, and excitation of the fine structure levels in O compete for the available secondary electron energy. Electron heating results in an enhanced electron temperature, and the temperature profile derived for our model is shown in Figure 9. Heat conduction plays a relatively minor role below about 300 km; the structure in the temperature profile below this level is principally due to local effects. The electron and ion energy equations are coupled because a fraction of the energy from hot electrons is shared with the ambient ions rather than the neutral gas. The ion temperature profile is also shown in Figure 9. Computation of the temperature profiles requires a boundary condition at the 1000 km level. We have assumed that there is no net charge flowing across this boundary between the ionosphere and the magnetosphere, the downward electron flux being balanced by an upward flowing reverse current (Rees *et al.*, 1971). Different boundary conditions yield different temperature profiles, but all viable assumptions lead to an enhanced electron temperature in the auroral ionosphere.

Aurorally produced ionization profoundly affects the electrical conductivity profiles and, hence, the current flowing in the ionosphere and magnetosphere. Profiles

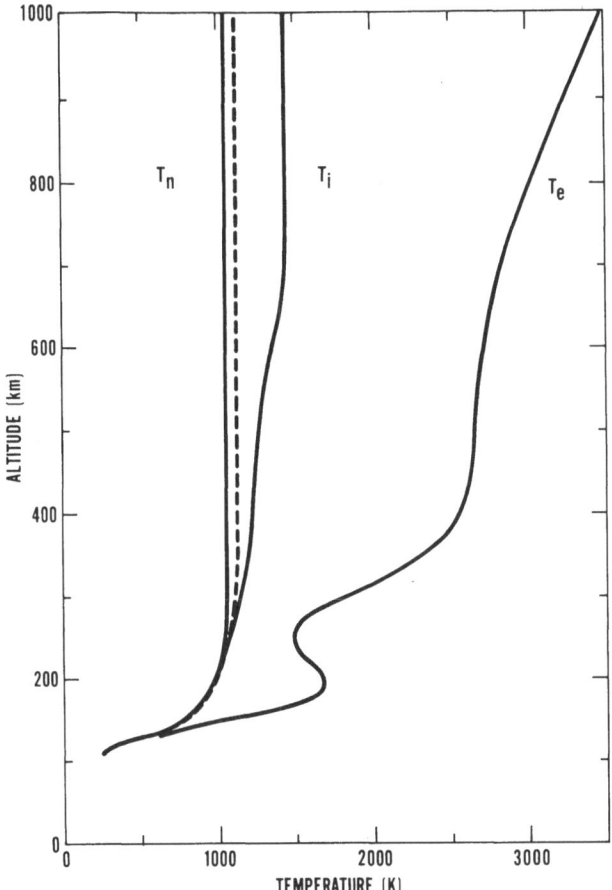

Fig. 9.   Profiles of electron and ion temperatures in our model aurora. The dashed curve represents the neutral gas temperature after 1 h of auroral heating.

appropriate to our model aurora are given in Figure 10. These show a large enhancement in the magnitudes over those found for unperturbed conditions. We observe that the relatively soft electron flux of the model contributes to a Pedersen component that is higher than the Hall component above 120 km: the Hall conductance is 13.5 mhos and the Pedersen conductance is 47.2 mhos. As indicated in Figure 1, enhanced ionospheric conductivities together with a convection electric field lead to field aligned currents (Kennel and Rees, 1972) that may be heavily involved in auroral particle acceleration mechanisms.

3. Discussion

I consider these initial model computations as a test for our overall understanding of the physical and chemical processes that follow electron bombardment of the atmosphere. Several derived parameters may be measured, some *in situ*, others remotely. Only rarely has it been possible, through coordinated experiments, to observe several auroral quantities simultaneously in a spatially and temporally well-defined event. A

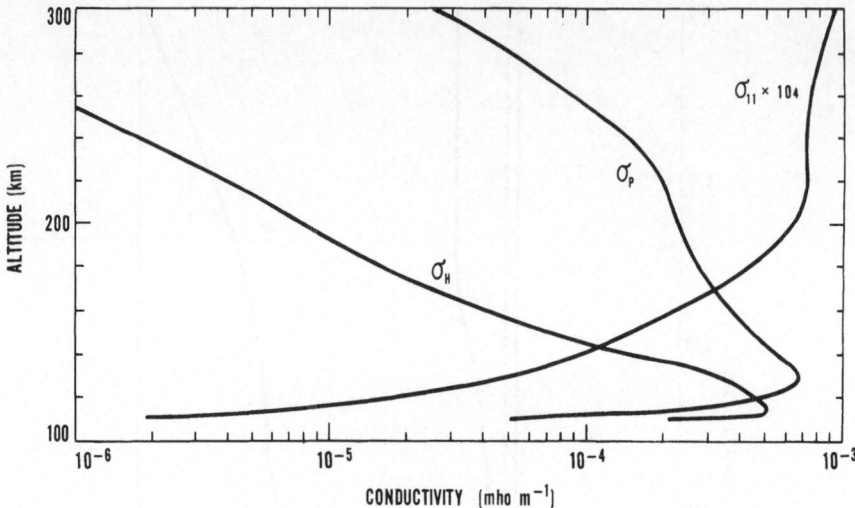

Fig. 10. Conductivity profiles in the ionosphere appropriate to the aurora under investigation. The Pedersen and Hall components are labeled σ_P and σ_H, respectively, and the conductivity parallel to the field lines is $\sigma_{\parallel}$.

few such experiments have borne out the general validity of the model, but there are persistent discrepancies that we do not understand as yet.

Observations generally yield abundance ratios of NO^+/O_2^+ that are larger than those predicted using currently accepted values of relevant rate coefficients and species abundances. Perhaps the vibrationally excited N_2 molecules formed in the auroral atmosphere react faster with O_2^+ ions to form NO^+ and NO than is predicted by laboratory measurements performed with ground state molecules. I strongly suspect that several rate coefficients measured under laboratory conditions may not be applicable to ionospheric species that have excess kinetic and internal energy. The importance of highly excited states and resulting EUV radiation is just beginning to be appreciated. Horizontal winds may affect the ion composition when the time constant for transport is less than that for chemical reaction.

A few measurements of auroral electron spectra have been made in the energy range of a few eV to some 100 eV. The observed decrease in flux with respect to energy agrees with the model computations up to about 20 eV, but above about 20 eV the observed flux vs. energy curve falls off less steeply than the model for some, but not all, observations. Primary low energy electrons may, somehow, find their way to altitude levels where the measurements are made, but the mechanism is not evident. Below about 100 eV, the model takes into account only secondary electrons.

The mechanism and the degree to which the ionospheric conductivity controls magnetospheric convection and particle acceleration are unanswered questions at this time. Ionosphere-magnetosphere current systems must be better understood before we can make progress in this area.

Although Figure 1 illustrates the notion that auroral processes are coupled, it does not show the dynamic character of the phenomenon. Auroral electron fluxes exhibit large temporal fluctuations in magnitude, energy spectrum, and angular distribution. Each of the processes shown in the Figure has a characteristic time constant that affects the other processes. It is not surprising to me, therefore, that the model does not reproduce all types of observations all the time. The challenge is to identify the important shortcomings. The numerical results presented here are based on a precipitation event of 20 min duration, invariant in electron flux characteristics. Most of the parameters reach a steady state in this interval. No electric fields have been introduced. While a derived electron temperature has been incorporated in the temperature dependent rate coefficients, a completely self-consistent solution does not seem reasonable at this time. As new information becomes available, this model can be tested and updated.

Acknowledgments

It is a pleasure to acknowledge the assistance given by Diane Luckey during various stages of this work. Financial support was provided, in part, by the Atmospheric Sciences Section of the National Science Foundation under grant GA-16290.

References

Banks, P. M., Chappell, C. R., and Nagy, A. F.: 1974, *J. Geophys. Res.* **79**, 1459.
Berger, M. J., Seltzer, S. M., and Maeda, K.: 1970, *J. Atmospheric Terrestr. Phys.* **32**, 1015.
Berger, M. J., Seltzer, S. M., and Maeda, K.: 1974, *J. Atmospheric Terrestr. Phys.* **36**, 591.
Cole, K. D.: 1971, *Planetary Space Sci.* **19**, 59.
Ferguson, E. E.: 1967, *Rev. Geophys.* **5**, 305.
Gattinger, R. L. and Vallance Jones, A.: 1973, *J. Geophys. Res.* **78**, 8305.
Hays, P. B., Jones, R. A., and Rees, M. H.: 1973, *Planetary Space Sci.* **21**, 559.
Jones, R. A. and Rees, M. H.: 1973, *Planetary Space Sci.* **21**, 537.
Kennel, C. F. and Rees, M. H.: 1972, *J. Geophys. Res.* **77**, 2294.
Lin, C. L. and Kaufman, F.: 1971, *J. Chem. Phys.* **55**, 3760.
Opal, C. B., Peterson, W. K., and Beatty, E. C.: 1971, *J. Chem. Phys.* **55**, 4100.
Rees, M. H.: 1963, *Planetary Space Sci.* **11**, 1209.
Rees, M. H. and Jones, R. A.: 1973, *Planetary Space Sci.* **22**, 1213.
Rees, M. H. and Luckey, D.: 1974, *J. Geophys. Res.* **79**, 5181.
Rees, M. H. and Maeda, K.: 1973, *J. Geophys. Res.* **78**, 8391.
Rees, M. H., Jones, R. A., and Walker, J. C. G.: 1971, *Planetary Space Sci.* **19**, 313.
Zipf, E. C.: 1970, *J. Geophys. Res.* **31**, 6371.

ROCKETBORNE OBSERVATIONS OF ATMOSPHERIC
INFRARED EMISSIONS IN THE AURORAL REGION

A. T. STAIR, JR. and J. C. ULWICK

Air Force Cambridge Research Laboratories L. G. Hanscom Field Bedford, Ma. 01730, U.S.A.

and

K. D. BAKER and D. J. BAKER

Utah State University, Logan, Utah 84321 U.S.A.

Abstract. This paper briefly reviews the preliminary results of spectral measurements of IR atmospheric emissions in the wavelength region from 1.6 to 23 μm. These measurements were achieved using cryogenic spectrometers on six recent rocket flights conducted under the Air Force Cambridge/DNA ICECAP program for coordinated auroral measurements from Poker Flat, Alaska.

1. Measurement System

The IR spectral results presented here were obtained with cryogenically-cooled spectrometers carried aboard sounding rockets that penetrated the high altitude emitting regions of the atmosphere. The rocket probes are summarized in Table I according to type of IR measurements, auroral conditions penetrated, and altitude range of the measurements. As shown, the measurements were conducted during a variety of auroral conditions including nondisturbed cases. In each case the spectrometer viewed forward along the rocket axis with a 5° (full angle) field of view. In addition to the IR spectrometers, each of these six rocket payloads contained other instruments, and in addition an extensive array of ground-based instruments provided measurements prior to and during the rocket flights (Burt and Davis, 1974). These other measurements will be alluded to here only in a general way to define the auroral conditions probed by the IR instruments. Details of the composite measurements and instrumentation will be reported elsewhere.

A cross sectional diagram of a typical spectrometer is shown in Figure 1. Although this diagram shows the details of the instrument used for measurements in the long wave IR (LWIR) region of the spectrum, the short wave IR (SWIR) instrument is quite similar.

The wavelength scanning of the spectrometers was accomplished by use of a circular-variable interference filter (CVF) that was rotated in the optical path giving 2 scans s^{-1}. The spectral resolution of the instruments was in the range of from 3 to 4%. The entire optical train of each spectrometer was housed in a high-vacuum dewar and cooled cryogenically by liquid He for the LWIR system and by liquid N_2 for the SWIR system. A cold cover was removed in flight just prior to taking measurements. The LWIR spectrometer utilized an As doped Si detector, whereas the SWIR instrument operated with an In-Sb detector. The two semicircular segments of the CVF were separated by an opaque mask that allowed the use of a dc reset electronic system. This dc reset system achieves significantly better signal to

B. M. McCormac (ed.), Atmospheres of Earth and the Planets, 335–346. All Rights Reserved.

TABLE I

Summary of infrared auroral measurements

Rocket	Date	Alt. range of meas. (km)	Spectral range of IR meas. (μm)	Auroral cond.	Primary results
Black Brant VA (18.006 − 2)	22 Mar 73	40–185	7–23	Rocket penetrated post-breakup auroral glow IBC II	First longwave infrared spectral measurements in aurora. Altitude profiles of 9.6 μm (O_3) and 15 μm (CO_2) emissions.
Paiute-Tomahawk	24 Mar 73	70–211	1.6–5.3	Rocket penetrated very bright auroral breakup IBC III[+]	Aurorally enhanced 4.3 μm (CO_2), 2.8 and 5.3 μm (NO) emissions.
Black Brant VA (18.205 − 1)	27 Mar 73	68–181	1.6–5.3	Rocket ascent penetrated bright auroral arc IBC II[+] Descent into quiet region	Measurements of SWIR emissions from auroral arc
Black Brant VA (18.006 − 4)	13 Feb 74	65–199	7–23	Aurorally quiet	Background for auroral case
Black Brant VA (18.219 − 1)	27 Mar 74	80–195	1.6–5.6	Rocket penetrated IBC II aurora in region previously occupied by IBC II[+] auroral arc	Auroral IR emission profiles
Nike-Javelin (NJ 74 − 1)	11 Apr 74	56–118	1.6–5.3	Aurorally quiet	Background for auroral cases

noise threshold sensitivity than an equivalent chopped and demodulated approach since the wider electrical bandwidth is not required and the optical efficiency is higher (choppers blank off the detector 50% of the time). Details of the spectrometers have been reported elsewhere for reference (Wyatt, 1971, 1975; Jensen *et al.*, 1972; Stair *et al.*, 1973).

The Black Brant payloads were separated from the rocket motor and were oriented to within about 3° of the local vertical at the altitudes above 65 km. (The actual altitude of the stabilized portion of the flight trajectory varied on the several flights.) The other two rocket payloads (Paiute-Tomahawk and Nike-Javelin) remained with the rocket second stage motor which was spin stabilized to within 30° of the local vertical during the measurements interval.

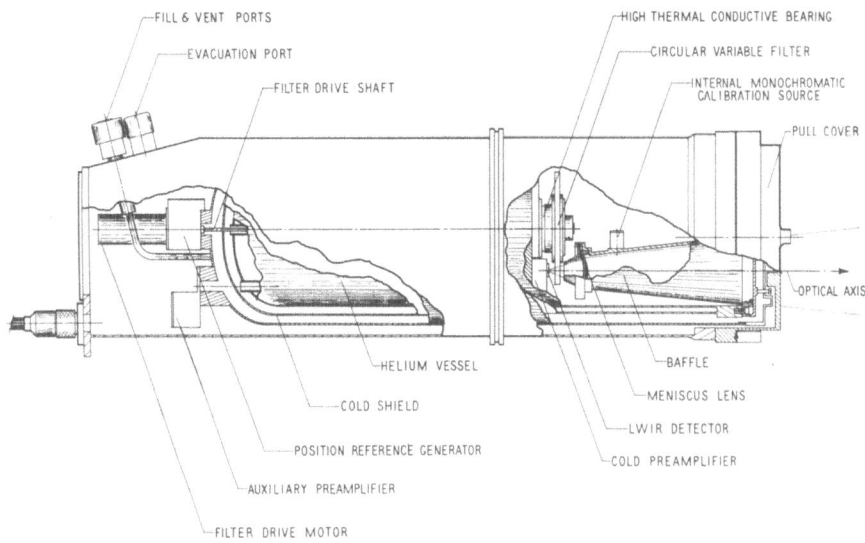

Fig. 1.   Cutaway view of CVF spectrometer (LWIR) rocket measurements of IR atmospheric emissions.

2. Results

All six of the IR spectrometer rocket flights were successful in providing high quality measurements in the various auroral conditions probed. The data analyses for these flights are still in process and absolute calibrations of instruments are still being finalized. Accordingly, the results presented here should be viewed as preliminary with some refinement and further analysis to be expected.

2.1. SWIR RESULTS

Sample spectral scans in the SWIR region from the Paiute-Tomahawk flown into an intense auroral breakup are shown in Figure 2. These are radiance spectra as viewed vertically (overhead) from the rocket. The obvious features in the spectra are a strong peak at 4.3 μm which is attributed to $CO_2(\nu_3)$ and lesser peaks at about 5.3 and 2.8 μm assumed to be due to emission from excited NO ($\Delta v = 1$ and $\Delta v = 2$, respectively). The wavelength of the peak at about 5.3 μm needs to be viewed with caution since the instrument response is falling off very rapidly at 5.3 μm.

The peak spectral radiance of the 4.3 μm feature observed in the Paiute-Tomahawk flight is plotted in Figure 3 as a function of altitude along with similar data from three other flights. The 4.3 μm peak spectral radiance viewed in the zenith shows a large enhancement (126 MR μm^{-1}) at about 92 km. Although these data have not been corrected for viewing angle, they approximately represent zenith spectral radiances. In viewing the profile and particularly the increase of spectral radiance with altitude from 75 to 92 km, it should be borne in mind that the ν_3 band of CO_2 is optically thick at this wavelength at these atmospheric pressures. As a result, the

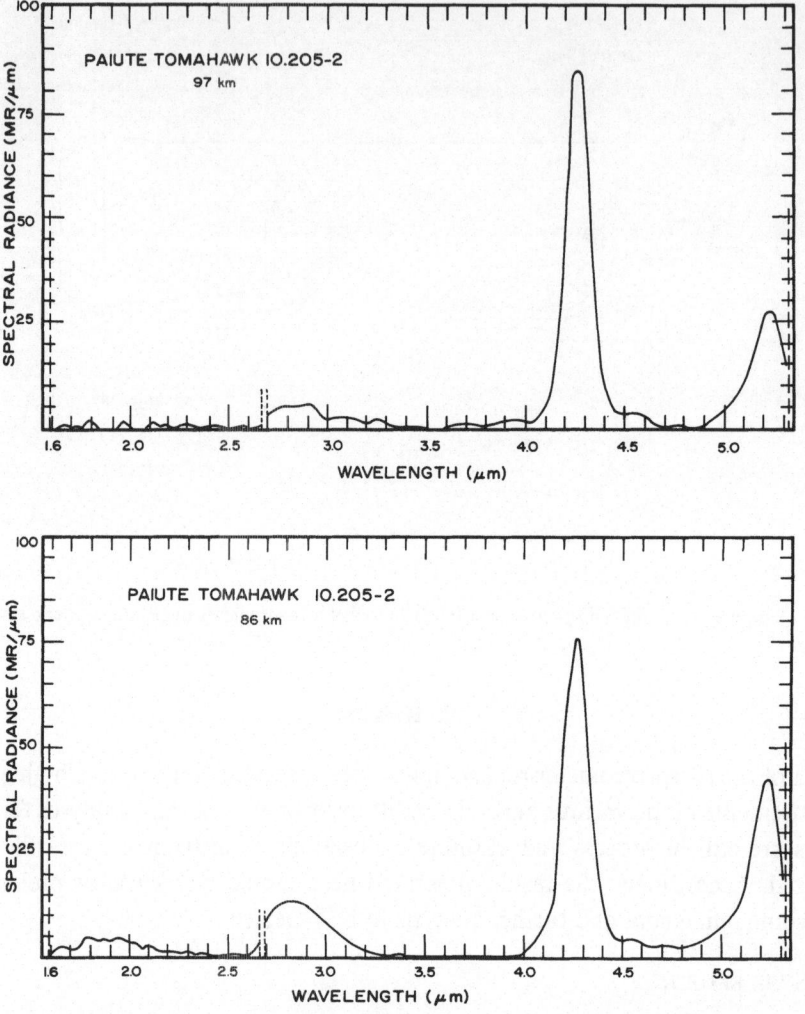

Fig. 2. Sample spectral scans from SWIR spectrometer aboard Paiute-Tomahawk rocket launched from Poker Flat, Alaska, 24 Mar 73. Although uncorrected for actual rocket aspect, the data approximate the zenith spectral radiance. The lower curve was observed at 86 km while the upper was at 97 km, both on rocket ascent.

observed values are not necessarily coming from radiation integrated over the total optical path above the rocket. This rocket penetrated a very bright visible region at the altitude of the peak IR radiance. Both ground photometers and an onboard photometer indicated over 200 kR of N_2^+ (3914 Å) emission from the region penetrated. An incoherent scatter radar operated by the Stanford Research Institute, Chatanika, observed electron densities of about 2×10^6 cm^{-3} from the vicinity of the rocket at an altitude of about 90 km (Baron and Chang, 1974). Comparison of the peak value at 4.3 μm for this breakup event with the value from the quiet background (Nike-Javelin) at 92 km shows an enhancement of about a factor of 50 in observed radiance due to the auroral activity.

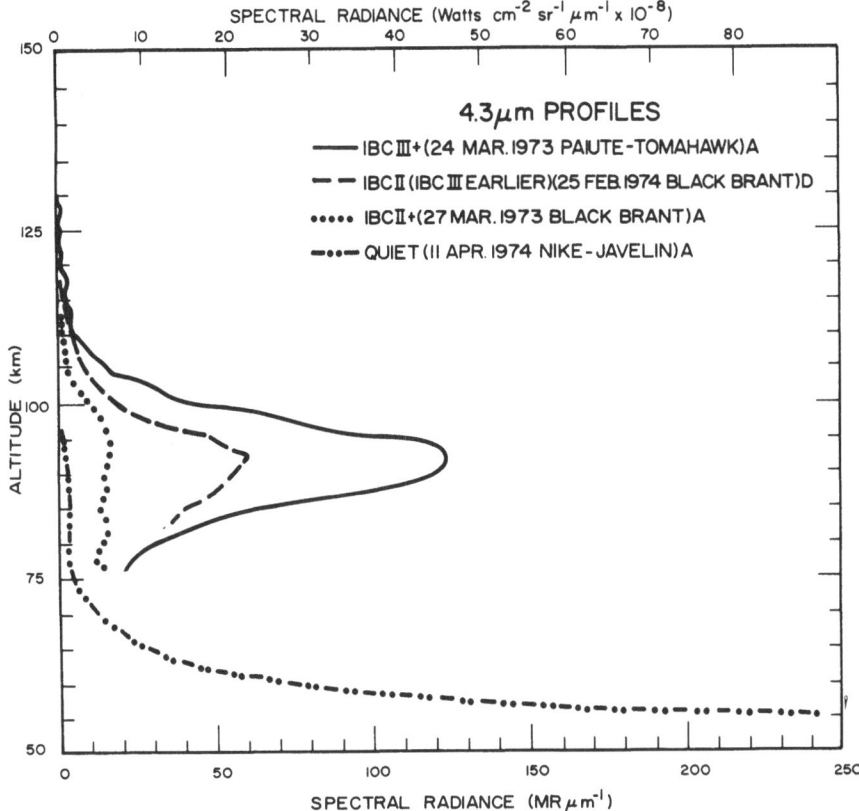

Fig. 3. Zenith peak spectral radiance at 4.3 μm measured with SWIR spectrometers flown on four different rockets under various auroral conditions.

The peak spectral radiances at 4.3 μm measured by the two Black Brant rockets show considerable enhancement over the background but not nearly as pronounced as the bright breakup case. This general pattern is to be expected since the auroral conditions for these probes were moderately bright but substantially less than the extreme case. An interesting feature is the fact that the 1974 Black Brant shows a much brighter region of 4.3 μm emission than the 1973 Black Brant, even though the visible aurora was considerably less bright as the rocket traversed the aurora. However, this can be accounted for as a bright auroral arc of IBC III intensity had occupied the region shortly before the rocket payload arrived. This was verified by ground station photometric and all sky camera coverage. When the rocket arrived at the altitude of the auroral energy deposition, though, the region showed slightly less than 10 kR of 5577 Å emission. In the 1973 case a 30 kR (5577 Å) auroral arc (IBC II$^+$) remained relatively stationary for the rocket penetration.

The above observations lead to the conclusion that the IR emission does not follow the visible emissions in close time correlation but rather depends on the time history of the energy deposition in the region. This conclusion was borne out by measurements during auroral breakup made on the descent portion of the Paiute-Tomahawk rocket flight. Although the visible auroral emission had dimmed more than a factor of 5 from rocket ascent to descent, the 4.3 μm emission profile on descent remained almost identical to the ascent curve.

Preliminary analysis of the 4.3 μm data is given by Kumer (1975). The highlights of this analysis are that the observed temporal and spatial behaviour of the 4.3 μm auroral data can be explained by a mechanism involving vibrational excitation of nitrogen to $N_2^{\ddagger}$, followed by a collisional resonance $v - v$ transfer of the vibration to excite CO_2 (001), followed by emission by the CO_2 at 4.3 μm. This mechanism is complicated by repeated transfers of the vibration back and forth between N_2 and CO_2 via $N_2^{\ddagger} + CO_2 \rightleftarrows N_2 + CO_2$ (001), and, since the CO_2 is optically thick (James and Kumer, 1973; Kumer and James, 1974) the CO_2 4.3 μm photons may be repeatedly emitted and absorbed via CO_2 (001) $\rightleftarrows CO_2 + hv$. This mechanism calls for an efficiency ε for the production of $N_2^{\ddagger}$ by particle precipitation which is in the range $3 \lesssim \varepsilon \lesssim 18$ vibrational quanta produced per ionization event. This value for ε is essentially consistent with the theoretical estimate by Stolarski (1968). Kumer also shows that emission by NO^+ ($\Delta v = 1$) at 4.3 μm does not significantly contribute to the observed 4.3 μm aurora. However, measurements employing better spectral resolution would be useful in more accurately determining the NO^+ contribution.

The zenith peak spectral radiance values measured at 2.8 and 5.3 μm are shown in Figure 4 for the auroral cases. A large enhancement is obvious at these wavelengths again for the bright auroral breakup (IBC III$^+$), although the altitude of the maximum radiation is observed at a lower altitude (84 km) than for the 4.3 μm emission. The IBC II and the IBC II$^+$ auroral arc profiles (1973 and 1974 Black Brants) show magnitudes at 5.3 μm that are essentially at the quiet background levels except for slight auroral enhancements due to the arcs observed from about 90 to 110 km. Since the spectrometer range does not allow total coverage of the NO ($\Delta v = 1$) band sequence, which may extend to over 6 μm, the spectral radiance values shown here do not give comprehensive information on the total band structure or strength. However, the measured spectral radiance values should be representative of the relative intensity of NO emissions for comparison sake.

Results under quiet conditions showed a 5.3 μm profile similar to those under auroral conditions except without the slight bulges at auroral altitudes, whereas there was no detectable emission at 2.8 μm. The principal mechanism for NO excitation under nonauroral conditions is due to earthshine and possibly O atom interchange (Degges, 1971). The O atom interchange reaction is not energetic enough to excite vibrational levels above $v = 1$ and at the collision-limited densities above 100 km most of the ambient nitric oxide is in $v = 0$; consequently, resonance fluorescence excites primarily the $v = 0 \rightarrow 1$ levels. Under auroral activity the NO is excited by nitrogen-oxygen processes that can excite both the $\Delta v = 1$ and $\Delta v = 2$ sequences. A principal precursor for these excitation processes is $N(^2D)$ which is produced by various electron and ion reactions (Reidy et al., 1974). The absence of the 2.8 μm band under nonauroral conditions coupled with the similarity of the altitude profiles of this emission to that at 5.3 μm leads to the conclusion that excited NO ($\Delta v = 2$) is responsible for the auroral enhancement at 2.8 μm.

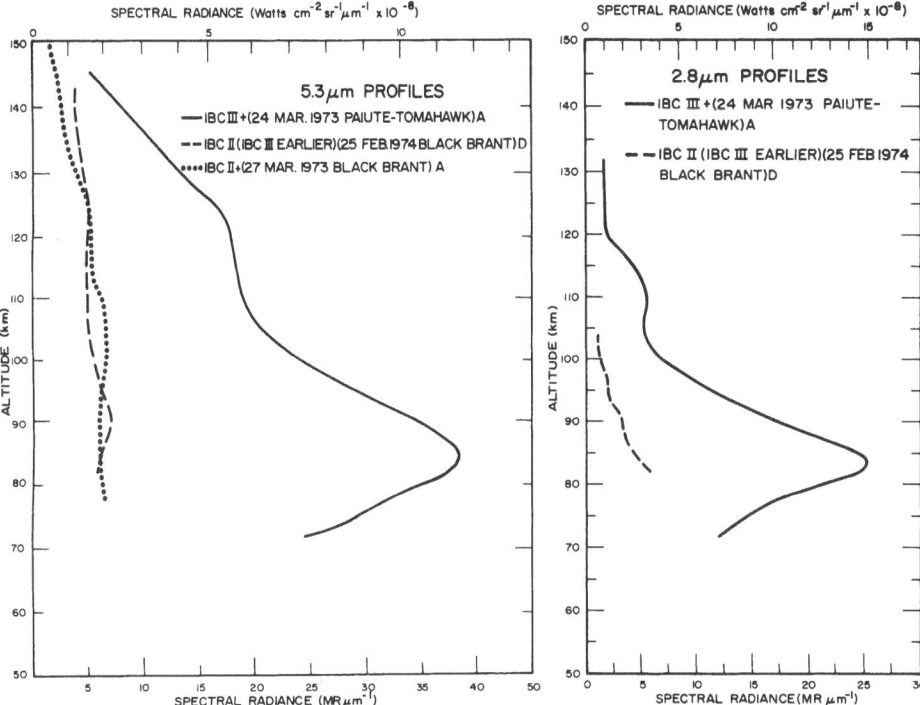

Fig. 4.   Zenith peak spectral radiance measured with SWIR spectrometers flown on several rockets under
various auroral conditions.
(a) 5.3 μm data   (b) 2.8 μm data

2.2. LWIR RESULTS

One of the measured radiance spectra (zenith) from the LWIR instrument flown
on 1973, March 22 is shown in Figure 5. The two pronounced spectral peaks at
9.6 and 15 μm are due to $O_3(v_3)$ and $CO_2(v_2)$, respectively (Stair *et al.*, 1974). This
payload penetrated a region of postbreakup auroral glow which had persisted at
the IBC II level for nearly 10 min before the rocket reached the region. The actual
intensity at the time of penetration was 13 kR of 5577 Å emission (zenith).

The results obtained on this flight illustrated the need for refinement of earlier
radiance models. The measured values at 9.6 μm are in reasonable agreement with
the models of Corbin *et al.* (1970) and Degges (1972) at altitudes below 75 km but
give values two orders of magnitude larger than these models at the higher altitudes.

Comparison between the measurements and a more recently calculated radiance
profile (Degges, 1974) utilizing the $[O_3]$ profile of Roble and Hays (1974) gives much
better agreement as shown in Figure 6. Degges results in W cm^{-2} sr^{-1} are shown
in comparison to the observed radiance in W cm^{-2} sr^{-1}. By inspection of Figure 5 one might judge to first order that the full width of the 9.6 μm feature is
about 0.3 μm. In this case the agreement between the model and the data is within

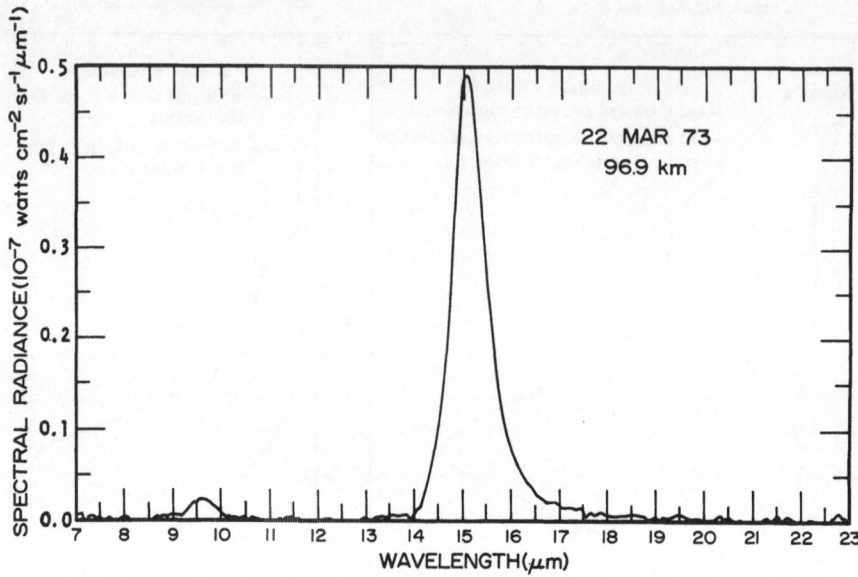

Fig. 5.   Sample spectral scans (vertical) from LWIR spectrometer aboard Black Brant rocket flown from Poker Flat, Alaska on 22 Mar 73. The rocket altitude is about 97 km on rocket descent. The spectral radiance is given in this case in 10^{-7} W cm^2 s^{-1} μm^{-1}.

a factor of five in the 60 to 100 km range. In the model the principal mechanisms for producing $O_3^{\ddagger}$ vibrationally excited in the v_3 mode are radiation transport and resonance scattering of earthshine $hv + O_3 \rightleftarrows O_3^{\ddagger}$, and thermal collisions $O_3 + M \rightleftarrows O_3^{\ddagger} + M$. Thus accurate modeling of the 9.6 μm altitude radiance profile requires accurate knowledge of the altitude profiles of T, $[M]$ and $[O_3]$. Not all of these quantities were measured in the 22 March 1973 rocket flight. Degges (1974) describes the atmospheric model he used for T and $[M]$ in computing the 9.6 μm radiance profile shown in Figure 6. Agreement between the data and predictions is good if one considers the uncertainties introduced in the calculation by the necessity to model the atmospheric parameters T, $[M]$ and $[O_3]$. The 3 body mechanism $O + O_2 + M \rightarrow O_3^{\ddagger} + M$ ($k = 3.2 \times 10^{-35}$ $e^{1.7/RT}$ cm^6 s^{-1}), which is not included in the model calculations shown in Figure 6, is too slow to contribute significantly to the emission above 80 km but does become important in the 70 to 80 km region.

Two model calculations for the CO_2 15 μm radiance profile in W cm^{-2} sr^{-1} μm^{-1} are compared with the measured results in Figure 7. These are the Degges (1974) model (D) and the results (KJ) of an independent calculation that were presented by James and Kumer (1974) and Kumer (1974) at the 1974 annual spring meeting of the American Geophysical Union. In these models the principal mechanisms for producing CO_2 vibrationally excited in the v_2 state $CO_2^{\ddagger}$ are radiation transport $hv + CO_2 \rightleftarrows CO_2^{\ddagger}$ and thermal collisions $CO_2 + M \rightleftarrows CO_2^{\ddagger} + M$. Accurate modeling requires a knowledge of the altitude profile of the atmospheric parameters T, $[M]$ and $[CO_2]$. These were modeled in the 2 cases as described by Degges (1974) and Kumer (1974). Differences in the D and KJ calculations largely reflect the differences in the atmospheric models employed in these 2 cases. The CO_2 mixing ratio in the D model remains constant at essentially 3.1×10^{-4} above 100 km whereas the KJ model utilizes the smaller CO_2 mixing ratio above 100 km which was introduced by James and Kumer (1973). Neither model accounts for the enhanced scattering of earthshine in the region above 100 km which should result from the CO_2 band and line broadening which should occur due to the increased temperature in the thermosphere. These models are in excellent agreement at the lower altitudes with the measurements but exhibit discrepancies increasing with altitude.

Indirect evidence supports the claim that the experimental measurements are accurate to within a factor of 2. First of all, upon extrapolation the rocket data joins smoothly with balloon measurements (Murcray et al., 1972; Stair et al., 1974). Secondly, satellite measurements tend to support the low altitude data (Conrath et al., 1970; Barnett et al., 1973). Finally, the 1974 rocket measurements under ambient conditions give roughly the same intensity as under auroral conditions (Stair et al., 1974). This was expected since neither band is predicted to be enhanced by aurora to any significant extent (Bishop et al., 1973).

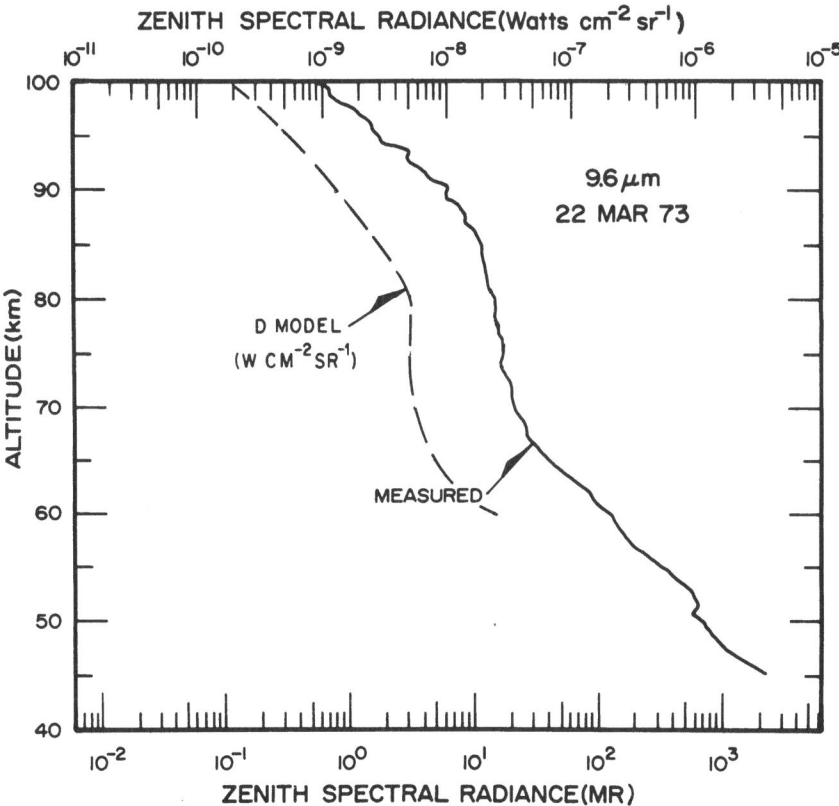

Fig. 6.   Zenith peak spectral radiance at 9.6 μm measured with LWIR spectrometer aboard rocket flown at Poker Flat, Alaska on 22 Mar 73 compared to theoretical model. The solid curve is the measured profile while the dashed curve shows the D model calculations in W cm^{-2} sr^{-1}.

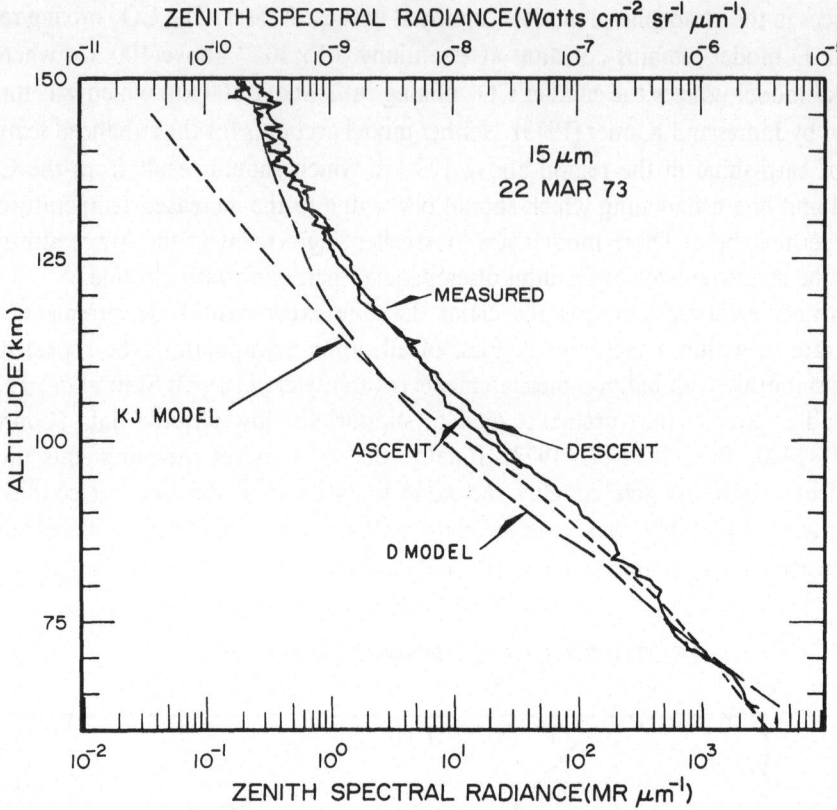

Fig. 7.   Zenith peak spectral radiance at 15 μm measured with LWIR spectrometer aboard rocket flown at Poker Flat, Alaska on 22 Mar 73 compared to theoretical model. The solid curve is the measured profiles (ascent and descent) while the dashed curves show the model calculations of Kumer and James (KJ) and Corbin and Degges (D).

3. Summary of Results

Infrared spectra from 1.6 to 23 μm have been measured under a variety of auroral conditions ranging from aurorally quiet to intense auroral breakup. The main IR emission features originating from NO, CO_2, and O_3 show varying degrees of enhancement associated with the aurora. The general features are summarized in Table II. The most significant auroral enhancement observed occurred in the 4.3 μm band of CO_2. The strength of this band was observed to be enhanced above background by an amount that depended on the total energy deposited within the region some minutes prior to the observation. The extreme case of the intense auroral breakup showed an enhancement of nearly two orders of magnitude over the non-disturbed case.

Whether or not enhancement of the long-wave emissions occurs in aurora is not clear at this time. Results of one observation in a rather long duration IBC II auroral glow did not produce marked evidence for enhanced emissions.

TABLE II

Summary of primary infrared emissions observed in auroral zone flights

Wavelength (μm)	Species	Maximum altitude of observed emission (km)	Auroral association
2.8	NO $(\Delta v = 2)$ OH $(\Delta v = 1)$?	130	Moderate enhancement in bright aurora (IBC III)
4.3	CO_2 (v_3) NO$^+$ $(\Delta v = 1)$?	130	Strong enhancement in bright aurora (IBC III)
5.3	NO $(\Delta v = 1)$	150	Moderate enhancement in bright aurora (IBC III) Slight enhancement in (IBC II) auroral arcs
9.6	O_3 (v_3)	110	No marked enhancement in IBC II
15	CO_2 (v_2)	150	No marked enhancement in IBC II

Acknowledgments

The successful measurements program which provided the results described herein are the culmination of the efforts of a large team of dedicated professionals conquering what appeared to be innumerable insurmountable obstacles. The staffs of Air Force Cambridge Research Laboratories, Defense Nuclear Agency, Utah State University, University of Alaska, Northeastern University, Oklahoma State University, Space Date Corporation, NASA Wallops, Stanford Research Institute, White Sands Missile Range/MEWTA, and Lockheed Palo Alto Research Laboratory, contributed heavily to the success of the program. Although by no means an exhaustive list, the authors would like to specifically acknowledge and thank the dedicated efforts of their colleagues who played particularly key roles in the program: Clair Wyatt and David Burt who were the principal creators of the rocket instruments and payloads, and Glenn Allred, Tom Condron, Gary Frodsham, Bill Grieder, Ron Huppi, John Kemp, Ed McKenna, Herb Mitchell, James Rogers, Ned Wheeler, and Raymond Wilton. This program was sponsored by the Defense Nuclear Agency under Sub-Task L25AAXHX632, Work Unit 20.

References

Barnett, J. J., Houghton, J. T., Morgan, C. G., Pick, D. R., Rogers, C. D., Williamson, E. J., Cross, M. J., Flower, D., Peckham, G., and Smith, S. D.: 1973, *Nature* **245**, 141.

Baron, M. and Chang, N. J.: 1974, 'ICECAP 73A – Chatanika Radar Results', Technical Report 4, 130 pp., Contract DNA 001-74-C-0167, Stanford Research Institute, Menlo Park, Calif.

Bishop, R. H., Han, R. Y., Shaw, A. W., and Megill, L. R.: 1973, 'Infrared Radiance Model for the Aurorally Disturbed Atmosphere', Final Report, AFCRL-TR-73-0527, 138 pp., Contract No. F19628-71-C-0257, Utah State University, Logan.

Burt, D. A. and Davis, C. S.: 1974, 'Rocket Instrumentation for ICECAP 73A Auroral Measurements Program – Black Brant 18.205-1', SSL Sci. Rep. No. 3, HAES Report No. 3, AFCRL-TR-74-0195, 147 pp., Contract No. F19628-72-C-0255, Utah State University, Logan.

Conrath, B. J., Hanel, R. A., Kunde, V. G., and Prabhakara, C.: 1970, *J. Geophys. Res.* **75**, 5831.

Corbin, V. L., Dalgarno, A., Degges, T. C., House, F. B., Lilienfeld, P., Ohring, G., and Oppel, G. E.: 1970, 'Atmospheric Radiance Models for Limbviewing Geometry in the 5–25 μ Spectral Region', Honeywell, Inc. Sci. Rep.No. 1, AFCRL Contract F19628-69-C-0268.

Degges, T. C.: 1971, *Appl. Opt.* **10**, 1856.

Degges, T. C.: 1972, 'A High Altitude Radiance Model', AFCRL-72-0273, Visidyne Corp., Burlington, *Mass.*

Degges, T. C.: 1974, 'A High Altitude Radiance Model AFCRL-74-0606, Visidyne Corp., Burlington, Mass.

James, T. C. and Kumer, J. B.: 1973, *J. Geophys Res.* **78**, 8320.

James, T. C. and Kumer, J. B.: 1974, *EOS* **55**, 375.

Jensen, L. L., Kemp, J. C., and Bell, R. J.: 1972, 'Small Rocket Instrumentation for Measurements of Infrared emissions – Astrobee D 30.205-3 and Astrobee D 30.205-4', SSL Sci. Rep. No. 3, AFCRL 72-0691, 89 pp., Contract No. F19628-70-C-0302, Utah State University, Logan.

Kumer, J. B.: 1974, *EOS* **55**, 375.

Kumer, J. B.: 1975, this volume, p. 347.

Kumer, J. B. and James, T. C.: 1974, *J. Geophys. Res.* **79**, 638.

Murcray, D. G., Barker, D. B., Brooks, J. M., Kosters, J. J., Murcray, F. H., and Williams, W. J.: 1972, 'Atmospheric Emission at High Latitudes', Final Report, Contract No. F19628-71-C-0171, University of Denver, Denver, Colo.

Reidy, W. P., Degges, T. C., Manley, O. P., Smith, H. J., Carpenter, J. W., Stair, A. T., Ulwick, J. C., and Baker, K. D.: 1974, 'Analysis of HAES Results: ICECAP 72 – HAES Report No. 2', Final Report, DNA 3247F, 232 pp., Contract No. DNA 001-73-0020.

Roble, R. G. and Hayes, P. B.: 1974, *Planetary Space Sci.* **22**, 1974.

Stair, A. T., Jr., Ulwick, J. C., Baker, D. J., Wyatt, C. L., and Baker, K. D.: 1974, *Geophys. Res. Letters* **1**, 117.

Stair, A. T., Jr., Wheeler, N. B., Baker, D. J., and Wyatt, C. L.: 1973, 'Cryogenic IR Spectrometers for Rocketborne Measurements', IEEE/NEREM 1973 Record Part 3 Infrared.

Stolarski, R. S.: 1968, *Planetary Space Sci.* **16**, 1265.

Wyatt, C. L.: 1971, 'Infrared Helium-Cooled Circular-Variable Spectrometer, Model HS-1', Final Report, AFCRL-71-0340, Contract No. F19628-67-C-0322, Utah State University, Logan.

Wyatt, C. L.: 1975, *Appl. Opt.* submitted.

SUMMARY ANALYSIS OF 4.3 μm DATA

J. B. KUMER

Lockheed Palo Alto Research Laboratory, 3251 Hanover Street,
Palo Alto, Calif. 94304 U.S.A.

1. Introduction

A preliminary analysis of zenith radiance data at 4.3 μm obtained as part of the DNA/AFCRL ICE CAP auroral measurements program at Poker Flats, Alaska (65° N, 147.5° W) from the following three rocket flights is presented.

Code	Rocket	Launch Date/Time (UT)
AD72	Astrobee D	March 9, 1972; 1052:19
PT73	Paiute	March 24, 1973; 1031:42
BB73	Black Brandt	March 27, 1973; 0937:45

Zenith spectral radiance data were obtained in all three shots via a circular variable filter (CVF) in the approximate wavelength range 2 μm $\lesssim \lambda \lesssim$ 5 μm. The CVF zenith radiance data we discuss in this presentation were obtained by integrating over the emission feature observed near 4.3 μm as discussed by Baker *et al.* (1973). Zenith radiance radiometer 4.3 μm data were also obtained on the BB73 rocket flight.

Energy input data were also obtained for these 3 auroras in the coordinated ICE CAP auroral measurements program. The energy input measurements included, for example, ground based optical measurements by photometers, scanning photometers, all-sky cameras and television, electron density measurements by the Chatanika radar, and radar and optical inferences on wind data as reported by Sears and Evans (1975). Rocket borne instrumentation such as photometers, electron spectra analyzers, and others also obtained energy input data.

Here we report our work in which we relate the spatial and temporal history of the energy input to the measured 4.3 μm zenith radiance in the three auroras. The three auroras were quite dissimilar in the intensity, temporal history and spatial extent of the energy input. We were able, however, to consistently relate the energy input to the 4.3 μm output on the basis of a mechanism involving vibrationally excited N_2 and CO_2 [$N_2^+ \rightleftarrows CO_2(001)$] that will be discussed in detail in Section 3 below. Kumer (1974) presents a detailed analysis of the three auroras. In this presentation we shall discuss in detail just the PT73 flight since we believe the comparison of the data with our analysis based on the $N_2^+ \rightleftarrows CO_2(001)$ mechanism is best illustrated by the March 24, 1973 aurora.

B. M. McCormac (ed.), Atmospheres of Earth and the Planets, 347–358. All Rights Reserved.
Copyright © 1975 by D. Reidel Publishing Company, Dordrecht-Holland.

2. Qualitative Analysis

The PT73 flight exemplifies most clearly the qualitative characteristics of the 4.3 μm aurora. The up and downleg 4.3 μm and 3914 Å vertical looking radiance altitude profiles for the PT73 flight are presented in Figure 1.

The 4.3 μm up and downleg are essentially identical above 90 km, but the 3914 Å up and downleg radiance altitudes differ markedly from one another. The 3914 Å channel was saturated with signal > 200 kR below about 100 km on the upleg and had a maximum of $\simeq$ 45 kR on the downleg. Furthermore, the 4.3 μm radiance profiles show an *increase* in radiance in going from 80 to 90 km on both legs which is certainly not reflected in the unsaturated 3914 Å downleg profile. Finally, the slopes in the 3914 Å profiles above 90 km differ a great deal from those of the 4.3 μm profiles.

To dispel any doubts that the PT73 4.3 μm data are of auroral origin, we also show

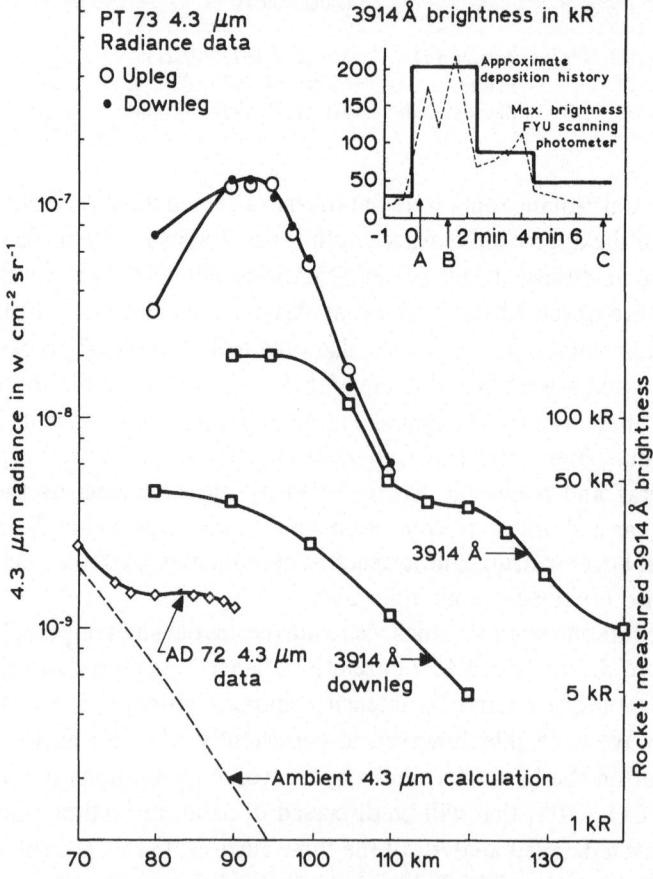

Fig. 1. The 3914 Å and 4.3 μm radiance altitude profiles for the PT73 flight. The solid curves through the 4.3 μm data from PT73 are an eyeball fit to the data. A calculation for the undisturbed 4.3 μm radiance altitude profile as well as 4.3 μm radiance obtained in a weaker aurora AD72 are also shown. The insert is the ground based scanning photometer maximum brightness. The points *A*, *B* and *C* are launch time and the times of rocket up and downleg penetration, respectively.

in Figure 1 a March model atmosphere 4.3 μm ambient radiance calculation in which thermal collisions and the scattering of earthshine are the excitation mechanisms for producing $CO_2(001)$ which can then radiate in the vibration rotation band $001 \rightarrow 000$ with band origin near 4.26 μm. We also show 4.3 μm radiance data from AD72 in a weak ($\simeq 5$ kR 3914 Å) aurora. One sees that the 4.3 μm radiance data obtained from PT73 in the strong aurora are much brighter than in the ambient calculation or that obtained in the weak aurora from AD72, thus we conclude the PT73 data are indeed auroral phenomena; however, with temporal and spatial (vertical) characteristics that are much different from those of the visual (3914 Å) aurora.

The insert in the upper right hand corner of Figure 1 shows the ground-based 3914 Å deposition history of the maximum brightness reading of the FYU (Fort Yukon) meridian scanning photometer (perforated line) (other ground-based visible observations were degraded by clouds). The points A, B and C on the insert time scale are the approximate times of rocket launch and of rocket penetration at 80 km on the up and downlegs respectively. We have adopted the square wave labeled 'approximate deposition history' to represent the deposition history for both the upleg and the downleg. This approximation agrees in general with the scanning photometer results and with the onboard rocket 3914 Å measurements at points B and C, and with all-sky camera montages for the PT73 flight. By comparing the time histories of the 3914 Å data with the 4.3 μm data, we conclude that the time constant for the duration of 4.3 μm emission due to particle deposition must be of the order of more than 5 min.

We believe that both the spatial and temporal differences in the behavior of the 4.3 μm vs. the 3914 Å data argue against the emission of 4.3 μm radiation by NO^+ formed in a vibrationally excited state NO^{+*} which may then radiate with fundamental band origin near 4.3 μm. Since the NO^+ emission lines are not coincident with the atmospheric CO_2 absorption lines and since the ratio of CO_2 line spacing to upper atmospheric Doppler e-fold line width is of the order of 680, the atmosphere is optically thin above 80 km for NO^+ 4.3 μm radiation. Therefore, we would expect the 4.3. μm radiance altitude profile due to this NO^+ emission mechanism to decrease monotonically as a function of altitude z in the same way the radiance profile for the optically thin 3914 Å emission behaves. This is in stark contrast with the 4.3 μm radiance data which increases with altitude in the 80 to 90 km region. Furthermore, although the time constant required to explain the 4.3 μm data must be $\gtrsim 5$ min one can show that the time constant τ which is of the order of $\tau \bar{\simeq} (\alpha n_e)^{-1}$ for the formation of NO^+ in this intense PT73 aurora is less than $\simeq 10$ s in the altitude range $90 \lesssim z \lesssim$ $\lesssim 120$ km. This may be derived from Chatanika radar electron density n_e measurements and the value for $\alpha(z)$ the recombination coefficient as may be inferred from Biondi (1969).

In summary, most of the ionization formed by deposition eventually (via a chain of chemical reactions which occurs on a time scale $\tau \lesssim 10$ s for this intense PT73 aurora) results in the formation of NO^+. The formation of this NO^+ in vibrationally excited states (i.e. NO^{+*}), followed by emission at 4.3 μm via radiative decay of the

NO^{++}, is probably not responsible for the bulk of the observed 4.3 μm radiance since it would then produce a volume emission which is approximately proportional to the deposition (and 3914 Å volume emission) and would be essentially a prompt emission (10 s is essentially prompt considering the time scales involved in the problem). Thus, by this hypothesis, if NO^+ is formed vibrationally hot followed by emission at 4.3 μm, we would expect the observed 4.3 μm radiance profiles to have essentially the same spatial and temporal dependence as the prompt 3914 Å radiance, but the data in Figure 1 show this is definitely not the case.

3. The $N_2^{\ddagger} \rightleftarrows Co_2$ (001) Mechanism

A second mechanism for the production of 4.3 μm auroral emission involves auroral excitation of vibrationally excited nitrogen $N_2^{\ddagger}$. The basic processes involved in this mechanism are:

$$P(N_2^{\ddagger}) = \varepsilon q(z, t), \tag{1}$$

$$N_2^{\ddagger} + CO_2 \rightleftarrows N_2 + CO_2(001); \ k_2, \tag{2}$$

$$CO_2(001) \rightleftarrows CO_2 + h\nu, \tag{3}$$

$$N_2^{\ddagger} + O \rightleftarrows N_2 + O; \ k_{10}, \tag{4}$$

$$N_2^{\ddagger} + O_2 \rightleftarrows N_2 + O_2^{\ddagger}; \ k_{e11}, \tag{5}$$

and

$$CO_2(001) + M \rightleftarrows CO_2(v_1 v_2) + M; \ k_1. \tag{6}$$

The selection and notation for the rate coefficients that are used in the calculation are discussed by Kumer and James (1974). The rate constant values we used for the calculation were somewhat modified from those used in the calculation reported by Kumer and James in accord with the latest paper by Taylor (1974) which deals with temperature dependent approximations for the rate constants. The reaction (1) expresses the basic assumption that $N_2^{\ddagger}$ is produced in proportion to energy deposit $q(z, t)$ measured in ionizations $cm^{-3} s^{-1}$. Reaction (2) is a resonance vibration transfer reaction since the spacing of the N_2 vibrational level is approximately $(4.3 \ \mu m)^{-1}$. Reaction (3) denotes multiple scattering of photons in the CO_2 001→000 band. Reactions (4) and (5) are quenching reactions of $N_2^{\ddagger}$ and reaction (6) is the quenching reaction for $CO_2(001)$. This set of reactions and other reactions important for this problem are discussed in more detail by Kumer and James.

It is necessary to calculate $[001]_A$ the number density in units cm^{-3} of $CO_2(001)$ in order that we may calculate the 4.3 μm auroral radiance. The auroral time dependent radiation transport problem involving solution for $[N_2^{\ddagger}]_A$ and $[001]_A$ may be formulated as follows

$$\frac{d[001]_A}{dt} = k_2[CO_2][N_2^+]_A - k_2[N_2][001]_A - k_1[M][001]_A$$

$$- [001]_A A_{TR} + A_{4.26} \int \sigma[CO_2]' \, dz' H(\sigma|N' - N|)[001]'_A. \quad (7)$$

$$\frac{d[N_2^+]_A}{dt} = \varepsilon q(z, t) + k_2[N_2][001]_A - k_2[CO_2][N_2^+]_A -$$

$$- k_{10}[O][N_2^+]_A - k_{e11}[O_2][N_2^+]_A. \quad (8)$$

In this discussion we strictly follow the notation developed by Kumer and James. We also assume the reader is familiar with the analysis and terminology employed in that paper.

If one employs $J+1$ altitude mesh points, Equations (7) and (8) represent $2(J+1)$ coupled linear differential equations. These present a formidable numerical problem, and although the problem is solvable, the scope and time scale of our effort did not permit us to obtain the exact solution of Equations (7) and (8). Fortunately, approximate solutions may be obtained in the escape function approximation.

In the escape function approximation, Equation (7) may be rewritten

$$\frac{d[001]_A}{dt} \cong k_2[CO_2][N_2^+]_A - k_2[N_2][001]_A - k_1[M][001]_A -$$

$$- [001]_A AE_T, \quad (9)$$

where $AE_T = A_{4.26}E(z) + \frac{1}{2}(A_{10.4} + A_{9.4})$, where $E(z)$ is the probability the photons in the $001 \rightarrow 000$ band will escape to space from altitude z, and $A_{10.4} \simeq 0.357$ s^{-1} and $A_{9.4} \simeq 0.373$ s^{-1} are the Einstein coefficients for radiative decay in the fluorescent bands $001 \rightarrow 100$ and $001 \rightarrow 020$. One notes that the factor $\frac{1}{2}$ accounts in the escape function approximation for integral transport terms for the fluorescent bands that were neglected in Equation (7).

Now Equations (8) and (9) represent 2 coupled equations for $[N_2^+]_A$ and $[001]_A$ at each altitude mesh point z_j. This simplification in the mathematical formulation of the problem is gained at the expense of some physical reality, spatial photon diffusion.

Solutions for coupled linear, first order, inhomogeneous Equations (8) and (9) may be obtained via methods described by Frazer *et al.* (1938). In the notation of Frazer *et al.* Equations (8) and (9) take the matrice form

$$f(D) \, y = (ID - u) \, y = \eta, \quad (10)$$

where

$$ID = \begin{pmatrix} 1 & 0 \\ 0 & 1 \end{pmatrix} \frac{d}{dt}, \quad (11)$$

and

$$y = \begin{pmatrix} [N_2^+]_A \\ [001]_A \end{pmatrix}, \quad (12)$$

and

$$\eta = \begin{pmatrix} \varepsilon q(z, t) \\ 0 \end{pmatrix}, \tag{13}$$

and

$$u_{11} = -k_2[CO_2] - k_{10}[O] - k_{e11}[O_2],$$
$$u_{12} = k_2[N_2],$$
$$u_{21} = k_2[CO_2],$$
$$u_{22} = -k_2[N_2] - k_1[M] - AE_T.$$

From the discussion in Frazer *et al.*, one sees that the solution for Equation (10) and for boundary conditions $t_0 = -\infty$ and $y(t_0) = 0$ is given by

$$y = \sum_{r=1}^{2} y^r, \tag{14}$$

where

$$y^r = \frac{F(\lambda_r)}{\Delta^{(1)}(\lambda_r)} \int_{-\infty}^{t} e^{-\lambda_r(t'-t)} \, dt' \eta(t'),$$

where the λ_r are solutions of the determinental equation

$$|I\lambda - u| = \lambda^2 - (u_{11} + u_{22})\lambda + u_{11}u_{22} - u_{21}u_{12} = 0. \tag{15}$$

The solutions are

$$\lambda = (u_{11} + u_{22})/2 \mp \sqrt{[(u_{11} + u_{22})/2]^2 + u_{11}u_{22} - u_{12}u_{21}}, \tag{16}$$

and the r index on the λ_r is 1 or 2 for a $-$ sign or $+$ sign occurring in Equation (16) respectively. The reader may verify that

$$|\lambda_1| \simeq \left(\frac{k_2[N_2]}{AE_T + Q_T} + 1\right)^{-1} k_2[CO_2] + k_{10}[O] + k_{e11}[O_2]. \tag{17}$$

The remaining quantities in Equation (14) are given by:

$$\Delta^{(1)}(\lambda_r) = \frac{\partial}{\partial \lambda_r} |I\lambda_r - u|, \tag{18}$$

and

$$F(\lambda_r) = -(I\lambda_p - u); \quad p \neq r. \tag{19}$$

Inspection of the numerical solution shows $|y^1| \gg |y^2|$ by orders of magnitude at all altitudes, hence $1/|\lambda_1|$ is the dominant decay time. The quantity $\tau(z) = (1/|\lambda_1|)\left(\frac{1}{60}\right)$ is then the lifetime (in minutes) in the escape function approximation for an N_2 vibration quantum produced at altitude z. The quantities $\tau(z)$, $A_{4.26}E(z)$, $Q_T(z)$ and

$k_2[N_2]$ (z) are plotted versus altitude in Figure 2. The maximum in τ which occurs when $Q_T = A_{4.26}E$ and the minimum in τ occurring when $A_{4.26}E = k_2[N_2]$ are easily understood by inspection of Equation (17). The maximum near $z \simeq 115$ km is a result of the increasing value for k_{10} with temperature in the $z \simeq 100$ to 120 km region. Of course, $\tau(z)$ will begin to increase again in isothermal regions of the atmosphere at higher altitudes.

Billingsley (1973) has calculated $A \cong 14$ and 26 s^{-1} for the NO$^+$ $(1 \rightarrow 0)$ and $(2 \rightarrow 1)$ bands. One notes from inspection of Figure 2 that if the rate constant k^* for the resonance vibration transfer reaction

$$N_2^* + NO^+ \rightleftarrows N_2 + NO^{+*}, \tag{20}$$

were 2 orders of magnitude larger than k_2, then the absence of NO$^+$ emission near 4.3 μm in the data might well be explained since in this case the N$_2$ would quench the NO^{+*} before it could radiate in the altitude region $z \lesssim 120$ km. From Table I one sees that the NO^{+*} will be formed predominantly in the region $z \lesssim 120$ km in the PT73

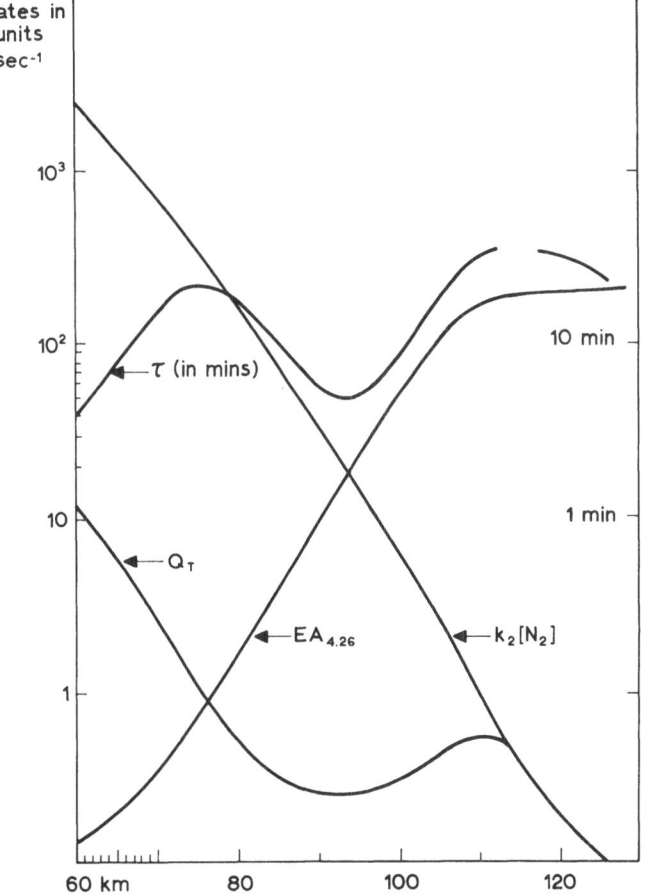

Fig. 2. The quantities τ, $EA_{4.26}$, Q_T and $k_2[N_2]$ as described in the text are plotted vs. altitude.

aurora. This hypothesis calls for $k^* \simeq 10^2 k_2 \simeq 5 \times 10^{-11}$ cm^3 s^{-1}. This may not be too far-fetched since one might expect a much stronger interaction between ion and neutral [reaction (20)] than between neutral and neutral [reaction (2)]. Furthermore the NO^{++} may be formed with more than ambient translational energy which would also tend to increase k^*. A measurement or calculation for k^* would be very useful for assessing this hypothesis.

In order to get $q(z, t)$ for the PT73 flight, we used electron density data n_e obtained from the SRI Chatanika radar data appropriate for upleg penetration and the re-combination coefficient α (Biondi, 1969). The resultant values for $\alpha n_e^2 = C^{-1} q(z)$ are tabulated below in Table I. The value $c = 1.4$ was found by requiring $\int dz c \alpha n_e^2 =$ $= 4 \times 10^{12}$ cm^{-2} s^{-1} in accordance with the 200 kR 3914 Å observed on the upleg.

The approximate deposition history depicted in the upper right hand corner insert in Figure 1 provides us with $g(t)$ for both the upleg and downleg if we assume both up and downleg penetration points had approximately similar deposition histories; this point was discussed in Section 2 above. Now, with this value for $g(t)$ and with $q(z) = c \alpha n_e^2$, we get $q(z, t) = g(t) q(z)$.

TABLE I

Pertinent data for the calculation of αn_e^2

z(km)	$\alpha(10^{-7}$ cm^3 s$^{-1})$	$n_e(10^5$ cm$^{-3})$	$\alpha n_e^2(10^5$ cm^{-3} s$^{-1})$
80	8.5	2.0	0.34
90	7.5	10.0	7.50
95	7.0	16.0	17.9
100	6.2	14.0	12.2
105	5.6	10.0	5.6
112	5.0	8.0	3.6
118	3.6	6.0	1.3
125	2.9	4.0	0.5
145	1.65	2.0	0.07

In order to approximately allow for spatial diffusion, we can use the stationary escape function approximation to obtain

$$R = (A_{TR} + Q_T) [001]_A (1 - \Omega[1 - E]). \tag{21}$$

The reader may verify that

$$R \simeq \varepsilon q(z) f \int_\infty^0 dt' \tau^{-1} e^{t'\tau^{-1}} g(t'),$$

where the upper limit 0 in the integration refers to points B and C on the insert shown in Figure 1 for up and downlegs, respectively, and $f \simeq k_2 [CO_2]/(k_2 [CO_2] + k_{10} [O])$.

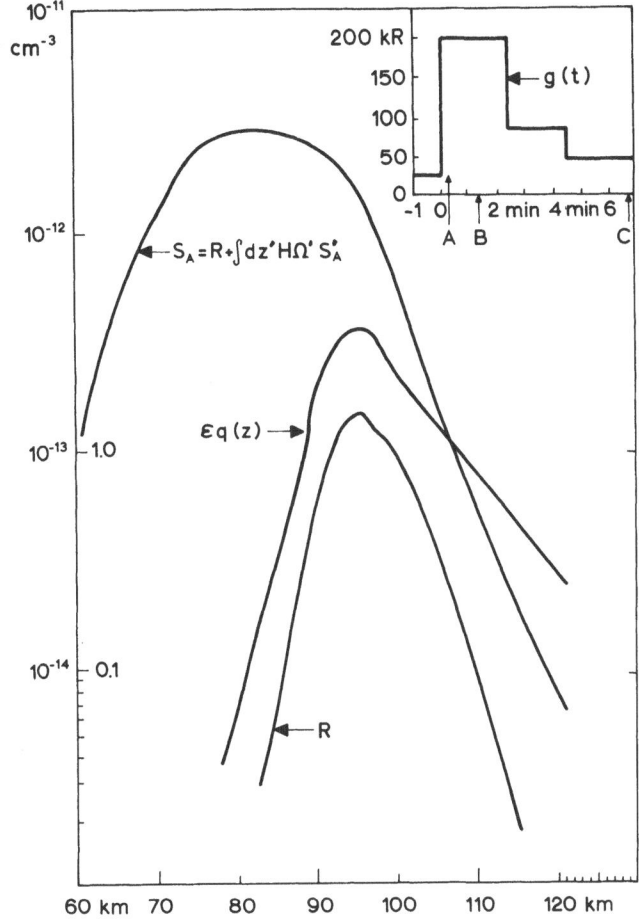

Fig. 3.   The quantities $\varepsilon q(z)$, R and S_A for the PT73 aurora are plotted vs. altitude. The insert shows our model for $g(t)$.

Then we may solve the stationary transport equation

$$S_A(z) = R(z) + \int dz' H(z, z')\, \Omega(z')\, S_A(z').$$
(22)

The quantities $S_A(z)$, $\varepsilon q(z)$ and $R(z)$ are shown in Figure 3.

4. Comparison of Theory and Data for the PT73 Aurora

From the solution $S_A(z)$ of the transport equation, we may obtain the auroral radiance I_A via

$$4\pi I_A(z) = \int_z^\infty dz' T(z, z')\, \Omega(z')\, S_A(z')),$$
(23)

where the transmission function $T(z, z')$ is the fraction of the emission at z' directed towards the rocket-borne detector at z that is transmitted through the ambient absorbing CO_2. The transmission function is discussed in detail by Kumer and James (1974). The excellent quality of the fit is shown in Figure 4. The consistency between up and downleg results is remarkable with proportionality constant $\varepsilon = 16.3$ and 14.4 required on the up and downlegs, respectively. We believe that the relatively

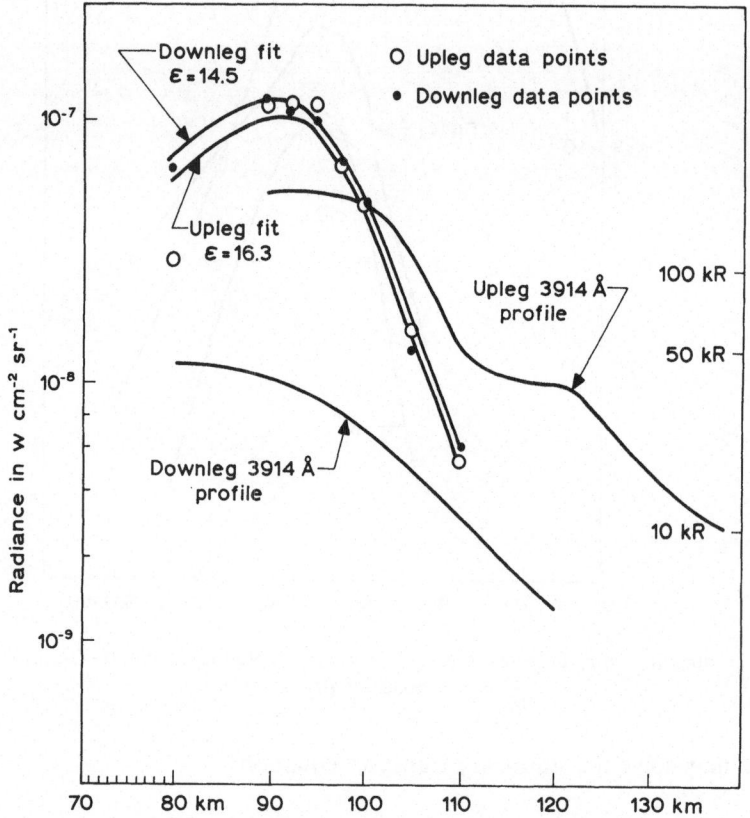

Fig. 4. The result of fitting $I_A(z)$ to the upleg and downleg 4.3 μm radiance data for the PT73 aurora is shown. The 3914 Å radiance profiles are also shown to emphasize the dissimilarity in the spatial and temporal nature of the 3914 Å and the 4.3 μm auroras.

poor theoretical fit to the upleg data at 80 km is at least partially the result of using the modified escape function approximation via Equations (8), (9), (20), and (21) to obtain $I_A(z)$ via Equation (22) as opposed to solving the more accurate time dependent radiation transport Equations (7) and (8) in order to obtain $S_A(z)$. The 3914 Å profiles are shown again in Figure 4 to re-emphasize the dissimilarity in the spatial and temporal characteristics of the 3914 Å and 4.3 μm radiance profiles.

5. Other Auroral Data

Analysis of the data taken from the AD72 and BB73 auroras is reported by Kumer (1974). We briefly report the highlights of this analysis.

Analysis of the AD72 data indicates that the $N_2^+ \rightleftarrows CO_2(001)$ mechanism is dominant over the NO^{+*} mechanism in producing the observed 4.3 μm aurora and that uncertainties in the energy input model may cause a factor 2 uncertainty in the value for ε that is required to fit the data. Values for ε which fit the data for this aurora are $3.5 < \varepsilon < 6.4$.

Analysis of the BB73 data again showed the $N_2^+ \rightleftarrows CO_2(001)$ mechanism dominant. This aurora also pointed up that transport of N_2^+ by winds, lateral radiative transport, and a more exact treatment of the time dependence via solution of Equations (7) and (8) are necessary to gain satisfactory understanding of the BB73 aurora.

6. Conclusions

Our main conclusion from the analysis of the three flights cases cited above is that our 4.3 μm predictions on the basis of the $N_2^+ \rightleftarrows CO_2(001)$ mechanism with the N_2^+ formed proportional to 3914 Å volume emission are in good qualitative agreement with the spatial and temporal features of the data in all cases. This is in striking contrast with the NO^{+*} 4.3 μm radiance mechanism which was shown to be unsatisfactory, qualitatively, in explaining the spatial and temporal features of the 4.3 μm data. It is intriguing that our theory for N_2^+ resident time $\tau(z)$ and the data for the case of the BB73 downleg indicate that transport of N_2^+ by wind plays a crucial role in forming the structure of the 4.3 μm aurora.

We obtained values for ε in the range $3 \lesssim \varepsilon \lesssim 18$ via analysis of the data. This large range in ε is due in part to a factor 3 in uncertainty in the absolute calibration of the IR instrumentation. The nominal calibration of the absolute calibration of instrumentation is uncertain within a factor f such that $0.5 \lesssim f \lesssim 1.5$. Uncertainty in the input models for $q(z, t)$ also contributes about a factor 2 to the uncertainty in ε.

Acknowledgments

I would especially like to thank D. M. Hunten, T. M. Donahue, F. Kaufman, F. Gilmore, and C. B. Leovy for their comments and questions which considerably enhanced the value of this presentation at the Liège summer study. This work was supported partly by the AFCRL Contract F19628-73-C-0288 under the DNA ICE CAP program and partly by Lockheed independent research. The success of this work is largely due to the generous attitude amongst other participating scientists in providing data and useful suggestions. For this, I would like to thank A. T. Stair, R. D. Sears, T. C. James, W. R. Pendleton, R. Hegblom, I. Kofsky, M. Baron, B. M. McCormac, R. E. Meyerott and others.

References

Baker, D. J., Kemp, E. C., Bruce, M., and Ulwick, J. C.: 1973, AGU Washington, D.C. April Meeting.

Biondi, M.: 1969, *Can. J. Chem.* **47**, 10.

Frazer, R. A., Duncan, W. J., and Collar, A. R.: 1938, *Elementary Matrices*, University Press, Cambridge.

Kumer, J. B.: 1974, 'Analysis of 4.3 μm ICE CAP Data', Draft Report on Contract F19628-73-C-0288, July.

Kumer, J. B. and James, T. C.: 1974, *J. Geophys. Res.* **79**, 638.

Sears, R. D. and Evans, John, E.: 1975, this volume, p. 125.

Taylor, R. L.: 1974, *Can. J. Chem.*, in press.

PART VII

ATMOSPHERES OF OTHER PLANETS

THE ATMOSPHERES OF MARS AND VENUS – A COMPARISON

JOHN C. McCONNELL

Physics Dept. and CRESS, York University Downsview M3J 1P3, Ont., Canada

1. Introduction

In this short review I will outline briefly the main areas of current interest for CO_2 aeronomy, i.e., the aeronomy of the atmospheres of Mars and Venus. The discussion will be an attempt to distill the pertinent information from the many recent reviews containing information on CO_2 atmospheres (e.g. Hunten, 1971, 1974; Marov, 1972; McConnell, 1973; Prinn, 1973a, b; Whitten and Colin, 1974). In addition I have included a summary of the most recent results (at the date of this meeting) which have been stimulated by the recent fly-by of Mariner 10 past Venus (Broadfoot *et al.*, 1974).

On Venus the discussion will be limited to the region above the cloud tops.

2. The Lower Atmosphere

Table I lists the mixing ratios of gases on Mars and Venus on which most attention has focused in recent years. The major component of both atmospheres is CO_2 which is photodissociated right to the ground on Mars and to the cloud tops on Venus. Thus the low abundance of CO and O_2, noted in Table I, in the atmospheres presents a puzzling problem. On Mars it has been noted that there is a correlation between high H_2O mixing ratios and low O_3 densities (Barth and Leovy, 1974). At present COS has not been detected, however its abundance is an essential parameter in our understanding of the composition of the Venus clouds (e.g. Prinn, 1973a, b; Wofsy and Sze, 1975).

As noted above the stability of a CO_2 atmosphere is a major problem. At present the concensus is that CO_2 recombination in the atmospheres of Mars and Venus is catalysed by odd hydrogen $HO_x = H + OH + HO_2 + H_2O_2$ (McElroy and Donahue, 1972; Parkinson and Hunten, 1972; McElroy *et al.*, 1973; Liu and Donahue, 1974; Sze and McElroy, 1974). The major reaction reforming CO_2 from the initial photolysis products CO and O is

$$CO + OH \rightarrow CO_2 + H. \tag{1}$$

The reaction of CO and O to form CO_2 is much too slow to be important. The catalysis is completed, i.e., the OH reformed, via two major paths

$$H + O_2 + M \rightarrow HO_2 + M \tag{2}$$
$$HO_2 + O \rightarrow OH + O_2 \tag{3}$$

(I)

and

$$2(H + O_2 + M \rightarrow HO_2 + M) \tag{2}$$

TABLE I

Mixing ratios*

Constituent	Mars	Venus
CO_2	0.75–1.0	0.95–1.0
CO	8(−4)	5(−5)
O_2	1(−3)	<1(−6) (a)
H_2O	∼1(−3) V	∼1(−6) (b)
O_3	2–60 μ atm V (d)	<2 μ atm (c)
HCl	<1(−7)	6(−7)
COS	<6(−7)	<1(−7)

* Mixing ratios above cloud tops from McConnell (1973), Prinn (1973a, b).
(a) Traub and Carleton (1973).
(b) Fink et al. (1972).
(c) Owen and Sagan (1972).
(d) Barth (1974).

$$2HO_2 \rightarrow H_2O_2 + O_2 \qquad (II) \tag{4}$$

$$H_2O_2 + h\nu \rightarrow 2\,OH. \tag{5}$$

The recombination via I or II is equivalent to the stoichiometric equations $CO + O \rightarrow CO_2$ and $CO + 1/2\,O_2 \rightarrow CO_2$ respectively.

For Mars the main sources of odd hydrogen are photolysis of H_2O and reaction with $O(^1D)$ with H_2 and H_2O (e.g. McConnell, 1973). On Venus the main source of odd hydrogen is photolysis of HCl

$$HCl + h\nu \rightarrow H + Cl \tag{6}$$

followed by reaction of Cl with H_2

$$Cl + H_2 \rightarrow HCl + H \tag{7}$$

of which the net effect is production of 2H from H_2. Such schemes require large sources of H_2. On both planets the reaction

$$H + HO_2 \rightarrow H_2 + O_2 \tag{8}$$

seems to be a major source of H_2. On Venus an additional source of H_2 may be required to balance the destruction by Cl. This may be provided by the reaction

$$CO + H_2O \rightarrow CO_2 + H_2 \tag{9}$$

which would occur in the deep atmosphere near the surface where the pressure ∼95 atmospheres and the temperature ∼750 K (McElroy et al., 1973). On Venus the reaction

first suggested by Prinn (1971) appears to give but a minor contribution in the CO_2 recombination scheme (Sze and McElroy, 1974).

In both atmospheres the stability is due to the catalytic effects of very dilute reactive radicals. The efficacy of the catalytic schemes depends sensitively on the abundance of the radicals which in turn depend on the sources mentioned earlier and on the radical removal processes. At present the rates of the radical scavenging reactions are not well determined, and this has import as to the relative importance of various recombination schemes such as I or II. This type of information may in turn reveal knowledge on transport processes in the lower atmosphere of Mars (e.g. McConnell, 1973). The main HO_x scavenging reactions in both atmospheres are

$$OH + HO_2 \rightarrow H_2O + O_2 \tag{11}$$

$$H + HO_2 \rightarrow H_2 + O_2 \tag{12}$$

$$\rightarrow H_2O + O \tag{13}$$

and to a lesser extent

$$OH + H_2O_2 \rightarrow H_2O + HO_2. \tag{14}$$

In addition on Venus the reactions

$$OH + HCl \rightarrow H_2O + Cl \tag{15}$$

$$Cl + HO_2 \quad HCl + O_2 \tag{16}$$

are important sinks for HO_x radicals. Of the above reaction rates only the rates of Equations (14) and (15) are well determined.

From a stoichiometric point of view when CO_2 is dissociated a CO to O_2 ratio of two is expected. On Mars the ratio is $\sim 1/2$. This can be readily accounted for by an extra source of O_2 which arises from photolysis of H_2O followed by H escape (McElroy and Hunten, 1969a). However on Venus, $CO/O_2 \geqslant 50$, and the problem is where is the O_2? Since CO has a mixing ratio of 5×10^{-5} then the lost O component must have a similar O mixing ratio. McElroy $et~al.$ (1973) surmise that the O_2 is tied up in Cytherian H_2O. However the observed mixing ratio of water above the cloud tops appears to be in the ppm range (Fink $et~al.$, 1972; Gull $et~al.$, 1974) rather than in the 10^{-4} range required by continuity. McElroy $et~al.$ (1973) suggest that H_2O is present in adequate quantities below the cloud deck and such an abundance does not violate any observational constants on H_2O to date. Thus according to McElroy $et~al.$'s (1973) dry model reactions such as Equations (11, 14, 15) produce H_2O which then flows down to below the cloud deck. H_2 flows up and the cycle is completed by reaction (7) which is the source of HO_x from which H_2O is formed.

On both planets the main sources of O_2 are

$$O + O + M \rightarrow O_2 + M \tag{17}$$

$$OH + O \quad \rightarrow H + O_2 \tag{18}$$

while the main sinks are

$$O_2 + h\nu \rightarrow 2O \tag{19}$$

and the net effect of *II*. On Venus there is also a contribution from the reactions

$$Cl + O_3 \rightarrow ClO + O_2 \tag{20}$$

$$ClO + O \rightarrow Cl + O_2 \tag{21}$$

(Sze and McElroy, 1974), the same ones which have attracted so much attention recently because of their effects on odd oxygen in the stratosphere (Molina and Rowland, 1974).

3. The Upper Atmosphere

There has been relatively little new information on the upper atmosphere of Mars in the time since the last meeting (McConnell, 1973). On Mars, interpretation of the UV spectrometer data and the ionospheric data on Mariners 6, 7 and 9 indicates that the upper atmosphere is mainly CO_2, in spite of the relatively rapid dissociation of CO_2 to CO and O. The low mixing ratios of CO and O observed ($\sim 1\%$ at the ionospheric peak) indicate rapid transport mechanisms in order to remove the CO_2 dissociation products. In terms of a vertical eddy diffusion coefficient K, a $K \sim 10^8$ cm^2 s^{-1} is required. Recent work by Leovy (1974) on the effects of mixing due to tidal motions indicates that K's of this magnitude are not unreasonable.

The problem of the atmospheric heat budget on Mars has still not been resolved.

Much work has been done on the Cytherian upper atmosphere since the last meeting (Prinn, 1973b). The status of the composition of the upper atmosphere is not clear. The main items of data which must be reconciled with a consistent picture of the upper atmosphere are shown in Table II. The first information on the temperature structure in the upper atmosphere came from the radio-occultation data (Kliore *et al.*, 1967) and the Lα experiment (Barth *et al.*, 1967). The theoretical calculations of McElroy (1968) were consistent with an exospheric temperature ~ 700 K and an *F*1 type ionosphere with less than 5% O at the ionospheric peak. Assuming a 700 K exospheric temperature, interpretation of the Lα airglow indicated the presence of both H and D in the upper atmosphere (Donahue, 1969; McElroy and Hunten,

TABLE II

Data relating to the upper atmospheric structure of Venus

Temperature	Composition	K[a]	References
700 K	H + D	low	McElroy and Hunten (1969b), Donahue (1969)
350 K	H + hot H		Barth (1968), Kumar and Hunten (1974)
700 K	<5% O[b]	mod	McElroy (1968)
350 K	<1% O	high	Kumar and Hunten (1974)
	≥10% CO, O	low	Rottman and Moos (1973), Strickland (1973)
400 K	H	high	Liu and Donahue (1974), Sze and McElroy (1974)
380 K	100% O	low	Bauer and Hartle (1974)

[a] description refers to Earth's thermosphere as standard where $K \sim 2 \times 10^6$ cm^2 s^{-1}.
[b] O and CO mixing ratios refer to ionospheric peak.

1969b). Recently Kumar and Hunten (1974) have shown that a thermospheric model with $\sim 1\%$ O and an exospheric temperature of 350 K could also provide agreement with the Mariner 5 ionospheric results. This in turn would imply that the Lα data of Mariner 5 could be explicable in terms of a two-temperature model, i.e., H and hot H atoms as originally suggested by Barth (1968). Barth (1968) suggested photodissociation H$_2$

$$H_2 + L\alpha \rightarrow 2H \tag{22}$$

as the source of hot H. It was later shown that the required abundance of H$_2$ was much too large to be consistent with the cold Lα component (Donahue, 1969; McElroy and Hunten, 1969b). However in the Kumar and Hunten (1974) model the sources of hot H suggested are different. They suggest

$$O^+ + H_2 \rightarrow OH^+ + O \tag{23}$$

$$OH^+ + e \rightarrow O + H. \tag{24}$$

It appears that this source of hot H is not adequate to provide an adequate source of hot H to maintain the two-temperature model (Sze and McElroy, 1974). The Mariner 10 Lα data also suggest a 400 K exospheric temperature, but in this case there is no second Lα component such as was observed by Mariner 5 (Broadfoot et al., 1974).

The Mariner 5 and 10 Lα data can be used to provide strong constraints on vertical mixing in the atmosphere or the presence of efficient non-thermal escape processes (Liu and Donahue, 1974; Sze and McElroy, 1974). The lower limit to the observed mixing ratio of total hydrogen $(2H_2O + HCl)$ in the lower atmosphere, plus a moderate value of K $(\sim 10^6 \text{ cm}^2 \text{ s}^{-1})$ imply a thermal escape flux $\sim 10^8$ atoms cm^{-2} s^{-1} (cf. Hunten, 1973). The observed H density together with an exospheric temperature 350 to 700 K yield a Jeans escape flux $\sim 10^2 - 10^6$ cm^{-2} s^{-1}. To account for the observations one requires either a large value of $K \sim 10^8$ cm^2 s^{-1} or else an efficient nonthermal escape process for H from Venus.* There have been several suggestions to date for nonthermal escape processes however none seems adequate to provide the required flux (Sze and McElroy, 1974). The high K required to maintain a low H escape flux also would suppress O and CO densities as is the case for Mars. In fact for $K = 10^8$ the O/CO$_2$ ratio is 10^{-3} at the ionospheric peak (Liu and Donahue, 1974). Such a low O/CO$_2$ ratio and an exospheric temperature of 400 K would not be able to fit the ionospheric data, since the O ionization source in the 200 km region required for the low temperature ionospheric fit would be suppressed (cf; Kumar and Hunten, 1974).

To further confuse the situation there are two other items of information which suggest that O is abundant in the upper atmosphere rather than sparse. Rottman and Moos (1973) have measured the Cytherian airglow spectrum in the 1200 to 1900 Å region and distinguish Lα, O (1304, 1356) and CO fourth positive systems. The 1356

---

* It should be noted that the theory which will yield high K's on Mars will not hold on Venus due to its slower rotation period.

and 1304 lines have been analyzed by Strickland (1973). Allowing for uncertainties in excitation parameters one can conclude that the 1356/1304 ratio measured indicates that optically thick conditions prevail for the 1304 line and that $O/CO_2 > 10\%$ at the ionospheric peak. Also the Mariner 10 ionospheric data have been analyzed by Bauer and Hartle (1974). The conclude that the atmosphere could be predominantly O at the ionospheric peak. They revive the solar wind interaction theory of Cloutier *et al.* (1967) to account for the suppression of an $F2$ peak which would be expected to form in a predominantly O atmosphere. The high O atmosphere would require low K's which would in turn require an efficient nonthermal escape process for H.

The evidence regarding the solar wind interaction theory is of course quite speculative at present. In addition, the airglow data may be accounted for by a suggestion due to Gutcheck and Zipf (1973, 1974). It is that the source of 1304 and 1356 is not due to resonance scattering and electron excitation as is thought presently, rather the excitation mechanism is

$$CO_2^{+\prime} + e \rightarrow CO'' + O''', \tag{25}$$

where the primes indicate excited states and either CO'' or O''' can contribute to 1304 and 1356 radiations. This suggestion of course has its attendant difficulties such as finding an adequate source of highly vibrationally excited CO_2^+ energetic enough to produce required photons on recombination.

If the exospheric temperatures are ~ 400 K then the specter of unbalanced energy budget in thermosphere rears its head on Venus as well as Mars. For example Dickenson and Ridley (1972) calculate exospheric temperatures ~ 600 corresponding in location to the 'measured' temperature ~ 400 K.

4. Conclusions

CO_2 recombination on Mars and Venus occurs via catalysis by HO_x. The source of HO_x is H_2O and H_2 on Mars and HCl and H_2 on Venus. A more definitive analysis is at present hindered by lack of rate data. In the upper atmosphere of Mars rapid transport of the dissociation products of CO_2 is required to maintain low mixing ratios of O and CO. On Venus similar conditions may prevail. However this implies O densities which conflict with the ionospheric data and possibly airglow data also, although the H escape flux problem is satisfied. Alternatively the upper atmosphere may be mainly O. This requires a solar wind interaction model to maintain agreement with the ionospheric observations and a process to maintain a large nonthermal H escape flux. However the airglow constraint would be satisfied. In both atmospheres there appears to be a thermospheric heat budget problem.

Many of the problems may be resolved with the further interpretation of the Mariner 10 data.

Acknowledgments

This work was supported by the National Research Council of Canada.

References

Barth, C. A.: 1968, *J. Atmospheric Sci.* **25**, 564.

Barth, C. A.: 1974, *Ann. Rev. Astron. Astrophys.*, to be published.

Barth, C. A. and Leovy, C. B.: 1974, private communication.

Barth, C. A., Pearce, J. B., Kelly, K. K., Wallace, L., and Fastie, W. G.: 1967, *Science* **158**, 1675.

Bauer, S. J. and Hartle, R. E.: 1974, *Geophys. Res. Letters* **1**, 7.

Broadfoot, A. L., Kumar, S., Belton, M. J. S., and McElroy, M. B.: 1974, *Science* **183**, 1315.

Clouter, P. A., McElroy, M. B., and Michel, F. C.: 1969, *J. Geophys. Res.* **74**, 6215.

Dickenson, R. E. and Ridley, E. C.: 1972, *J. Atmospheric Sci.* **29**, 1557.

Donahue, T. M.: 1969, *J. Geophys. Res.* **73**, 1128.

Fink, U., Larsen, H. P., Kuiper, G. P., and Poppen, R. F.: 1972, *Icarus* **17**, 617.

Gull, T. R., O'Dell, C. R., and Parker, R. A. R.: 1974, *Icarus* **21**, 213.

Gutcheck, R. A. and Zipf, E. C.: 1973, *J. Geophys. Res.* **78**, 5429.

Gutcheck, R. A. and Zipf, E. C.: 1974, private communication.

Hunten, D. M.: 1971, *Space Sci. Rev.* **12**, 539.

Hunten, D. M.: 1973, *J. Atmospheric Sci.* **30**, 1481.

Hunten, D. M.: 1974, *Rev. Geophys.* **12**, 529.

Kliore, A., Levy, G. S., and Cain, D. L.: 1967, *Science* **158**, 1683.

Kumar, S. and Hunten, D. M.: 1974, *J. Geophys. Res.* **79**, 2529.

Leovy, C.: 1974, private communication.

Liu, S. C. and Donahue, T. M.: 1974, *J. Atmospheric Sci.*, submitted.

McConnell, J. C.: 1973, in B. M. McCormac (ed.), *Physics and Chemistry of Upper Atmospheres*, D. Reidel Publishing Co., Dordrecht-Holland, p. 309.

McElroy, M. B.: 1968, *J. Atmospheric Sci.* **25**, 574.

McElroy, M. B. and Donahue, T. M.: 1972, *Science* **177**, 986.

McElroy, M. B. and Hunten, D. M.: 1969a, *J. Geophys. Res.* **74**, 5807.

McElroy, M. B. and Hunten, D. M.: 1969b, *J. Geophys. Res.* **74**, 1720.

McElroy, M. B. and McConnell, J. C.: 1971, *J. Atmospheric Sci.* **28**, 879.

McElroy, M. B., Sze, N. D., and Yung, Y. L.: 1973, *J. Atmospheric Sci.* **30**, 1437.

Marov, M. Ya.: 1972, *Icarus* **16**, 415.

Molina, M. J. and Rowland, F. S.: 1974, *Nature* **249**, 810.

Owen, T. and Sagan, C.: 1972, *Icarus* **16**, 557.

Parkinson, T. M. and Hunten, D. M.: 1972, *J. Atmospheric Sci.* **29**, 1380.

Prinn, R. G.: 1971, *J. Atmospheric Sci.* **28**, 1058.

Prinn, R. G.: 1973a, *Science* **182**, 1132.

Prinn, R. G.: 1973b, in B. M. McCormac (ed.), *Physics and Chemistry of Upper Atmospheres*, D. Reidel Publishing Co., Dordrecht-Holland, p. 335.

Rottman, G. J. and Moos, H. W.: 1973, *J. Geophys. Res.* **78**, 8033.

Strickland, D. J.: 1973, *J. Geophys. Res.* **78**, 2827.

Sze, N. D. and McElroy, M. B.: 1974, *J. Atmospheric Sci.*, submitted.

Traub, W. A. and Carleton, N. P.: 1973, 'A Search for H_2O and O_2 on Venus', paper presented at the American Astronomical Society Third Annual Meeting, Tucson, Arizona.

Whitten, R. C. and Colin, L.: 1974, *Rev. Geophys.* **12**, 155.

Wofsy, S. C. and Sze, N. D.: 1975, this volume, p. 369.

VENUS CLOUD MODELS

STEVEN C. WOFSY and NIEN DAK SZE

Center for Earth and Planetary Physics, Harvard University, Cambridge, Mass. 02138, U.S.A.

Abstract. Remote observations of Venus are reviewed. The strongest inferences of cloud properties can be drawn from the polarization data which provide information about the cloud particles near 68 km. Particle properties are not as well determined at higher and lower levels. If the clouds are generated photochemically from reduced sulfur species, the supply of O_2 may be an important constraint on cloud production. Finally, vapor-pressure data are reviewed. Deep clouds cannot be H_2SO_4–H_2O aerosols unless the mixing ratios of both H_2SO_4 and H_2O approach 10^{-3} below 50 km.

1. Introduction

The clouds of Venus are now widely believed to be droplets composed of H_2SO_4–H_2O solutions. We would like to be able to explain how such clouds can be created and maintained in the atmosphere. In a preliminary model, Prinn (1973) suggested that photochemical oxidation of carbonyl sulfide (OCS) provides the source of H_2SO_4 to the clouds. If they are indeed generated photochemically, then the clouds and the complex photochemistry of the upper atmosphere must be closely linked. In the present review we shall examine some possible relationships between cloud composition, physical properties, and photochemistry.

2. Observations of Cloud Properties

Earth-based observations provide virtually all of the information from which several authors have inferred the composition of Venus' clouds (Young and Young, 1973; A. T. Young, 1973; Sill, 1972). The data are necessarily indirect and therefore these inferences cannot be established unambiguously. At best we can state that a number of independent pieces of evidence support the proposed cloud composition, and that no clearly contradictory observations have been reported.

Young (1973) has discussed models in which all the optically observable cloud phenomena arise from particles composed of about 75% H_2SO_4, 25% H_2O, by weight, and has discussed models in which the cloud extends from the upper atmosphere to as low as ~ 30 km. Cloud particles extend for at least three scale heights, as indicated by various optical phenomena. We shall review three altitude regions separately, since the inferences as to composition are not equally strong at the various altitudes. We shall show that the composition of particles and atmosphere must vary considerably with height if H_2SO_4–H_2O clouds extend over the full height range discussed by Young (1973).

2.1. THE HIGHEST OBSERVABLE HAZE

The existence of a very tenuous upper haze layer was inferred from limb photometry (cf. Goody, 1967). Goody (1967) analyzed data from solar transits by Venus to estimate the pressure at the level of maximum refraction. He concluded that particles existed very high indeed, with optical depth ~0.01 at ~7 mb (79 km) (uncertainty at least a factor of 2). By combining the refraction level with the 'level of line formation' ($\tau \sim 20$ at 150 mb [64 km]; Belton *et al.*, 1968), which is again uncertain by at least a factor of 2, Goody (1967) inferred that the particle scale height was $\lesssim 4$ km, somewhat smaller than atmospheric.

The estimate for $\tau \simeq 1$ from the polarization analysis (see below) is consistent with this inference, but the uncertainties are large, as shown in Figure 1. In this figure, we have shown for comparison the optical depth at each altitude which would result if vertically uniform mixing ratios $f (v/v)$ of gaseous H_2SO_4 were condensed into H_2SO_4–H_2O aerosols with radius $\simeq 1$ μm. Since the particles are larger than visible wavelengths, the extinction cross section was taken as $\sim 2\pi r^2$, which leads to the following expression for the optical depth τ in terms of the atmospheric column number density, $N(z)$,

$$\tau \simeq 1.5 \times 10^{-22} \frac{fN(z)}{r}. \tag{1}$$

Equation (1) might be a slight over-estimate for some of the IR measurements discussed below. Figure 1 shows that, in accord with Goody (1967), the particles

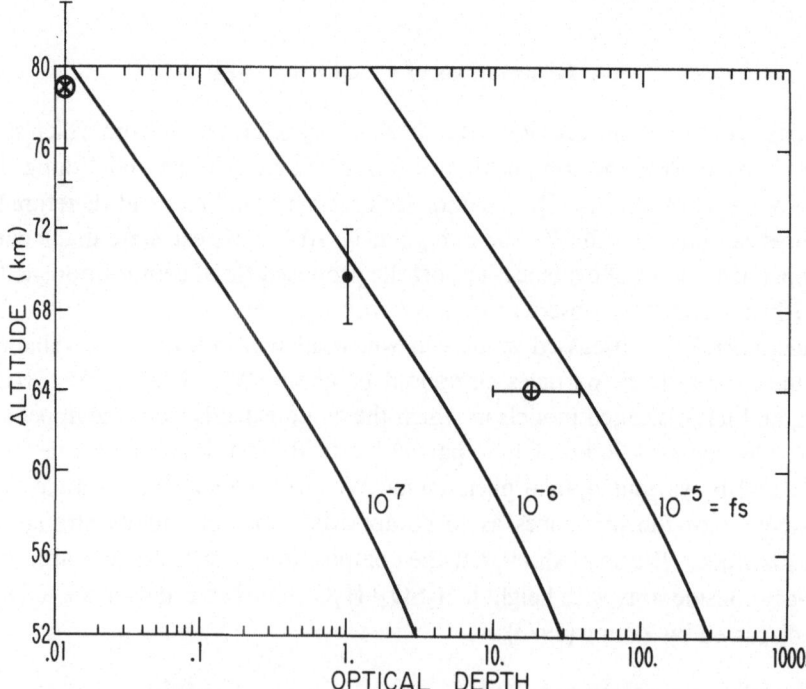

Fig. 1.   Optical depth is shown as a function of height, for particles of radius 1 μm, with various mixing ratios (fs, v/v) of gaseous H_2SO_4 condensed into the cloud particles. The upper experimental point is from Goody (1967), the middle point is from Hansen and Hovenier (1974), and the lowest point from Belton *et al.* (1968).

do not appear to maintain a mixed distribution, but a quantitative interpretation is evidently difficult (cf. Prinn, 1974a). Any increase of r with decreasing altitude would decrease still further the scale-height of the aerosol concentration.

Mariner 10 observed a limb haze with thickness ~ 8 km (Murray *et al.*, 1974) which probably corresponds to the haze discussed by Goody (1967). Preliminary estimates placed the haze top at about 78 km. Stratification is evident in the published photographs. Hopefully, quantitative information about the particle distribution can be obtained from these observations.

2.2. The 'Polarization Clouds'

Coffeen (1969) and Coffeen and Gehrels (1969) measured polarization of light reflected from Venus as a function of wavelength and phase angle. Narrow-band filters were used, covering the spectrum from the near UV through the visible. The data reveal very detailed structure in the phase-angle dependence of Venus polarization. The polarization is affected principally by singly-scattered photons, with lesser contributions from doubly and triply-scattered photons, hence the measurements sample the atmosphere near the level $\tau \simeq 1$.

Coffeen (1969) analyzed the polarization data using a single-scattering calculation, and Hansen and Arking (1971) and Hansen and Hovenier (1974) used multiple scattering. In all cases a *homogeneous*, plane-parallel atmosphere was assumed. The results indicate a very narrow size distribution of spherical particles, radius $r = 1$ μm and index of refraction $n = 1.44 \pm 0.015$ ($\lambda = 5500$ Å). In the UV data, a contribution from gaseous Rayleigh scattering was inferred which yielded an estimate of 50 ± 25 mb (70 ± 2 km) for the level where $\tau \simeq 1$ (Hansen and Hovenier, 1974). These conclusions are based on detailed agreement with the complex structure exhibited by the data. Hansen and Hovenier (1974) place quite narrow error estimates on their numbers, but their use of a homogeneous model complicates an analysis of the true uncertainty. Telescopic and Mariner-10 observations (Murray *et al.*, 1974) have shown significant horizontal inhomogeneities over the planet, particularly in the near UV where the planet exhibited distinct markings variable in time. It is difficult to estimate the effects of such variations on analysis of the polarization data.

Deductions regarding size, sphericity (suggests liquid particles), and index of refraction restrict the possible constituents of these particles near 50 mb, where the temperature $T \sim 230$ K (NASA, 1972). Young and Young (1973), Young (1973), and Sill (1972) suggested H_2SO_4–H_2O droplets, 75–90% H_2SO_4. To date no other suggestion has been able to meet the optical constraints as satisfactorily as H_2SO_4 solutions. Numerous earlier suggestions (H_2O ice, HCl–H_2O, $FeCl_2$, etc.) failed to reproduce the index of refraction and freezing point. The vapor pressures of H_2SO_4 and H_2O over such droplets appear to be consistent with observations of atmospheric composition on Venus; this question is discussed further below.

2.3. Reflecting Clouds

The clouds below 70 km can be observed principally through measurements of reflected sunlight at IR wavelengths. CO_2 and minor constituents (CO, HCl) have IR absorption bands which provide a measure of the quantity of gas traversed by reflected photons. Chamberlain (1965) developed the theory generally used to determine the 'depth of line formation' for these bands (Belton, 1968; Belton *et al.*, 1968; L. D. G. Young, 1972) from the variation of equivalent width with line strength and phase angle. This level occurs on Venus at about 140 mb (64 km) and corresponds to $\tau \sim 20$ (cf. Gierasch and Goody, 1970). The rotational temperature deduced from these observations (~ 250 K) is consistent with the atmospheric temperature at this level, which adds confidence to the inferred altitude. Unfortunately, in order to quantitatively account for the phase curve, complicated inhomogeneous cloud models are required (cf. Hunt, 1972). Thus significant doubt remains as to the cloud optical depth at 64 km.

Pollack *et al.* (1974) observed the near IR reflection spectrum of Venus at low resolution, and concluded that the spectrum exhibits a feature at 3μm which may be attributed to H_2SO_4–H_2O aerosols. They attributed the unexplained feature at 3.8μm reported by Beer *et al.* (1971) to absorption by $CO^{16}O^{18}$, for which other bands have been observed at high resolution (L. D. G. Young, 1972). Venus also exhibits low reflectivity between 1.2 and 2.3μm and in the near UV. These absorptions are not present in the H_2SO_4–H_2O spectrum and unknown impurities must be invoked to explain them (Pollack *et al.*, 1974).

Evidently the data do not provide nearly as much information on the higher and lower cloud deck as on the 'polarization clouds' near 68 km. Virtually no information is available about deeper clouds, which apparently persist down to ~ 35 km (Lacis and Hansen, 1974). Therefore it cannot be demonstrated whether or not all the optical phenomena arise from particles of the same composition, and there are significant uncertainties with respect to the altitudes at which the various optical phenomena are produced. On the other hand, no òbservational data exclude the possibility that all the cloud layers are composed of H_2SO_4–H_2O aerosols. We will therefore examine the implications of such a composition for the atmospheric environment.

3. Proposed Photochemical Mechanism

Prinn's (1973) photochemical model is the only proposed mechanism for the production of H_2SO_4–H_2O clouds on Venus. The following chemical reactions are involved:

$$OCS + hv \rightarrow CO + S(^1D) \quad \begin{matrix} J_2 = 10^{-5}\ s^{-1},\ z > 67\ km \\ J_2 = 0\ s^{-1},\ z < 67\ km \end{matrix} \quad (2)$$

$$CO_2 + S(^1D) \rightarrow CO_2 + S \quad k_3 = 1.7 \times 10^{-11}\ cm^3\ s^{-1} \quad (3)$$

$$S + O_2 \rightarrow SO + O \quad k_4 = 2 \times 10^{-12}\ cm^3\ s^{-1} \quad (4)$$

$$SO + OH \rightarrow SO_2 + H \quad k_5 = 1.2 \times 10^{-10}\ cm^3\ s^{-1} \quad (5)$$

$$SO_2 + HO_2 \rightarrow SO_3 + OH \quad k_6 = 3 \times 10^{-18}\ cm^3\ s^{-1} \quad (6)$$

$$SO_3 + H_2O -- \rightarrow H_2SO_4\ aerosol \quad (7)$$

$$Aerosol + CO \rightarrow H_2O,\ CO_2,\ OCS \quad lifetime = 10^5\ s \quad (8)$$

$$below\ 50\ km$$

TABLE I

Composition of the Venus atmosphere

Gas	Mixing ratio (v/v)	Method	Reference
CO_2	0.97 ± 0.04	Venera 4, 5, 6	Vinogradov et al. (1968, 1970)
N_2 (including inert gases)	$< 2 \times 10^{-2}$	Venera 4, 5, 6	Vinogradov et al. (1968, 1970)
H_2O	1.1×10^{-2}	Venera 4, 5, 6	Vinogradov et al. (1968, 1970)
H_2O	10^{-6}	Venera 4, 5, 6	Vinogradov et al. (1968, 1970)
H_2O	$\sim 7 \times 10^{-7}$	Spectroscopic	Kuiper (1969); Kuiper and Forbes (1968)
H_2O	10^{-6}	Spectroscopic	Fink et al. (1972)
O_2	$< 10^{-3}$	Venera 5, 6	Vinogradov et al. (1970)
O_2	$< 10^{-5}$	Spectroscopic	Belton and Hunten (1968); Owen (1968)
O_2	$< 10^{-6}$	Spectroscopic	Traub and Carleton (1973)
CO	5×10^{-5}	Spectroscopic	Connes et al. (1968)
HCl	6×10^{-7}	Spectroscopic	Connes et al. (1968)
HF	10^{-9}	Spectroscopic	Connes et al. (1968)
CH_4	$< 10^{-6}$	Spectroscopic	Connes et al. (1968)
CH_3Cl	$< 10^{-6}$	Spectroscopic	Connes et al. (1968)
CH_3F	$< 10^{-6}$	Spectroscopic	Connes et al. (1968)
C_2H_2	$< 10^{-6}$	Spectroscopic	Connes et al. (1968)
HCN	$< 10^{-6}$	Spectroscopic	Connes et al. (1968)
O_3	$< 10^{-8}$	Spectroscopic	Jenkins et al. (1969)
SO_2	$< 3 \times 10^{-8}$	Spectroscopic	Cruikshank and Kuiper (1967)
COS	$< 10^{-6}$	Spectroscopic	Cruikshank (1967)
COS	$< 10^{-8}$	Spectroscopic	Kuiper (1969); Kuiper and Forbes (1968)
C_3O_2	$< 5 \times 10^{-7}$	Spectroscopic	Cruikshank and Sill (1967)
H_2S	$< 2 \times 10^{-4}$	Spectroscopic	Cruikshank (1967)
NH_3	$< 3 \times 10^{-8}$	Spectroscopic	Kuiper (1969); Kuiper and Forbes (1968)

According to Prinn (1973), sulfur is provided to the stratosphere by vertical transport of OCS, which is expected to be present in the lower atmosphere, (Lewis, 1970) $f_{OCS} \sim 5 \times 10^{-5}$ v/v. Of the 11 reactants in this scheme, only H_2O and CO have been observed. Searches for sulfur bearing gases (SO_2, H_2S and OCS) have produced negative results (see Table I).

In Prinn's calculation the column destruction rate of OCS is about 10^{13} cm^{-2} s^{-1} (Prinn, 1974b), and the cloud production time is constrained to be equal to the photolysis lifetime of OCS, 10^5 s. He calculated a cloud bottom at ~ 57 km where he assumed that the particles evaporated with a mean lifetime of 10^5 s. His cloud optical depth was about 100.

Prinn's (1973) assumption of the cloud bottom at 57 km was based on the level where water vapor exceeded 10^{-4} mixing ratio for particles made of 75% H_2SO_4. According to Figures 6 and 7 (see discussion below), such particles would not evaporate until much lower levels (~ 44 km); the H_2SO_4 and SO_3 vapor pressures remain quite low, hence in Prinn's model the particles would lose H_2O, but not evaporate, at 57 km. Of course, the chemical instability of SO_3 might also produce a cloud bottom, as could dynamical processes (see below).

According to the above scheme, the atmosphere must supply one OCS, one H_2O, and $3/2O_2$ in order to generate each H_2SO_4 in the photochemical clouds. The lower atmosphere proposed by Lewis (1970) could perhaps supply H_2O and OCS with the necessary fluxes, but O_2 is virtually absent and could not be supplied. If insufficient O_2 were present at cloud level, the S atom produced by photolysis of OCS could recombine with CO or with other S-atoms, and not participate in H_2SO_4 formation. Possible oxygen sources might include the energetically allowed reaction

$$S(^1D)+CO_2 \rightarrow SO+CO \tag{9}$$

or the slightly endothermic reaction*

$$S+CO_2 \rightarrow SO+CO, \tag{10}$$

but no laboratory evidence exists for such reactions.

The most likely oxygen source appears to be CO_2 photolysis in the upper atmosphere. We shall now briefly review photochemical models of the upper atmosphere to estimate how much oxygen could be supplied for cloud generation. We shall see that it is difficult to create a thick cloud by means of chemistry in the thin atmosphere above the clouds.

If CO_2 photolysis is the source of O_2 in Prinn's (1973) model, then the CO left behind is a measure of how much O_2 is bound up in H_2SO_4. Coincidentally, the observed CO abundance is about the same as Lewis' (1970) prediction for OCS in the lower atmosphere, which is in turn slightly more than the H_2SO_4 abundance inferred for particles near $\tau = 1$ (Hansen and Hovenier, 1974). Since OCS destruction also leaves behind a CO molecule, we conclude that the H_2SO_4 mixing ratio must be less than the observed CO abundance, 5×10^{-5}, if the assumptions are correct.

Solar radiation photolyzes CO_2 at the rate of $\sim 10^{13}$ cm^{-2} s^{-1} on Venus,

$$CO_2+h\nu \rightarrow CO+O. \tag{11}$$

If CO_2 were the only component of the atmosphere, aeronomic models would predict abundances of CO, O and O_2 substantially greater than the observations shown in Table I. Moreover, solar radiation in a pure-CO_2 model produces CO and O in equal amounts, but the observations show that O, $O_2 \ll$ CO. Explanations for the observed atmospheric composition have concentrated on catalytic cycles, initiated by photochemistry of H_2, H_2O and HCl, which enhance the rate of CO recombination with O and O_2 (McElroy, 1967, 1968a, b; Donahue, 1968, 1971; Prinn, 1971; McElroy et al., 1973; Sze and McElroy, 1974).

Our present purposes are quite different. We would like to know how much oxygen can be provided by the upper atmosphere to maintain a photochemical cloud. For Prinn's (1973) proposed cloud mechanism, nearly all of the oxygen produced from photolysis is required for cloud formation.

Oxygen atoms are produced primarily above 72 km on Venus, as shown in Figure 2.

---

* According to the JANEF (1965) tables, reaction (10) is endothermic by about 2.5 kcal mole^{-1}. At 250 K, the equilibrium constant $K=[S][CO_2]/[SO][CO]=3.6$. If equilibrium could be attained, we would have $[SO]/[S] \simeq 5500$.

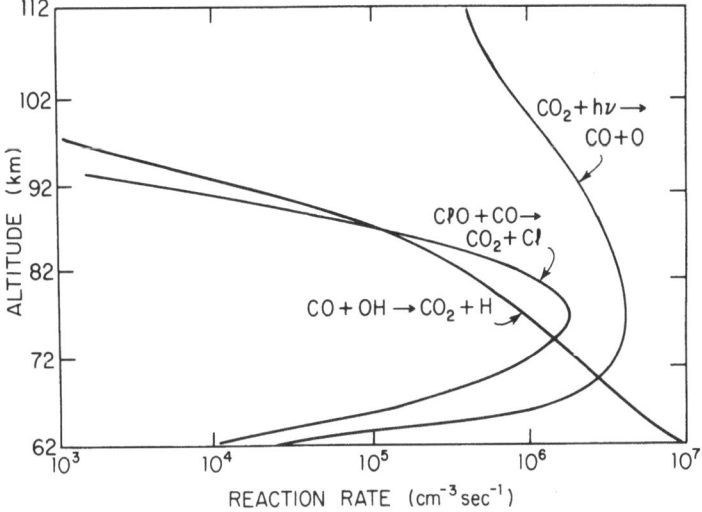

Fig. 2. Reaction rates for key reactions leading to production ($CO_2 + h\nu$) of CO and O and removal ($CO + OH$ and $ClO + CO$) of CO (Sze and McElroy, 1974).

According to recent calculations (Sze and McElroy, 1974), formation of O_2 is the principal sink for O atoms between 70 and 80 km. The process is catalyzed by Cl-atoms and by O_2, in the cycle

$$O + O_2 + CO_2 \rightarrow O_3 + CO_2 \tag{12}$$
$$Cl + O_3 \rightarrow ClO + O_2 \tag{13}$$
$$\underline{ClO + O \rightarrow Cl + O_2} \tag{14}$$
$$O + O \rightarrow O_2$$

HO_x radicals also contribute to O-atom removal through cycles terminating with (McElroy and Donahue, 1972; McElroy et al., 1973)

$$O + OH \rightarrow O_2 + H \tag{15}$$
$$O + HO_2 \rightarrow O_2 + OH. \tag{16}$$

Reactions (12) to (16) are much faster than the three-body removal processes

$$O + O + CO_2 \rightarrow O_2 + CO_2 \tag{17}$$

and

$$O + CO + CO_2 \rightarrow CO_2 + CO_2 \tag{18}$$

which would characterize a 'pure CO_2' model. The reaction rates for (14), (15), and (17) are shown in Figure 3. The O-atom chemical lifetime is $\sim 10^4$ s, which is too short for significant vertical transport of O in the absence of exceedingly vigorous vertical motions.

The conversion of O to O_2 above 70 km does not guarantee that O_2 is available to oxidize S to H_2SO_4. The reduced sulfur species (OCS and H_2S) are both rapidly photolyzed above 60 km on Venus. One therefore expects them to be absent from

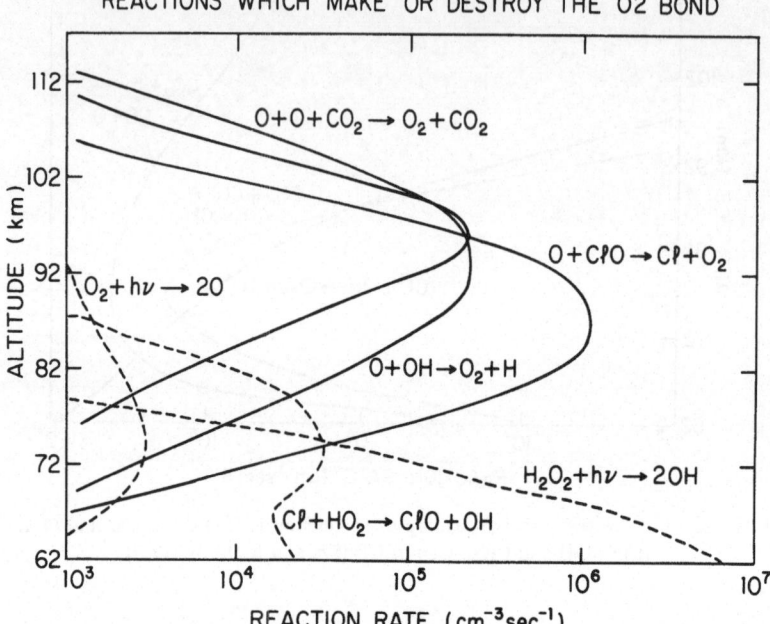

Fig. 3. Reaction rates for key reactions leading to formation of O—O bond ($O+ClO$, $O+OH$, and $O+O+CO_2$), and to the destruction of O—O bond (HO_2+HO_2, $O_2+h\nu$, and $Cl+HO_2$). The rate for Equation (26b) was assumed to be 10^{-12} cm^{-3} s^{-1} (Sze and McElroy, 1974).

the atmosphere above 70 km*, which implies that O_2 must be transported down to the altitudes where OCS and H_2S are photo-oxidized. A significant flux of O_2 from the CO_2 photolysis region is possible only if the vertical transport rate is faster than the rate of O_2 recombination with CO.

Direct reaction of O_2 with CO is negligibly slow, but several catalytic cyles have been suggested which could enhance CO oxidation on Venus. Prinn (1971) proposed the reaction

$$CO+ClO \rightarrow CO_2+Cl \tag{19}$$

and McElroy et al. (1973) argued for the importance of

$$CO+OH \rightarrow CO_2+H. \tag{20}$$

In both cases a prior reaction that breaks the O—O bond is crucial. Prinn (1971) suggested the reaction sequence

$$Cl+O_2+M \rightarrow ClOO+M \tag{21}$$
$$ClOO+CO \rightarrow ClO+CO_2 \tag{22}$$

* If OCS or H_2S were regenerated after photolysis, or if vertical transport were sufficiently rapid, traces of these gases might survive above 70 km. In that case, direct reactions with O-atoms might play a significant role in H_2SO_4 formation. For example, the reaction $OCS+O \rightarrow CO+SO$ (a) would be a principal sink for O atoms above 70 km if $[OCS] = 10^{-6}$ $[CO_2]$ and if the rate for (a) were 10 times faster than the analogous reaction $OCS+S \rightarrow CO+S_2$ (b), which has been measured recently by Klemm and Davis (1974). Reaction (b) is potentially a strong source of elemental sulfur, as noted by Prinn (1974b).

in which Equation (22) is the reaction which breaks the O—O bond. However, collisional dissociation of ClOO,

$$ClOO + M \rightarrow Cl + O_2 + M, \tag{23}$$

appears to reduce the concentration of ClOO to negligible amounts, making Equation (22) ineffective (cf. Watson, 1974).

In the reaction sequence proposed by McElroy et al. (1973) (also see Parkinson and Hunten, 1972), the O—O bond is broken by

$$H_2O_2 + h\nu \rightarrow OH + OH. \tag{24}$$

The H_2O_2 is formed by disproportionation of HO_2,

$$HO_2 + HO_2 \rightarrow H_2O_2 + O_2. \tag{25}$$

Another potentially important reaction is

$$Cl + HO_2 \rightarrow ClO + OH \tag{26a}$$

but the rate of Equation (26a) is unknown. Sze and McElroy (1974) adopted a value of 10^{-12} cm^{-3} s^{-1}, which makes Equation (26a) moderately important. The HO_x cycles could be quite efficient in catalyzing O_2 recombination with CO, as shown in Figure 3. However, the cycles may be subject to interference from processes which terminate the HO_x radical chains. Sze and McElroy (1974) explicitly considered the effect of the reaction

$$Cl + HO_2 \rightarrow HCl + O_2, \tag{26b}$$

which was also mentioned by Prinn (1971), for which rate measurements have not been reported. Figure 4 shows the downward flux of O_2 as a function of the rate for Equation (26b). The maximum yield of O_2 ($\sim 3 \times 10^{12}$ cm^{-2} s^{-1}) is obtained when Equation (26b) is fast enough to effectively eliminate HO_x catalysis. If Equation (26b) is slow, however, most of the O_2 recombines with CO, leaving a much smaller flux available for photo-oxidation of OCS.

Sze and McElroy (1974) also examined the effects of vertical transport on the supply of O_2. The downward flux of O_2 varies directly with the rate of vertical transport. If Prinn (1974a) is correct in his assertion that vertical motions are weak in the 60 to 80 km region, then a relatively fast rate for Equation (26b) is required for an appreciable downward flux of O_2.

In summary, we find that current aeronomic models provide at most a flux of $\sim 3 \times 10^{12}$ cm^{-2} s^{-1} O_2 molecules for cloud formation, which permits an efficiency of ~ 0.3 for Prinn's (1973) photo-oxidation mechanism. The O_2 flux might be very much smaller if catalytic oxidation of CO is rapid.

4. Physical Chemistry of H_2SO_4—H_2O Aerosols

4.1. VAPOR PRESSURES

Solutions of H_2SO_4 and H_2O exist in equilibrium with gaseous H_2SO_4, H_2O, and

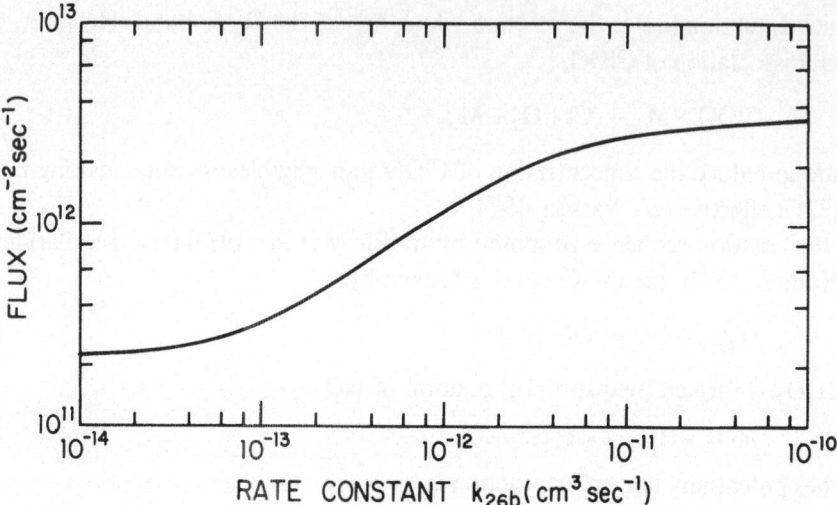

Fig. 4.   Downward flux of O_2 at 62 km as a function of the rate constant for the reaction $Cl + HO_2 \rightarrow$
$\rightarrow HCl + O_2$ (Sze and McElroy, 1974).

SO_3, with the vapor pressures strongly dependent on temperature. The properties
of this system are not as well determined as one might wish, but they appear to be
known adequately for our purpose. We have used the data collected by Luchinski
(1956) and Pickering (1890) supplemented by values from Timmermans (1960) where
necessary. For purposes of meaningful extrapolation, we have plotted the logarithm
of the vapor pressure vs. reciprocal temperature (cf. Toon and Pollack, (1973). The
results are shown in Figures 5 and 6.

Vapor pressures for H_2O over various solutions of H_2SO_4–H_2O are shown in
Figure 5, along with the freezing point curve and the conditions on Venus for several
assumed H_2O abundances. Our analysis must remain incomplete until the actual
water vapor profile on Venus is determined. Observations of H_2O IR bands yield
measurements ranging from $f = 10^{-6}$ (Fink et al., 1972) to higher than 10^{-4} (Belton
et al., 1968) and Venera probes reported values in excess of 10^{-3} in the lower atmosphere (Vinogradov et al., 1970). We therefore consider water vapor mixing ratios
from 10^{-6} to 10^{-4} for the altitudes above 60 km.

4.2. PARTICLE COMPOSITION AND FREEZING CHARACTERISTICS

Figure 6 shows the vapor pressures of H_2SO_4 and SO_3 on Venus (expressed as a
mixing ratio) in equilibrium with aerosols* of H_2SO_4–H_2O, with the water vapor
mixing ratio taken at 10^{-4}, 10^{-5}, and 10^{-6}. Below about 56 km the vapor pressures
of H_2SO_4 and SO_3 sharply increase, from negligible to quite large values. The abrupt-
ness of this increase is a consequence of the strong dependence of the vapor pressures
on temperature, so that varying the ambient H_2O pressure by a factor of 100 dis-
places the curves by less than a scale height ($\Delta T \simeq 50$ K).

---

* The small increases in vapor pressures due to surface curvature have been neglected.

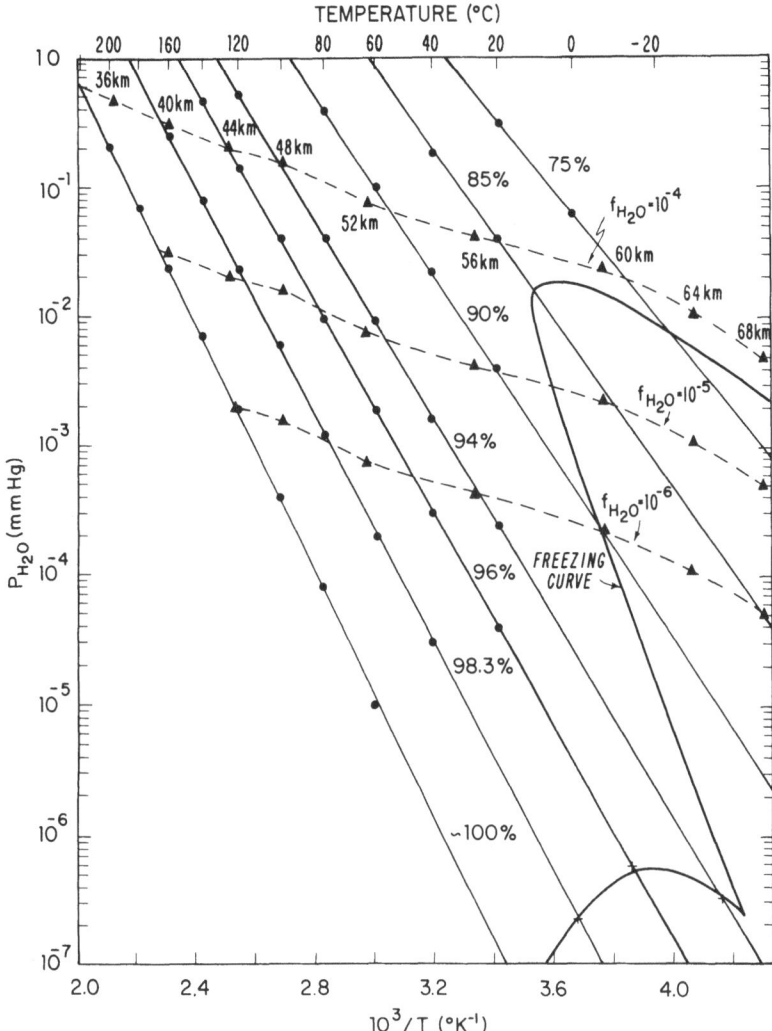

Fig. 5. The equilibrium vapor pressure of H_2O is shown for solutions of various concentration of H_2SO_4 (wt %) in water, as a function of temperature. The data are from Luchinski (1956) and Timmermans (1960). The freezing point curve according to Pickering (1890) for the solutions is shown by the heavy line. Conditions on Venus are shown, by the dotted lines, using the NASA (1972) atmosphere, for H_2O mixing ratios of 10^{-4}, 10^{-5}, and 10^{-6}. The triangles (Δ) designate the given altitude on Venus.

We may interpret Figure 6 as showing the mixing ratios of gaseous sulfur compounds $(H_2SO_4 + SO_3)$ which produce saturation at a given level on Venus. The figure thus shows the lowest altitude where a H_2SO_4 cloud can exist if the sulfur mixing ratio on Venus were specified. For example, Lewis (1970) has suggested a sulfur abundance of 5×10^{-5} in Venus' lower atmosphere, in the form of OCS. If this value applied near the surface and minerals were the ultimate sulfur source, then 5×10^{-5} would be the maximum abundance that could occur for sulfur bearing gases higher in the atmosphere. The H_2SO_4 cloud bottom could not then be lower than 48 km for $f_{H_2O} = 10^{-6}$, or 43 km for $f_{H_2O} = 10^{-4}$. Note, however, that the cloud

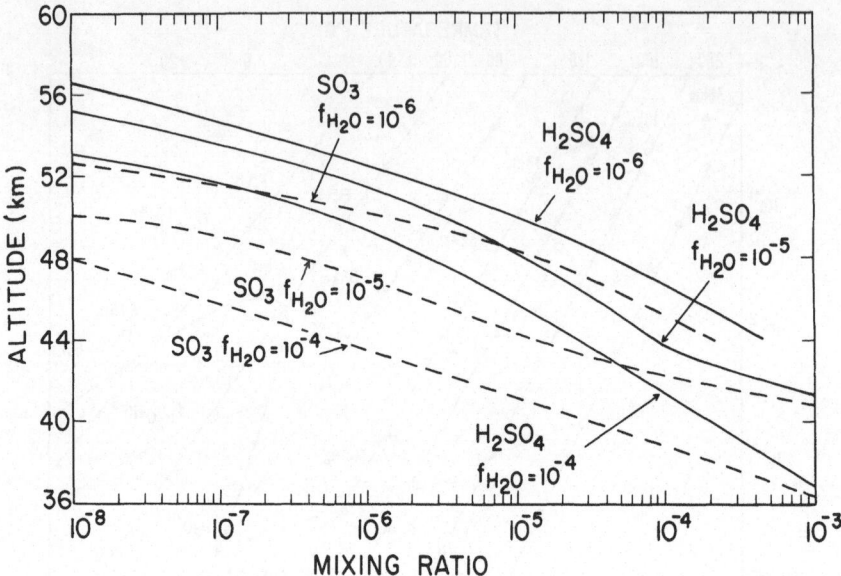

Fig. 6. The equilibrium vapor pressures of H_2SO_4 and SO_3 (expressed as mixing ratios) as a function of altitude in the Venus atmosphere. Profiles are given for assumed background H_2O abundances of 10^{-4}, 10^{-5}, and 10^{-6} mixing ratio (v/v).

bottom could occur higher, due to dynamical effects such as rapid downward mixing below some level, or to incomplete conversion of OCS to H_2SO_4 and SO_3.

Young (1973) proposed that much deeper clouds could be composed of H_2SO_4–H_2O aerosols. His models imply very large abundances of *both* H_2O and sulfur ($\gtrsim 10^{-3}$ v/v) in the lower atmosphere. Lewis' (1970) discussion of surface minerals is inconsistent with such abundances, but in the absence of data the disagreement cannot be resolved. Such deep H_2SO_4 clouds evidently are not possible if the sulfur abundance on Venus is $\lesssim 10^{-4}$.

We further note that high ambient temperature below 50 km produces substantial mixing ratios of SO_3 at the expense of H_2SO_4. SO_3 is quite reactive photochemically. The observed CO abundances, and Lewis' (1970) model, suggest that Venus' lower atmosphere is likely to be somewhat reducing. The maintenance of sufficient SO_3 therefore poses difficulties for models in which H_2SO_4 clouds extend below 50 km. The presence of very large ($\sim 10^{-3}$) H_2O abundances would be required to suppress SO_3 formation thermochemically. If such large H_2O abundances occurred at 50 km, the low H_2O density observed above the clouds would be quite difficult to explain, since the H_2SO_4 in the gas phase would be much too small to 'dry out' the atmosphere.

Extrapolation of the vapor pressures shows that freezing of pure H_2SO_4–H_2O particles may occur at levels where interpretation of the polarization data suggest spherical particles, i.e. at temperatures between 232 and 217 K (cf. Hansen and Hovenier, 1974).

Figure 7 shows the equilibrium composition for droplets in the Venus atmosphere

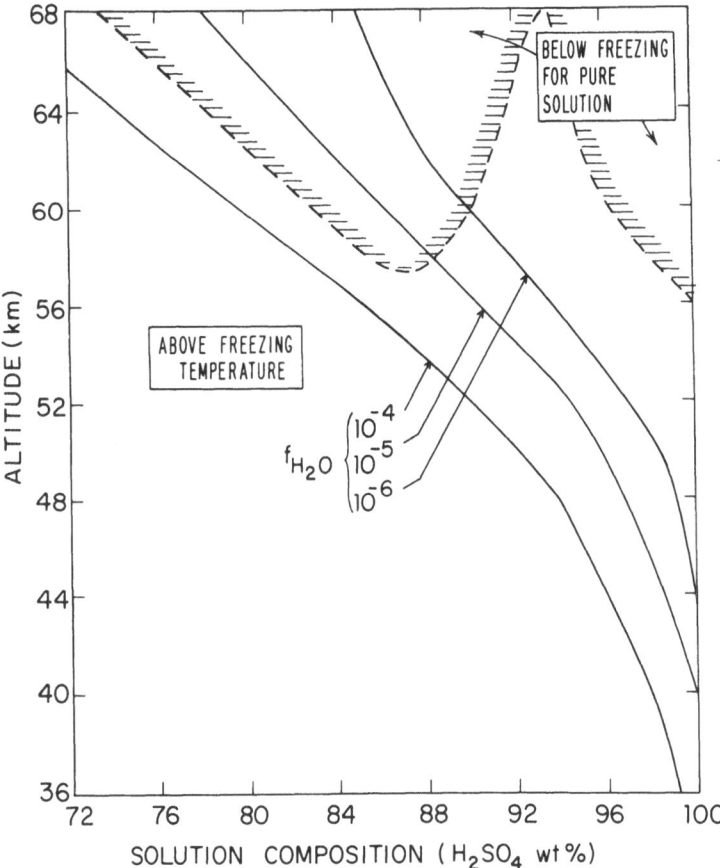

Fig. 7. Solution composition profiles are shown for Venus with assumed H_2O abundances 10^{-4}, 10^{-5}, and 10^{-6}. The regions where pure solutions could freeze are shown by the hatched areas. This figure is constructed from Figure II.2 using the NASA (1972) atmosphere.

and the region in which freezing of the pure solutions could occur. There are several ways that liquid particles can be maintained at 68 km: if $f_{H_2O} < 7 \times 10^{-5}$, either (a) the particles must be supercooled or (b) impurities must be present which lower the freezing points of the solutions. (c) If $f_{H_2O} \sim 10^{-4}$, pure solutions could not freeze. At present either (a), (b) or (c) could provide an explanation. If (a) or (b) were correct, some frozen particles would probably be present. If (c) were correct, then the H_2O abundances must be larger than the recent measurements indicate, and the solutions would contain more H_2O than currently assumed.

If the upper atmosphere must supply O_2 to the clouds via direct photodissociation of CO_2, then the cloud depth could well be limited by this supply. Figure 8 shows the time required to produce clouds of different thicknesses, compared to the time required for various dynamical loss processes. Here we have used our vapor-pressure curves and assumed that the particles evaporate at the level of saturation, with subsequent net transport of vapor down to the lower atmosphere where it is converted back to OCS. The cloud is replenished by a flux of $\sim 10^{12}$ cm^{-2} s^{-1} H_2SO_4 mole-

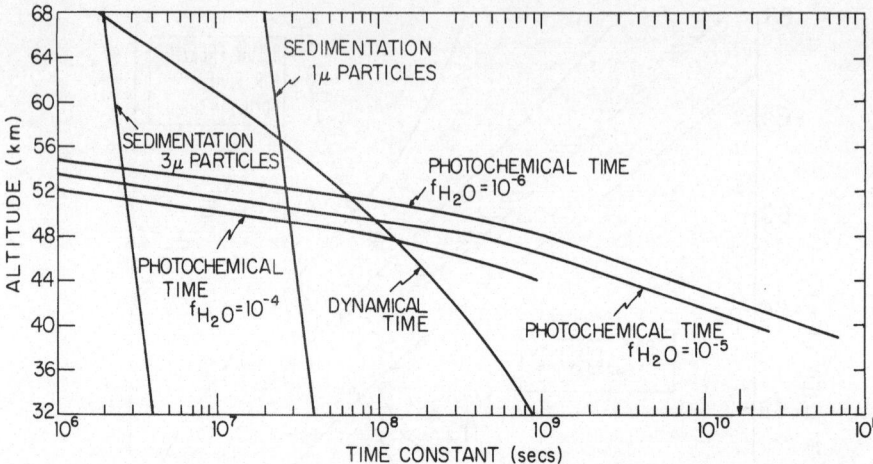

Fig. 8.   Time constant profiles are plotted vs. the cloud bottom altitude, for cloud-related processes on Venus. The photochemical time is derived from Figure II.5 (NASA, 1972) by assuming a well-mixed cloud above the level of saturation, with the concentration at that level from Figure II.5. The photochemical time is then given by the cloud H_2SO_4 content (cm^{-2}) divided by the flux of O_2 into the cloud ($\sim 10^{12}$ cm^{-2} s^{-1}) (see text). The sedimentation time is calculated from the formula given by Byers (1965), assuming viscous flow. The 'dynamical' time is a rough estimate of the transport time for large scale motions; the curve shown is the time required to heat the atmosphere to the ambient temperature at each altitude, using 5% of the solar constant is assumed to be absorbed at that altitude.

cules, generated by Prinn's (1973) scheme but limited by the supply of O_2 from the upper atmosphere. The figure shows that the cloud bottom will occur at ~ 50 km, regardless of the H_2O vapor pressure, because very large mixing ratios of H_2SO_4 must be maintained in order for the particles to exist at the high temperatures below 50 km. The time required to produce clouds near 50 km is seen to be 10^7 to 10^8 s, which is the same order as (or larger than) the coagulation time (assuming $\sim 10^2$ particles cm^{-3}). Thus we might expect that the size distribution inferred from the polarization data would be unlikely to persist down to the cloud bottom.

We may summarize our conclusions from the vapor pressure data for H_2SO_4–H_2O aerosols:

(1) If $f_{H_2O} \cong 10^{-6}$ between 60 and 70 km, then the particles would be $\sim 20°$ below the freezing point for pure solutions, and they would be composed of 88 to 92% H_2SO_4, rather than 75% H_2SO_4 as favored by Young (1974).

(2) If $f_{H_2O} \simeq 10^{-4}$ near 60 km, some mechanism must exist for drying the atmosphere to $f = 10^{-6}$ above 60 km, or else the data of Fink et al. (1972) are erroneous. H_2SO_4 particles seem to have insufficient density to be an effective drying agent ($f_{H_2SO_4} \sim 3 \times 10^{-6}$, cf. Gierasch and Goody, 1970 and Hanson and Hovenier, 1974).

(3) H_2SO_4–H_2O aerosols will evaporate between 45 and 50 km unless both H_2SO_4 and H_2O have very large abundances ($\sim 10^{-3}$).

(4) As a consequence of point (3) and the exponential dependence of density on height, an exceedingly long time is required ($\gg 10$ yr) to generate clouds down to 35 km by Prinn's (1973) mechanism.

Acknowledgments

This work was supported jointly by the Atmospheric Sciences Division of the National Science Foundation, and the National Aeronautics and Space Administration, under grant numbers GA 33990X and NGR 22-007-067 respectively to Harvard University. The authors thank M. B. McElroy, R. M. Goody, and D. M. Hunten for helpful discussions and suggestions.

References

Beer, R., Norton, R. H., and Matonchik, J. V.: 1971, *Astrophys. J. Lett.* **68**, L121.
Belton, M. J. S.: 1968, *J. Atmospheric Sci.* **25**, 596.
Belton, M. J. S. and Hunten, D. M.: 1968, *Astrophys. J.* **154**, 797.
Belton, M. J. S., Hunten, D. M., and Goody, R. M.: 1968, *The Atmospheres of Mars and Venus*, Gordon and Breach, New York, 69.
Byers, H. R.: 1965, *Elements of Cloud Physics*, University of Chicago Press, Chicago, Illinois.
Chamberlain, J. W.: 1965, *Astrophys. J.* **141**, 1184.
Coffeen, D. L.: 1969, *Astron. J.* **74**, 446.
Coffeen, D. L. and Gehrels, T.: 1969, *Astron. J.* **74**, 433.
Connes, P., Connes, J., Kaplan, L. D., and Benedict, W. S.: 1968, *Astrophys. J.* **152**, 731.
Connes, P., Connes, J., Benedict, W. S., and Kaplan, L. D.: 1967, *Astrophys. J.* **147**, 1230.
Cruikshank, D. P.: 1967, *Univ. Ariz. Commun. Lunar Planetary Lab.* **6**, 199.
Cruikshank, D. P. and Kuiper, G. P.: 1967, *Univ. Ariz. Commun. Lunar Planetary Lab.* **6**, 195.
Cruikshank, D. P. and Sill, G. T.: 1967, *Univ. Ariz. Commun. Lunar Planetary Lab.* **6**, 302.
Donahue, T. M.: 1968, *J. Atmospheric Sci.* **25**, 568.
Donahue, T. M.: 1971, *J. Atmospheric Sci.* **28**, 895.
Fink, V., Larson, H. P., Kuiper, G. P., and Poppen, R. F.: 1972, *Icarus* **17**, 617.
Goody, R. M.: 1967, *Planetary Space Sci.* **15**, 1817.
Gierasch, P. and Goody, R. M.: 1970, *J. Atmospheric Sci.* **27**, 224.
Hansen, J. E. and Arking, A.: 1971, *Science* **171**, 669.
Hansen, J. E. and Hovenier, J. W.: 1974, *J. Atmospheric Sci.* **31**, 1137.
Hunt, G.: 1972, *J. Quant. Spectr. Radiative Transfer* **12**, 387.
JANEF Tables: 1965, compiled by Dow Chemical Co., Midland, Michigan, under PB-168370.
Jenkins, E. B., Morton, D. C., and Sweigart, A. V.: 1969, *Astrophys. J.* **157**, 913.
Klemm, R. B. and Davis, D. D.: 1974, *J. Phys. Chem.* **78**, 1137.
Kuiper, G. P.: 1969, in C. Sagan, T. C. Owen, and H. J. Smith (eds.), 'Planetary Atmospheres', *IAU Symp.* **40**, 91.
Kuiper, G. P. and Forbes, F. F.: 1968, *Univ. Ariz. Commun. Lunar Planetary Lab.* **6**, 177.
Lacis, A. A. and Hansen, J. E.: 1974, *Science* **184**, 979.
Lewis, J. S.: 1970, *Earth Planetary Sci. Letters* **10**, 73.
Luchinski, G. P.: 1956, *J. Fiz. Khim.* **30**, 1208.
McElroy, M. B.: 1967, *Astrophys. J.* **150**, 1125.
McElroy, M. B.: 1968a, *J. Geophys. Res.* **73**, 1513.
McElroy, M. B.: 1968b, *J. Atmospheric Sci.* **25**, 574.
McElroy, M. B. and Donahue, T. M.: 1972, *Science* **171**, 986.
McElroy, M. B., Sze, N. D., and Yung, Y. L.: 1973, *J. Atmospheric Sci.* **30**, 1437.
Murray, B. C., Belton, M. J. S., Danielson, G. E., Davis, M. E., Gault, D., Hapke, B., O'Leary, B., Storn, R. G., Suomi, V., Trask, N.: 1974, *Science* **183**, 1307.
NASA: 1972, *Models of the Venus Atmosphere*, SP-8077.
Owen, T.: 1968, *J. Atmospheric Sci.* **25**, 583.
Parkinson, T. D. and Hunten, D. M.: 1972, *J. Atmospheric Sci.* **29**, 1380.
Pickering, S. V.: 1890, *J. Chem. Soc.* **1890**, 331.
Pollack, J. B., Erikson, E. F., Witteborn, F. C., Chackerian, C., Summers, A. L., Van Camp, W., Baldwin, B. J., Augason, G. C., and Caroff, L. J.: 1974, 'Aircraft Observation of Venus' Near-Infrared Reflection Spectum', paper presented at AMS-DPS meeting, March 1973.

Prinn, R. G.: 1971, *J. Atmospheric Sci.* **28**, 1058.
Prinn, R. G.: 1973, *Science* **182**, 1132.
Prinn, R. G.: 1974a, *J. Atmospheric Sci.* **31**, 1691.
Prinn, R. G.: 1974b, private communication.
Sill, G. T.: 1972, *Commun. Lunar Planetary Lab.* **9**, 191.
Sze, N. D. and McElroy, M. B.: 1974, *Planetary Space Sci.*, in press.
Timmermans, J.: 1960, *The Physico-Chemical Constants of Binary Mixture in Concentrated Solution*, Vol. 14, Interscience, New York.
Toon, O. B. and Pollack, J. B.: 1973, *J. Geophys. Res.* **78**, 7051.
Traub, W. A. and Carleton, N. P.: 1973, 'A Search for H_2O and O_2 on Venus', presented at American Astronomical Society, third annual meeting, Tucson, Arizona.
Vinogradov, A. P., Surkov, Yu. A., Florensky, K. P., and Andreichikov, B. M.: 1968, *Soviet Phys. Doklady* **13**, 176.
Vinogradov, A. P., Surkov, Yu. A., and Andreichikov, B. M.: 1970, *Soviet Phys. Doklady* **15**, 4.
Watson, R. T.: 1974, *JCS Far. Trans.*, in press.
Young, A. T.: 1973, *Icarus* **18**, 564.
Young, A. T.: 1974, *Science* **183**, 407.
Young, L. D. G.: 1972, *Icarus* **17**, 632.
Young, L. D. G. and Young, A. T.: 1973, *Astrophys. J.* **179**, L39.

THE IONOSPHERE AND UPPER ATMOSPHERE OF VENUS

SHAILENDRA KUMAR*

Kitt Peak National Observatory, P.O. Box 26732, Tucson, Ariz. 85726, U.S.A.

1. Introduction

The Mariner 10 flyby of Venus has provided us a second close look at the upper atmosphere of a planet which has intrigued aeronomers for a long time. The Mariner 10 payload included an objective grating spectrometer designed to measure airglow radiations from Venus and Mercury in the spectral range 200 to 1700 Å at discrete wavelengths corresponding to the resonance radiations of H, He, C, O, Ar, Ne and CO (Broadfoot *et al.*, 1974). The data on the ionosphere and the neutral atmosphere were obtained from the radio-occultation experiment (Howard *et al.*, 1974) whereas an electrostatic analyzer was employed to make *in situ* measurements of the charged particle environment around Venus (Bridge *et al.*, 1974). A highly sensitive dual magnetometer system was employed to measure magnetic fields around Venus, especially in the bow shock region (Ness *et al.*, 1974). A number of questions raised by the Mariner 5 flyby in 1967 have been answered while many more new ones have evolved. Far UV spectra (1200–1900 Å) of Venus with moderate spectral resolution (~ 20 Å) have also been obtained from a number of rocket experiments (Moos *et al.*, 1969; Moos and Rottman, 1971; Rottman and Moos, 1973). These spectra have lead to the identification of several minor constituents in the upper atmosphere. This paper summarizes the current understanding of the upper atmosphere and ionosphere of Venus and its interaction with solar wind.

2. Exosphere and Thermosphere

2.1. COMPOSITION

A model for the upper atmosphere of Venus is shown in Figure 1 and the various parameters are listed in Table I. The far UV spectrum of Venus obtained from a recent rocket flight is shown in Figure 2 (Rottman and Moos, 1973). CO_2 is undoubtedly the primary constitutent in the atmosphere of Venus (Hunten, 1971; Marov, 1972). Although minor species such as CO, H_2O, HCl and HF have been identified from ground based spectroscopy with mixing ratios (Belton, 1968) 2×10^{-4}, 10^{-4}–10^{-6}, 10^{-6} and 2×10^{-8} respectively, these mixing ratios correspond to the cloud tops (~ 60 to 70 km altitude) and may not necessarily be the same in the upper atmosphere due to the destruction of these molecules by photodissociation and chemical reactions. Except for CO_2 and possibly N_2, if present, the mixing ratio of all other constituents probably varies throughout the atmosphere. The altitude of the

* Now at Jet Propulsion Laboratory, 4800 Oak Grove Dr. Pasadena, Calif., 91103, U.S.A.

B. M. McCormac (ed.), Atmospheres of Earth and the Planets, 385–399. All Rights Reserved.
Copyright © 1975 by D. Reidel Publishing Company, Dordrecht-Holland.

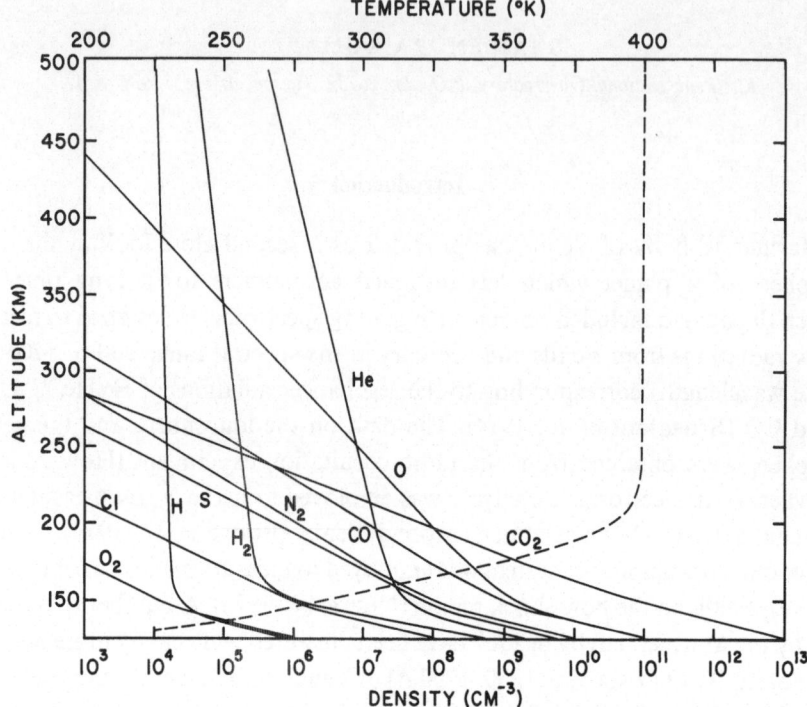

Fi-. 1. Model for the neutral atmosphere of Venus for an exospheric temperature of 400 K. The mixing ratios are $CO_2(0.98)$, $CO(0.01)$, $O(0.01)$, $N_2(10^{-3})$, $H_2(2 \times 10^{-6})$, $He(2 \times 10^{-4})$, $S(5 \times 10^{-5})$, $Cl(10^{-5})$ and $O_2(10^{-7})$. Total density is $\simeq 1.0 \times 10^{13}$ cm^{-3} at 125 km. Eddy diffusion coefficient $K = 10^8$ cm^2 s^{-1}; height of the turbopause is 145 km.

turbopause is not known. As a matter of fact, it is not clear whether a well defined homopause or turbopause exists at all on Venus. However, defining the turbopause at a level above which the distribution of neutral constituents is determined purely by molecular diffusion and below which atmospheric turbulence or eddy mixing keep the atmosphere uniformly mixed (Hunten, 1975), the altitude of the turbopause on Venus is estimated to be about 120 km for $K \simeq 10^6$ cm^2 s^{-1} and about 145 km for $K \simeq 10^8$ cm^2 s^{-1}. The lower value for K is appropriate for the Earth's atmosphere (Hunten, 1975) while the higher value is needed to explain the Mars observations from Mariner 6, 7 and 9 (Strickland et al., 1972, 1973). Although the recent models (Kumar and Hunten, 1974; Sze and McElroy, 1974; Donahue and Liu, 1974) of Venus atmosphere favor higher K value, the low K can not be excluded (see McConnell, 1975, for a discussion).

In addition to CO_2, the positively identified constituents in the upper atmosphere of Venus are H, He, O, C and CO. Other constituents indicated in Table I and shown in Figure 1 are either predicted or represented by marginal evidence for them. Hydrogen has been identified on Venus by a number of experiments (Barth et al., 1967; Moos et al., 1969; Wallace et al., 1971; Moos and Rottman, 1971; Rottman and Moos, 1973) and the H Ly-α emission was again measured by Mariner 10 (Broadfoot et al., 1974). Helium has been identified on Venus for the first time by the Mariner 10

TABLE I

Upper atmosphere of Venus

Exospheric temperature	~ 400 K
Thermopause	~ 220 km
Mesopause	~ 100 km ($T \sim 200$ K)
Turbopause	130–145 km ($K_{eddy} = 10^6$–10^8 cm^2 s)
Mixing ratios at the turbopause:	
CO_2	0.95 ± 0.05
CO	$\sim 1\%$
O	$< 20\%$
N_2	$\sim 1\%$
O_2	$< 10^{-6}$
He	$< 2 \times 10^{-4}$
H	$\sim 10^{-5}$
Cl	10^{-6}–10^{-4}
S	10^{-5}–10^{-4}

UV experiment and it appears to be a major constituent in the exosphere. The measured density of He is, however, at least a factor of ten lower than that shown in Figure 1 (Kumar and Broadfoot, 1975).

The source of He is probably the radioactive decay from U and Th in the planetary crust. The presence of radioactivity was indicated by the Venera 8 landing probe (Vinogradov et al., 1973). Argon and Ne have not yet been detected. Argon, however, may also be present due to the radioactive decay of K^{40}, as on the Earth.

Atomic oxygen was identified on Venus by means of scanning far UV spectrometers flown on rockets (Moos and Rottman, 1971; Rottman and Moos, 1973) and the O1304 Å resonance emission was also measured by Mariner 10. However, the O abundance and distribution are as yet uncertain. Although attempts have been made to derive O abundance from the measured O1304 Å and 1356 Å emission rates (Strickland, 1973; Rottman and Moos, 1973), the process is complicated since a number of excitation mechanisms (photoelectron excitation, CO_2 dissociative excitation, resonance scattering, various chemical reactions, etc.) are involved and some of the excitation parameters such as the g values for electron excitation, are not well known. Additional complication resulted due to the high intensity (~ 17 kR) of the 1304 Å emission measured by the Mariner 10 UV instrument (compared to 5.7 kR measured by the rocket UV experiment). This difference, however, could be due to the different viewing geometry and calibration differences between the two instruments. For instance, the rocket measurement of 5.7 kR is an average over the bright disc (phase angle $\sim 90°$), which implies a subsolar nadir O1304 Å brightness of ~ 13 kR (Rottman and Moos, 1973). The O abundances in the dayside upper atmosphere may be significantly affected by thermospheric circulation (Dickinson and Ridley, 1972). Also a number of mechanisms describing the recycling of O and CO to form CO_2 have been proposed (Stewart, 1971; Prinn, 1971, 1972; McElroy and Donahue, 1972; Parkinson and Hunten, 1972; McElroy et al., 1973). The problem of O distribution on Venus is, however, far from understood. From the present estimates, the O mixing ratio could be up to 20% at 145 km altitude.

Atomic carbon has emission lines in the far UV at 1261, 1277, 1329, 1432, 1470, 1561, 1657, and 1750 Å wavelengths. The Venus spectrum shown in Figure 2 indicates positive signal at all of these wavelengths. A-X fourth positive bands of CO, found to be the brightest far UV features on Mars (Barth *et al.*, 1972), lie in the 1450 to 1850 Å range and there could be a significant overlap between the C and CO emissions. The Venus far UV spectrum shown in Figure 2 requires the presence of both C and CO emissions. Considering various excitation mechanisms, Rottman and Moos (1973) deduced a CO composition of 1 to 10% at a level where the column density of CO_2 is 4×10^{16} cm^{-2} (roughly ~ 155 km). The Mariner 10 UV data also indicate bright emission at 1657 Å, although a significant signal in this channel could be due to the fourth positive (0, 2) band of CO. The intensity at 1657 Å is more than an order of magnitude larger than an intensity computed by McElroy and McConnell (1971). The discrepancy could be reduced appreciably if CO were a major constituent of the upper atmosphere. The actual abundances of C and CO however still remain uncertain.

Rottman and Moos's spectrum may also contain evidence of Cl and S in the upper atmosphere. Chlorine is known to be present on Venus in the form of HCl (Connes *et al.*, 1967) and there is some evidence for H_2SO_4 (Young, 1973, 1974). These molecules would be dissociated by the solar UV at higher altitudes and give rise to Cl and S. Atomic Cl may be detectable via the transitions at 1336 and 1351 Å, which can be resonantly excited by the 1336 Å solar line of C$^+$. An upper limit to the Cl mixing ratio of $\sim 10^{-4}$ is obtained if about half of the 1356 Å emission in Figure 2 is due to Cl (Kumar, 1974a). The Cl mixing ratio probably lies in the range 10^{-6} to 10^{-4}. If the 1400 Å feature in Figure 2 is identified as the ^3P–$^3S°$ transition of S that

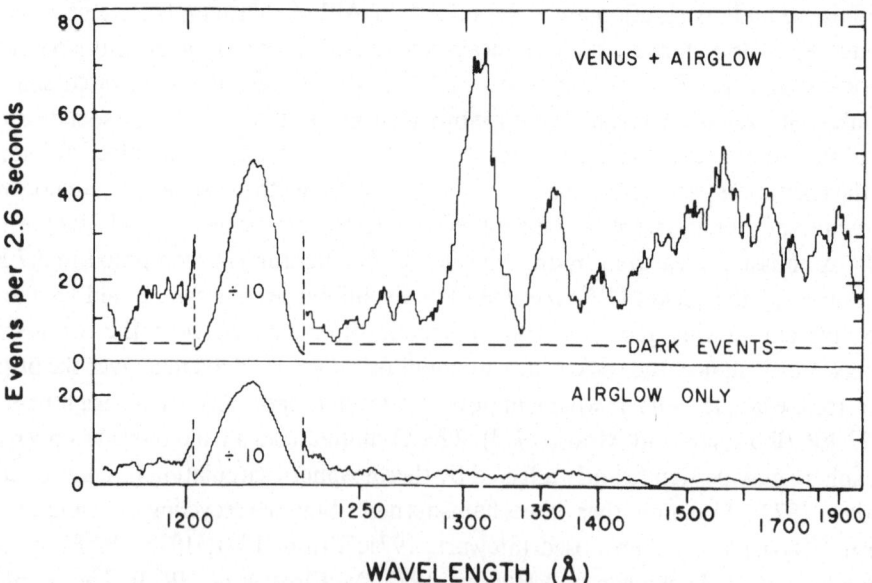

Fig. 2. (Top) The far UV spectrum of Venus (including terrestrial airglow) obtained from a rocket experiment. (Bottom) Terrestrial airglow only. After Rottman and Moos (1973).

forms the multiplet at 1401, 1409, and 1413 Å, the measured brightness would correspond to a S mixing ratio in the range of 10^{-5} to 10^{-4} (Herman, 1974). The laboratory data on the radiative recombination of vibrationally excited CO_2^+ (Zipf and Unger, 1974) may explain several features in the Venus for UV spectrum, therefore Cl and S mixing ratios may be significantly lower than those quoted here.

Molecular H is important in the detailed understanding of photochemistry, cloud formation and H corona on Venus. Although H_2 has not yet been identified, the absence of the $v' = 6$ Lyman band of H_2 in Rottman and Moos's spectrum indicates that there is less than 10^8 cm^{-3} of H_2 at 120 km (Herman, 1974). An upper limit on the H_2 density of $\sim 10^6$ cm^{-3} at ~ 200 km was derived by McElroy and Hunten (1969) to keep the concentration of H atoms (resulting from H_2 photodissociation) consistent with the Mariner 5 Ly-α observations. These H_2 concentrations are well above the mixing ratio of a few ppm assumed in the stratospheric model of Prinn (1971), and Kumar and Hunten's (1974) model of 'hot H' atoms in the Venus corona.

The most abundant gases on the Earth, N_2 and O_2 have not yet been identified on Venus. The upper limit to O_2 mixing ratio is ~ 1 ppm (Carleton and Traub, 1972) near the cloud-top level. In the upper atmosphere Sze and McElroy (1971) require an O_2 mixing ratio of $\sim 10^{-4}$ in their model. The mixing ratio of N_2 could be as high as a few percent.

2.2. Thermal Structure and H Ly-α Emission

The thermosphere is heated by the solar EUV radiation (in analogy with the Earth). The absorption of most of the solar energy takes place near 150 km. No direct measurement of temperature is available above ~ 70 km, therefore the temperature determination in the thermosphere and exosphere is mainly model dependent. The mesopause altitude and temperature are quite uncertain although the estimated mesopause is at ~ 100 km, temperature ~ 200 K. The exosphere temperature (T_∞) can be determined by measuring the airglow emission of H and He as a function of altitude, and also by measuring the electron density profile by the radio occultation method. The latter method is, however, model dependent and requires *a priori* knowledge of the neutral composition.

One of the objectives of the Mariner 10 UV experiment was to resolve the problem of two scale heights for the Ly-α emission from Venus which was indicated by the Mariner 5 observations (Barth *et al.*, 1967, 1968; Wallace, 1969). The Mariner 5 Ly-α data (Barth *et al.*, 1967) could be interpreted in terms of a two-temperature hydrogen exosphere at 350 K and 700 K (Wallace, 1969; Kumar and Hunten, 1974) or a two component exosphere of H and D at 700 K (McElroy and Hunten, 1969; Donahue, 1969). The data obtained from Mariner 10 (Broadfoot *et al.*, 1964) on the H corona around Venus are shown in Figure 3. For comparison, the Ly-α observations of Venus from Mariner 5 are also included in this figure. It appears that the intensity of Ly-α was lower during the Mariner 10 encounter by at least a factor of 2 over the height range 1000 to 3000 km. Since the solar Ly-α flux does not vary by more than 50% as a function of solar activity (Hinteregger, 1970), the observed variation in Venus Ly-α

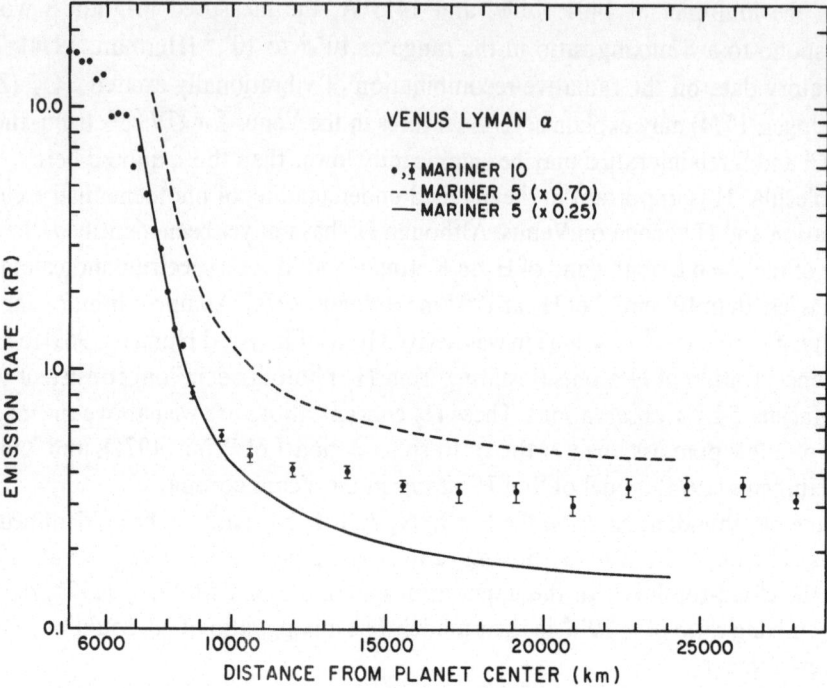

Fig. 3. H Ly-α emission rate versus minimum distance of the line of sight from the center of the planet. Also shown are the data obtained by Mariner 5 scaled down by a factor of 4 for direct comparison at 7000 to 10000 km, and by a factor of 0.7 to allow for a postflight recalibration of the Mariner 5 instrument.

brightness cannot be accounted for in terms of the variation in the solar Ly-α flux only. The calibration differences between the Mariner 5 and Mariner 10 instruments could not account for this variation either because the Ly-α sky background measurements made from the two instruments are compatible (Broadfoot *et al.*, 1974). Evidently the two sets of data suggest a significant variation in the total H content in the exosphere of Venus. The Mariner 10 data however show no indication of the extensive component observed by Mariner 5. Either this extensive component was too faint or non-existent at the time of Mariner 10 encounter. If the observed Ly-α emission from Venus results from the resonance scattering by H, the Mariner 10 data are consistent with an exospheric temperature of 380 K. The thermal escape rate for at this temperature is about 10^4 cm^{-2} s^{-1}, some four orders of magnitudes less than escape rates for the Earth and Mars.

The determination of exospheric temperature from the Ly-α observations is not straightforward because a 700 K D component will show an essentially similar Ly-α distribution as the 380 K H component. Since the H (1215.72 Å) and D (1215.39 Å) Ly-α components differ by only 0.33 Å, the low resolution Ly-α data obtained from Mariner 5 and 10 cannot rule out the possibility of D. Wallace *et al.* (1971), however, attempted to separate the two components from a rocket experiment and did not find any D emission. Unless the D component is highly variable, this experiment in combination with the Mariner 5 Ly-α observations suggested a low exospheric temperature on Venus. The He 584 Å data obtained from Mariner 10 (Kumar, 1974b; Broadfoot *et al.*, 1974) provide the conclusive evidence on the exospheric temperature. These data show that the He distribution is consistent with an exospheric temperature of about 350 K.

What, then, was the dayside exospheric temperature on Venus at the time of Mariner 5 encounter? Although a 700 K temperature was usually taken for granted before Mariner 10, Kumar and Hunten (1974) recently argued that the evidence of a 700 K exosphere was ambiguous and showed that the Mariner 5 ionosphere observations were consistent with a 350 K exosphere as well. The solar activity was somewhat different for the two Mariner encounters (10.7 cm solar flux ~ 75 W m^{-2} Hz^{-1} for Mariner 10, ~ 120 W m^{-2} Hz^{-1} for Mariner 5). However, using the formulae developed for a 'slab approximation' by Gross (1972), we find that the exospheric temperature on Venus during the Mariner 10 encounter would not be more than 10% lower than the value at the time of Mariner 5 encounter. Using the same formula for the Earth's atmosphere we estimate that the corresponding variation in the temperature is $\sim 15\%$, in accord with the CIRA (1972) models. It appears therefore that T_∞ at the time of Mariner 5 encounter was also near 400 K. The two scale heights for Ly-α observed from Mariner 5 should therefore be interpreted in terms of two temperature components of H, the 'normal' component at ~ 350 K and a 'hot' component at ~ 700 K.

Why the temperature should be so low on Venus is puzzling. A similar problem however exists for Mars which also has a CO_2 predominant atmosphere. The analysis of the Mariner 6 and 7 observations of Mars (Stewart, 1972) indicated a temperature considerably less than that predicted earlier from theoretical considerations. Stewart had to postulate several mechanisms for the loss of solar EUV energy, namely radiation in the airglow emissions of CO (Cameron bands), CO_2^+ (A–X and B–X bands) and O (^{3}P–^{1}S, 2972 Å; ^{1}D–^{1}S, 5577 Å), to explain the observed low temperature on Mars. A recalibration of the Mariner instrument, however, revealed that even these emissions would not be enough to explain the low temperature on Mars. The loss of energy in the 15 μm band emission of CO_2 appears to affect the temperature structure in the Earth's atmosphere. A similar phenomenon may also take place on Venus. Further studies will be needed to understand the physics behind the thermal structure of these planets.

3. The Ionosphere and Its Interaction with Solar Wind

3.1. DAYSIDE IONOSPHERE

Figure 4 shows the dayside ionosphere profiles obtained from Mariner 5 (Mariner Stanford Group, 1967, Fjeldbo and Eshelman, 1969) and Mariner 10 (Howard *et al.*, 1974). Both sets of data show a prominent peak near 145 km identified as the main F_1 layer formed by the absorption of solar EUV radiation in the atmosphere. A secondary peak is also apparent in the Mariner 10 data near 187 km which may be identified as an F_2 layer. The observed ledge, however, could also be due to horizontal advection and fluctuations in the ionosphere which cannot be modeled. The signature

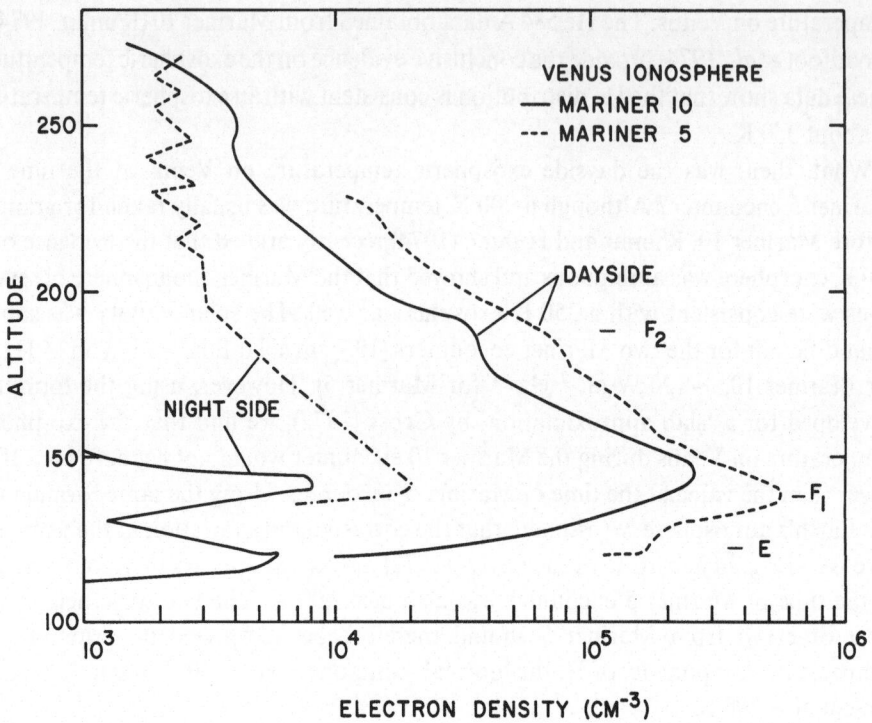

Fig. 4.   Dayside and nightside ionosphere profiles deduced from Mariner 10 (Howard *et al.*, 1974) and
Mariner 5 (Mariner Stanford Group, 1967) measurements.

of a similar peak, if at all, is quite faint in the Mariner 5 data. On the other hand
Mariner 5 data show a noticeable feature at ~125 km, identified as the E layer, which
is not so apparent in the Mariner 10 data. This layer is formed by the absorption of
solar X-rays in the atmosphere (see Figure 4). Overall, the two profiles appear quite
similar below 200 km and the observed differences are probably due to different
occultation points (solar zenith angle 40° for Mariner 5 and 70° for Mariner 10) and
different solar fluxes.

A 350 K exosphere model (Kumar and Hunten, 1974) for the dayside ionosphere is
given in Figure 5, where it is compared with the Mariner 5 data. Taking a temperature of 400 K would not substantially alter this model. The ion chemistry is faster than
diffusion below ~210 km. The main layer is formed by the photoionization of CO_2:

$$CO_2 + h\nu \rightarrow CO_2^+ + e$$
$$\rightarrow \ O^+ + CO + e \tag{1}$$
$$\rightarrow CO^+ + \ O + e.$$

However CO_2^+ would not be the major ion in the presence of more than 1% O,
which may be the case (see Section 2.1). O_2^+ instead would be the dominant ion due
to the reaction,

$$CO_2^+ + O \rightarrow O_2^+ + CO. \tag{2}$$

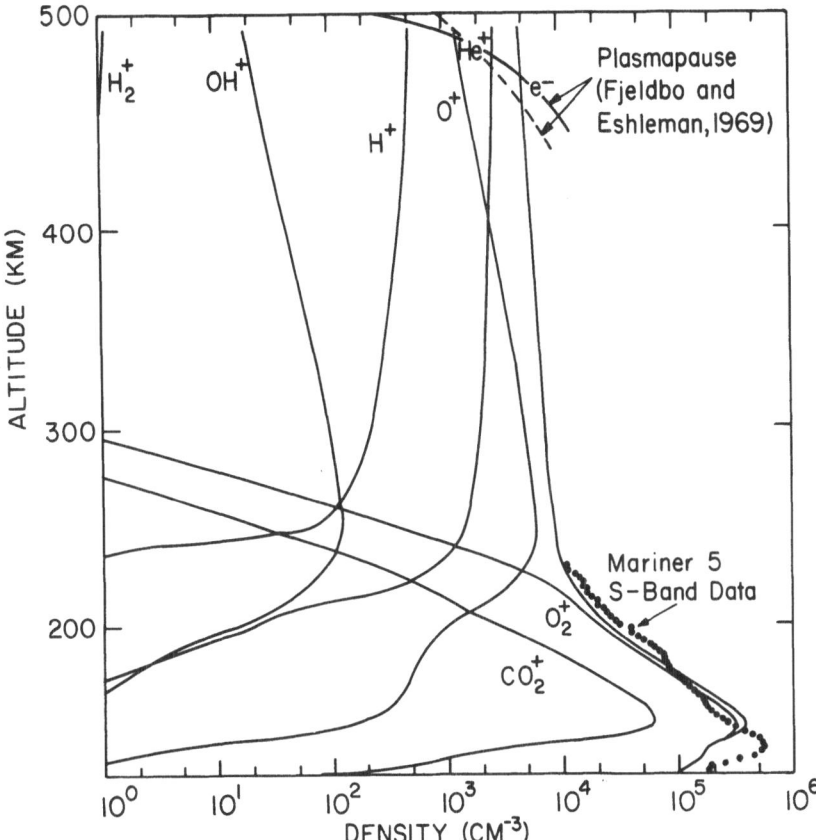

Fig. 5. Ionosphere model for the neutral atmosphere shown in Figure 1. Also shown are the electron density profile obtained from the Mariner 5 S band experiment and the plasmapause profile deduced from Mariner 5 49.8-MHz data (Fjeldbo and Eshleman, 1969).

In the previous models (McElroy, 1968a, 1968b, 1969; Stewart, 1968) a pure CO_2 atmosphere at 700 K was found to fit the Mariner 5 ionosphere data. For a cool exosphere however, photoionization of CO_2 alone is not enough to explain the electron scale height observed. The additional source near 220 km is provided by ionization of O; much of the O^+ is then converted to O_2^+ by the fast reaction:

$$O^+ + CO_2 \rightarrow O_2^+ + CO. \tag{3}$$

Up to 1.5% O could be tolerated without forming a dense F_2 layer, while up to 5% O_2, N_2 or CO in this model does not affect the electron densities noticeably.

Reaction (3) ceases to be important above ~ 250 km, therefore the O^+ is the dominant ion at higher altitudes. He^+ is also produced in abundance since He is a dominant neutral constituent above 260 km. The major loss of He^+ is via slow reaction with O

$$He^+ + O \rightarrow O^+ + He \tag{4}$$

therefore He^+ is also expected to be a major ion in the topside ionosphere. The He^+

density however will be less than that indicated in Figure 5 because of the lower He mixing ratio measured from Mariner 10.

H^+ and H_2^+ will be insignificant because of the rapid reactions:

$$H^+ + CO_2 \rightarrow COH^+ + O \qquad (5)$$

$$H_2^+ + O \rightarrow OH^+ + H. \qquad (6)$$

$$COH^+ + e \rightarrow CO + H \qquad (7)$$

$$OH^+ + e \rightarrow O + H. \qquad (8)$$

Reactions (6) and (8), and charge exchange reaction

$$O^+ + H_2 \rightarrow OH^+ + H \qquad (8)$$

could provide a source of the 'hot H' atoms needed to explain the Mariner 5 Ly-α data (Kumar and Hunten, 1974). In the model shown in Figure 5, these reactions produce 8×10^6 cm^{-2} s^{-1} H atoms in the exosphere. Sze and McElroy (1974) have suggested that an even more important reaction is

$$H^+ + O \rightarrow H + O^+ \qquad (10)$$

which may provide a source of $\sim 5 \times 10^7$ hot H atoms. The sink of hot atoms may be estimated as follows. The lifetime is of the order of twice the time for free fall over a scale height, or approximately 1000 s for a velocity distribution represented by a temperature of 1000 K. A column density of 10^{11} cm^{-2} (model 4 of Wallace [1969]) therefore requires a source of 10^8 cm^{-2} s^{-1}. Alternatively, we may take Wallace's base density, 840 cm^{-3}, and calculate the flux across a surface, again 10^8 cm^{-2} s^{-1}. These estimates are probably high by a factor of 10 or more, because H atoms are slowed inefficiently by collisions with heavier particles. Reduction of the energy by 1/e requires 23 collisions with CO_2, or 9 with 0. The source provided by the photo-chemical reaction in the ionosphere may therefore be enough. McElroy and Hunten (1969) did not include the last factor, and also used a higher estimate for the density of the hot component.

Above ~ 200 km, the Mariner 5 and Mariner 10 ionosphere profiles are considerably different. The Mariner 5 data suggest that at ~ 500 km altitude, a sudden transition occurs from the relatively cool ionospheric plasma ($T_i \sim 3 \times 10^3$ K) of relatively high density ($\sim 10^4$ cm^{-3}) to interplanetary conditions dominated by energetic protons ($T_p \sim 3 \times 10^5$ K) at a density of $\lesssim 10$ cm^{-3}. No data were obtained between 230 km (density $\sim 10^4$ cm^{-3}) and 450 km due to formation of caustics although a height independent electron density of $\sim 10^4$ cm^{-3} is usually taken at these altitudes. On the other hand, Mariner 10 measurements show that the electron density continues to decrease above 200 km to $\sim 10^3$ cm^{-3} at ~ 270 km with only slightly larger scale height than that below 200 km.

This difference in the topside ionospheric structure may be understandable in terms of a dynamic interaction of solar wind with the planetary ionosphere. Due to the absence of a planetary magnetic field ($<4\,\gamma$ at the surface), the solar wind interacts directly with the Venus atmosphere. From the *in situ* measurements of magnetic field (Ness *et al.*, 1974) and electron density, temperature and energy distribution (Bridge *et al.*, 1974) made from Mariner 10, it is fairly certain that a bow shock exists; however, the nature of the obstacle responsible for the deflection of the solar wind flow continues to elude identification. Four types of models have been suggested to describe the nature of this obstacle: (1) Existence of a boundary surface, an iono-sphere, in the form of a tangential discontinuity at the interface between the solar wind and the ionosphere separating the two plasmas of solar wind and planetary ionosphere (Spreiter *et al.*, 1970), (2) A direct interaction of the solar wind with the atmosphere in which the solar wind is decelerated by the mass loading due to plan-etary photoions and subsequently penetrates deep into the ionosphere at subsonic velocities (Cloutier *et al.*, 1969, originally for Mars; Cloutier, 1970), (3) The inter-planetary magnetic field is compressed against the highly conducting ionosphere to induce a magnetic field on the dayside strong enough to cause a standing shock wave (Dessler, 1968; Johnson and Midgeley, 1969; Michel, 1971; Cloutier and Daniell, 1973), and (4) A comet type interaction leading to a nearly shockless deceleration and deflection of the solar wind flow (Wallis, 1972, 1973). A combination of (1) and (3) was discussed by Bauer *et al.* (1970) and Herman *et al.* (1971), by assuming He^+ as the dominant ion in the topside ionosphere. The Mariner 10 magnetic field data are essentially consistent with all four models (Ness *et al.*, 1974) because a relatively broad pulsation type bow shock was detected and the observed position of the bow shock can be explained by adjusting the various free parameters in these models even though the nature of the obstacle is different in each case.

In the first model, the ionopause (or plasmapause) on a magnetic field free planet would occur at the level where the solar wind ram pressure equals the ionospheric plasma pressure. The ionopause boundary is characterized by H/r_0 in this model, where H is the plasma scale height and r_0 is the planetocentric distance of the iono-pause in the subsolar direction. Spreiter *et al.* (1970) obtained a best representation of the Mariner 5 bow shock with $H/r_0 = 0.25$. Mariner 10 data, however, require a much sharper obstacle than this. As noted by the Mariner 10 experimenters (Bridge *et al.*, 1974), even the Mariner 5 data would require a sharper obstacle. Due to this reason and the fact that during Mariner 10 encounter the solar wind ram pressure was about a factor of 1.5 greater than that during Mariner 5 encounter, the two sets of data appear consistent. The best fit to the Mariner 10 data is characterized by $H/r_0 = 0.01$. A sharp obstacle required by the Mariner 10 plasma science experiment is also consistent with the compressed ionosphere observed. The compressed ions are perhaps squeezed laterally into the tail thereby reducing the plasmapause al-titude.

Although the hydrodynamic model discussed above provides a reasonable explana-tion of the solar wind interaction, it does not include any possible solar wind diffusion across the ionopause. Model 2 (Cloutier *et al.*, 1969) allows a subsonic solar wind flow deep into the atmosphere. A one dimensional interaction of this type can be described by a downward transport of ionospheric plasma with a constant velocity (Bauer, 1973). In this model, the solar wind suppresses the ionosphere such that the ion scale height is reduced to half as much in the photochemical equilibrium only. Bauer and Hartle (1974) have suggested that the Mariner 10 topside ionosphere may be understood in terms of this model. They interpret the F_2 ledge consisting of O^+ if the number density ratio $[O]/[CO_2] = 60$ at this level for a 380 K exosphere. This ratio, however, corresponds to an exceedingly large O mixing ratio of $\sim 60\%$ at 145 km. The main layer in this model can still be described as a dynamically unaffected F_1 layer with neutral temperature corresponding to 380 K and consisting of O_2^+ and CO_2^+, in the same manner as described in Figure 5. It should be emphasized that although some of the observed characteristics of the topside ionosphere may be understood in terms of the simple models mentioned above, it appears that the interaction of the solar wind with Venus is a more complex phenomenon than previously anticipated. For example, the topside ionosphere may be in rapid motion while the Mariner 5 and Mariner 10 ionospheric profiles are simply two snapshots of a highly dynamic system. The existing models are only one dimensional, and a two dimensional or three dimensional treatment may be necessary to describe the observed phenomena.

3.2. The night ionosphere

The night ionosphere profiles obtained from Mariner 5 and Mariner 10 are also shown in Figure 3. Both sets of data showed a major peak near 140 km with a density $\sim 3\%$ of its dayside counterpart. Mariner 10 measurements also showed a secondary peak at ~ 120 km almost as pronounced as the main peak. Although significant data are available, the nightside ionospheric processes are much less understood than on the dayside. Due to the long rotation period of Venus and in the absence of any known ionization source at night it has been suggested (Mariner Stanford Group, 1967; McElroy and Strobel, 1969) that the main layer on the nightside may arise from the rapid transport of He^+ and H_2^+ from the dayside and then charge exchange of these ions with CO_2. H_2^+, however, is unlikely in view of the dayside model shown in Figure 5. The transport of O^+ and then conversion to O_2^+ via reaction (3) should also be considered. Although horizontal winds of ~ 200 m s^{-1} required in this model are not unlikely (Dickinson, 1971; Dickinson and Ridley, 1972; Prinn, 1972; Murray et al., 1974), the presence of two pronounced peaks may require additional sources. The photoionization of CO_2 and O by the nightglow 584 Å emission observed from Mariner 10 (Broadfoot et al., 1974) may be important. A similar suggestion has been made for the nightside ionosphere of the earth (Ogawa and Tohmatsu, 1966; Strobel et al., 1974). If a sufficient amount of N_2 is present on Venus, then the photoionization of NO by nightglow H Ly-α emission may also be significant. Of course, a sporadic E phenomenon as on the earth may also appear on Venus.

In addition to the main layer, the Mariner 5 data showed that electron densities of $\sim 10^2$–10^3 cm^{-3} extended out to ~ 10000 km in the wake region. Banks and Axford (1970) suggested that for solar zenith angle $\psi > 70°$, the ions in the topside ionosphere will be pushed toward the tail by the solar wind such that as viewed from behind the planet topside ionosphere near the terminator would appear as a ring shaped source of ions streaming into the tail. H^+ and D^+ ions considered in their model are, however, unlikely due to a rapid charge exchange with CO_2 (Banks, 1971) and should be replaced by He^+ and O^+. Although this transport mechanism appears quite attractive, detailed calculations have not yet been performed. A related problem may be the relatively bright UV emission detected on the nightside which extends to thousands of kilometers (Broadfoot *et al.*, 1974) from the surface. These data have not yet been analyzed and the nature and identification of this emission are unknown.

4. Conclusion

The temperature in the exosphere of Venus is ~ 400 K for low solar activity applicable at the time of Mariner 5 and Mariner 10 encounters. In addition to CO_2, the upper atmosphere of Venus contains H, He, O, C, CO and probably also N_2, Cl and S. The H content in the exosphere appears highly variable although the thermal escape rate is only $\sim 10^4$ cm^{-2} s^{-1}, some four orders of magnitude less than that on the Earth and Mars. Atomic O mixing ratio is still uncertain and may be as high as 20%. The ionosphere of Venus has a prominent peak near 140 km identified as an F_1 layer, which appears both on the day and nightside. An E layer near 125 km and, possibly, an F_2 layer near 170 km are also present. Although the dayside ionosphere can be explained in terms of the absorption of solar radiation by CO_2, O and He, the source of nightside ionosphere is not well understood. The transport of ions from day to nightside may be important; however, an additional source may be needed to explain the nightside counterpart of the E layer. Observational evidence exists from Mariner 5 and Mariner 10 that the solar wind interacts directly with the atmosphere of Venus and results in the formation of a bow shock. A number of models for this interaction have been proposed. Some of the observed phenomena can be explained in terms of a pressure balance at the boundary, namely the ionopause, between the solar wind ram pressure and the planetary plasma pressure just below the ionopause. The Mariner 10 dayside ionosphere, however, seems to require a subsonic solar wind flow deep into the ionosphere. Most of these models are one dimensional and at least a two dimensional or three dimensional treatment including non thermal phenomena at the boundary will be needed to fully explain the observations.

References

Banks, P.: 1971, *J. Geophys. Res.* **76**, 8455.
Banks, P. M. and Axford, W. I.: 1970, *Nature* **225**, 924.
Barth, C. A.: 1968, *J. Atmospheric Sci.* **25**, 564.
Barth, C. A., Pearce, J. B., Kelly, K. K., Wallace, L., and Fastie, W. G.: 1967, *Science* **158**, 1675.
Barth, C. A., Hord, C. W., and Steward, A. I.: 1972, *Science* **175**, 309.
Bauer, S. J.: 1973, *Physics of Planetary Ionospheres*, Springer-Verlag, New York, Heidelberg, Berlin.
Bauer, S. J. and Hartle, R. E.: 1974, *Geophys. Res. Letters* **1**, 7.
Belton, M. J. S.: 1968, *J. Atmospheric Sci.* **25**, 596.
Bridge, H. S. *et al.*: 1974, *Science* **183**, 1293.
Broadfoot, A. L., Kumar, S., Belton, M. J. S., and McElroy, M. B.: 1974, *Science* **183**, 1315.

CIRA: 1972, *COSPAR International Reference Atmosphere*, North-Holland Publishing Co., Amsterdam.
Cloutier, P. A., McElroy, M. B., and Michel, F. C.: 1969, *J. Geophys. Res.* **74**, 6215.
Cloutier, P. A.: 1970, *Radio Sci.* **5**, 387.
Cloutier, P. A. and Daniell, R. E.: 1973, *Planetary Space Sci.* **21**, 463.
Connes, P., Connes, J., Benedict, W. S., and Kaplan, L. D.: 1967, *Astrophys. J.* **147**, 1230.
Dessler, A. J.: 1968, in J. C. Brandt and M. B. McElroy (ed.), *The Atmospheres of Venus and Mars*, Gordon and Breach Science Publishers, Inc.
Dickinson, R. E.: 1971, *J. Atmospheric Sci.* **28**, 885.
Dickinson, R. E. and Ridley, E. C.: 1972, *J. Atmospheric Sci.* **29**, 1557.
Donahue, T. M.: 1969, *J. Geophys. Res.* **74**, 1128.
Fjeldbo, G. and Eshleman, V. R.: 1969, *Radio Sci.* **4**, 879.
Gross, S. H.: 1972, *J. Atmospheric Sci.* **29**, 214.
Herman, J. R.: 1974, preprint.
Herman, J. R., Hartle, R. E., and Bauer, S. J.: 1971, *Planetary Space Sci.* **19**, 1971.
Hinteregger, H. E.: 1970, *Ann. Geophys.* **26**, 547.
Howard, H. T. *et al.*: 1974, *Science* **183**, 1297.
Hunten, D. M.: 1971, *Space Sci. Rev.* **12**, 539.
Hunten, D. M.: 1975, this volume, p. 59.
Johnson, F. S. and Midgeley, J. E.: 1969, *Space Res.* **9**, 760.
Kumar, S.: 1974a, *Geophys. Res. Letters* **1**, 153.
Kumar, S.: 1974b, Paper, Amer. Astron. Soc.-Div. for Plan. Sci.
Kumar, S. and Hunten, D. M.: 1974, *J. Geophys. Res.* **79**, 2529.
Kumar, S. and Broadfoot, A. L.: 1975, Amer. Astron. Soc. Div. for Plan. Sci.
Mariner Stanford Group: 1967, *Science* **158**, 1678.
Marov, M. Y.: 1972, *Icarus* **16**, 415.
McConnell, J. C.: 1975, this volume, p. 361.
McElroy, M. B.: 1968a, *J. Geophys. Res.* **73**, 1513.
McElroy, M. B.: 1968b, *J. Atmospheric Sci.* **25**, 574.
McElroy, M. B.: 1969, *J. Geophys. Res.* **74**, 29.
McElroy, M. B. and Donahue, T. M.: 1972, *Science* **177**, 986.
McElroy, M. B. and Hunten, D. M.: 1969, *J. Geophys. Res.* **74**, 1720.
McElroy, M. B. and McConnell, J. C.: 1971, *J. Atmospheric Sci.* **28**, 879.
McElroy, M. B. and Strobel, D. F.: 1969, *J. Geophys. Res.* **74**, 1118.
McElroy, M. B., Sze, N. D., and Yung, Y. L.: 1973, *J. Atmospheric Sci.* **30**, 1437.
Michel, F. C.: 1971, *Rev. Geophys. Space Sci.* **9**, 427.
Moos, H. W. and Rottman, G. J.: 1971, *Astrophys. J. Letters* **169**, L127.
Moos, H. W., Fastie, W. G., and Bottema, M.: 1969, *Astrophys. J.* **155**, 887.
Murray, B. C. *et al.*: 1974, *Science* **183**, 1307.
Ness, N. F., Behannon, K. W., and Lepping, R. P., Whang, Y. C., and Schatten, K. H.: 1974, *Science* **183**, 1301.
Ogawa, T. and Tohmatsu, T.: 1966, *Rept. Ionosphere Space Res. Japan* **20**, 395.
Parkinson, T. D. and Hunten, D. M.: 1972, *J. Atmospheric Sci.* **29**, 3180.
Prinn, R. G.: 1971, *J. Atmospheric Sci.* **28**, 1058.
Prinn, R. G.: 1972, Paper, Advanced Study Inst.-Physics & Chem. of Upper Atmos., Orleans, France.
Rottman, G. J. and Moos, H. W.: 1973, *J. Geophys. Res.* **78**, 8033.
Spreiter, J. R., Summers, A. L., and Rizzi, A. W.: 1970, *Planetary Space Sci.* **18**, 1281.
Stewart, A. I.: 1972, *J. Geophys. Res.* **77**, 54.
Stewart, R.: 1968, *J. Atmospheric Sci.* **25**, 578.
Stewart, R.: 1971, *J. Atmospheric Sci.* **28**, 1069.
Strickland, D. J.: 1973, *J. Geophys. Res.* **78**, 2827.
Strickland, D. J., Thomas, G. E., and Sparks, P. R.: 1972, *J. Geophys. Res.* **77**, 4052.
Strickland, D. J., Stewart, A. I., Barth, C. A., Hord, C. W., and Lane, A. L.: 1973, *J. Geophys. Res.* **78**, 4547.
Strobel, D. F., Young, T. R., Meier, R. R., Coffey, T. P., and Ali, A. W.: 1974, *J. Geophys. Res.* **79**, 3171.
Sze, N. O. and McElroy, M. B.: 1974, *Planetary Space Sci.*, submitted.
Traub, W. A. and Carleton, N. P.: 1972, *Bull. Amer. Astron. Soc.* **4**, 371.
Vinogradov, A. P., Surkov, Y. A., Kirnozov, F. F., and Glazov, V. N.: 1973, *Dokl. Akad. Nauk SSSR* **208**, 576.

Wallace, L.: 1969, *J. Geophys. Res.* **74**, 115.
Wallace, L., Stuart, F. E., Nagel, R. H., and Larson, M. D.: 1971, *Astrophys. J. Letters* **168**, L29.
Wallis, M.: 1972, *Cosmic Electrodyn.* **3**, 45.
Wallis, M.: 1973, *Planetary Space Sci.* **21**, 1647.
Young, A. T.: 1973, *Icarus* **18**, 564.
Young, A. T.: 1974, *Science* **183**, 407.
Zipf, E. C. and Unger, G.: 1974, private communication.

OUTER SOLAR-SYSTEM AERONOMY

DARRELL F. STROBEL

Naval Research Laboratory, Code 7750, Washington, D.C. 20375, U.S.A.

1. Introduction

With the close encounter to Jupiter by Pioneer 10, the exploration of the outer planets has begun. Generally, the outer planets with their strong gravitational fields have retained their original composition and their present atmospheric composition should be indicative of solar abundances. In many respects, their reducing atmospheres are similar to our present speculations on the nature of the Earth's primitive atmosphere. Exploration of the outer solar-system is essential if we are to understand the formation and evolution of our own atmosphere.

Some satellites of the outer planets also possess atmospheres that lead to rather interesting aeronomical processes. For example, a light gas may escape the gravitational field of the satellite but not the stronger planetary field. This leads to the formation of a gaseous toroidal cloud around the planet that can alter the net escape rate of atmospheric constituents from the satellite (McDonough and Brice, 1973a, b; Judge and Carlson, 1974).

The photochemistry of these atmospheres is considerably more complex than the terrestrial analogues. It is well known that hydrocarbons in the presence of UV radiation form complex hydrocarbons known as polymers.

2. Composition

In Table I the planets and satellites with atmospheres in the outer solar-system are listed with their physical properties and composition, if known. For comparison purposes the physical properties of the Earth's moon are also tabulated. Those constituents which are likely to be present in the atmosphere but have not been detected with spectroscopic certainty are followed by a question mark. With negligible escape of H and H_2, the outer planets have retained their primitive atmospheres and we would expect the fully saturated hydrides of the most abundant reactive atoms (CH_4, NH_3, H_2O) plus H_2 and H_e to be the principal atmospheric constituents. Molecular hydrogen is undoubtedly the major constituent and the H_e/H_2 ratio should be given by solar abundances, 0.11 by volume (Hunten and Münch, 1973). For Jupiter the spectroscopically derived abundance for CH_4 is consistent with the solar abundance of C, $C/H \sim 3.5 \times 10^{-4}$ by number (McElroy, 1969). The water is frozen out deep in these atmospheres and the NH_3 is also saturated over an extended altitude range. As a consequence their concentrations in the upper atmospheres will be extremely sensitive to the respective cold trap temperatures.

Since atmospheric escape is possible from the satellites, their atmospheric composi-

B. M. McCormac (ed.), Atmospheres of Earth and the Planets, 401–408. All Rights Reserved.
Copyright © 1975 by D. Reidel Publishing Company, Dordrecht-Holland.

TABLE I

Physical properties and atmospheric composition

Planet	Radius (R_E)[a]	Mass (M_e)[b]	Mean density $(g\ cm^{-3})$	Atmospheric composition
Jupiter	11.2	318	1.4	H_2, CH_4, NH_3, CH_3D, HD, H, C_2H_6, C_2H_2, PH_3, He
Saturn	9.5	95.2	0.7	H_2, CH_4, C_2H_6, He?, NH_3? PH_3?
Uranus	3.8	14.6	1.6	H_2, CH_4, He?, NH_3?
Neptune	3.5	17.2	2.5	H_2, CH_4, He?, NH_3?
Titan	0.4	0.023	2.1	CH_4, C_2H_6, H_2?
Io	0.28	0.015	3.5	Na, H
Ganymede	0.4	0.025	2.0	?
Moon	0.27	0.012	3.3	--

[a] in units of Earth radii.
[b] in units of Earth masses.

tion cannot be inferred from solar abundance arguments. Direct spectroscopic observations are required. On Titan large amounts of CH_4 are present (Kuiper, 1944) with observable amounts of C_2H_6 (Gillett *et al.*, 1973; Danielson *et al.*, 1973). For the other satellites the principal atmospheric constituents remain unknown.

3. Thermal Structure

A distinctive feature of the outer solar-system is the near IR thermal emission observed from their atmospheres: Jupiter (Gillett *et al.*, 1969), Saturn (Gillett and Forrest, 1974), and Titan (Gillett *et al.*, 1973). In Figure 1 the Saturn spectrum is presented as representative of these spectra. Of most interest are the pronounced emission peaks at 7.7 μm (v_4 band of CH_4) and at 12.2 μm (v_9 band of C_2H_6). The most plausible interpretation of these emission spectra is the presence of a temperature inversion in their stratospheres similar to the Earth's O_3 layer (Gillett *et al.*, 1969; Danielson *et al.*, 1973). Gillett *et al.* (1969) suggested that the temperature inversion is due to the near IR solar heating in 3.3 μm CH_4 band with cooling in the 7.7 μm CH_4 band and the 12.2 μm C_2H_6 band for thermal energy balance. The most detailed thermal calculations confirm the temperature inversion model for Jupiter (Wallace *et al.*, 1974). Additional heat sources are the weak CH_4 bands in the region 0.7 to 2.3 μm (Wallace *et al.*, 1974) and absorption of visible and near UV solar radiation by fine dust particles (Axel, 1972). In an analysis of the visible and UV geometric albedo measurements of Jupiter, Axel (1972) found it necessary to postulate the presence of fine absorbing particles in the stratosphere. The heating rate associated with these particles can be substantially larger than the energy absorbed by CH_4 in the 3.3 μm band. Suggested sources of these particles include hydrocarbon photochemical products, meteoric dust, and N_2H_4 condensation.

In the vicinity of the mesopause the upper atmosphere is cooled by IR active hydrocarbons. For mesopause temperatures ~ 150 K, C_2H_2 is most effective in cooling

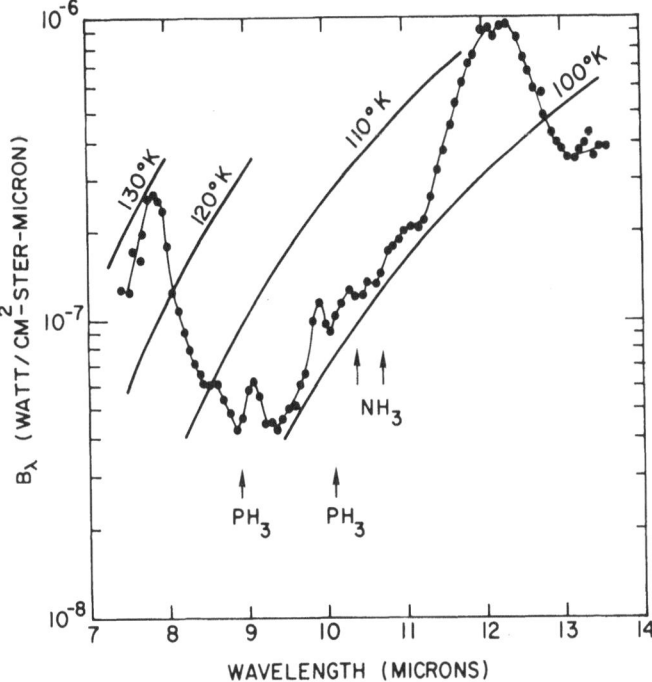

Fig. 1.   Surface brightness of Saturn as a function of wavelength. For Titan the NH_3 and PH_3 features are not evident. For Jupiter the NH_3 features are more pronounced and mask the 12.2 μm C_2H_6 peak. The locations of the Q-branches of the ν_2 band of NH_3 and the ν_2 and ν_4 bands of PH_3 are indicated (after Gillett and Forrest, 1974).

the upper atmosphere by radiating IR energy in its 13.7μm band (Strobel and Smith, 1973). At higher temperatures (~ 200 K) CH_4 is the most important IR radiator. At very cold temperatures (~ 100 K) Strobel (1974a) has noted the potential importance of CH_3 which can radiate at 16.5 μm (Tan *et al.*, 1973).

Strobel and Smith (1973) concluded that the globally averaged vertical temperature contrast in the thermospheres of the outer planets is $\leqslant 15$ K. These small temperature gradients are due to the small solar EUV flux (at these distances from the Sun) and the high thermal conductivity of H_2 as pointed out by Gross and Rasool (1964).

4. Photochemistry of H_2

Molecular hydrogen, the major constituent of the outer planets, has a dissociation continuum below 845 Å and an ionization continuum below 804 Å (Cook and Metzger, 1964). Discrete absorption in the Lyman and Werner bands can lead to fluorescent dissociation of H_2 (cf. Field *et al.*, 1966; Stecher and Williams, 1967). The deposition of solar EUV radiation in a H_2 atmosphere results initially in the production of primarily H_2^+ ions, which react with H_2 to produce $H_3^+ + H$ and break a H_2 bond. If He is present the production of He^+ ions will result in the dissociation of H_2 by ion-molecule reactions. For each H_2 molecule and He atom ionized two H atoms will be produced (Gross and Rasool, 1964). Three body recombination of H atoms is exceedingly slow in the ionosphere and consequently there is a large downward flux of H atoms from this region. For Jupiter the globally averaged downward H atom flux is $\sim 7 \times 10^8$ cm^{-2} s^{-1} (Strobel, 1973c). Appropriate scaling of the solar flux yields estimates for the other planets.

The planetary albedo at Ly-α is in simple terms a measure of the H atom column density above the absorbing CH$_4$ layer. This column density is a sensitive function of the eddy diffusion coefficient, K in the vicinity of the turbopause (Hunten, 1969). Wallace and Hunten (1973) placed an upper limit of 10^6 cm^2 s^{-1} on this coefficient for Jupiter based on the rocket measurement of 4.2 kR($+2.1$, -1.4) Ly-α brightness by Rottman et al. (1973). Recent Ly-α measurements by Pioneer 10 (Judge and Carlson, 1974) and the *Copernicus* satellite (Jenkins et al., 1974) however are <1 kR and imply that mixing is considerably more vigorous near the turbopause than previously anticipated ($K > 10^8$ cm^2 s^{-1}). Measurements of the Ly-α brightness around the satellites will provide important information on processes that produce the gaseous toroidal clouds discussed in the Introduction.

The formation of an ionosphere in a H$_2$ dominated atmosphere has been most recently discussed by Atreya et al. (1974) and Capone and Prasad (1973). Of particular importance are the major sources of H$^+$ ions as a H$^+$ plasma can only recombine radiatively at a slow rate ($\sim 7 \times 10^{-12}$ cm^3 s^{-1}). The H$_3^+$ ions produced by the reaction H$_2^+ +$ H$_2$ dissociatively recombine rapidly ($\sim 4 \times 10^{-7}$ cm^3 s^{-1}). Thus H$^+$ will be the major ion down to a level where the three body reaction H$^+ +$ H$_2 +$ H$_2 \rightarrow$ H$_3^+ +$ H$_2$ proceeds rapidly. This behavior is illustrated in Figure 2 for the Jovian ionosphere. The possible presence of Na complicates matters by leading to the formation of a low altitude ionization layer (Atreya et al., 1974).

5. Photochemistry of Hydrocarbons

The principal hydrocarbon in the outer solar system is CH$_4$. C$_2$H$_6$ has been detected in the atmospheres of Jupiter, Saturn, and Titan (Ridgway, 1974; Gillett and Forrest, 1974; Gillett et al., 1973). C$_2$H$_2$ has been identified only in the Jovian atmosphere (Ridgway, 1974). Hydrocarbons in the presence of UV radiation can form polymers. Based on photochemical models for the outer planets Strobel (1974c) concluded that only a small percentage of dissociated CH$_4$ molecules are converted to complex hydrocarbons. To first order a closed photochemical model can be constructed; the principal reactions are schematically presented in Figure 3. Approximately 70% of the solar photons which dissociate CH$_4$ are at Ly-α where the primary processes are (Rebbert and Ausloos, 1972; Ausloos, 1972)

$$CH_4 + h\nu(\text{Ly-}\alpha) \rightarrow {}^1CH_2 + H_2 \quad 92\%$$
$$CH + H + H_2 \quad 8\% \tag{1}$$

where ^{1}CH$_2$ denotes the singlet state of CH$_2$.

The only chemical means (other than energetic photons) for breaking the bond of two C atoms is the reaction sequence: $H + C_2H_4 \rightarrow C_2H_5$, $H + C_2H_5 \rightarrow 2CH_3$. For this destruction to be important a large H atom concentration is required at pressures ~ 10 mb. As a consequence there is some production of higher hydrocarbons. To

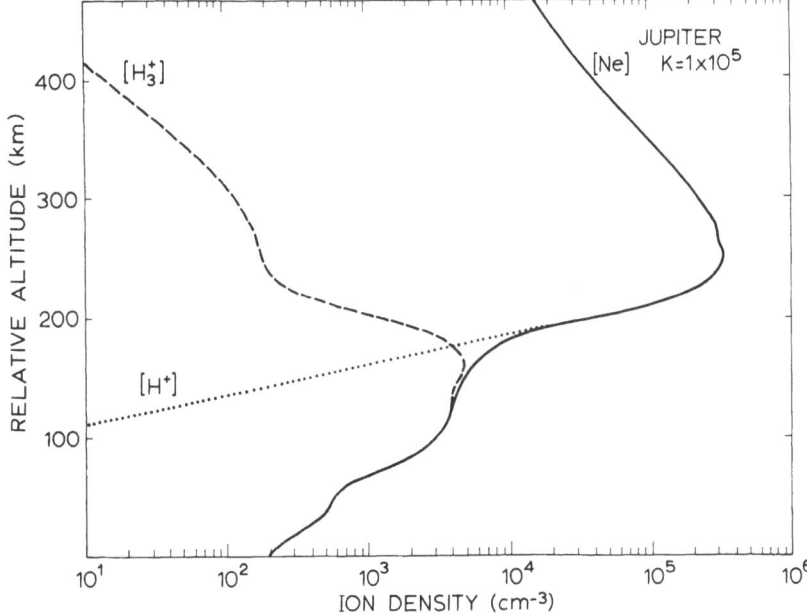

Fig. 2.   Model for Jupiter's ionosphere with eddy diffusion coefficient $K = 1 \times 10^5$ cm^2 s^{-1} and He mixing ratio 0.24. The maximum HeH$^+$, H$_e^+$, and H$_2^+$ densities are less than 10^1 cm^{-3} (after Atreya *et al.*, 1974).

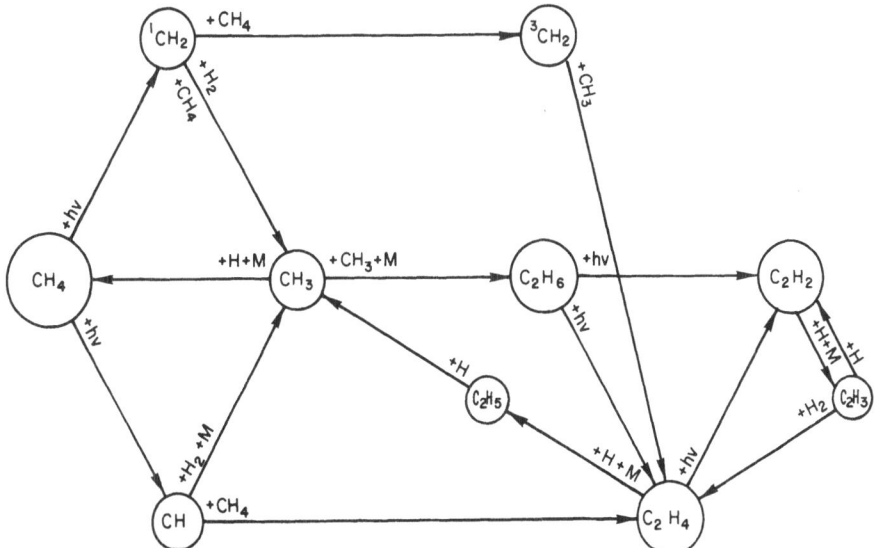

Fig. 3.   Principal reactions of hydrocarbon photochemistry in the outer solar system atmospheres
(after Strobel, 1974c).

conserve C atoms a downward flow of C_2H_6 and C_2H_2 is balanced by an upward flow of CH_4. It is postulated that a deep circulation is present in the major planets that transports higher hydrocarbons to the hot, dense interior where they undergo thermal decomposition to produce fresh CH_4 which is transported upwards to replenish the CH_4 destroyed in photolysis.

The photochemical model is most sensitive to the $[CH_4]/[H_2]$ mixing ratio, the escape rate of H atoms from the atmosphere, and the atmospheric mixing rate (eddy diffusion coefficient). From Figure 3 it is evident that as the $[CH_4]/[H_2]$ ratio increases, the production rates of C_2H_4 and C_2H_2 will increase. The fate of the CH_3 radical depends on the $[H]/[CH_3]$ ratio and determines the rate at which CH_4 is recycled. Although the escape rate of H atoms from the major planets is negligible, it can be substantial from their satellites and actually control the H atom density distribution. A large escape rate depresses the H concentration and results in large conversion rates of CH_4 to C_2H_6 and C_2H_2 (>90% for Titan (Strobel, 1974a)). Also large concentrations of C_2H_2 will efficiently remove H atoms by catalytic recombination as illustrated in Figure 3, $H + C_2H_2 \rightarrow C_2H_3$, $H + C_2H_3 \rightarrow H_2 + C_2H_2$ (Strobel, 1973a).

The most abundant hydrocarbons produced in CH_4 photolysis are C_2H_6 and C_2H_2. They have vertical density distributions of the form (Strobel, 1974b):

$$n = c \, \text{exo}\left(\frac{-z}{H_{av}}\right) + \frac{\phi_0}{K}\left(\frac{1}{H_{av}} - \frac{1}{H_k}\right)^{-1}, \qquad (2)$$

where $K = K_0 \exp(z/H_k)$, H_{av} is the atmospheric scale height, c is an integration constant, and ϕ_0 is the downward flux approximately equal to the column production rate. Typical hydrocarbon densities are illustrated in Figure 4 for the Jovian atmosphere. C_2H_6 is a stable molecule in a cold, reducing atmosphere. If the interior conditions do not require rapid downward flow, then the C_2H_6 density profile is given approximately by the first term of (2), i.e., C_2H_6 is mixed. For C_2H_2 some chemical removal occurs when the ratio $[H]/[H_2]$ is very small. A large downward flux is required to balance this loss. Consequently its density profile is represented by the second term of Equation (2). The rapid decrease in the H concentration below 200 km is due to catalytic removal by C_2H_2 (Strobel, 1973a). The predicted C_2H_6 and C_2H_2 abundances are in reasonable agreement with the observational estimates given by Ridgway (1974). Photochemical models for Titan predict observable amounts of C_2H_6 and C_2H_2 (Strobel, 1974a).

6. Photochemistry of NH_3

The NH_3 photochemistry is considerably simpler than its hydrocarbon counterpart. The only product of NH_3 photolysis is $NH_2(\tilde{X}^2B_1)$ radicals since UV radiation shortward of 1600 Å is absorbed by CH_4. The essential chemical reactions are outlined in Figure 5. For pressures characteristic of the stratosphere, the reaction of NH_2 with itself is competitive with the reaction $H + NH_2$. Unless $[H] \gg [NH_2]$

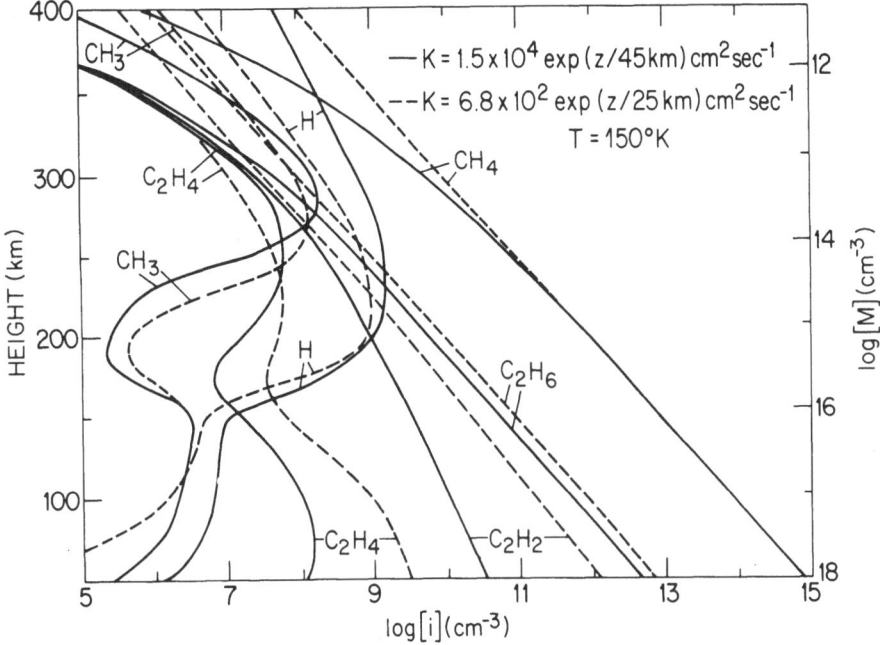

Fig. 4. Hydrocarbon density profiles for the Jovian atmosphere with indicated eddy diffusion profile. Solid line: $K \propto [M]^{-0.5}$ dashed line: $K \propto [M]^{-1}$, where $[M]$ = number density of atmospheric gas (after Strobel, 1974b).

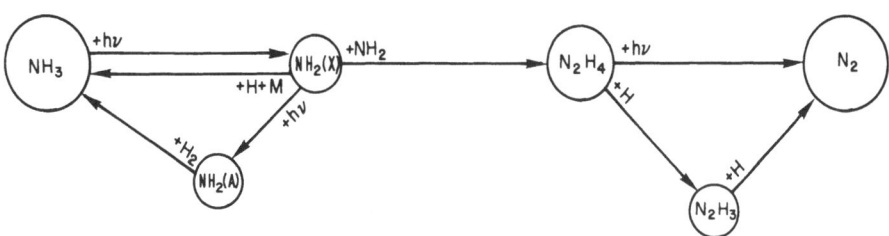

Fig. 5. Principal reactions of NH_3 photochemistry in outer solar system atmospheres (after Strobel, 1973b).

and/or the eddy mixing rates are large compared to the dissociation rate, there will be significant irreversible photochemical destruction of NH_3. Once N_2H_4 is formed it is either chemically converted to N_2 and $2H_2$ or it will condense out at low temperatures to form a photochemical smog.

The CH_4 dissociation rate is sufficiently slow that it is essentially mixed below the turbopause. For NH_3 which absorbs out at 2300 Å, its dissociation rate can be competitive with atmospheric mixing rates. The NH_3 density distribution is thus described by two relevant vertical scales: H_{av} and $(K/J_{net})^{1/2}$, the 'photomechanical' scale height, where J_{net} is the net destruction rate of NH_3 (Strobel, 1973b). In the limit of fast mixing NH_3 approaches a mixed distribution, while for slow mixing substantial photochemical destruction of NH_3 occurs and the NH_3 density profile departs significantly from a mixed distribution. For Jupiter a typical value for J_{net} is $\sim 3 \times 10^{-7}$ s^{-1} (Strobel, 1973b). Thus for $K \sim 2 \times 10^4$ cm^2 s^{-1}, the appropriate NH_3 scale height is ~ 3 km, while $H_{av} \sim 17$ km. Calculations of the UV albedo at 2000 to 2300 Å as a function of K and comparison with the observed albedo indicate that a model with $K \sim 2 \times 10^4$ cm^2 s^{-1} best describes the NH_3 density profile and mixing conditions in the lower stratosphere (Strobel, 1973b; Tomasko, 1974). In the other atmospheres of the outer solar system NH_3 is probably frozen out deep in the atmosphere. Consequently the NH_3 mixing ratios are sufficiently small that NH_3 does not play an important aeronomic role in their upper atmospheres.

Acknowledgment

This work was supported by the Office of Naval Research.

References

Atreya, S. K., Donahue, T. M., and McElroy, M. B.: 1974, *Science* **184**, 154.
Ausloos, P.: 1972, private communication.
Axel, L.: 1972, *Astrophys. J.* **173**, 451.
Capone, L. A. and Prasad, S. S.: 1973, *Icarus* **20**, 200.
Cook, G. R. and Metzger, P. H.: 1964, *J. Opt. Soc. Am.* **54**, 968.
Danielson, R. E., Caldwell, J. J., and Larach, D. R.: 1973, *Icarus* **20**, 437.
Field, G. B., Somerville, W. B., and Dressler, K.: 1966, *Ann. Rev. Astron. Astrophys.* **4**, 207.
Gillett, F. C. and Forrest, W. J.: 1974, *Astrophys. J. (Letters)* **187**, L37.
Gillett, F. C., Low, F. J., and Stein, W. A.: 1969, *Astrophys. J.* **157**, 925.
Gillett, F. C., Forrest, W. J., and Merrill, K. M.: 1973, *Astrophys. J. (Letters)* **184**, L93.
Gross, S. H. and Rasool, S. I.: 1964, *Icarus* **3**, 311.
Hunten, D. M.: 1969, *J. Atmospheric Sci.* **26**, 826.
Hunten, D. M. and Münch, G.: 1973, *Space Sci. Rev.* **14**, 433.
Jenkins, E. B., Drake, J. F., and Wallace, L.: 1974, private communication.
Judge, D. L. and Carlson, R. W.: 1974, *Science* **183**, 317.
Kuiper, G. P.: 1944, *Astrophys. J.* **100**, 378.
McDonough, T. R. and Brice, N. M.: 1973a, *Nature* **242**, 513.
McDonough, T. R. and Brice, N. M.: 1973b, *Icarus* **20**, 136.
McElroy, M. B.: 1969, *J. Atmospheric Sci.* **26**, 798.
Rebbert, R. E. and Ausloos, P.: 1972, *J. Photochem.* **1**, 167.
Ridgway, S. T.: 1974, *Astrophys. J.* **187**, L41.
Rottman, G. J., Moos, H. W., and Freer, C. S.: 1973, *Astrophys. J.* **184**, L89.
Stecher, T. P. and Williams, D. A.: 1967, *Astrophys. J.* **149**, L29.
Strobel, D. F.: 1973a, *J. Atmospheric Sci.* **30**, 489.
Strobel, D. F.: 1973b, *J. Atmospheric Sci.* **30**, 1205.
Strobel, D. F.: 1973c, in B. M. McCormac (ed.), *Physics and Chemistry of Upper Atmospheres*, D. Reidel Publ. Co., Dordrecht, Holland, p. 345.
Strobel, D. F.: 1974a, *Icarus* **21**, 466.
Strobel, D. F.: 1974b, *Astrophys. J.* **192**, L47.
Strobel, D. F.: 1974c, *Rev. Geophys. Space Phys.*, in press.
Strobel, D. F. and Smith, G. R.: 1973, *J. Atmospheric Sci.* **30**, 718.
Tan, L. Y., Winer, A. M., and Pimental, G. C.: 1973, *J. Chem. Phys.* **57**, 4028.
Tomasko, M. G.: 1974, *Astrophys. J.* **187**, 641.
Wallace, L. and Hunten, D. M.: 1973, *Astrophys. J.* **182**, 1013.
Wallace, L., Prather, M., and Betton, M. J. S.: 1974, *Astrophys. J.* **193**, 481.

THE ATMOSPHERE AND IONOSPHERE OF JUPITER

MICHAEL B. McELROY

Center for Earth and Planetary Physics, Harvard University, Cambridge, Mass. 02138, U.S.A.

Abstract. Selected topics in Jovian research are reviewed with an emphasis on recent results from Pioneer 10. The Pioneer 10 results for the temperature field in the lower atmosphere are not consistent with other observational constraints and must be rejected. The results for the ionosphere suggest that the thermosphere may be surprisingly hot, $\sim 10^3$ K. It is suggested that this high temperature may be maintained by upward propagation of inertia-gravity waves.

1. Introduction

This paper gives a brief review of our present understanding of Jupiter. The treatment is necessarily limited. The choice of topics, and the emphasis with which they are treated, reflects personal prejudice rather than absolute priority. An attempt is made to focus on areas of contention and to identify problems requiring further work.

Special attention is paid to the thermal structure of Jupiter. Results derived from an analysis of data obtained by the radio occultation experiment on Pioneer 10 (Kliore *et al.*, 1974; 1975) are in marked disagreement with previous models. We explore possible sources for the discrepancy, reviewing in some detail the basis for the earlier models. We discuss also the chemistry of Jupiter's atmosphere, and review the processes which are expected to play a role in shaping the structure of the planet's ionosphere.

2. General Considerations

Jupiter is the largest planet in the solar system, with a total mass equal to approximately 318 times that of the Earth. The planet has an equatorial radius of 71 600 km (Hubbard and Van Flandern, 1972), and a mean density of 1.314 g cm^{-3}. Its density is such that hydrogen and helium must be major atmospheric constituents, and most models implicitly assume that the planet is composed of material of essentially solar composition.

Molecular hydrogen was first detected as an atmospheric species by Kiess *et al.* (1960). Helium was observed only recently (Judge and Carlson, 1974) although quantitative estimates for its abundance had been given earlier by Baum and Code (1953) and by Elliot *et al.* (1974). Its abundance remains uncertain. Present data would allow mixing ratios, $[He]/[H_2]$, as low as 0.06, and as high as 0.60, with a preferred value of about 0.2 (Carlson and Judge, 1974). Consideration of solar abundances suggests a ratio of 0.11 (Hunten and Munch, 1973).

A summary of the best current information on composition is given in Table I. With the exception of the He results, the table is based on analyses of data obtained with Earth-based telescopic equipment. The results for H_2, CH_4, NH_3 and CH_3D were derived from studies of reflected sunlight, which dominates Jupiter's spectrum at

B. M. McCormac (ed.), Atmospheres of Earth and the Planets, 409–423. All Rights Reserved.
Copyright © 1975 by D. Reidel Publishing Company, Dordrecht-Holland.

TABLE I

Composition of Jupiter's atmosphere

Gas	Abundance[a]	Mixing ratio[b]	Wavelength[c]
H_2	67	1	0.82
He		0.2	0.06
CH_4	45	$\sim 10^{-3}$	1.1
NH_3	13	$\sim 10^{-4}$	1.1
CH_3D	2.6	$\sim 10^{-7}$	4.6
C_2H_6	10^{-4}	4×10^{-3} [d]	12
C_2H_2	2×10^{-6}	8×10^{-5} [d]	13
PH_3	[e]	[e]	
H_2O	[f]	[f]	5

[a] The abundance of H_2 is expressed in km atm, while the abundance of CH_3D is given in cm atm. Abundances for all other species are in m atm.
[b] Mixing ratio by volume with respect to H_2.
[c] Approximate wavelength (μm) at which the gas was detected.
[d] These abundances and ratios are uncertain. See discussion in the text, Section 4.
[e] Observed by S. T. Ridgway, but abundance not determined as yet.
[f] Positive detection reported by R. Treffers et al.

wavelengths less than 5 μm. The results for C_2H_6, C_2H_2 and PH_3 were obtained from studies of the emission spectrum at longer wavelengths.

Interpretation of the reflection spectrum is complicated due to effects of multiple scattering. For general discussions of line formation in scattering atmospheres the reader is referred to papers by Chamberlain and Kuiper (1965), Belton (1968), McElroy (1969, 1971), and Chamberlain (1970). For specific discussions of the Jovian problem, he is directed to papers by Danielson and Tomasko (1969), Axel (1972) and Hunt (1972). The spectroscopic data appear to require at least two distinct cloud layers. The upper layer is optically thin, and must extend upwards in the atmosphere from levels where the temperature is about 150 K, to perhaps as high as the tropopause where the temperature could be as low as 115 K. The upper cloud is most probably composed of frozen crystals of NH_3. Thermochemical studies by Lewis (1969a, b) suggest that the lower cloud may be composed of ammonium hydrosulfide (NH_4SH).

Lewis argues that Jupiter's atmosphere should contain additional layers of clouds at altitudes below the NH_4SH cloud. He predicts clouds of water ice, aqueous ammonia, ammonium chloride, and silicates extending over a range of altitudes through which the temperature varies from about 250 K (water ice) to about 1600 K (silicates). The high temperatures at depth in Jupiter's atmosphere are a consequence of the relatively large flux of energy which must be transported, mainly by convection, from the deep interior of the planet.

Jupiter emits somewhere between 2.0 and 2.5 times the amount of energy it absorbs from the sun (Aumann et al., 1969; Chase et al., 1974). The excess emission is probably supplied by gravitational contraction, and Smoluchowski (1967) has argued that a radial contraction as small as 1 mm yr^{-1} would suffice to explain the observations.

Jupiter is in rapid rotation. Clouds within 10° of the equator rotate with a period of $9^h 50^m 30^s.003$, defining a system of coordinates known as System I. Clouds at higher latitudes rotate more slowly, with a period of $9^h55^m40\overset{s}{.}632$ (System II), while the rotation period indicated by decimeter and decameter studies of the magnetosphere (System III) is $9^h55^m29\overset{s}{.}75 \pm 0.04$ s.

Jupiter has a strong magnetic field. The Pioneer 10 results imply a dipole of moment $4.0 \, GR_j^3$, displaced a distance $0.11 \, R_J$ from the center of Jupiter, and tilted an angle of $11°$ with respect to the rotation axis (Smith *et al.*, 1974). The magnetosphere is extensive, populated by large fluxes of high energy protons and electrons, as well as thermal plasma (cf. series of papers in the Pioneer-10 issue of *Science*, **183** (January 25, 1974). Synchrotron emission from trapped electrons provides a strong source of radiation at decimetric wavelengths (0.1 to 1.0 m), making Jupiter one of the brightest objects in the sky at radio wavelengths. The planet also emits strongly, though sporadically, at decametric wavelengths (between 75 and 100 m). The nature of the decametric radiation is not well understood, although there are theories which account for some of its characteristics, including the remarkable modulating effect exercised by Io (see Newburn and Gulkis, 1973, for a more complete discussion).

3. Thermal Structure

Present knowledge (?) of Jupiter's thermal structure is summarized in Figure 1. The

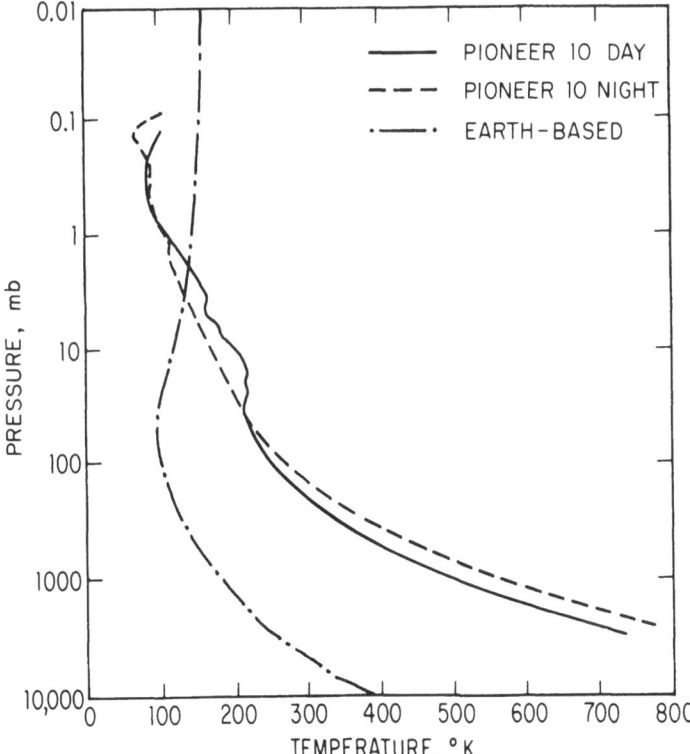

Fig. 1.   Temperature vs. pressure in Jupiter's atmosphere derived from analysis of the Pioneer-10 results (Kliore *et al.*, 1975) and for comparison a model based on a variety of Earth-based measurements (Divine, 1970), supplemented by calculations of thermal structure (Wallace *et al.*, 1974).

confusion is obvious. Analysis of the Pioneer:10 radio occultation experiment
(Kliore *et al.*, 1974, 1975) indicates remarkably high temperatures at modest depths in
Jupiter's atmosphere. For example, the temperature at levels where the pressure is
about 1 atmosphere should be about 500 K according to Pioneer 10, but no more than
200 K according to other sources of information summarized by the model in Figure
1b. What is the nature of the discrepancy? Which of the various models in Figure 1
most closely represents actual conditions in Jupiter's atmosphere?

The radio occultation technique has an impressive history as a tool for remote
sensing of planetary atmospheres. Experience with various Mariner experiments
(Kliore *et al.*, 1965, 1967, 1969) lends confidence to the Pioneer results. On the other
hand, the pre-Pioneer model in Figure 1 also has credibility and is based on a wide
range of observational constraints. After reviewing the evidence, we are led to decide
in favor of the early model. The Pioneer-10 data can be reconciled with spectroscopic
data only if one is prepared to make a series of *ad hoc* assumptions regarding the
structure and optical properties of Jupiter's clouds. The spacecraft results appear to
be totally inconsistent with ground-based measurements of the planetary emission
at radio wavelengths. We conclude that the occultation experiment should be re-
examined. Further analysis of the data may provide useful information on the distribution and scale of turbulent elements in Jupiter's stratosphere and upper troposphere. It is unlikely, however, to shed much light on the details of the temperature
field in the deep atmosphere

A summary of various temperature determinations is given in Figure 2. The results
shown here refer to brightness temperature, defined as follows. We assume that
Jupiter may be approximated by a plane parallel slab, characterized by a state of local
thermodynamic equilibrium (LTE).* The intensity of radiation emitted at frequency
and zenith angle $\cos^{-1} \mu$ is given then by (Chandrasekhar, 1950)

$$I_\nu(\mu) = \int_0^\infty B_\nu(\tau) \exp\left(-\frac{\tau}{\mu}\right) \frac{d\tau}{\mu}, \tag{1}$$

where $B_\nu(\tau)$ denotes the Planck function at optical depth τ. The brightness temperature, T_B, is defined by Equation (1), with the additional relation

$$I_\nu(\mu) = B_\nu(T_B). \tag{2}$$

At longer wavelengths, such that $h\nu \ll kT$, we may write

$$B_\nu(T) \simeq \frac{2\nu^2 kT}{c^2}. \tag{3}$$

Then

$$T_B = \int_0^\infty T(\tau) \exp\left(-\frac{\tau}{\mu}\right) \frac{d\tau}{\mu}, \tag{4}$$

* The LTE approximation should be reasonably reliable at wavelengths longward of 5 μm, where solar
radiation plays a negligible role.

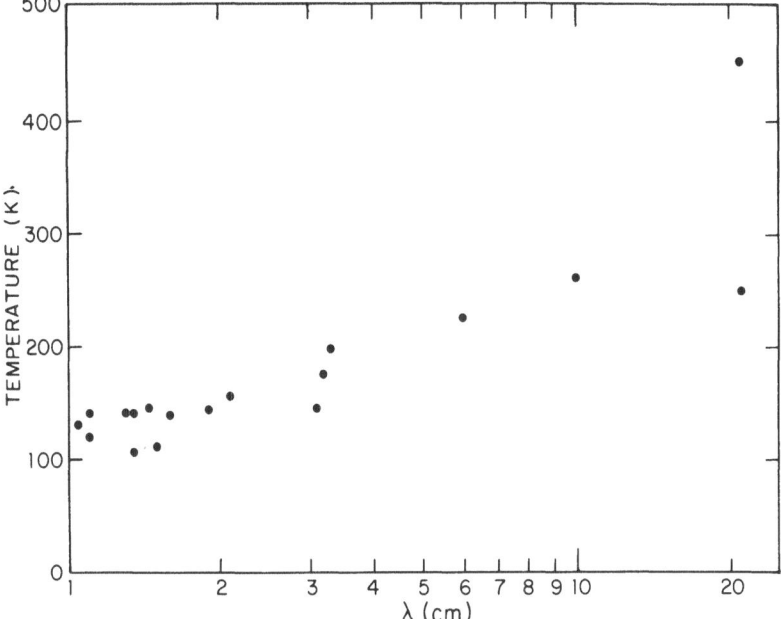

Fig. 2.   Brightness temperatures vs. wavelength for Jupiter, from a variety of sources listed by Newburn
and Gulkis (1973).

where τ denotes optical depth, a function of both frequency and height, given by

$$\tau(v, z) = \int_{z}^{\infty} k_v(z')\, \varrho(z')\, dz'. \tag{5}$$

Here k_v is the mass absorption coefficient and ϱ is the mass density. The brightness
temperature, T_B, is an approximate measure of the temperature at $\tau = \mu$.

The opacity over much of the spectral range covered by Figure 2 is dominated by
H_2, and is due primarily to pressure-induced translational and rotational-transla-
tional tansitions (Trafton, 1967). Absorption by NH_3 is important at wavelengths
longward of 0.5 cm, where the inversion transition centered near 23.5 GHZ plays a
predominant role. Absorption by NH_3 is important also between 9 and 12 μm (see for
example the observations by Aiken and Jones, 1972), while CH_4 plays a role near
7.8 μm (Gillett et al., 1969).

We note that the brightness temperature is less than 450 K for all wavelengths included in Figure 2. It follows, if the Pioneer-10 data are assumed valid, that the emission over this entire spectral range must originate at levels where the pressure is less
than 1 bar. Indeed, if we exclude the datum at 21 cm, the emission must originate at
pressures below 0.1 bar. Absorption by atmospheric gases listed in Table I cannot
account for the necessary opacity $(\tau_v \sim 1)$. It may be possible to explain the data by
invoking a high altitude cloud. This cloud must, however, include significant numbers
of particles with radii of the order of 1 cm in order to account for the radio observa-

tions, an unlikely situation in view of the exceedingly low values which must apply for the background atmospheric pressure. The results in Figure 2 are clearly in serious conflict with any simple interpretation of the Pioneer-10 results.

It is difficult also to reconcile the Pioneer 10 results with various spectroscopic observations at shorter wavelengths. Fink and Belton (1969) found a rotational temperature of 145 ± 20 K, from an analysis of quadrupole lines in the 3-0 vibration-rotation band of H_2 near 8200 Å. Their analysis indicates that this temperature should apply at levels where the pressure is about 2 bar. Studies of the $3v_3$ band of CH_4 at 1.1 μm give a similar indication. The temperature inferred here (Margolis and Fox, 1969) is somewhat higher than the value found by Fink and Belton, but significantly less than temperatures implied by the Pioneer-10 data.

Inspection of the brightness temperatures at radio wavelengths reveals further difficulties.* If we accept the interpretation of the Pioneer 10 results, it follows that the radio signal from the spacecraft must have had the capability to probe at least to levels in the atmosphere where the temperature is as high as 700 K. It follows that these levels can contribute also to passive radio observations at the same wavelength (~ 13 cm). There is no indication, however, of any such contribution in the data shown in Figure 2, and it is difficult to escape the conclusion that the interpretation of the Pioneer-10 data, as presented in Figure 1, must be in error.

The radio occultation technique relies on very precise measurements of the changes in phase, frequency, and amplitude of the telemetry wave, and on the ability of the experimeter to isolate those changes which are due to the passage of the wave through the refractive planetary atmosphere. The phase change associated with the planetary atmosphere is given by

$$\Delta \Phi = \frac{f}{c} (\Delta r + \Delta l), \tag{6}$$

where f denotes frequency, c the velocity of light, Δr the increase in ray path due to propagation delay, and Δl the increase in ray path due to refractive bending of the wave.

The increase in path due to propagation delay is given by

$$\Delta r = 2 \int_{s}^{\infty} (n-1) \, ds \tag{7}$$

where n is the refraction index and the integral extends over the actual ray path. It is customary to assume that n is a function only of r, distance from the center of the planet. In that case, at least for small bending angles, we have

---

* The argument presented here is apparently due to S. Gulkis and was first brought to my attention by D. Hunten.

$$\Delta r \simeq 2 \int_{R}^{\infty} \frac{(n-1)\,dr}{\sqrt{1-\left(\dfrac{R}{r}\right)^2}}, \tag{8}$$

where R is the distance of closest approach.

The increase in path length due to refractive bending is given by

$$\Delta l = \frac{R_S R_E}{R_S + R_E}(1-\cos\varepsilon), \tag{9}$$

for small values of the bending angle ε, where R_S and R_E define distances of the occulting limb from the spacecraft and Earth, respectively. In general, $R_E \gg R_S$ and

$$\Delta l \simeq R_S \frac{\varepsilon^2}{2}. \tag{10}$$

Assuming spherical symmetry as before, we find

$$\varepsilon = -2n(R)\,R \int_{R}^{\infty} \frac{n'(r)\,dr}{n(r)\,[n^2 r^2 - n^2(R)\,R^2]^{1/2}}, \tag{11}$$

where $n'(r)$ denotes the derivative of n with respect to r.

Hunten (1974) has made the interesting suggestion that the difficulties encountered in the application of the Pioneer-10 data may be related to the assumption of spherical symmetry, as implied by the use of Equations (8) and (11) in the analysis of the spacecraft data. The assumption might not be justified if small scale turbulence were to play a role in the determination of ε and if refractive bending of the radio ray were to play a major role in the determination of the residual phase shift. We note that Δl is proportional to R_S. The value of R_S appropriate for Pioneer-10 is significantly larger, by a factor of order 10^2, than values appropriate for the earlier Mariner missions, and possible effects of turbulence would be amplified accordingly. Hunten's suggestion deserves further attention and may provide a resolution to the intriguing puzzle posed by the Pioneer temperature profile.

4. Chemistry

We restrict our attention here to photochemical processes, which occur primarily in regions of the atmosphere where the pressure is less than about 2 bar. The chemistry of CH_4 and NH_3 is of special interest, reflecting in part the comparatively high abundance of these gases in the Jovian atmosphere, in part the relatively rich suite of possible photochemical products.

Dissociation of CH_4 is energetically possible at wavelengths below 2700 Å, but proceeds to a significant extent only at wavelengths shortward of 1600 Å.* Approxi-

---

* The discussion of CH_4 chemistry given here relies heavily on work by Strobel (1969, 1973a, 1974). The summary of NH_3 chemistry is patterned after Strobel (1973b).

mately 70% of the total CH_4 photolysis takes place at Ly-α. The primary paths are

$$hv + CH_4 \rightarrow {}^1CH_2 + H_2 \tag{12}$$

and

$$hv + CH_4 \rightarrow CH + H + H_2 \tag{13}$$

for which the branching ratios are 92% and 8%, respectively (Strobel, 1975).

The radical ${}^1CH_2$ is removed mainly by

$$^1CH_2 + H_2 \rightarrow CH_3 + H \tag{14}$$

and CH is removed by

$$CH + CH_4 \rightarrow C_2H_4 + H \tag{15}$$

or

$$CH + H_2 + M \rightarrow CH_3 + M. \tag{16}$$

Reaction (15) is more important in regions of the atmosphere where the density of third bodies is less than 10^{15} cm^{-3}. Reaction (16) is dominant at lower altitudes. The radical CH_3 is removed primarily by

$$CH_3 + H + M \rightarrow CH_4 + M, \tag{17}$$

with an additional contribution due to

$$CH_3 + CH_3 + M \rightarrow C_2H_6 + M. \tag{18}$$

Ethane (C_2H_6) is a relatively stable compound in the Jovian environment. Ethylene (C_2H_4), on the other hand, may photolyze to give acetylene (C_2H_2) which in turn may polymerize by reactions initiated by sequences such as

$$hv + C_2H_2 \rightarrow C_2H_2^* \tag{19}$$

followed by

$$C_2H_2^* + H_2 \rightarrow C_2H_2 + H_2. \tag{20}$$

Detailed photochemical calculations by Strobel (1973a, b) suggest that the most abundant hydrocarbons, in order of decreasing importance, are CH_4, C_2H_6, C_2H_2, C_2H_4, and CH_3. The mixing ratios of C_2H_6 and C_2H_2 are estimated to be of the order of 10^{-5} and 10^{-7}, respectively, somewhat less than the values reported by Ridgway (1974). Strobel, however, does not view this discrepancy as serious.

Ethane, acetylene, and hydrogen produced in Jupiter's mesosphere and stratosphere must be transported downward, where they will be converted eventually to CH_4 at the high temperatures and pressures which must prevail below the visible cloud deck. The downward fluxes of C_2H_6, C_2H_2 and H_2 will be balanced by a corresponding upward flux of CH_4.

Photolysis of NH_3 takes place below 2300 Å but above 1600 Å where the primary reaction is

$$hv + NH_3 \rightarrow NH_2 + H. \tag{21}$$

Reaction (21) is followed either by

$$NH_2 + NH_2 + M \rightarrow N_2H_4 + M \tag{22}$$

or by

$$NH_2 + H + M \rightarrow NH_3 + M. \tag{23}$$

In any event there must be significant net decomposition of NH_3. Strobel (1973b) drew attention to the interesting possibility that hydrazine (N_2H_4), formed by Equation (22), may condense in Jupiter's stratosphere and contribute to the high altitude haze discussed by Danielson and Tomasko (1969), Axel (1972) and Hunt (1972). He argued also that photolysis should lead to significant depletion of high altitude NH_3 and might allow one to reconcile conflicting estimates for the NH_3 abundance derived on the basis of UV and IR observations discussed by Anderson *et al.* (1969) and by Tomasko (1974).

5. Ionosphere

Most theories for Jupiter's ionosphere (Rishbeth, 1959; Zabriskie, 1960; Gross and Rasool, 1964; Hunten, 1969; Shimizu, 1971; Prasad and Capone, 1971; Capone and Prasad, 1973; McElroy, 1973; Tanaka and Kirao, 1973; Atreya *et al.*, 1974; Prasad and Tan, 1974) assume that H^+ is the major ion. It is produced in part by photo-ionization of H, in part by dissociative ionization of H_2, with additional minor contributions due to reactions of He^+ with H_2. Protons are removed mainly by downward diffusion, and by reactions with H_2,

$$H^+ + H_2 + M \rightarrow H_3^+ + M, \tag{24}$$

followed by recombination,

$$H_3^+ + e \rightarrow H_2 + H. \tag{25}$$

The various reactions thought to play a role in Jupiter's ionosphere are listed in Table II. The table includes also a number of reactions involving hydrocarbons. The potential importance of hydrocarbons in the chemistry of Jupiter's ionosphere appears to have been first recognized by Prasad and Tan (1974), who drew attention to the potential role of CH_3^+ produced by ionization of CH_3 at wavelengths between 1000 and 1260 Å. Ionization of CH_3 in Jupiter's ionosphere would be analogous to ionization of NO in the Earth's atmosphere, a process which plays an important role in the chemistry of the D region. It could lead to formation of a thin layer of intense low altitude ionization on Jupiter, and Prasad and Tan (1974) surmised that this ionization may have been detected by Pioneer 10.

The Pioneer 10 results (Fjeldbo *et al.*, 1974) are presented in Figure 3, which includes also a pre-Pioneer model for the ionosphere (Atreya *et al.*, 1974). The ionosphere is located at significantly higher altitudes than might be expected on the basis of the earlier theoretical work. The vertical profile of electron density derived from analysis of the Pioneer-10 data exhibits a variety of structural features not included in

TABLE II

Reactions of importance in Jupiter's ionosphere. Rate constants are in units of cm^3 s^{-1} for two-body reactions, cm^6 s^{-1} for three-body reactions (from Atreya *et al.*, 1974; Prasad and Tan, 1974)

Reaction number	Reaction		Rate constant
(a) Ion production			
p1	$H_2 + h\nu$	$\rightarrow H_2^+ + e$	
p2	$H_2 h\nu$	$\rightarrow H^+ + H + e$	
p3	$H_2 + e$	$\rightarrow H_2^+ + 2e$	
p4	$H_2 + e$	$\rightarrow H^+ + H + 2e$	
p5	$H + h\nu$	$\rightarrow H^+ + e$	
p6	$H + e$	$\rightarrow H^+ + 2e$	
p7	$He + h\nu$	$\rightarrow He^+ + e$	
p8	$He + e$	$\rightarrow He^+ + 2e$	
p9	$CH_4 + h\nu$	$\rightarrow CH_4^+ + e$	
p19	$CH_3 + h\nu$	$\rightarrow CH_3^+ + e$	
(b) Ion exchange			
e1	$H_2^+ + H_2$	$\rightarrow H_3^+ + H$	$\sim 5 \times 10^{-10}$
e2	$H_2^+ + H$	$\rightarrow H_2^+ + H^+$	$\sim 1 \times 10^{-10}$
e3	$He^+ + H_2$	$\rightarrow He + H_2^+$	
e4	$He^+ + H_2$	$\rightarrow HeH^+ + H$	sum $< 1 \times 10^{-13}$
e5	$He^+ + H_2$	$\rightarrow He + H + H^+$	
e6	$H^+ + H_2 + H_2 \rightarrow H_3^+ + H_2$		3.2×10^{-29}
e7	$HeH^+ + H_2$	$\rightarrow H_3^+ + He$	1.85×10^{-9}
e8	$He^+ + CH_4$	$\rightarrow CH_2^+ + H_2 + He$	9.3×10^{-10}
e9		$\rightarrow CH^+ + H_2 + H + He$	2.4×10^{-10}
e10		$\rightarrow CH_3^+ + H + He$	6.0×10^{-11}
e11		$\rightarrow CH_4^+ + He$	4.0×10^{-11}
e12	$H^+ + CH_4$	$\rightarrow CH_4^+ + H$	1.5×10^{-9}
e13		$\rightarrow CH_3^+ + H_2$	2.3×10^{-9}
e14	$H_2^+ + He$	$\rightarrow HeH^+ + H$	1.0×10^{-10}
e15	$H_3^+ + CH_4$	$\rightarrow CH_5^+ + H_2$	2.4×10^{-9}
e16	$CH^+ + H_2$	$\rightarrow CH_2^+ + H$	1.01×10^{-9}
e17	$CH_2^+ + H_2$	$\rightarrow CH_3^+ + H$	7.2×10^{-10}
e18	$CH_3^+ + CH_4$	$\rightarrow C_2H_5^+ + H_2$	8.9×10^{-10}
e19	$CH_4^+ + H_2$	$\rightarrow CH_5^+ + H$	4.1×10^{-11}
e20	$CH_4^+ + H$	$\rightarrow CH_3^+ + H_2$	1.0×10^{-11}
e21	$CH_4^+ + CH_4$	$\rightarrow CH_5^+ + CH_3$	1.11×10^{-9}
e22	$CH_5^+ + H$	$\rightarrow CH_4^+ + H_2$	1.0×10^{-11}
(c) Ion removal/electron-ion recombination			
r1	$H_3^+ + e$	$\rightarrow H_2 + H$	3.8×10^{-7}
r2	$H_2^+ + e$	$\rightarrow H + H$	1×10^{-8}
r3	$HeH^+ + e$	$\rightarrow He + H$	1×10^{-8}
r4	$H^+ + e$	$\rightarrow H + h\nu$	6.6×10^{-12}
r5	$He^+ + e$	$\rightarrow He + h\nu$	6.6×10^{-12}
r6	$CH^+ + e$	$\rightarrow C + H$	3.0×10^{-7}
r7	$CH_2^+ + e$	$\rightarrow$ Products	3.0×10^{-7}
r8	$CH_3^+ + e$	$\rightarrow$ Products	4.0×10^{-7}
r9	$CH_4^+ + e$	$\rightarrow$ Products	1.5×10^{-6}
r10	$CH_5^+ + e$	$\rightarrow$ Products	1.9×10^{-6}
r11	$C_2H_5^+ e$	$\rightarrow$ Products	1.9×10^{-6}

the theoretical model, and it is clear that the model tends to seriously underestimate the value for the topside ionospheric scale height. Several of these difficulties could be avoided if one were to assume that the neutral thermosphere were significantly hotter than might be expected on the basis of the pre-Pioneer work.

The earlier models considered only that heating of the atmosphere which would occur due to direct absorption of sunlight. It is difficult in this case to rationalize temperatures much larger than 150 K (Strobel and Smith, 1973). On the other hand, energy generated at lower levels in Jupiter's atmosphere could be transported upwards by vertically propagating waves, and could, in principle, dominate the upper atmospheric energy budget. There is some evidence for these waves in light curves measured during the occultation of β-Scorpii (Veverka *et al.*, 1974), and French and Gierasch (1974) argue that the stellar data may be interpreted to provide evidence for the presence of inertia-gravity waves in Jupiter's lower thermosphere. They estimate that the associated energy flux could be as large as 3 ergs cm^{-2} s^{-1} and could lead to a significant perturbation in the structure of Jupiter's upper thermosphere.

The possible magnitude of the resulting perturbation in thermospheric temperature can be estimated in an approximate manner as follows. The energy absorbed above

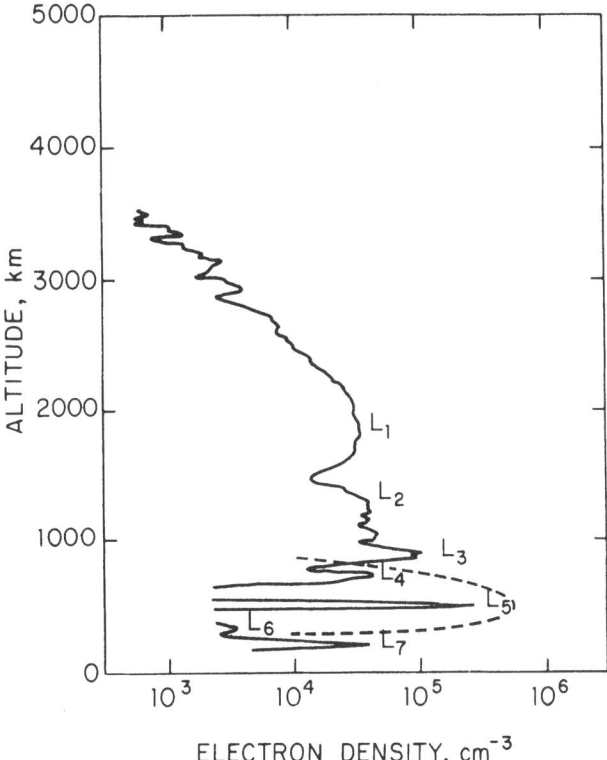

Fig. 3.   Electron densities obtained from an analysis of the Pioneer-10 immersion data (Fjeldbo *et al.*, 1974). The measurements were taken in late afternoon at 26° N when the solar zenith angle was approximately 81°. The height scale is referenced to zero at a planetocentric distance of 70 300 km and the observations are compared with theoretical calculations reported earlier by Atreya *et al.* (1974).

a level z in the thermosphere must be conducted downward to the vicinity of the mesopause, where it can be radiated by IR active species such as C_2H_2 (Strobel and Smith 1973). The heat flux $\mathscr{F}$ at z is given by

$$\mathscr{F}(z) = k(z) \frac{dT}{dz}(z), \tag{26}$$

where $k(z)$ defines the thermal conductivity at z. If we assume that the bulk of the wave energy is dissipated above z, then the temperature at z should be given approximately by

$$\Delta T = \frac{H(z)\, \Delta z}{k(z)}, \tag{27}$$

where ΔT and Δz are respectively the difference in temperature and height which should apply between the mesopause and z: $H(z)$ defines the heat absorbed above z. Taking H equal to 3 erg cm^{-2} s^{-1}, and $k(z)$ equal to $69\, T(z)$ erg cm^{-1} s^{-1} K^{-2} (Strobel and Smith, 1973), we estimate that ΔT should be of the order of 10^3 K. In making this estimate we assumed that Δz was approximately equal to 150 km, a reasonable choice in light of the results in Figure 3.

Fjeldbo et al. (1974) suggest that the structure in Figure 3 may indicate the presence of heavy ions in Jupiter's ionosphere. Layering, however, could be a natural consequence of the interaction between relatively long-lived chemical species such as H^+, and the propagating vertical waves discussed above. A change in sign for the horizontal component of the wave velocity could result in either a clustering or a dispersal of ions near the velocity null point. A similar mechanism leads to clustering of ions in the Earth's ionosphere, where it is thought to be responsible for the phenomenon of sporadic E.

In the foregoing discussion we assumed that the observational data in Figure 3 were representative of conditions in Jupiter's ionosphere. As emphasized by Fjeldbo et al. (1974), however, and underscored by the remarks in Section 3 of this paper, interpretation of the Pioneer radio data is a complex task, and the results for Jupiter may be significantly less certain than were the earlier results obtained for Mars and Venus. The results in Figure 3, and the following discussion, should be interpreted with caution in light of these remarks.

6. Concluding Remarks

We have attempted to give a brief review of selected topics in contemporary Jovian research. The review was purposely limited in scope.

We did not include any discussion of present ideas regarding the structure of Jupiter's interior. Neither did we attempt to summarize the various chemical and dynamical models which have been introduced recently in order to account for the visual image of Jupiter – its belts, its zones, its Great Red Spot, and various other structures. We omitted all discussion of models for the magnetosphere, and resisted the temptation to comment on possible interactions between Jupiter and its satellites. The reader is invited to pursue these matters further, and a brief bibliography is appended to this paper which may serve as an introduction to these subjects.

We discussed a variety of topics, gave some answers but raised many more questions. We dismissed high temperatures derived from analysis of the Pioneer-10 data, on the grounds that these results could not be reconciled with other more direct information. We speculated on possible sources for the error. Is the error restricted to interpretation of the data pertaining to the lower atmosphere, or are there similar problems with the ionospheric results? We discussed the possible role of dynamical processes, in particular inertia-gravity waves, with regard to the temperature and structure of Jupiter's ionosphere. Can these speculative ideas stand up to a more careful scrutiny? What is the nature and excitation mechanism of the waves?

There are many other questions. Further analyses of the Pioneer-10 data may provide some answers. Pioneer 11 and Mariner 1977 will help, but definitive results will surely require *in situ* measurments from probes and orbiting vehicles. The scientific exploration of Jupiter is as yet at a youthful stage. The journey towards a fuller appreciation of Jupiter's qualities and characteristics promises to be both exciting and scientifically rewarding.

Acknowledgments

I am indebted to S. Wofsy, Y. L. Yung, and Max Havlick for invaluable assistance during the preparation of this manuscript. This work was supported by the Atmospheric Sciences section of the National Science Foundation under grant no. GA-33990X to Harvard University.

References and Additional Bibliography

A. REFERENCES

Aiken, D. K. and Jones, B.: 1972, *Nature* **240**, 230.
Anderson, R. C., Pipes, J. G., Broadfoot, A. L., and Wallace, L.: 1969, *J. Atmospheric Sci.* **26**, 874.
Atreya, S. K., Donahue, T. M., and McElroy, M. B.: 1974, *Science* **184**, 154.
Aumann, H. H., Gillespie, C. M., and Low, F. J.: 1969, *Astrophys. J.* **157**, L69.
Axel, L.: 1972, *Astrophys. J.* **173**, 451.
Baum, W. A. and Code, A. D.: 1953, *Astron. J.* **58**, 108.
Belton, M. J. S.: 1968, *J. Atmospheric Sci.* **25**, 596.
Capone, L. A. and Prasad, S. S.: 1973, *Icarus* **20**, 200.
Carlson, R. W. and Judge, D. L.: 1974, *J. Geophys. Res.* **79**, 3623.
Chamberlain, J. W.: 1970, *Astrophys. J.* **159**, 137.
Chamberlain, G. W. and Kuiper, G. P.: 1956, *Astrophys. J.* **124**, 399.
Chandrasekhar, S.: 1950, *Radiative Transfer*, Clarendon Press, Oxford.
Chase, S. C., Ruiz, R. D., Munch, G., Neugebauer, Schroeder, M., and Trafton, L. M.: 1974, *Science* **183**, 315.
Danielson, R. E. and Tomasko, M. G.: 1969, *J. Atmospheric Sci.* **26**, 889.
Divine, N.: 1970, NASA SP-8069.
Elliot, J. L., Wasserman, L. H., Veverka, J., Sagan, C., and Liller, W.: 1974, *Astrophys. J.* **190**, 719.
Fink, U. and Belton, M. J. S.: 1969, *J. Atmospheric Sci.* **26**, 952.
Fjeldbo, G., Kliore, A., Seidel, B., Sweetnam, D., and Cain, D.: 1974, to be published.

French, R. G. and Gierasch, P. J.: 1974, *J. Atmospheric Sci.* **32**, 1707.
Gillett, F. C., Low, F. J., and Stein, W. A.: 1969, *Astrophys. J.* **157**, 925.
Gross, S. H. and Rasool, S. I.: 1964, *Icarus* **3**, 311.
Hubbard, W. B. and Van Flandern, T. C.: 1972, *Astron. J.* **77**, 65.
Hunt, G. E.: 1972, Paper, AAS Meeting, Kona, Hawaii.
Hunten, D. M.: 1969, *J. Atmospheric Sci.* **26**, 826.
Hunten, D. M.: 1974, private communication.
Hunten, D. M. and Munch, G.: 1973, *Space Sci. Rev.* **14**, 433.
Judge, D. L. and Carlson, R. W.: 1974, *Science* **183**, 317.
Kiess, C. C., Corliss, C. H., and Kiess, H. K.: 1960, *Astrophys. J.* **131**, 221.
Kliore, A., Cain, D. L., Levy, G. S., Eshleman, V. R., Fjeldbo, G., and Drake, F. D.: 1965, *Science* **149**, 1243.
Kliore, A., Levy, G. S., Cain, D. L., Fjeldbo, G., and Rasool, S. I.: 1967, *Science* **158**, 1683.
Kliore, A., Fjeldbo, G., Seidel, B. L., and Rasool, S. I.: 1969, *Science* **166**, 1393.
Kliore, A., Cain, D. L., Fjeldbo, G., Seidel, B. L., and Rasool, S. I.: 1974, *Science* **183**, 323.
Kliore, A., Cain, D. L., Fjeldbo, G., Seidel, B. L., and Rasool, S. I.: 1975, to be published.
Lewis, J. S.: 1969a, *Icarus* **10**, 365.
Lewis, J. S.: 1969b, *Icarus* **10**, 393.
Margolis, J. S. and Fox, K.: 1969, *Astrophys. J.* **157**, 935.
McElroy, M. B.: 1969, *J. Atmospheric Sci.* **26**, 798.
McElroy, M. B.: 1971, *J. Quant. Spectr. Radiative Trans.* **11**, 813.
McElroy, M. B.: 1973, *Space Sci. Rev.* **14**, 460.
Newburn, R. L. and Gulkis, S.: 1973, *Space Sci. Rev.* **3**, 179.
Prasad, S. S. and Capone, L. A.: 1971, *Icarus* **15**, 45.
Prasad, S. S. and Tan, A.: 1974, to be published.
Ridgway, S. T.: 1974, *Astrophys. J.* **187**, L41.
Rishbeth, H.: 1959, *Austral. J. Phys.* **12**, 466.
Shimizu, M.: 1971, *Icarus* **14**, 273.
Smith, E. J., Davis, L., Jones, D. E., Coleman, P. J., Colburn, D. S. Dyal, P., Sonett, C. P., and Frandsen, A. M. A.: 1974, *J. Geophys. Res.* **79**, 3501.
Smoluchowski, R.: 1967, *Nature* **215**, 691.
Strobel, D. F.: 1969, *J. Atmospheric. Sci.* **26**, 906.
Strobel, D. F.: 1973a, *J. Atmospheric Sci.* **30**, 489.
Strobel, D. F.: 1973b, *J. Atmospheric Sci.* **30**, 1205.
Strobel, D. F.: 1974, *Astrophys. J.*, in press.
Strobel, D. F.: 1975, this volume, p. 401.
Strobel, D. F. and Smith, G. R.: 1973, *J. Atmospheric Sci.* **30**, 718.
Tanaka, T. and Kirao, K.: 1973, *Planetary Space Sci.* **21**, 751.
Tomasko, M. G.: 1974, *Astrophys. J.* **187**, 641.
Trafton, L. M.: 1967, *Astrophys. J.* **147**, 765.
Treffers, R., Larson, H. P., Fink, U., and Gautier, T. N.: 1975, paper presented at meeting of the Division for Planetary Sciences, American Astronomical Society, Columbia, Maryland.
Veverka, J., Wasserman, L. H., Elliot, J., Sagan, C., and Liller, W.: 1974, to be published.
Wallace, L., Prather, M., and Belton, M. J. S.: 1974, *Astrophys. J.*, in press.
Zabriskie, F. R.: 1960, Ph.D. Dissertation, Princeton Univ.

B. ADDITIONAL BIBLIOGRAPHY

1. *Dynamics*

Gierasch, P. J. and Goody, R. M.: 1969, *J. Atmospheric Sci.* **26**, 979.
Gierasch, P. J. and Stone, P. H.: 1968, *J. Atmospheric Sci.* **25**, 1169.
Gierasch, P. J., Goody, R. M., and Stone, P. H.: 1970, *Geophys. Fluid. Dyn.* **1**, 1.
Hide, R.: 1969, *J. Atmospheric Sci.* **26**, 841.
Ingersoll, A. P.: 1969, *J. Atmospheric Sci.* **26**, 744.
Ingersoll, A. P. and Cuzzi, J. N.: 1969, *J. Atmospheric Sci.* **26**, 000.
Peek, B. M.: 1958, *The Planet Jupiter*, Faber and Faber.
Stone, P. H.: 1967, *J. Atmospheric Sci.* **24**, 642.

Stone, P. H.: 1971, *Geophys. Fluid Dyn.* **2**, 147.
Stone, P. H. and Baker, D. H.: 1968, *Quart. J. Roy. Meteoral Soc.* **94**, 576.

2. *Interior*

Hubbard, W. B.: 1968, *Astrophys. J.* **152**, 745.
Hubbard, W. B.: 1969, *Astrophys. J.* **155**, 333.
Hubbard, W. B.: 1970, *Astrophys. J.* **162**, 687.
Smoluchowski, R.: 1970, *Phys. Rev. Letters* **25**, 693.
Smoluchowski, R.: 1971, *Astrophys. J.* **166**, 435.

3. *Magnetosphere*

See papers in *Science* **183**, No. 4122, 1974; also *J. Geophys. Res.* **79**, 25. 1974.
Goldreich, P. and Lynden-Bell, D.: 1969, *Astrophys. J.* **156**, 59.
Gurnett, D. A.: 1972, *Astrophys. J.* **175**, 525.

SPECTROSCOPY OF JUPITER AND SATURN

G. E. HUNT

Meteorological Office, London Road, Bracknell, Berkshire, England

1. Introduction

There is considerable cosmological interest in determining the composition of the atmospheres of Jupiter and Saturn. These huge planets are low density, rapidly rotating bodies whose atmospheres are apparently banded, with rapidly changing colors caused by complicated chemical reactions.

At the present time most of our information on their composition has been obtained by instrumentation either on or in the neighborhood of the Earth. Space exploration of the major planets has just begun with the successful Pioneer 10/11 flybys of Jupiter, so that we may anticipate rapid developments in our knowledge of the outer solar system with further missions.

There are considerable problems associated with obtaining the composition of Jupiter and Saturn from Earth-based observations. The planets are covered by layers of cloud whose precise composition and structure are unknown. The rapid rotations result in considerable atmospheric motions so that we must expect variations in the abundances determined in the neighborhood of the cloud tops. Seasonal changes in the atmospheric composition, while likely to be small for Jupiter, may be considerable for Saturn whose rotational axis is inclined at 26° to the equatorial plane.

TABLE I

Basic physical properties

Parameters	Earth	Jupiter	Saturn
Mean distance from Sun	1 AU	5.203 AU	9.523 AU
Mass	5.97×10^{24} kg	317.9[a]	95.1[a]
Equatorial radius	6378.16 km	11.19[a]	9.47[a]
Mean density	5.517 g cm^{-3}	1.33 g cm^{-3}	0.69 g cm^{-3}
Equatorial surface gravity	978 cm s^{-2}	2620 cm s^{-2}	1120 cm s^{-2}
Rotation period sidereal	$23^h56^m04^s$	$9^h55^{m[b]}$	$10^h14^{m[c]}$
Length of year	365 days	11.86 yr	29.46 yr
Inclination of equator to orbit	23°45′	3°	26°
Albedo	35%	45%	61%
Cloud cover	50%	100%	100%
Satellites	1	13	10

[a] Earth = 1
[b] System III
[c] Visible spots at the equator

B. M. McCormac (ed.), Atmospheres of Earth and the Planets, 425–432. All Rights Reserved.

Situated beyond the asteroid belt in the outer solar system, we see only one face of these planets. The phase angle – the angle between the Sun, planet and Earth – varies by only 12° for Jupiter and 6° for Saturn. Consequently, angular characteristics of the atmosphere, which are important for determining the structure of the clouds, may only be obtained from center to limb observations.

Progress toward more accurate information on the atmospheric composition of Jupiter and Saturn depends upon the development of realistic interpretative techniques for the problem of line formation in a cloudy atmosphere. In this article we review our present knowledge of the composition of these atmospheres, the type of observations used and methods available for the interpretation of the data.

2. Composition of the Atmospheres of Jupiter and Saturn

2.1. HYDROGEN

Hydrogen is the major constituent of these atmospheres. It is a homopolar molecule and does not, therefore, possess a dipole spectrum. It does however have a quadrupole moment, which gives rise to the (3–0) and (4–0) absorption features which occur at $\sim 12\,200$ cm^{-1} and $\sim 15\,700$ cm^{-1}, respectively, (see for example Hunt, 1973b; Margolis and Hunt, 1973). To interpret the lines of this spectrum one is faced with the additional complication of the collision narrowing phenomenon exhibited by these lines before the more usual pressure broadening takes place. McKellar (1974) has recently investigated the effects of pressure shifts on the interpretation of these lines. They are negligible for Jupiter and small for Saturn so there is no need to make any revision to the present interpretations.

Hunt and Margolis (1973) have shown that simple approximations or the neglect of the effects of collision narrowing will invalidate the interpretation: Fortunately, these effects may be accurately represented by the Galatry profile and computed by the algorithms suggested by James (1969), Hunt and Margolis (1973) and Herbert (1974).

Hunt and Bergstrahl (1974a, b) suggest the abundance of H_2 in the line of formation region of Jupiter is extremely variable. They estimate values of 32 to 120 km amagats (am) for the (4–0) S(1) line and 30 to 75 km am for the stronger (3–0) S(1) line for the period of their observations. Clearly we cannot base our understanding on an analysis of a single observation.

Owen (1969) suggested a H_2 abundance of 190 km am in the Saturn atmosphere. Trafton's (1971, 1974) observations indicate an abundance of ~ 40 km am. These are the two extremes of a limited number of observations suggesting that we must expect some variability, which can only be fully understood by a patrol of these absorption features.

2.2. HELIUM

Helium, the second most abundant element in the solar system, has been positively detected for the first time in Jupiter's atmosphere by the dayglow observations of the resonance line at 584 Å by Pioneer 10's UV photometer (Judge and Carlson, 1974). It is a minor constituent, and the data suggest a ratio of He to H of 0.18 approximately which is close to the solar value. The determination of the proportion of He in the Jovian atmosphere will remove one free parameter from the interpretative models.

The amount of He in the Saturn atmospheres is still unknown and we shall have to wait for the Pioneer 11 flyby in 1979 for the first information. At present we assume the He/H is the same for Jupiter and Saturn.

2.3. METHANE

Methane is a minor constituent of both the Jovian and Saturn atmospheres (see for example, Newburn and Gulkis, 1973; Trafton, 1973). It is also a possible cloud condensate at the cooler levels of the Saturn troposphere.

Ideally one would like to be able to study individual rotational lines in a band so that a temperature and an abundance can be derived at the same time. The combination bands of CH_4 that are more prominent in the photographic IR are not suitable for this purpose since their extremely complex structure has not been analyzed. The overtones of the fundamental frequencies are considerably easier to study, and the R branch of $3v_3$, occurs in a relatively clear region of the spectrum at 1.1 μm where observations at high resolution are possible.

Margolis and Fox (1969) have determined an abundance of 30 m-atm of CH_4 while Hunt's (1973b) analysis is consistent with $CH_4/H_2 \sim (7 \pm 1) \times 10^{-4}$ for Jupiter.

For Saturn, Bergstrahl (1973) finds an abundance of 86 ± 14 to 51 ± 11 m amagats of CH_4, while Trafton's (1973) analysis yields $\sim 153 \pm 16$ m am.

Trafton's values gives CH_4/H_2 in the range 0.4×10^{-3} to 0.7×10^{-3}.

2.4. AMMONIA

The abundance and distribution of NH_3 is of special interest because it condenses at pressures and temperatures existing in Jupiter's atmosphere (see for example Newburn and Gulkis, 1973; Owen, 1970). It has long been thought to be the dominant constituent of the visible clouds and its role as a solvent is of critical importance in developing ideas about inorganic and organic chemistry in the atmosphere. Although NH_3 has a rich dipole spectrum, its condensation properties severely complicate the interpretation of observations. Attempts have been made to analyse the $5v_1$ and $3v_1$ bands which occur in the photographic infrared. Owen and Mason (1968) estimated an abundance of 13 m-atm by comparing laboratory and planetary data, Anderson et al. (1969) found only 2×10^{-5} atm from the UV data near 2000 Å, Savage and Danielson (1968) inferred an abundance of ~ 0.2 m atm from their observations near 1 μm. Poynter and Gulkis (1972) derived a mixing ratio of $(1.5 \pm 1) \times 10^{-4}$ from their centimeter wavelength observations. All these values should be considered uncertain until we can make more accurate interpretations of the observations.

The presence of NH_3 in the atmosphere of Saturn is a controversial topic, (Newburn and Gulkis, 1973; Teifel, 1974). Some authors have reported its presence while others have failed to detect it. However, there are several factors that must be taken into account in explaining this apparent conflict. The Saturn atmosphere is expected to show seasonal effects. The appearance and disappearance of traces of NH_3 in the line formation region may be related to the Saturnian season and the position of the planet on its long orbit around the Sun. Also, any NH_3 in the atmosphere is likely to form a substantial cloud layer. These spectroscopic observations may therefore refer to the cloud top region, which when viewed from the Earth with limited spacial resolution, may observe an abundance of NH_3 which is close to the limit of detectability.

2.5. TRACE GASES

In addition to the gases discussed in the previous section several trace gases have been observed in these atmospheres and are summarized in Table II. The importance of these gases to the atmospheric chemistry and origin of these planetary atmospheres has been emphasized by several authors (Owen, 1970; Newburn and Gulkis, 1973).

3. Interpretative Models

In the preceding section we have briefly outlined the main gases that have been detected in these atmospheres. Now we will consider the procedures to follow in order to interpret the observations. The models will be discussed in the order of increasing complexity, indicating in each case the assumptions and limitations of each technique.

We should be reminded that there are two similar types of high resolution observational material available to the spectroscopist. They are observations of the equivalent width of a spectral line (see for example Margolis and Fox, 1969) and the extremely high resolution observations of the profile of an individual spectral line (see for example Carleton and Traub, 1974). Since the EW is simply the integral of the line profile, the spectroscopic information obtained from these observations may be similar within the limitations of our knowledge of the planet. However, the line profile studies do have one distinct advantage. They are very sensitive to changes in the vertical opacity of the atmosphere, so that they may be used to indicate the posi-

TABLE II

Trace gases which have been detected in the atmospheres of Jupiter and Saturn

Jupiter	
CH_3D	Beer and Taylor (1973)
HD	Trauger *et al.* (1973)
$C^{13}H_4$	Fox *et al.* (1972)
C_2H_6 }	
C_2H_2 }	Ridgway (1974)
PH_3 }	
Saturn	
C_2H_6	Gillett and Forrest (1974)

tions of scattering layers which complicate the interpretation of data from cloudy atmospheres, such as Jupiter, Saturn and Venus.

3.1. REFLECTING LAYER MODEL

This is the simplest model and should therefore be the starting point for the analysis of the observations. It neglects scattering by cloud particles and assumes that the absorption lines are formed at a lambertian reflecting surface. Generally it is assumed to be white although an albedo $A(\leqslant 1)$ can be prescribed to the surface. Certainly this model is simple to use. In spite of its gross assumptions it can be useful for interpreting observations in an atmosphere where the lines are formed in a region bounded by an opaque cloud layer. It cannot of course predict the angular characteristics of spectral lines, such as their center to limb variation.

3.2. HOMOGENOUS CLOUD LAYER

This assumes the whole atmosphere is cloudy, and that an effective (p, t) can be obtained to represent the atmospheric characteristics. The scattering medium may either be isotropic or scaled through the similarity relations to be anisotropic (Hansen, 1969). This type of model may only be used when the line formation region is completely cloudy. If the clouds form distinct layers then this model may predict the wrong angular characteristics (see Belton (1968); Fink and Belton (1969); McElroy (1969)) for discussions of this type of model.

3.3. INHOMOGENEOUS MODEL ATMOSPHERE

This is the most sophisticated model developed so far, and the procedures to be followed in developing this approach are described in a number of papers by Hunt (1972a, b, 1973a, b). It is only suitable for use when sufficient information is available about a planetary atmosphere since it requires details of the atmospheric profile and cloud positions. The latter could of course be adjusted in fitting the observations. We know enough about the cloudy atmospheres of Venus, Jupiter and Saturn so that this model must be used to interpret their planetary spectra. Clearly the model has several free parameters associated with the cloud. However the number of degrees of freedom may be reduced by constraining the model to fit the wavelength dependence of the planetary albedo which controls the characteristics of the scattering medium. The application of this model to interpreting planetary spectra is given in Hunt (1973b), Hunt and Bergstralh (1974a, b).

So far all our models have been one dimensional, taking into account the variation of the physical variables in the vertical only. We will need a sophisticated 2-dimensional model in order to accurately interpret the centre to limb variations of observations of the major planets where the tremendous horizontal inhomogeneities are evident in the planetary photographs.

4. Structure and Composition of the Clouds

In order to construct an inhomogeneous model for the analysis of spectroscopic observations, we need to know the structure of the clouds. There are no complete or accepted models for the Jovian or Saturnian clouds, although we do have some observations to assist us to postulate possible structures.

There is observational evidence from the eclipse data of Price and Hall (1971), the theoretical studies of Axel (1972) and the Pioneer 10 S band occultation data of Kliore *et al.* (1974) of an aerosol layer in the Jovian stratosphere which causes an additional heating, and therefore a temperature inversion at this region of the atmosphere. The occultation data suggests the layer extends between 20 and 100 mb. Since NH_3 is a constituent of the Jovian atmosphere it will condense into a layer of cloud at some tropospheric level. Indeed, we believe that NH_3 cirrus mark the top of ascending fluid which corresponds to the *bright zones*, while the regions of cool descending fluid may be obscured by a haze and appear as *dark belts*. For the nominal model of Divine (1971) the NH_3 cloud will extend from ~ 140 K to the tropopause at 113 K, although the level of the cloud base will depend upon the NH_3 mixing ratio. The cirrus layer may be optically thin with an isotropic optical thickness of 1.5 to 3 at visible wavelengths (Hunt, 1973b) and ~ 0.5 in the IR (Taylor and Hunt, 1972).

Below this diffuse NH_3 cloud is a dense layer whose top may be characterized by $p \sim 2$ atm, $T \sim 240$ K (Hunt, 1973a, b). Since there is insufficient NH_3 for saturation below the base of the diffuse upper layer, the gap between the clouds may be essentially free of particles. A two cloud model of this type for the visible layer has been suggested before by Owen (1969), Danielson and Tomasko (1969), Axel (1972) and successfully employed to interpret Jovian observations by Hunt (1973b) and Taylor and Hunt (1972). It is the variability of this cloud structure which results from the atmospheric motions, that causes the large variations in spectroscopic abundances reported by Hunt and Bergstralh (1974a, b).

We do not have any further information on the cloud structure at deeper levels of the Jovian (or Saturnian) atmospheres, apart from that postulated by Lewis (1969) in his geochemical analyses. Entry probes will be required to improve our knowledge beyond its present position.

It is probable that Saturn has a similar cloud structure to that described for Jupiter. The NH_3 cloud may be more extensive and there is no clear evidence that optical measurements penetrate below this layer. A further complication is caused by the role of the tropospheric, stratospheric aerosols. Trafton (1974) does not require it in order to interpret his observations, while Macy (1974) finds it mandatory. Good spatial resolution observations of the continuum and absorption lines are required to resolve this problem.

5. Conclusion

Determining the properties of an atmosphere is a spectral jigsaw. Spacecraft missions exploit this by carrying instrument packages that make measurements at various wavelength ranges. This could also be achieved for ground based observations of the planets by introducing a *spectral planetary patrol*. By this procedure, where simultaneous observations of the same portion of a planet are carried out at different wavelengths, the total information may provide more insight into the atmospheric processes contained in the individual measurements. Temporal patrols are required to study the time variability of the spectroscopic abundances when spectral lines are formed in dynamically active regions, such as in the case of Jupiter and Saturn. Furthermore, we urgently need simultaneous observations of the continuum and absorption lines with high spatial resolution to test the validity of our present set of models used to analyze the observations in a cloudy planetary atmosphere. There is still a great deal to be done with our computers, laboratory spectroscopes and telescopes to improve our knowledge of the composition and structure of the atmospheres of Jupiter and Saturn.

References

Anderson, R. C., Pipes, J. G., Broadfoot, A., and Wallace, L.: 1969, *J. Atmospheric Sci.* **26**, 874.
Axel, L.: 1972, *Astrophys. J.* **173**, 451.
Beer, R. and Taylor, F. W.: 1973, *Astrophys. J.* **179**, 309.
Belton, M. J. S.: 1968, *J. Atmospheric Sci.* **25**, 596.
Bergstrahl, J.: 1973, *Icarus* **18**, 605.
Carleton, N. and Traub, W.: 1974, *Bull. Amer. Astron. Soc.*, in press.
Danielson, R. E. and Tomasko, M. G.: 1969, *J. Atmospheric Sci.* **26**, 889.
Divine, N.: 1971, NASA SP-8069.
Fink, U. and Belton, M. J. S.: 1969, *J. Atmospheric Sci.* **26**, 252.
Fox, K., Owen, T., Mantz, A. W., and Rao, K. N.: 1972, *Astrophys. J.* **176**, L76.
Gillett, F. C. and Forrest. W. J.: 1974. *Astrophys. J.* **187**, L37.
Hansen, J. E.: 1969, *Astrophys. J.* **158**, 337.
Herbert, F.: 1974, *J. Quant. Spectr. Radiative Transfer* **14**, 943.
Hunt, G. E.: 1972a, *J. Quant. Spectr. Radiative Transfer* **12**, 387.
Hunt, G. E.: 1972b, *J. Quant. Spectr. Radiative Transfer* **12**, 1023.
Hunt, G. E.: 1973a, *Icarus* **18**, 637.
Hunt, G. E.: 1973b, *Monthly Notices Roy. Astron. Soc.* **161**, 347.
Hunt, G. E. and Bergstrahl, J. T.: 1974a, *Nature* **249**, 635.
Hunt, G. E. and Bergstrahl, J. T.: 1974b, in A. Woszczyk and C. Iwaniszewska (eds.), 'Exploration of the Planetary System', *IAU Symp.* **65**, 385.
Hunt, G. E. and Margolis, J.: 1973, *J. Quant. Spectr. Radiative Transfer* **13**, 417.
Jones, T. C.: 1969, *J. Opt. Soc. Amer.* **59**, 1602.
Judge, D. L. and Carlson, R. W.: 1974, *Science* **183**, 317.
Kliore, A. J., Cain, D. L., Fjeldbo, G., Seidel, B. L., and Rasool, S. I.: 1974, *Science* **183**, 323.
Lewis, J. S.: 1969, *Icarus* **10**, 365.
Macy, W.: 1974, *Bull Amer. Astron. Soc.*, in press.
Margolis, J. and Fox, K.: 1969, *Astrophys. J.* **157**, 935.
Margolis, J. and Hunt, G. E.: 1973, *Icarus* **18**, 593.
McElroy, M. B.: 1969, *J. Atmospheric Sci.* **2**, 798.
McKellar, A. R. W.: 1974, *Icarus* **22**, 212.
Newburn, R. and Gulkis, S.: 1973, *Space Sci. Rev.* **13**, 179.
Owen, T.: 1970, *Science* **167**, 1675.
Owen, T.: 1969, *Icarus* **10**, 355.
Owen, T. and Mason, H. P.: 1968, *Astrophys. J.* **154**, 317.
Poynter, R. and Gulkis, S.: 1972, *Bull. Am. Astron. Soc.* **6**, 62.
Price, M. J. and Hall, J. S.: 1971, *Icarus* **14**, 3.
Ridgway, S. T.: 1974, *Astrophys. J.* **187**, L41.
Savage, B. D. and Danielson, R. F.: 1968, in P. J. Brancazio and A. G. W. Cameron (eds.), *Infrared Astronomy*, Gordon and Breach, N.Y.
Taylor, F. W. and Hunt, G. E.: 1972, *Bull. Am. Astron. Soc.* **6**, 67.
Trafton, L.: 1971, *Bull. Am. Astron. Soc.* **5**, 291; *Astrophys. J.* **182**, 615 (1973).

432   G. E. HUNT

Trafton, L.: 1973, *Astrophys. J.* **182**, 615.
Teifel, V.: 1974, in A. Woszczyk and C. Iwaniszewska (eds.), 'Exploration of the Planetary System', *IAU Symp.* **65**, 415.
Trauger, J. T., Roesler, F. L., Carleton, N. P., and Traub, W. A.: 1973, *Astrophys. J.* **184**, L137.

USEFUL ADDITIONAL REFERENCES

1. 'The Atmospheres of the Major Planets', 1969, *J. Atmospheric Sci.* **26**, 795.
2. 'Jupiter and the Other Planets', 1969, *Icarus* **10**, 353.
3. Hunten, D. M.: 1971, *Space Sci Rev.* **12**, 539.
4. Hunt, G. E.: 1974, *Endeavour* **32**, 43.
5. 'Outer Solar System Exploration – An Overview', 1973, *Space Sci. Rev.* **14**, 347.
6. Pioneer 10 Report: 1974, *Science* **183**, 301.

SODIUM 'AIRGLOW' FROM Io

T. D. PARKINSON*

Kitt Peak National Observatory, P.O. Box 26732, Tucson, Ariz. 85726, U.S.A.

1. Introduction

Io is the innermost of the four Galilean satellites of Jupiter. In size, density, and distance from its parent planet it is quite similar to our own Moon. Most of its other physical properties are different from our lunar experience. To gain some familiarization with this satellite, the following brief discussion will highlight some of its observed physical properties. An excellent review has recently been published by Morrison and Cruikshank (1974).

The orbital period of Io is 42-$\frac{1}{2}$ h. The nearly circular orbit lies close to the Jovian equatorial plane (Allen, 1963). At opposition Io can be followed through about 90° of its orbit in a night of observing. On successive nights the satellite has generally changed from one side of Jupiter to the other.

At mean opposition Io is 0.9″ diameter and has a visual magnitude of 5.0 (Jupiter is -2.5). It exhibits a $\pm 13\%$ intensity variation with orbital position, the leading face being brighter than the trailing face (Morrison and Cruikshank, 1974). Direct imaging shows darker polar cap regions and a lighter equatorial region (Danielson and Tomasko, 1971; Kuiper, 1973; Minton, 1973). Spectrophotometric studies show that it has a very high albedo in the red and IR. At 5980 Å the albedo of Io is about 0.66 [Johnson and McCord (1971) normalized by Morrison and Cruikshank (1974)]. From 1.2 to 3.4 μm the albedo of Io remains high while for the other three Galilean satellites the albedos decrease dramatically and show considerable structure in the higher resolution studies (Morrison and Cruikshank, 1974). The high resolution spectra taken by Pilcher *et al.* (1972) and Fink *et al.* (1973) show features due to water ice on Ganymede and Europa. Pilcher *et al.* found a weak indication for ice on Callisto. Thus it seems likely that all of the Galilean satellites except Io have some reasonable fraction of their surface covered with simple ice.

Polarimetric studies indicate that all but Callisto (the outermost of the Galilean satellites) have surfaces that are covered by a substance having the structure of frost (Veverka, 1971). Thermal IR measurements at 10 and 20 μm show brightness temperatures of 140 K and 152 K for Ganymede and Callisto, respectively. For Io the 10 μm measurements give a temperature of 138 K, about 10° higher than the 20 μm observations; Europa is 130 K at 10 μm, also about 10° higher than it is at 20 μm (Morrison and Cruikshank, 1974).

Being innermost of these satellites, Io is the most exposed to the Jovian radiation belts. Io is also known to modulate the Jovian decametric radio emission (Bigg, 1964;

* Present address Gould Laboratories, Rolling Meadow, Illinois 60008.

Warwick, 1967). Carr and Gulkis (1969) review the phenomenon in some detail.
Jovian decametric radio bursts at about 18 MHz show three longitude zones (System
III) from which the emission apparently originates. The majority of noise bursts
from the earliest of these longitude zones occurs when Io's orbital position is about
93° measured from superior *geocentric* conjunction (Bigg, 1964). Not so strong a
correlation exists between noise bursts from the most active zone and an orbital
position of about 246° for Io. The mechanism responsible for the decametric radiation, and Io's influence on it, is not well understood although many ideas have
been suggested (Newburn and Gulkis, 1973). Goldreich and Lynden-Bell (1969) discuss perhaps the most complete model which details the interaction of Io with the
Jovian magnetic field.

2. Atmosphere

The gravitational attraction of the Galilean satellites is low, so like our own moon
no atmosphere would be expected for these bodies were it not for their low temperature. Perhaps the first positive indication of an atmosphere on any of these satellites
was Binder and Cruikshank's (1964) observation of Io in which they observed a
post-eclipse brightening that persisted for some 15 min following emergence. They
suggested that during eclipse an atmospheric constituent precipitated leaving a light
frost on the surface which then sublimed upon exposure to the Sun. Not all observers
who look for this phenomenon have been successful, so at best the effect can only
be considered intermittent.

Sinton (1973) has further developed Binder and Cruikshank's idea using an atmosphere containing 0.5 cm atm of NH_3. This amount is consistent with the upper
limits set by Fink *et al.* (1973) from the IR spectra. Sinton argues that this much NH_3
will explain the intermittent post-eclipse brightening as a seasonal phenomenon, reconcile the 10 and 20 μm temperature differences, and explain some aspects of
thermal recovery.

A stellar occultation by Io occurred in 1971 (Taylor *et al.*, 1971). The observations
give an upper limit of the bulk atmosphere. Smith and Smith (1972) place that limit
at 9×10^{-8} bar for an N_2 atmosphere. More recently the Pioneer 10 spacecraft was
Earth-occulted by Io. The radio science team measured both day and night ionospheres. Peak intensities were about 6×10^4 cm^{-3} and 10^4 cm^{-3}, respectively (Kliore
et al., 1974a, b). Inferences from such a large ionosphere place lower limits on the
surface pressure which are compatible with the occultation results.

3. Sodium D Line Emission

Just prior to the Pioneer 10 flyby, Brown (1974) announced the observation of Na
D line emission from Io. The emission was intense and intermittent. Figure 1 shows
four spectra of Io taken by Brown and Chaffee (1974). The second of these shows
no measurable emission, while the other three show 3 MR, 900 kR, and 800 kR of
emission assuming that this emission comes from very close to the satellite. The

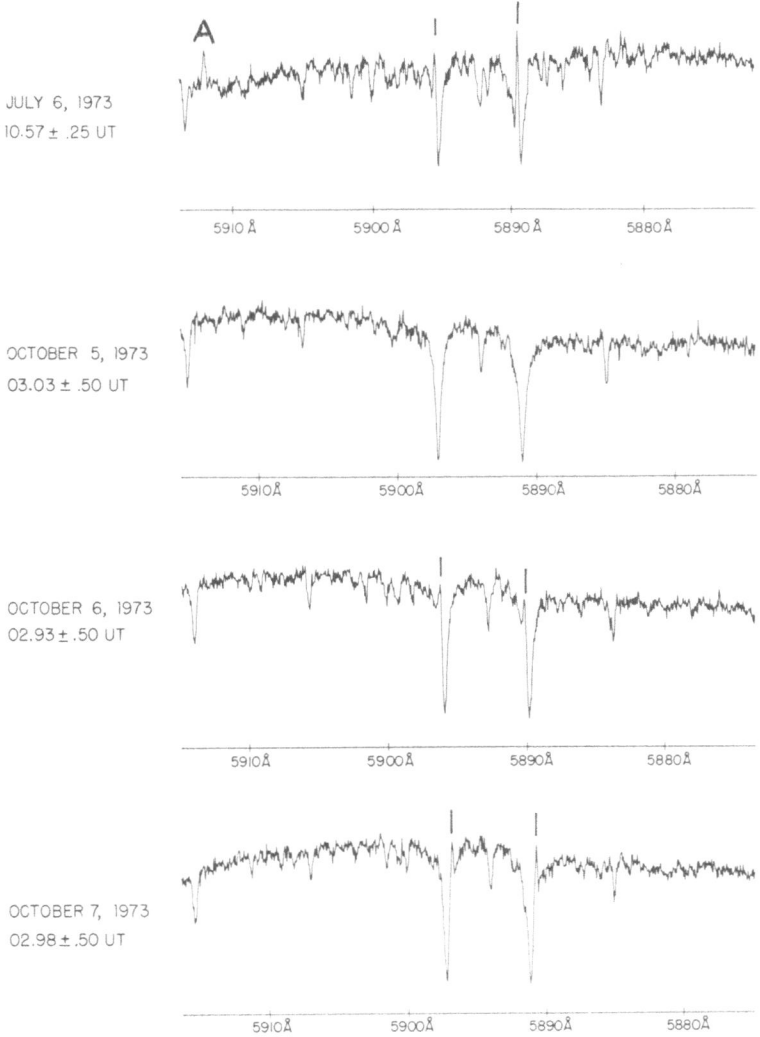

JULY 6, 1973
10·57 ± .25 UT

5910 Å   5900 Å   5890 Å   5880 Å

OCTOBER 5, 1973
03.03 ± .50 UT

5910Å   5900Å   5890Å   5880Å

OCTOBER 6, 1973
02.93 ± .50 UT

5910Å   5900Å   5890Å   5880Å

OCTOBER 7, 1973
02.98 ± .50 UT

5910Å   5900Å   5890Å   5880Å

Fig. 1.   Echelle spectra of Io (from Brown and Chaffee, 1974). The D line emission features are indicated. The feature marked A in the first spectrum is from the calibration lamp.

D_2/D_1 ratios observed are 1.4, 1.2, and 1.6. These numbers indicate considerable optical depth. Note that the Doppler shifts caused by the satellite's orbital motion (Swings effect) moves the position of the emission from one side of the reflected solar Fraunhofer features to the other.

The intensity of the emission rules out resonant scattering of the solar flux; there simply is not enough energy available. McElroy *et al.* (1974) suggested a model in which metastable N_2 excited the D lines. They argue that photolysis of NH_3 and subsequent chemistry would lead to N_2 and H_2. Their estimate of the average production rate of N_2 is a healthy 2×10^{11} cm^{-2} s^{-1}. Excitation of N_2 would be through electron collisions, as in terrestrial aurorae (Hunten, 1965). This feature could explain the intermittent nature of the emission. One feature of this model is the lower density of Na required since the particular collision processes involved are very efficient. McElroy *et al.* distribute the Na in two layers, a hotter source layer underneath a cooler scattering layer. They are able to obtain the low D_2/D_1 ratio observed. McElroy and Yung (1975) discuss further details of this and other models.

Shortly after Brown's announcement Trafton *et al.* (1974) found that the emission came not only from Io, but also from a large area surrounding the satellite. This region extended more than 24 satellite radii from Io, well beyond the Jupiter-Io equilibrium point. The spectra taken by the author (from Trafton *et al.*) simultaneously measured the D_2 and D_1 intensities. These spectra are shown in Figure 2. Also shown in the figure are the two emission intensities, the D_2/D_1 ratio, and the observing geometries for each spectrum. The intensities given for when Io was in the aperture (spectra A and E) are averages over the whole aperture. Were the emission to be considered to originate only near the satellite surface, these intensities would increase by factors of 50 and 30, respectively.

The high D_2/D_1 ratios observed in this cloud indicate that it is optically thin. It is difficult to imagine an efficient collision type source so far from the satellite surface. The lower intensities mean that solar resonant scattering may be viable for excitation of this cloud. We first ask what portion of the solar flux available would be necessary to supply the energy in the observed emission. The expected surface brightness I can be approximated by

$$4\pi I \leqslant \Delta\lambda f_{\text{Io}},$$

where f_{Io} is 1.2×10^{12} photons cm^{-2} s^{-1} Å^{-1}. The Doppler shift due to orbital motion causes f_{Io} as given above to be about half of the continuum flux. The minimum Doppler width $\Delta\lambda$ is about 15 mÅ, which would correspond to a minimum temperature of 1600 K.

The demands of this simple process are not excessive. McElroy *et al.* infer temperatures of 8000 K from a high resolution spectrum by Brown and Chaffee (1974). The column density N can be easily calculated, for the optically thin case, from

$$4\pi I = gN,$$

where $g = 0.183$ s^{-1} for the D_2 line, and includes the transition probability and properly adjusted solar flux for Io's distance and relative motion. The 16 kR D_2 emission shown in spectrum F of Figure 2 implies a column density of 8.8×10^{10} cm^{-2}. The assumption of optical thinness appears justified. Optical depth unity occurs at approximately 10^{11} cm^{-2} at 220 K so that at the higher temperatures involved here the density for $\tau = 1$ would be greater by at least a factor of 3. The above column density implies an average density of 13 cm^{-3} in a cloud with a radius of 24 satellite radii.

The D_2/D_1 intensity ratio also provides useful information. The high ratio observed in the extended cloud has been used to indicate an optically thin situation.

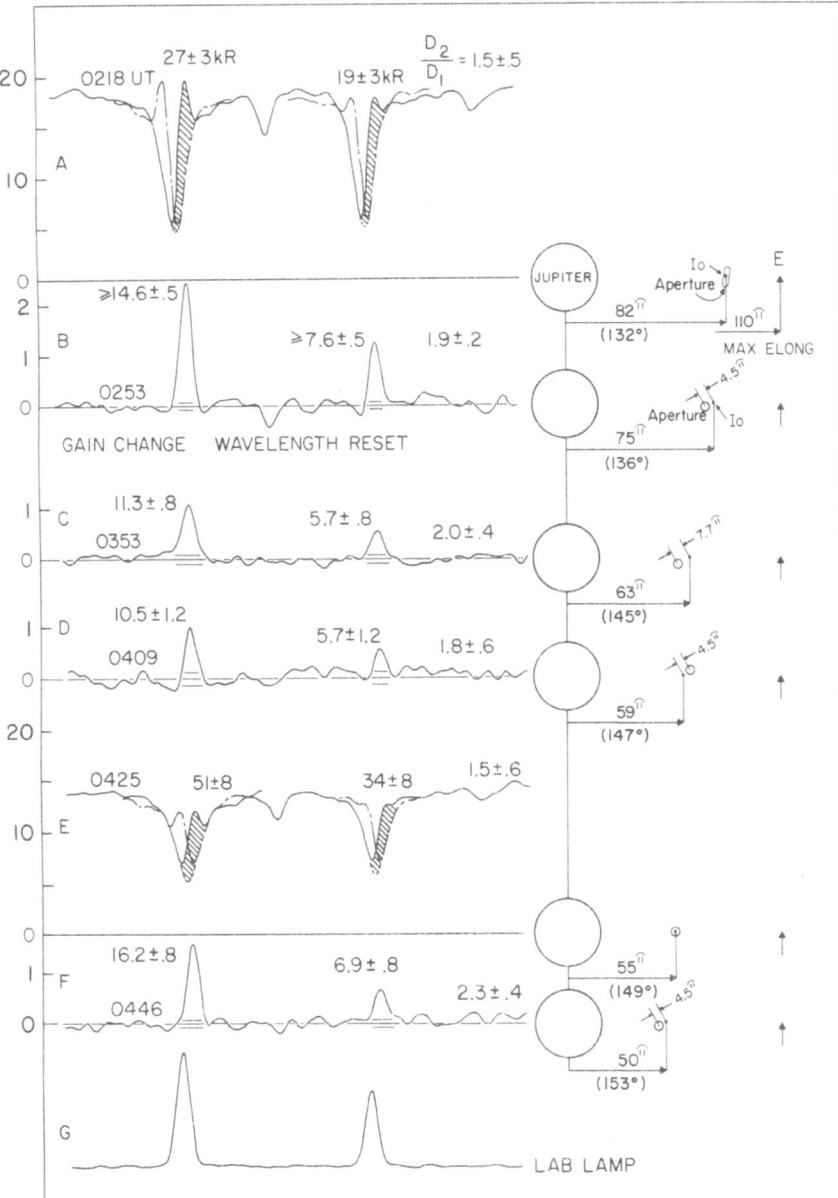

Fig. 2.   Spectra obtained at Kitt Peak National Observatory on November 16, 1973 (from Trafton *et al.*, 1974). Between Spectra B and C there was a grating malfunction and a gain change. Spectra A and B have been adjusted in wavelength. The small differences in the line locations in C, D, and F are easily accounted for by orbital motion. Time of observation, intensity of each emission feature (averaged over the aperture), and D_2/D_1 intensity ratio are indicated for each spectrum. In spectra A and E the broken line shows the spectrum reversed about the center of each Fraunhofer feature. This facilitates estimation of the net emission which is indicated by the shaded area. Schematic drawings show the observing geometry and indicate orbital position measured from superior geocentric conjunction. The intensity scales were improperly labeled in the original illustration in Trafton *et al.*; proper relative intensities are given here.

It has been noted that the observed ratios are somewhat too high to be explained only by solar resonance scattering for which the simple picture would predict a ratio of 1.7 (Macy and Trafton, 1974; Parkinson, 1974). Correction is necessary for the phase function of the D_2 emission (Chamberlain, 1961). This 6% correction increases the expected ratio to 1.8. Further increase seems necessary, however, because of spectrum F in Figure 2 for which the D_2/D_1 ratio is 2.3.

As an aside it should be noted that such high ratios have been seen in comets. Evans and Malville (1967) report, without discussion, values up to 2.4 for the Sun grazing comet Ikeya-Seki. These values should not be unexpected, however, because of the Swings effect. A maximum Doppler shift of 4.7 Å is easily calculated for this comet, which means that during approach the D_1 line is shifted into the Fraunhofer features at 5893 Å, thus raising the D_2/D_1 ratio with no basic change necessary in the resonant scattering mechanism. Similarly, on recession low values could be observed.

For Io we have no such simple explanation. The large optical depth inferred from the D_2/D_1 ratio when Io is in the aperture implies there must be an intense surface or near source such as the one discussed by McElroy *et al.* This source would supplement the solar flux scattered from the very extensive scattering cloud (Parkinson, 1974). The intensity of the near surface source can be estimated from

$$\left.\frac{D_2}{D_1}\right|_{obs} = \frac{2[1.06 f_{\odot, D_2} + f_{Io, D_2}]}{f_{\odot, D_1} + f_{Io, D_1}}.$$

Correction has been made for the D_2 phase function for the solar flux. That phase function is assumed to integrate to 1 for the supplemental flux from Io. $f_{\odot, D_2}/f_{\odot, D_1}$ is 0.85 and $f_{Io, D_2}/f_{Io, D_1}$ is assumed 1.5. In this simplified single scattering model it is found that 3.3 MR of $D_2 + D_1$ emission from the near surface source is needed to give a ratio of 2.0 at a distance of 9.4 satellite radii. This is in very good agreement with spectra A and E assuming that much of the emission in those spectra is from the vicinity of the satellite.

It is very tempting to try to correlate the intermittent nature of the Na emission with the decametric radio burst probability. Before discovery of the large spatial extent, Brown's (1974) observations indicated maximum probability of emission shortly after the maximum probability for radio bursts. This could be the case if the two phenomena were related; the phase lag would be a consequence of the time required to populate the D line emission regions with atomic sodium. Any direct coupling of the radio bursts and D line emission would cause difficulty for the beaming theories of the decametric bursts. It is easy to understand that a beam could sweep the Earth causing the intermittent radio bursts, but the Na emission should still be observable even when such a beam is not directed toward Earth. At this time further discussion would be unsupported due to the lack of definition of the temporal behavior of the Na emission.

4. Sodium Source

At the present time the source of Na is not well understood. Fanale *et al.* (1974) suggest a surface composition with considerable salt deposits rich in Na and S. McElroy *et al.* (1974) discuss Na and K dissolved in solid NH_3. McElroy and Yung (1975) note that such solutions may provide needed electrical conductivity and may mask characteristic NH_3 ice absorbtion features. Sodium could also be resupplied by meteoritic impact. It is clear that some surface reservoir is needed.

To raise an extended cloud is not particularly easy. Sputtering is discussed by Matson *et al.* (1974) and seems most reasonable. Such a non-thermal mechanism seems necessary for populating the extended cloud because of the high 'temperature' required in order that the solar flux can provide sufficient energy and because of the apparent time variations of the phenomenon itself. A Na atom would take 3 h to get 16 radii from the satellite if it left from near the surface with a radial velocity of 3.3 km s^{-1}. An atom at escape velocity would take 9 h to get the same distance.

The Na leaving the satellite is still bound to Jupiter; hence a torus of Na, at Io's radius, would be expected to form (McDonough and Brice, 1973). The lifetime of that Na against photoionization would be about 15 days (Hunten, 1954). This is sufficient time for at least a large portion of a torus to accumulate, such as has been observed in Ly-α by Pioneer 10 (Judge and Carlson, 1974). An unsuccessful attempt has been made to observe this torus (Trafton *et al.*, 1974). Observation at the elongation point indicated less than 2 KR emission, which reduces to an average Na density less than 0.1 cm^{-3}.

5. Other Emissions

As briefly mentioned above, the Pioneer 10 UV photometer saw 200 R of Ly-α emission from a segment of a torus about 120° in extent (Judge and Carlson, 1974). McElroy and Yung (1975) infer a short H lifetime of 10^5 s from this observation. Trafton has unsuccessfully searched for Hα, Li, and K resonance lines, Mg, and O I 6300 (Trafton *et al.*, 1974). He also failed to find Na emission at 6160.7 Å and 8194.8 Å. The absence of these lines indicates selective excitation processes such as considered above. The author has briefly searched for N_2^+ 3914 Å emission which could be expected from the McElroy *et al.* mechanism.

6. Conclusions

Intense Na D line emission has been unexpectedly observed from Io, a satellite of Jupiter. The emission also comes from a very extensive cloud surrounding the satellite. Solar resonant scattering is the most likely excitation source within the extended emission cloud. Some other source is needed to excite the emission from close to the Ionian surface. An auroral type transfer of excitation from mestastable N_2 seems reasonable. The Na is most likely sputtered from the satellite surface by intermittent bombardment by low energy protons.

Less than a year has elapsed since announcement of this exciting phenomenon. Most of the observations reported have been hurriedly made during the last stages of Jupiter's 1973 apparition. The new observations from this season will hopefully pin down many aspects of the phenomenon, most important of which is the temporal behavior.

Acknowledgment

Kitt Peak National Observatory is operated by the Association of Universities for Research in Astronomy, Inc., under contract with the National Science Foundation.

References

Allen, C. W.: 1963, *Astrophysical Quantities*, 2nd ed., Athlone Press, London.
Bigg, E. K.: 1964, *Nature* **203**, 1008.
Binder, A. B. and Cruikshank, D. P.: 1964, *Icarus* **3**, 299.
Brown, R. A.: 1974, in A. Woszczyk and C. Iwaniszewska (eds.), 'Exploration of the Planetary System', *IAU Symp.* **65**, 527.
Brown, R. A. and Chaffee, F. H.: 1974, *Astrophys. J. Letters* **187**, L125.
Carr, T. D. and Gulkis, S.: 1969, *Ann. Rev. Astron. Astrophys.* **1**, 577.
Chamberlain, J. W.: 1961, *Physics of the Aurora and Airglow*, Academic Press, New York.
Danielson, R. E. and Tomasko, M. G.: 1971, *Bull. Am. Astron. Soc.* **3**, 243.
Evans, C. and Malville, J. McK.: 1967, *Publ. Astron. Soc. Pacific* **79**, 310.
Fanale, F. P., Johnson, T. V., and Matson, D. L.: 1974, *Science*, submitted.
Fink, U., Dekkers, N. H., and Larson, H. P.: 1973, *Astrophys. J. Letters* **179**, L155.
Goldreich, P. and Lynden-Bell, D.: 1969, *Astrophys. J.* **156**, 59.
Hunten, D. M.: 1954, *J. Atmospheric Terrestr. Phys.* **5**, 44.
Hunten, D. M.: 1965, *J. Atmospheric Terrestr. Phys.* **27**, 583.
Johnson, T. V. and McCord, T. B.: 1971, *Astrophys. J.* **169**, 589.
Judge, D. L. and Carlson, R. W.: 1974, *Science* **183**, 318.
Kliore, A., Cain, D. L. Fjeldbo, G., Seidel, B. L., and Rasool, S. I.: 1974a, *Science* **183**, 323.
Kliore, A. J., Cain, D. L., Fjeldbo, G., and Seidel, B. L.: 1974b, Paper, Amer. Astron. Soc.-Div. Planet. Sci. 5th Annual Meeting.
Kuiper, G. P.: 1973, *Sky Telesc.* **46**, 228.
Macy, W., Jr. and Trafton, L.: 1974, *Bull. Am. Astron. Soc.* **6**, in press.
Matson, D. L., Johnson, T. V., and Fanale, F. P.: 1974, Paper, Amer. Astron. Soc.-Div. Planet. Sci. 5th Annual Meeting.
McDonough, T. R. and Brice, N. M.: 1973, *Icarus* **20**, 136.
McElroy, M. B. and Yung, Y. L.: 1975, *Astrophys. J.* **196**, 227.
McElroy, M. B., Yung, Y. L., and Brown, R. A.: 1974, *Astrophys. J. Letters* **187**, L127.
Minton, R. B.: 1973, *Commun. Lunar Planetary Lab.* **10**, 35.
Morrison, D. and Cruikshank, D. P.: 1974, *Space Sci. Rev.* **15**, 641.
Newburn, R. L., Jr. and Gulkis, S.: 1973, *Space Sci. Rev.* **3**, 179.
Parkinson, T. D.: 1974, *J. Atmospheric Sci.*, submitted.
Pilcher, C. B., Ridgway, S. T., and McCord, T. B.: 1972, *Science* **178**, 1087.
Sinton, W. M.: 1973, *Icarus* **20**, 284.
Smith, B. A. and Smith, S. A.: 1972, *Icarus* **17**, 218.
Taylor *et al.*: 1971, *Nature* **234**, 405.
Trafton, L., Parkinson, T. D., and Macy, W., Jr.: 1974, *Astrophys. J. Letters* **190**, L85.
Veverka, J.: 1971, *Icarus* **14**, 355.
Warwick, J. W.: 1967, *Space Sci. Rev.* **6**, 841.

IO: RECENT OBSERVATIONS

YUK L. YUNG

*Center for Earth and Planetary Physics, Harvard University, Cambridge,
Mass. 02138, U.S.A.*

1. Introduction

Our knowledge of Io has progressed dramatically in the past year. Prior to 1973, observations of Io made at optical, IR and radio wavelengths have revealed that this innermost Galilean satellite of Jupiter is unusual in a number of ways. A list would include post-eclipse brightening, high near-IR albedo, puzzling features (or lack of features) in the reflection spectrum, discordant brightness temperatures measured at 10 and 20 μm, and modulation of Jovian decametric activity. We now understand that Io has an atmosphere and an ionosphere. Their presence and interaction with the Jovian magnetosphere may account for much of the unusual behavior of the enigmatic satellite.

The emphasis of this review will be on recent observations and a few specific problems associated with their interpretations. For a more general review the reader is referred to Parkinson (1975).

2. Sodium Emission

Brown (1973) discovered intense Na emission from Io. Trafton *et al.* (1974) reported emission coming from an extended region of space around Io. The total column abundance of Na is estimated to be $\sim 10^{11}$ cm^{-2} in the Na cloud, $\sim 10^{13}$ cm^{-2} in the atmosphere of Io. Estimates of the lifetime of Na atoms in the cloud indicate that Io must be supplying Na at the rate of $\sim 10^7$ atoms cm^{-2} s^{-1}. The Na is presumably present initially in bound form on Io's surface and may be released by the sputtering mechanism suggested by Matson *et al.* (1974). Escape of Na atoms from the atmosphere must primarily occur during periods of intense atmospheric heating generated by interaction of Io's atmosphere with energetic particles in the Jovian matnetosphere (McElroy *et al.*, 1974). More recently Bergstrahl *et al.* (1974) reported observations on the correlation of Na emission with orbital position of Io. The results indicate a temporal behavior consistent with solar resonance scattering. The remarkable stability of the cloud over a period of many weeks of observation is puzzling in view of the short photoionization lifetime of a Na atom. The observations were made with a field of view about 30 times the size of Io and consequently provide little information on the emission source close to the surface of the satellite.

3. Ly-α Airglow

The UV airglow experiment on Pioneer 10 detected Ly-α emission ~ 300 R from an

B. M. McCormac (ed.), Atmospheres of Earth and the Planets, 441–446. All Rights Reserved.

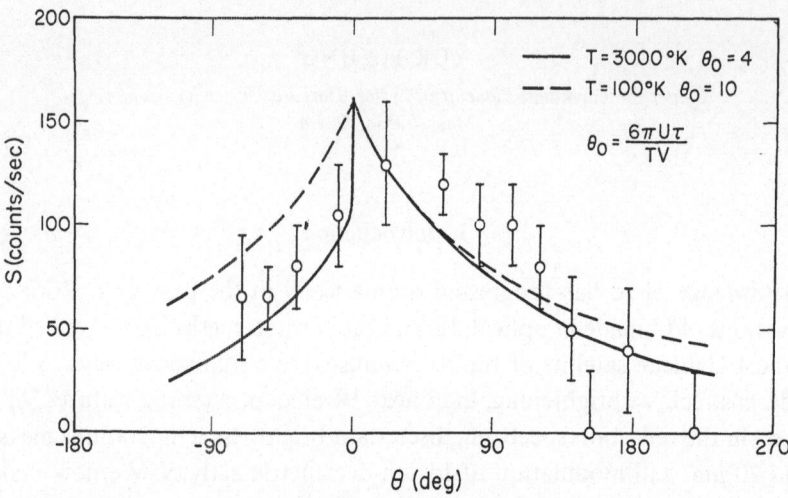

Fig. 1.   The observed H Ly-α intensity as a function of the orbital angle with respect to Io. The solid curve
shows calculation with $T = 3000$ K, the dashed curve shows results with
$T = 500$ K (Carlson and Judge, 1974).

extensive H torus surrounding Jupiter in the orbital plane of Io (Judge and Carlson, 1974; Carlson and Judge, 1974). The mean diameter of the torus is about equal to the diameter of the orbit of Io. The torus is not complete, however, but seems to subtend an effective angular width of about 150° as shown in Figure 1. The cloud was also observed while it was in eclipse and the lack of emission from that portion of the cloud which happened to lie in Jupiter's shadow suggests that the cloud is less than the size of Jupiter in vertical extent and that resonance scattering of sunlight is the primary excitation mechanism. The cloud is estimated to contain 10^{33} H atoms with a mean lifetime $\sim 10^5$ s. The source of toroidal H is most probably Io. Figure 1 shows the results of a calculation made by Carlson and Judge (1974) to model the distribution of H around Io. To account for the asymmetric distribution of the cloud about Io the escaping atoms must have a temperature exceeding 1000°. In order to maintain the cloud the satellite must be supplying H at a rate of 10^{11} atoms cm^{-2} s^{-1}. An escape flux of this magnitude, however, would exhaust the entire atmosphere on Io in a few years. It is clear that the surface of Io must be abundant in a H-rich material. The idea is consistent with Lewis' models for the composition of the satellites in the outer solar system (Lewis, 1971).

4. Ionosphere

The trajectory of Pioneer 10 took it behind Io. Such a trajectory gave rise to an opportunity to study Io's atmosphere by the method of radio occultation. Analysis of the S-band telecommunication signal by Kliore et al. (1974) established the presence of an ionosphere both on the dayside and the nightside. Electron densities observed on the dayside show a peak concentration of 6×10^4 cm^{-3} at an altitude of 100 km and extends for 750 km with a scale height of 200 km. On the nightside the maximum electron density is about 10^4 cm^{-3}. The peak occurs close to the surface and the

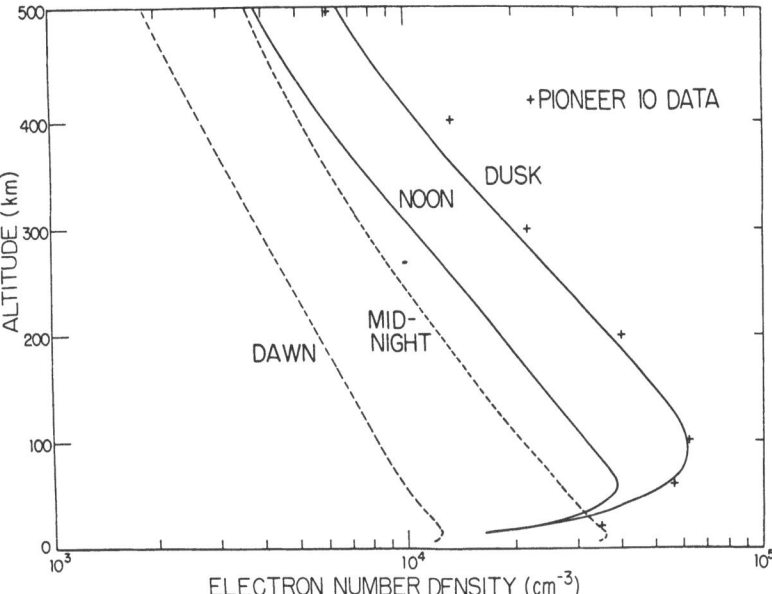

Fig. 2.   Electron density profiles for Io's ionosphere. Na^+ is the major ion. NH_3 is the major neutral species. In the daytime the surface number density of Na and NH_3 is taken to be 3.7×10^6 cm^{-3} and 5.0×10^{10} cm^{-3}, respectively. At night NH_3 is reduced by a factor of 5 and a vertical downward motion of 5 m s^{-1} is imposed on the atmosphere. The calculation is carried out for an isothermal atmosphere at 500 K (McElroy and Yung, 1975).

ionosphere appears to diminish rapidly above an altitude of 200 km. McElroy and Yung (1975) discussed a variety of models for the ionosphere. The most probable model requires Na^+ as the dominant ion. Sodium atoms have a low ionization potential ~ 5 eV and are readily ionized by UV sunlight:

$$Na + h\nu \rightarrow Na^+ + e.$$

Loss of ionization by radiative recombination

$$Na^+ + e \rightarrow Na + h\nu$$

is a slow process. The model includes details of molecular diffusion, diurnal variation, and vertical bulk motion. Results of this calculation are shown in Figure 2.

5. He Emission

Recently Cruikshank *et al.* (1974) observed the 10830 Å He resonance line from Io. The equivalent width of the emission is 11 ± 4 cm^{-1}. If the emission comes from an extensive cloud uniform over the field of view (17″ diameter), the mean intensity is 34 ± 12 kR. The emission is most probably due to resonance scattering of sunlight by metastable He atoms. Since each atom will scatter 0.62 photons s^{-1}, one must account for the presence of $\sim 5 \times 10^{10}$ cm^{-2} of He (2^3S). The most likely process for populating the triplet He states is excitation by electrons in excess of 20 eV:

$$He(1^1S) + e \rightarrow He(2^3S) + e.$$

This reaction has a cross section $\sim 5 \times 10^{-18}$ cm^2 near threshold and has been considered in detail by McElroy (1965) to account for the He emission in terrestrial airglow.

The observed intensity is surprisingly high, especially since, in view of Pioneer 10's failure to detect the 584 Å resonance emission, the total column abundance of He(^1S) cannot exceed 5×10^{12} cm^{-2} (Carlson, 1974). The natural lifetime and photoionization lifetime are 7.9×10^3 s (Drake, 1971) and 1.7×10^4 s, respectively. A consistent interpretation of the non-detection of He(1^1S) by Judge and Carlson (1974) and the detection of He(2^3S) by Cruikshank et al. (1974) may therefore require a flux of low energy (~ 20 eV) electrons of order 10^{11} cm^{-2} s^{-1} around Io.

The presence of an electron flux of this magnitude poses a serious problem. The He atoms can now be readily ionized by electron impact.

$$He + e \rightarrow He^+ + e + e.$$

The ionization lifetime is estimated to be about 10^6 s. If Io were to supply He atoms to maintain the cloud in a steady state, the required flux would be $\sim 10^9$ cm^{-2} s^{-1}. Since He is derived mainly from radiogenic sources, such a large production rate is inconsistent with current models of the Jovian satellites (Lewis, 1974). We are thus led to the conclusion that the He emission is probably sporadic.

6. Discussion

A consistent interpretation of recent observations is difficult. There are a number of puzzling features in our present understanding of Io. The observations by Bergstrahl et al. (1974) suggest that the Na cloud is remarkably stable. Since the cloud emission exhibits a temporal behavior consistent with resonance scattering, we can reliably estimate the total number of Na atoms in the vicinity of Io and their rate of destruction. These estimates have been made in Section 2. To account for the large escape flux of Na, the exospheric temperature must be in excess of 5000 K. Sputtering, as suggested by Matson et al. (1974), provides an attractive release mechanism for bound surface Na. But the flux of high energy protons, as measured by Pioneer 10 (Trainor et al., 1974), is probably too small for the purpose. Furthermore, the quantum yield is low for sputtering of Na by protons. The temperature of the upper atmosphere deduced from the electron density profile cannot exceed 500 K. If this quiescent state were to prevail at all times, no substantial amount of Na could escape from the satellite.

Interaction of Io with the Jovian magnetosphere provides clues in understanding many of the puzzling features. Io is known to modulate Jovian decametric activity (Bigg, 1964). The theory of Goldreich and Lynden-Bell (1969), illustrated by Figure 3, requires the existence of a current $\sim 10^6$ amp flowing in a loop between Io and Jupiter.

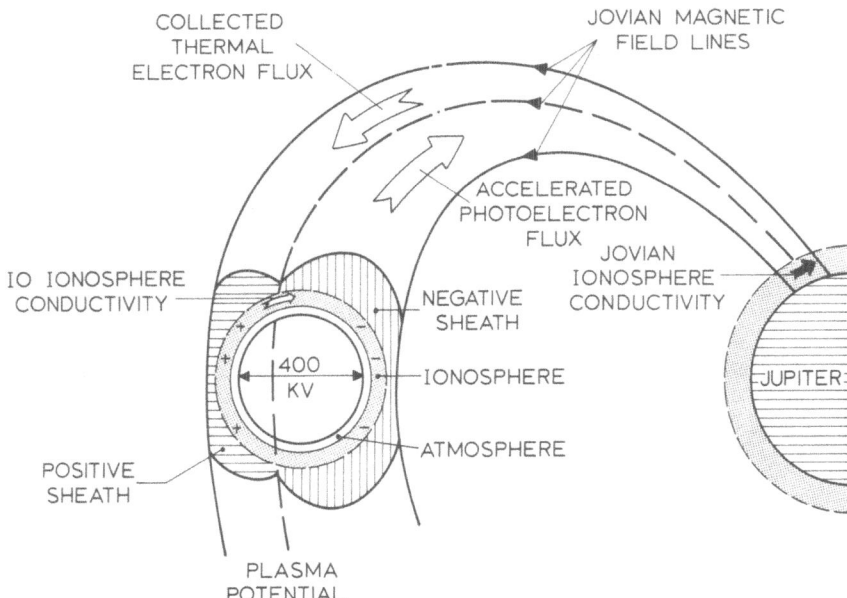

Fig. 3.   Basic Io sheath configuration and electron flux path (Shawhan *et al.*, 1974).

Gurnett (1972) has considered the effect of energetic electrons in the plasma sheath around Io. Further work by Shawhan *et al.* (1974) suggests that the electron fluxes directed into Io's atmosphere may be in the range 10^9 to 10^{10} electrons $cm^{-2} s^{-1}$. Pioneer 10 detected MeV proton and electron fluxes of the order of 10^6 and $10^7 cm^{-2} s^{-1}$, respectively (Trainor *et al.*, 1974; Fillius and McIlwain, 1974). We have no data on the fluxes of low energy charged particles. However from the distribution of Ly-α emission we may infer the existence of a flux of low energy protons $\sim 10^9 cm^{-2} s^{-1}$ (Carlson and Judge, 1974; McElroy and Yung, 1975). An analysis of the He 10830 Å emission shows that the low energy electron flux may be as high as $10^{11} cm^{-2} s^{-1}$.

The role played by large corpuscular fluxes in the atmosphere is complicated. McElroy and Yung (1975) and Shawhan *et al.* (1974) have discussed the possibility of ionization and auroral excitation. The energy flux is estimated to be as high as 10 erg $cm^{-2} s^{-1}$. Deposition of a fraction of this energy flux in the upper atmosphere will raise the exospheric temperature to thousands of degrees. Large numbers of atoms and molecules can escape from the upper atmosphere and form extensive clouds around Io.

7. Summary

Io may be covered with a layer of material rich in H and metallic Na. The satellite possesses an atmosphere of the order of 10^{-8} bar. Photoionization of Na atoms would lead to the formation of an ionosphere as observed by Pioneer 10. Interaction with energetic plasma in the Jovian magnetosphere could be responsible for the escape of large numbers of atoms and molecules from Io. These particles tend to remain in Io's orbit and give rise to an extensive cloud ~ 10 Ionian radii in size. Solar resonance scattering and electron impact excitation account for the large Na, H, and He emissions.

Acknowledgments

I would like to thank M. B. McElroy for many helpful suggestions and critical reading of the manuscript. I am indebted to J. T. Bergstrahl, R. W. Carlson, D. P. Cruikshank, T. V. Johnson, and D. L. Matson for discussion of their recent work. This work was supported by the Atmospheric Sciences section of the National Science Foundation under grant no. GA-33990X to Harvard University.

References

Bergstrahl, J. T., Matson, D. L., and Johnson, T. V.: 1974, Paper *IAU Colloq.*, No. 28, Aug. 18–21, 1974, Cornell University, in press.
Bigg, E. K.: 1974, *Nature* **203**, 1008.
Brown, R. A.: 1974, in A. Woszczyk and C. Iwaniszewska (eds.), 'Exploration of the Planetary System', *IAU Symp.* **65**, 527.
Carlson, R. W.: 1974, private communication.
Carlson, R. W. and Judge, D. L.: 1974, Paper, *IAU Colloq.*, No. 28, Aug. 18–21, in press.
Cruikshank, D. P., Pilcher, C. B., and Sinton, W. M.: 1974, Paper *IAU Colloq.*, No. 28, Aug. 18–21, Cornell University, in press.
Drake, G. W. F.: 1971, *Phys. Rev.* **A3**, 908.
Fillius, R. W. and McIlwain, C. E.: 1974, *Science* **183**, 314.
Goldreich, P. and Lynden-Bell, D.: 1969, *Astrophys. J.* **156**, 59.
Gurnett, D. A.: 1972, *Astrophys. J.* **175**, 525.
Judge, D. L. and Carlson, R. W.: 1974, *Science* **183**, 318.
Kliore, A., Cain, D. L., Fjeldbo, G., Seidel, B. L., and Rasool, S. I.: 1974, *Science* **183**, 323.
Lewis, J. S.: 1971, *Icarus* **15**, 174.
Lewis, J. S.: 1974, private communication.
Matson, D. L., Johnson, T. V., and Fanale, F. P.: 1974, Paper, Amer. Astron. Soc. Div. Planetary Science, 5th annual meeting.
McElroy, M. B.: 1965, *Planetary Space Sci.* **13**, 403.
McElroy, M. B. and Yung, Y. L.: 1975, *Astrophys. J.* **196**, 227.
McElroy, M. B., Yung, Y. L., and Brown, R. A.: 1974, *Astrophys. J. Letters* **187**, L127.
Parkinson, T. D.: 1975, this volume, p. 433.
Shawhan, S. D., Goertz, C. K., Hubbard, R. F., Gurnett, D. A., and Joyce, G.: 1974, Paper presented at 'The Magnetosphere of the Earth and Jupiter', Frascati, Italy, May 28–31. To be published in *Proceedings*, D. Reidel.
Trafton, L., Parkinson, T. D., and Macy, Jr., W.: 1974, *Astrophys. J. Letters* **190**, L85.
Trainor, J. H., Teegarden, B. J., Stilwell, D. E., McDonald, F. B., Roelof, E. C., and Webber, W. R.: 1974, *Science* **183**, 311.

GLOSSARY

Aerosol. Finely divided particles, liquid or solid, suspended in the atmosphere. The exact size range is arbitrary..

Auroral Absorption (AA). Measure of cosmic noise absorption through the ionosphere in the auroral oval.

Auroral Oval. Locus of auroras in latitude as a function of time, which has an oval shape.

Chemical symbols in brackets, e.g., $[N_2]$. Represents the concentration of the substance indicated. In this report, the units are frequently molecules cm^{-3}.

CIRA Model. COSPAR International Reference Atmosphere, published by North-Holland Publishing Co., Amsterdam (1965).

Cosmic Noise Absorption (CNA). Surface measurement of the absorption of cosmic noise, usually about 30 MHz, passing through the ionosphere.

D Region. Altitude region in which the density of ions rises steeply, from $\sim 10^2$ to $\sim 10^6$ ions cm^{-3}. The altitude is usually 50–90 km.

E Region. Altitude region next above D region with ion densities $\sim 10^6$ cm^{-3}. The altitude varies from day to night, usually 90–160 km in daytime.

Eddy Diffusion. A method of treating turbulent transport of minority constituents in the atmosphere in the mathematical formulation of classical molecular diffusion.

Electrojet. Refers to a current of electrons, on occasions reaching $\sim 10^6$ A moving horizontally from E to W in the early morning and much more weakly from W to E in the evening, at ~ 120 km altitude and $\sim 65°$ INLT.

ELF. Extremely low frequency and is from 30 to 3000 Hz.

EPR. Electron paramagnetic resonance.

Equatorial Electrojet. A current system in the ionosphere, flowing generally along the earth's equator.

EUV. Extreme ultraviolet, defined as starting below 1040 Å (the cutoff of the best window) and going down to 2 Å.

F Region. Altitude region next above the E region, characterized by a new rise in ion density.

F_0F_2. Maximum electron density in the E region of the ionosphere.

FP. Fabry-Pérot.

Hall Currents. Current flow perpendicular to both the electric and magnetic fields.

IMF. Interplanetary magnetic field.

IN Lat. Invariant latitude, Λ.

IN LT. Invariant local time.

IN Pole. Invariant pole, where $\Lambda = 90°$.

Invariant Coordinate System. McIlwain's B, L space magnetic coordinates.

Ionosphere. Altitude region where solar radiation is strongly effective in ionizing the atmosphere. Altitude limits usually given as 50–400 km, including the *D, E,* and *F* regions.

IR. Infrared radiation covering from about 7800 Å to 1000 μm.

IS. Ionospheric scatter.

Jacchia Model. Model atmosphere computed by L. G. Jacchia and staff of the Smithsonian Institute Astrophysical Observatory.

K_p. Quasi-logarithmic scale, from 0 to 9, measuring the range of activity of the most active component of the magnetic field within a 3 h interval.

L. McIlwain's invariant shell paramenter, whose units are expressed in R_E at the magnetic equator.

LBH. Lyman-Birge-Hopfield band system.

LT. Local time.

Ly-α. Stands for Lyman-alpha. Radiation (or absorption) of atomic H at 1216 Å, involving transition of the atomic electron between $n = 1$ to 2.

Magnetosphere. Region inside a surface surrounding the Earth at ~ 10 Earth radii inside of which the Earth's magnetic field exceeds external fields.

Mesopause. A narrow altitude region at the top of the mesosphere, often between 70 and 85 km altitude, in which the atmosphere is at its lowest temperature.

Mixing Ratio. Ratio of molecular density of a species to the total molecular density, sometimes alternatively the mass ratio or the specific volume ratio. Applies to atmospheric gases with good turbulent mixing.

MLT. Magnetic local time.

Molecular Diffusion. A term referring to interdiffusion of one gas into another in accordance with basic gas-kinetic theory.

$O(^1S)$ *and* $O(^1D)$. Two excited states of O atoms, ~ 4 and ~ 2 eV resp. above the ground energy states. Transitions between these states or to the ground state are quantum mechanically forbidden, hence the states are called metastable and have long lives, ~ 1 s for the first and ~ 145 s for the latter. These states only differ from each other and ground state by electron spin directions and the m quantum number.

$O_2(^1\Sigma_g^+)$ *and* $O_2(^1\Delta_g)$. Two excited states of O_2 molecules having ~ 1.6 and 1.0 eV energy resp. Transitions to the ground state are quantum mechanically forbidden leading to lifetimes ~ 12 s and ~ 3900 s resp. Basically, the states only differ from each other and from the ground state in electron spins.

Occultation. Important observations on planetary atmospheres are made during the brief period that a source of radiation is just disappearing or reappearing from behind the planet – hence occulting – and the radiation reaching the Earth observers passes through the planet atmosphere tangentially.

Oxygen Green Line. Radiation of wavelength 5577 Å resulting from transition in atomic O from ^1S to ^1D states.

PCA. Stands for Polar Cap Absorption but refers to an *event*, now usually written SPE for solar proton event. During the event, solar protons increase the ion density in the Earth's polar cap causing an increased absorption of radio frequencies.

Pedersen Current. Current flow along electric field which is perpendicular to the magnetic field.

Photochemical. Term referring to a chemical reaction caused directly by solar photons acting on the reagent.

Photoionization. Ionization of an atmospheric constituent by solar radiation.

Plasmapause. Boundary at about L of 3.5 to 4 inside of which the plasma density is much higher, but cooler.

Polar Cap. Region inside the auroral oval.

ppmv. Parts per million volume.

Pre-dawn Enhancement. Enhanced optical emission produced before normal sunrise behavior as a result of charge particles from the sunlit conjugate region.

Quenching. Metastable atoms and molecules 'normally' de-excite to a lower energy state spontaneously in a relatively long characteristic time interval after creation or excitation. Admixture of a different gas sometimes shortens this time; if so, the foreign gas is said to quench the metastable.

Rayleigh (R). Unit of brightness or radiative intensity, for example of an aurora or airglow, in simplest form amounting to 10^6 protons $cm^{-2} s^{-1}$ but complicated by thickness considerations.

Red Arc. Airglow at 6300 Å wavelength arising from the atomic O transition ^1D to ^3P (ground state).

Retrograde Motion. Most of the planets and satellites of the solar system rotate around the sun in the same direction with orbits nearly in the same plane (the ecliptic). They also revolve on their axes in the same directional sense. If either the orbit direction or the revolution is opposite to this sense, it is called retrograde motion.

Scale Height. Boltzmann's law of atmospheres for gases in equilibrium is $p = p_0 \exp(-mgh/kT)$. p is the pressure at height h for a gas of molecular weight m and constant temperature T. The quantity kT/mg is the scale height. It is thus also the value of the height h at which $p = p_0 e^{-1}$.

Schumann-Runge Band and Continuum. UV absorption by O_2 with a band system between ~ 2200 and 1759 Å. The continuum begins at 1759 Å and extends toward shorter wavelengths. It arises from dissociation of the O_2. The band system arises from transitions from the $X^3\Sigma_g^-$ ground state to the $B^3\Sigma_u^-$ excited state with lowest energy 6.1 eV.

SID. Sudden ionospheric disturbance.

Solar Wind. A steady flow or flux of particles escaping from the sun with high velocity up to 10^3 km s^{-1}. The particles are charged, ions and electrons, but the net flux is neutral. The flow can rise by a factor as large 10^2 during a solar storm.

SPA. Sudden phase anomaly in reflected radio waves.

SPE. Solar particle event.

Sporadic E. Irregular fluctuation in ion and electron density in the E region sometimes yielding very high densities. It occurs commonly in a thin layer at ~ 120 km altitude, leading to the term, sporadic E layer.

Stratopause. A narrow altitude band between the stratosphere and the mesophere marking the top of the constant and low temperature domain of the stratosphere, frequently at $\sim$25 km altitude in midlatitudes.

Superrotation. A condition where the atmosphere of Earth or other planets is moving in the same direction but faster than the planet itself is rotating.

UT. Universal time.

UV. Ultraviolet radiation and extends from 100 to 3800 Å.

Vibrational Temperature. The mean energy of molecular vibration, divided by Boltzmann's constant. It frequently differs from ordinary or 'translational' temperature because exchange of energy between vibration and translational motion may be extremely slow.

VK. Vegard-Kaplan band system.

VLF. Very low frequency and is from 3 to 30 kHz.

IN. First negative band system.

1P. First positive band system.

2P. Second positive band system.

SUBJECT INDEX

Air
  composition 7–9
    Ar 137–144
    electron 113, 160–162, 261, 263–264
    He 137–143
    Mars 361–367
    N_2 137–153, 165–167
    O 137–153, 164–167, 202
    O_2 165, 200–203
    Venus 361–367, 392–394
  neutral density 133–157, 161–162
  temperature 49–50, 90, 161–167, 171, 275–279, 331
  transport 269–270, 274–275
Alkali metals
  Na 434–446
Ar 67–68
  bulge 143–144
  density 137–144
Atmospheric absorption 75–77
Atmospheric chemistry
  Ca^+ 204
  CO_2, Venus 361–367
  experimental techniques 186–195, 220–223
  Fe 204
  Fe^+ 204–205
  H 198–199, 219–232
  H^+ 199–200
  H_2O, Venus 369–384
  H_2SO_4, Venus 369–384
  Jupiter 415–417
  Mars 361–367
  Mg^+ 204
  N^+ 202
  N_2 199–200
  N_2^+ 201–202
  neutral 21–43
  NO 203
  NO^+ 203, 205–207
  O 211–218, 222
  O^+ 198–200, 203
  O_2 201–203, 213–214, 216
  O_2^- 207–208
  O_2^+ 205–207
  O_3 203, 212–213, 216
  OH 226–231
  reaction kinetics 179–186
  reaction rates 9–13, 179–195, 223–232
  S^+ 203

Venus
  CO_2 361–367
  H_2O 369–384
  H_2SO_4 369–384
Atmospheric electric fields 87–97
Atmospheric emissions 323–333
  CO_2 337–358
  equatorial 286–288
  infrared 335–358
  midlatitude 285–286
  NO 304–306
  OI 125
  $O(^1S)$ 289–307
  O_3 341–345
  Venus 385–399
  UV 309–317
Atmospheric energy source 73–86
Atmospheric-ionosphere interactions
  midlatitude 145–153
  high latitude 153–156
Atmospheric models 5–7
  ionosphere 245–268
  neutral 235–243, 252–253
  Venus 385–399
Atmospheric physics 47–58
Atmospheric scale height 47–52
Atmospheric tides 89–90
Atmospheric transport 59–72
Atmospheric waves 169–170, 240–241
Atmospheric winds 87–97
Aurora
  electric fields 101, 104–105, 115–116, 128–130
  heating 99–109
  motions 125–130
  N 310, 316
  N_2 310–317, 319–322
  O 310, 315–316
  oval 283–284
  particle precipitation 107–108, 112–115
  storms 284–285
  UV emission 309–317

Barometric law 47–51

Ca^+ 204
CH_4 23–24, 32–33
  eddy diffusion 67–69
  Jupiter 402–406, 415–417, 427
  Saturn 402–406, 427

ASTROPHYSICS AND SPACE SCIENCE LIBRARY

Edited by

J. E. Blamont, R. L. F. Boyd, L. Goldberg, C. de Jager, Z. Kopal, G. H. Ludwig, R. Lüst,
B. M. McCormac, H. E. Newell, L. I. Sedov, Z. Švestka, and W. de Graaff

on the Symposium on the Magellanic Clouds, held in Santiago de Chile, March 1969, on the Occasion of the Dedication of the European Southern Observatory. 1971, XII + 189 pp.

24. B. M. McCormac (ed.), *The Radiating Atmosphere. Proceedings of a Symposium Organized by the Summer Advanced Study Institute, held at Queen's University, Kingston, Ontario, August 3–14, 1970.* 1971, XI + 455 pp.

25. G. Fiocco (ed.), *Mesopheric Models and Related Experiments. Proceedings of the 4th ESRIN-ESLAB Symposium, held at Frascati, Italy, July 6–10, 1970.* 1971, VIII + 298 pp.

26. I. Atanasijević, *Selected Exercises in Galactic Astronomy.* 1971, XII + 144 pp.

27. C. J. Macris (ed.), *Physics of the Solar Corona. Proceedings of the NATO Advanced Study Institute on Physics of the Solar Corona, held at Cavouri-Vouliagmeni, Athens, Greece, 6–17 September 1970.* 1971, XII + 345 pp.

28. F. Delobeau, *The Environment of the Earth.* 1971, IX + 113 pp.

29. E. R. Dyer (general ed.), *Solar-Terrestrial Physics/1970. Proceedings of the International Symposium on Solar-Terrestrial Physics, held in Leningrad, U.S.S.R., 12–19 May 1970.* 1972, VIII + 938 pp.

30. V. Manno and J. Ring (eds.), *Infrared Detection Techniques for Space Research, Proceedings of the 5th ESLAB-ESRIN Symposium, held in Noordwijk, The Netherlands, June 8–11, 1971.* 1972, XII + 344 pp.

31. M. Lecar (ed.), *Gravitational N-Body Problem, Proceedings of IAU Colloquium No. 10, held in Cambridge, England, August 12–15, 1970.* 1972, XI + 441 pp.

32. B. M. McCormac (ed.), *Earth's Magnetospheric Processes. Proceedings of a Symposium Organized by the Summer Advanced Study Institute and Ninth ESRO Summer School, held in Cortina, Italy, August 30–September 10, 1971.* 1972, VIII + 417 pp.

33. Antonin Rükl, *Maps of Lunar Hemispheres.* 1972, V + 24 pp.

34. V. Kourganoff, *Introduction to the Physics of Stellar Interiors.* 1973, XI + 115 pp.

35. B. M. McCormac (ed.), *Physics and Chemistry of Upper Atmospheres. Proceedings of a Symposium Organized by the Summer Advanced Study Institute, held at the University of Orléans, France, July 31–August 11, 1972.* 1973, VIII + 389 pp.

36. J. D. Fernie (ed.), *Variable Stars in Globular Clusters and in Related Systems. Proceedings of the IAU Colloquium No. 21, held at the University of Toronto, Toronto, Canada, August 29–31, 1972.* 1973, IX + 234 pp.

37. R. J. L. Grard (ed.), *Photon and Particle Interaction with Surfaces in Space. Proceedings of the 6th ESLAB Symposium, held at Noordwijk, The Netherlands, 26–29 September, 1972.* 1973, XV + 577 pp.

38. Werner Israel (ed.), *Relativity, Astrophysics and Cosmology. Proceedings of the Summer School, held 14–26 August, 1972, at the BANFF Centre, BANFF, Alberta, Canada.* 1973, IX + 323 pp.

39. B. D. Tapley and V. Szebehely (eds.), *Recent Advances in Dynamical Astronomy, Proceedings of the NATO Advanced Study Institute in Dynamical Astronomy, held in Cortina d'Ampezzo, Italy, August 9–12, 1972.* 1973, XIII + 468 pp.

40. A. G. W. Cameron (ed.), *Cosmochemistry. Proceedings of the Symposium on Cosmochemistry, held at the Smithsonian Astrophysical Observatory, Cambridge, Mass., August 14–16, 1972.* 1973, X + 173 pp.

41. M. Golay, *Introduction to Astronomical Photometry.* 1974, IX + 364 pp.

42. D. E. Page (ed.), *Correlated Interplanetary and Magnetospheric Observations. Proceedings of the 7th ESLAB Symposium, held at Saulgau, W. Germany, 22–25 May, 1973.* 1974, XIV + 662 pp.

43. Riccardo Giacconi and Herbert Gursky (eds.), *X-Ray Astronomy.* 1974, X + 450 pp.

44. B. M. McCormac (ed.), *Magnetospheric Physics. Proceedings of the Advanced Summer Institute, held in Sheffield, U.K., August 1973.* 1974, VII + 399 pp.

45. C. B. Cosmovici (ed.), *Supernovae and Supernova Remnants. Proceedings of the International Conference on Supernovae, held in Lecce, Italy, May 7–11, 1973.* 1974, XVII + 387 pp.

46. A. P. Mitra, *Ionospheric Effects of Solar Flares,* 1974, XI + 294 pp.

50. Z. Kopal and R. W. Carder, *Mapping of the Moon: Past and Present,* 1974, VIII + 237 pp.